AgroINNOVATION: Innovative Solutions in the Agro-Industrial Complex (AgroINNOVATION 2021)

IOP Conference Series: Earth and Environmental Science
Volume 965

Volgograd, Russia
12 – 14 May 2021

ISBN: 978-1-7138-4947-6
ISSN: 1755-1307

Printed from e-media with permission by:

Curran Associates, Inc.
57 Morehouse Lane
Red Hook, NY 12571

Some format issues inherent in the e-media version may also appear in this print version.

Institute of Physics
Dirac House, Temple Back
Bristol BS1 6BE UK

Phone: 44 1 17 929 7481
Fax: 44 1 17 920 0979

techtracking@iop.org

Additional copies of this publication are available from:

Curran Associates, Inc.
57 Morehouse Lane
Red Hook, NY 12571 USA
Phone: 845-758-0400
Fax: 845-758-2633
Email: curran@proceedings.com
Web: www.proceedings.com

TABLE OF CONTENTS

IOP Conf. Series: Earth and Environmental Science **965** (2022) 011001 doi:10.1088/1755-1315/965/1/011001

Preface

The international conference **"AgroINNOVATION - innovative solutions in the Agro-Industrial Complex"** was successfully held online on May 12-14, 2021 in Volgograd based on the Volgograd State Agrarian University. The organizer of the conference is the federal state budgetary educational institution of higher education **"Volgograd State Agrarian University".** In connection with COVID-2019 and global and national bans and other coordinated restrictive measures related to the pandemic, the organizers decided to hold the conference in an integrated manner using a virtual format and providing access for all participants to the presented reports.
The safety and well-being of all conference participants is our priority. COVID-19 is unpredictable, so the postponement of the conference is fraught with uncertainty, while many scientists and researchers want to attend this long-awaited conference and exchange academic knowledge with their colleagues. But, fortunately, there are no obstacles for science, and we continue to work in our areas remotely, using modern technical means. The conference was held using a virtual mode for 2 days (May 12-14, from 9:30 to 16:00), including program and invitation speeches, the author's presentation session.
Audiovisual contact was carried out using the Zoom communication platform for synchronous communication. The conference was organized as a large-scale online meeting of the participants of thematic sessions with the opportunity for speakers to show slides and video files in accordance with the approved technical program. Each presenter had a 15 minute presentation and a 10 minute Q&A session. There were always 25-30 participants online, along with participants at a safe distance in the room. During the online presentation, participants discussed more than 60 issues on innovation for sustainable and environmentally friendly development of the agro-industrial complex. The conference was attended by representatives of the leading institutions of the Russian Federation and foreign countries:

Agroinnovation 2021

1. Volgograd State Agrarian University (Volgograd, Russia)
2. Don State Agrarian University (Rostov region, Russia)
3. Saratov State Agrarian University named after N. I. Vavilov (Saratov, Russia)
4. Nikitsky Botanical Garden – National Scientific Center of the Russian Academy of Sciences (Yalta, Republic of Crimea, Russia)
5. Almaty Technological University (Almaty, Kazakhstan)
6. Russian State Agrarian University - Moscow Timiryazev Agricultural Academy (Moscow, Russia)
7. Smolensk State Agricultural Academy (Smolensk, Russia)
8. Kuban State Agrarian University named after I. T. Trubilin (Krasnodar, Russia)

IOP Conf. Series: Earth and Environmental Science **965** (2022) 011001 doi:10.1088/1755-1315/965/1/011001

9. Volga Region Research Institute of Manufacture and Processing of Meat-and-Milk Production (Volgograd, Russia)
10. All-Russian Research Institute of Irrigated Agriculture (Volgograd, Russia)
11. Volgograd State Technical University (Volgograd, Russia)
12. Ryazan State Agrotechnological University Named after P. A. Kostychev (Ryazan, Russia)
13. Chuvash State University named after I. N. Ulyanov (Cheboksary, Chuvash Republic, Russia)
14. Chuvash State Pedagogical University named after I. Ya. Yakovlev (Cheboksary, Chuvash Republic, Russia)
15. Kabardino-Balkarian State Agrarian University named after V. M. Kokov (Nalchik, Kabardino-Balkar Republic, Russia)
16. Ingush State University (Magas, Republic of Ingushetia, Russia)
17. Kabardino-Balkarian State University named after H. M. Berbekov (Nalchik, Kabardino-Balkar Republic, Russia)
18. Lomonosov Moscow State University (Moscow, Russia)
19. Financial University under the Government of the Russian Federation (Moscow, Russia)
20. Caspian Agrarian Federal Scientific Research Centre of Russian Academy of Sciences (Astrakhan region, Russia)
21. Mikhailovsky College of Technology (Volgograd region, Russia)
22. Moscow Power Engineering Institute (Moscow, Russia)
23. Federal Scientific Centre of Agroecology, Complex Melioration and Protective Afforestation of the Russian Academy of Sciences (Volgograd, Russia)
24. Research Institute of Agriculture of Crimea (Simferopol, Republic of Crimea, Russia)
25. V. I. Vernadsky Crimean Federal University (Simferopol, Republic of Crimea, Russia)
26. Medical Academy Named after S.I. Georgievsky" (Simferopol, Republic of Crimea, Russia)
27. Research Institute for Hygiene, Toxicology and Occupational Pathology» at Federal Medical and Biological Agency (Volgograd, Russia)
28. Ulyanovsk State Agrarian University named after P. A. Stolypin (Ulyanovsk, Russia)
29. Samara State Agrarian University (Samara, Russia)
30. Russian State Economic University named after G.V. Plekhanov (Moscow, Russia)
31. Department of Veterinary and Production Control of JSC firm "Agrokompleks" named after N. I. Tkachev (Krasnodar Territory, Russia)

IOP Conf. Series: Earth and Environmental Science **965** (2022) 011001 doi:10.1088/1755-1315/965/1/011001

32. Poultry Department of LLC “Russian Grain Ufa” (Ufa, Republic of Bashkortostan, Russia)
33. Rostov State Medical University (Rostov-on-Don, Russia)
34. Krasnodar Research Centre for Animal Husbandry and Veterinary Medicine (Krasnodar, Russia)
35. Maikop State Technological University (Maykop, Republic of Adygea, Russia)
36. Volgograd State University (Volgograd, Russia)
37. Gustave Eiffel University (France)

The excellent presentation of all the speakers made this conference truly spectacular.

The conference addressed the issues of sustainable development of agriculture and rural areas through the development and research of innovative agricultural technologies, land reclamation, ecology, rational use of natural resources and environmental protection.

The main topics of the conference were:

1. Innovative agricultural technologies.
2. Modern aspects of agricultural land reclamation, water resources and water supply.
3. Scientific support for the sustainable development of animal husbandry.
4. Progressive biotechnologies for the sustainable food production.
5. Energy and resource saving technologies and technical systems in agriculture.
6. IT-technologies in agriculture: problems and prospects for the sustainable development of agriculture.

Organising Committee

Chairman: Vitaly Alekseevich Tseplyaev, Candidate of Technical Sciences, Associate Professor, Rector of the University, volgau@volgau.com, +7 (8442) 41-17-84

Co-chairman: Aleksey Anatolyevich Ryadnov - Honorary Worker of Education of the Russian Federation, Dr. Biol. Sci., Professor, head of Department, Vice-Rector for Research, radnov@mail.ru, +7 (8442) 41-11-65

Co-chairman: Fomin Sergey Denisovich - Honorary Worker of Higher Professional Education of the Russian Federation, Doctor of Technical Sciences, head of Center; Scientific Editor, Deputy Editor-in-Chief: Editorial office of the journal "*Proc. of Low Volga Agro-University Complex*" fsd_58@mail.ru +7 (927) 521-88-55

Nesmiyanov Ivan Alekseevich - Doctor of Technical Sciences, Vice-Rector for Academic Affairs, Professor: Department of Mechanics, Head of laboratory "*Robotic technologies and digital adaptation of agricultural machinery*" ivan_nesmiyanov@mail.ru +7 (8442) 41-10-94

Borodychev Viktor Vladimirovich - Honored Scientist of the Russian Federation, Doctor of Agricultural Sciences, Academician of the Russian Academy of Sciences, Professor, Director of the Branch of the All-Russian Research Institute of Hydraulic Engineering and Land Reclamation named after A.N. Kostyakov vkoniigim@yandex.ru +7 (8442) 41-15-05

Gorlov Ivan Fedorovich, Honored Scientist of the Russian Federation, Doctor of Agricultural Sciences, Academician of the Russian Academy of Sciences, Scientific Director of the Volga Region Research Institute of Production and Processing of Meat and Dairy Products niimmp@mail.ru +7 8442 39-10-48

Kashtanova Elena G., Doctor, Professor, Dean of the Faculty of Agriculture, Ecotrophology and Landscape Development, Anhalt University of Applied Sciences, Bernburg (Germany)

Petrovich Milan - Doctor of Agricultural Sciences, Deputy Director of the Institute of Livestock Breeding, Belgrade, Serbia

Khramova Valentina Nikolaevna - Doctor of Biological Sciences, Professor, Dean of the Faculty of Food Production Technology, VolgSTU
hramova_vn@mail.ru +7 (903) 370-33-93

Chamurliev Omariy Georgievich - Doctor of Agricultural Sciences, Senior Researcher, Head of the Department, Director of the *Research Institute for Advanced Research and Innovation in the Agroindustrial Complex*
attika.ge@yandex.ru +7 (8442) 40-30-94

Zelenev Alexander Vasilievich - Doctor of Agricultural Sciences, Professor zelenev.a@bk.ru +7 (8442) 41-12-20

Gapich Dmitry Sergeevich - Doctor of Technical Sciences, head Department gds-08@mail.ru +7 (8442) 41-13-70

Ryadnov Alexey Ivanovich - Honored Worker of Higher Education of the Russian Federation, Honorary Worker of Higher Professional Education of the Russian Federation, Doctor of Technical Sciences, Professor emtp46@mail.ru alex.rjadnov@mail.ru +7 (8442) 41-81-88

Shaprov Mikhail Nikolaevich - Doctor of Technical Sciences, Professor, Honored Worker of Higher School of the Russian Federation, Professor
m.shaprov@yandex.ru +7 (8442) 41-15-18

Ovchinnikov Alexey Semenovich - Honored Worker of Higher School of the Russian Federation, Academician of the Russian Academy of Sciences, Professor, Doctor of Agricultural Sciences, head of department; Ch. editor: Editorial office of the journal "*Proc. of Low Volga Agro-University Complex*"; Chairman of the Board: Regional Fund "*Agrarian University Complex*"

Bocharnikov Viktor Sergeevich - Doctor of Technical Sciences, Professor; Scientific Secretary: Scientific Council; bocharnikov_vs@mail.ru +7 (8442) 41-30-95

Pakhomov Alexander Alekseevich - Doctor of Technical Sciences, Professor; Scientific supervisor of the laboratory pahomoff.1954@yandex.ru +7 (8442) 41-81-53

Varakin Alexander Tikhonovich - Doctor of Agricultural Sciences, Professor varakinat58@mail.ru +7 (8442) 41-77-13

Chamurliev Nodari Georgievich - Doctor of Agricultural Sciences, Professor zootexnia@mail.ru +7 (8442) 41-77-13

Цифровой диктант

Peer review declaration

All papers published in this volume of IOP Conference Series: Earth and Environmental Science have been peer reviewed through processes administered by the Editors. Reviews were conducted by expert referees to the professional and scientific standards expected of a proceedings journal published by IOP Publishing

The reviews comment on the main provisions of the paper, give a generalized reasoned assessment and draw conclusions about the relevance, scientific novelty, scientific and practical significance and the possibility of publishing the paper in the journal.
During the peer review, the following aspects were assessed:
• compliance with the subject of the journal;
• correspondence of the title of the paper to the content;
• whether the abstract contains essential information about the paper, whether it is suitable for inclusion in the summarization service;
• the degree of originality and novelty of the research results;
• statement of the problem in general terms, its relevance;
• sequence of presentation;
• compactness and clarity of illustrative material: diagrams, figures, tables and captions;
• use of scientific terms;
• structuredness of the paper material;
• theoretical and practical value of the paper;
• the validity of the conclusions presented in the paper, their value.

Particular attention is paid to the completeness of the bibliographic list of the paper. The process of unfair citation (unjustified increase in citations, as well as self-citation), incorrect citation (unjustified by the content of the cited papers) is controlled.

In the Conclusion, the reviewers conclude:
1. The paper is recommended for publication without revision;
2. The paper is recommended for publication, taking into account the comments of the reviewer (without re-reviewing);
3. The paper is recommended for publication, subject to revision and re-review.
4. The paper was rejected.

• Type of peer review: Double-blind
• Conference submission management system: submissions were received and handled via e-mail
• Number of submissions received: 104
• Number of submissions sent for review: 82
• Number of submissions accepted: 64

IOP Conf. Series: Earth and Environmental Science **965** (2022) 011002 doi:10.1088/1755-1315/965/1/011002

- **Acceptance Rate (Number of Submissions Accepted / Number of Submissions Received X 100): 61.5%**
- **Average number of reviews per paper: 2**
- **Total number of reviewers involved: 43**
- **Any additional info on review process: All reviewers have academic status and are recognized scientific experts in their field.**

Contact person for queries:

Name : Fomin Sergey Denisovich, Deputy Editor-in-Chief, Head of the NAMSI Center, Doctor of Technical Sciences
Affiliation: Volgograd State Agrarian University

Email: fsd_58@mail.ru

IOP Conf. Series: Earth and Environmental Science **965** (2022) 012001 doi:10.1088/1755-1315/965/1/012001

Optimization of the energy consumption of the shredder of the tops for the sustainable development of agriculture

Y N Abramov, S N Borychev, M Y Kostenko, M B Uglanov, A A Tsymbal and Y A Yudaev

Ryazan State Agrotechnological University named after P A Kostychev, 1 Kostychev Avenue, Ryazan, 390044, Russia

E-mail: yu.yudaev@mail.ru

Abstract. To optimize energy costs during the cutting process, a mathematical model is constructed and the calculation results are studied to optimize the cutting process. The proposed model is built for the operation of the rotor of the haulm shredder, but the technique of its construction can be used for any systems in which rotational movements are used to cut plants. A method for constructing a rotation model of the haulm shredder rotor with respect to two centers of rotation is presented. The method is based on the solution of differential equations of motion, the theorem on the center of mass of the system and the energy conservation law. To obtain an analytical expression, differential equations were solved that related the deviation of the cutting element of the haulm shredder from a radial position. Modeling showed that in the operating mode, at an angular velocity ω of rotor rotation of more than 100 rad s^{-1}, the dependence is almost linear and can be replaced by the formula in the characteristic mass range of rotating knives 0.7-1.5 kg in relation to $T \approx 0.6\text{-}0.001\omega$.

1. Introduction

When harvesting and other technological operations, there is a need to cut plants or their parts: harvesting grain, corn, removing potato tops, pruning shrubs, fruit trees, etc. For all these operations, cutting elements are used, which for the most part use rotational energy movement of the rotor of a harvester or a combine. The quality of the operations performed depends on the speed of rotation, the mass of the cutting element and the area of contact with the object being cut.

To optimize the energy costs of the cutting process, it is necessary to build a mathematical model and study the results of calculations for extreme points to determine the maximum efficiency [1].

This article continues the cycle of work related to modeling of technological processes in agriculture [2, 3, 4, 5, 6].

2. Materials and methods

The model will be built for the rotor of the haulm shredder, but the method of its construction can be used for any systems in which rotational motions are used to cut plants (pruning) [7, 8, 9, 10].

The creation of the model is based on the joint solution of differential equations of motion, the theorem on the center of mass of the system and the law of energy conservation.

When modelling, the following assumptions were made:

- external and internal forces act on the system;

IOP Conf. Series: Earth and Environmental Science **965** (2022) 012001 doi:10.1088/1755-1315/965/1/012001

- all forces balance each other;
- friction forces are neglected;
- the force of gravity is not taken into account in calculations.

Symbols are introduced and the forces acting on the simulated system, both external and internal, are represented. External forces are $\overline{F}_k^{\,e}$ and internal forces are $\overline{F}_k^{\,i}$. If the movement occurs with linear or angular acceleration $\overline{a}_k$, then: $m_k\overline{a}_k = \overline{F}_k^{\,e} + \overline{F}_k^{\,i}$.

In this case, the creation of the model is based on the solution of a system consisting of n - equations, which in general form can be written as:

$$\begin{cases} m_1\overline{a}_1 = \overline{F}_1^{\,e} + \overline{F}_1^{\,i} \\ m_2\overline{a}_2 = \overline{F}_2^{\,e} + \overline{F}_2^{\,i} \\ m_n\overline{a}_n = \overline{F}_n^{\,e} + \overline{F}_n^{\,i} \end{cases}$$

Considering that acceleration is $a = \frac{dV}{dt}$ or $a = \frac{d^2S}{dt^2}$, then the system presents differential equations of motion in a vector form.

The forces that are written to the right, depend on time, coordinates and speed. Having decomposed the forces along the coordinate axes, one obtains projections of these forces and, as a result, systems suitable for analytical or numerical solutions.

The solution to the system will be to integrate the corresponding differential equations. This can be done analytically in very limited tasks and in some cases, when the number of points in the system is small.

To find the desired function, one cannot consider the movement of individual points, but consider the system as a single object.

For example, for the radius of the vector of the center of mass, one can write:

$$M\overline{r}_C = \sum m_k\overline{r}_k .$$

Differentiating parts of this equation in time twice, one gets:

$$M\,\frac{d^2\,\overline{r}_C}{dt^2} = \sum m_k\,\frac{d^2\,\overline{r}_k}{dt^2},$$

where $\overline{a}_C$ is the center of mass acceleration.

By definition, the sum of all the internal forces of the system is zero, one obtains $\sum \overline{F}_k^{\,i} = 0$ and finally:

$$M\overline{a}_C = \sum \overline{F}_k^{\,e} .$$

One applies the above methodology to simulate the operation of a haulm shredder with rotating knives. Figure 1 shows the location of the knives (front view).

When harvesting potatoes, a potato digging device is clogged with tops. This leads to a decrease in the process productivity.

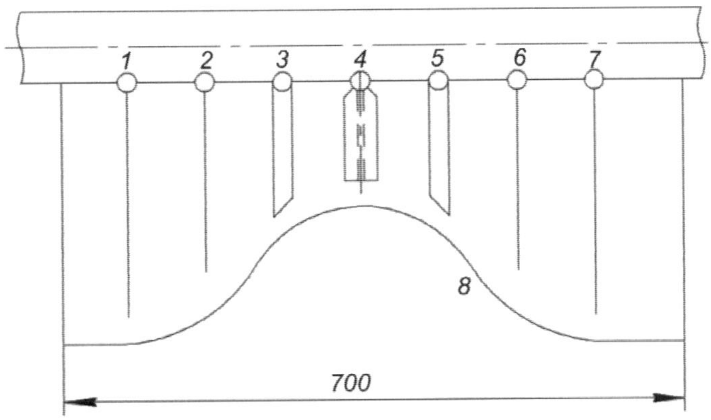

Figure 1. The arrangement of knives 1-7 on the shaft, 8 - a profile of a bed.

Knives 1-3 are flat. They perform the function of chopping potato tops and weeds between beds. To reduce the load on the shaft, such knives are set in pairs on opposite sides of the circle.

Knife 4 is spade-shaped. It provides cutting tops over the tuberous loculus.

A model of the movement of knives located on a rotating shaft will be constructed. The calculation will be carried out in accordance with the drawing (Figure 2).

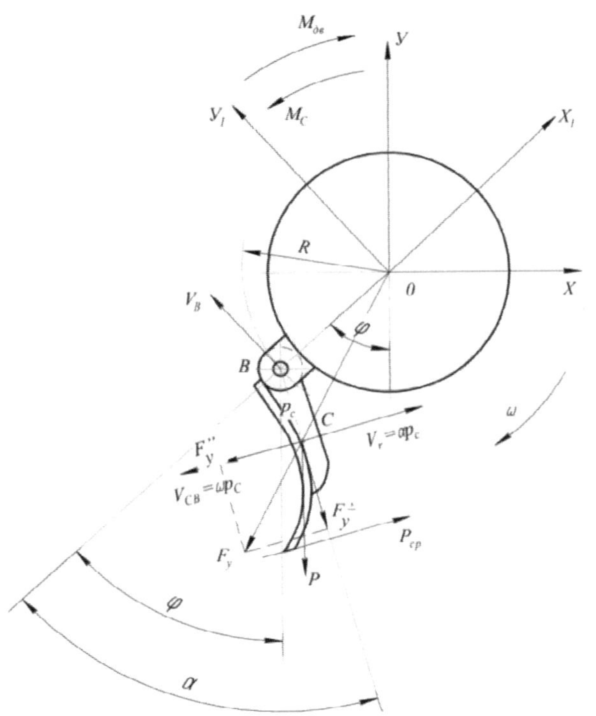

Figure 2. The scheme of the rotor of the haulm shredder with indication of the acting forces and direction of rotation.

It is convenient to describe the operation of the haulm shredder with the indication of the acting forces and the direction of rotation using the Lagrangian equations:

IOP Conf. Series: Earth and Environmental Science **965** (2022) 012001 doi:10.1088/1755-1315/965/1/012001

$$\frac{d(\partial T)}{dt(\partial \dot{\varphi})} - \frac{\partial T}{\partial \dot{\varphi}} = Q_1,$$

$$\frac{d(\partial T)}{dt(\partial \dot{\alpha})} - \frac{\partial T}{\partial \dot{\alpha}} = Q_2,$$

where T is kinetic energy; Q_1 and Q_2 are forces; φ and α are coordinates in accordance with Figure 2.

Kinetic energy depends on the angle of rotation or displacement of the knife from the radial position. Due to the fact that it consists of the energies of the rotor T_1 and the cutting element T_2, the total energy will be determined by the following formula:

$$T = T_1 + T_2.$$

Taking into account the moment of inertia of the rotor I and angular velocity ω:

$$T_1 = I \cdot \frac{\omega^2}{2}.$$

In the process of cutting, the knife takes part in rectilinear and rotational movements. Therefore, the total energy will be determined by these two components:

$$T_2 = \frac{1}{2} m V_c^2 + \frac{1}{2} I_c \omega_r^2,$$

where V_C is velocity of the center of mass; ω_r is angular velocity of rotation.

Having projected the velocity of the masses on fixed axes X, O, Y, one gets (Figure 2):

$$(V_c)_X^2 = -\omega R \cos\varphi - \omega \rho_c \cos(\alpha - \varphi) + \dot{\alpha} \rho_c \cos(\alpha - \varphi),$$

$$(V_c)_Y^2 = -\omega R \sin\varphi - \omega \rho_c \sin(\alpha - \varphi) + \dot{\alpha} \rho_c \sin(\alpha - \varphi),$$

$$V_c^2 = (V_c)_X^2 + (V_c)_Y^2,$$

Substituting values $(V_c)_X^2$ and $(V_c)_Y^2$ into the last equality and making the transformations, one gets:

$$V_c^2 = \omega^2 + R + \omega^2 \rho_c^2 + \dot{\alpha}^2 \rho_c^2 + [\cos\varphi \cos(\alpha - \varphi) - \sin\varphi \sin(\alpha - \varphi)] \cdot [2\omega^2 R \rho_c - 2\omega \dot{\alpha} R \rho_c] - 2\omega \dot{\alpha} \rho_c^2$$

Expression $\cos\varphi\cos(\alpha - \varphi) - \sin\varphi\sin(\alpha - \varphi)$ is the decomposed cosine of the difference of two angles $\cos[(\varphi + \alpha) - \varphi] = \cos\alpha$.

Thus, the expression will take the following form:

$$V_c^2 = \omega^2 R + \omega^2 \rho_c^2 + \dot{\alpha}^2 \rho_c^2 + 2\omega^2 R \rho_c \cos\alpha - 2\omega \dot{\alpha} R \rho_c \cos\alpha - 2\omega \dot{\alpha} \rho_c^2,$$

The kinetic energy of the system will be expressed as:

$$T = \frac{1}{2} m \dot{\varphi}^2 R^2 + \frac{1}{2} m \dot{\varphi}^2 \rho_c^2 + \frac{1}{2} m \dot{\alpha}^2 \rho_c^2 + m \dot{\varphi}^2 \rho_c \cos\alpha - m \dot{\varphi} \dot{\alpha} R \rho_c \cos\alpha - m \dot{\varphi} \dot{\alpha} \rho_c^2 + \frac{1}{2} I_c (\dot{\varphi} \pm \dot{\alpha})^2 + \frac{1}{2} I \dot{\varphi}^2$$

Let's proceed to the calculation of the components of the equation:

$$\frac{\partial T}{\partial \dot{\varphi}} = m \dot{\varphi} R^2 + m \dot{\varphi} \rho_c^2 + 2m \dot{\varphi} R \rho_c \cos\alpha - m \dot{\alpha} \rho_c \cos\alpha - m \dot{\alpha} \rho_c^2 + I_c (\dot{\varphi} \pm \dot{\alpha}) + I \dot{\varphi},$$

$$\frac{d(\partial T)}{dt(\partial \dot{\varphi})} = mR^2\ddot{\varphi} + m\rho_c^2\ddot{\varphi} + 2mR\rho_c \cos\alpha\ddot{\varphi} - 2mR\rho_c\dot{\varphi}\dot{\alpha}\sin\alpha - m\ddot{\alpha}R\rho_c \cos\alpha +$$
$$+ mR\rho_c\dot{\alpha}^2 \sin\alpha - m\ddot{\alpha}\rho_c^2 + I_c(\ddot{\varphi} \pm \ddot{\alpha}) + I\ddot{\varphi},$$

One makes transformations and denotes:

$$mR^2 + m\rho_c^2 + I_c + I = A,$$

$$m\rho_c^2 \pm I_c = B,$$

then one gets:

$$\frac{d(\partial T)}{dt(\partial \dot{\varphi})} = (A + 2C\cos\alpha)\ddot{\varphi} - (B + C\cos\alpha)\ddot{\alpha} + (\dot{\alpha} - 2\dot{\varphi})C\dot{\alpha}\sin\alpha,$$

Finally, one gets the first differential equation of motion:

$$(A + 2C\cos\alpha)\ddot{\varphi} - (B + CX\cos\alpha)\ddot{\alpha} + (\dot{\alpha} - 2\dot{\varphi})C\dot{\alpha}\sin\alpha = M_{\partial в} - M_c - PR\sin\varphi - P_{cp}R\cos\alpha$$

One calculates the components of the equation:

$$\frac{\partial T}{\partial \dot{\alpha}} = m\dot{\alpha}\rho_c^2 - m\dot{\varphi}R\rho_c \cos\alpha - m\dot{\varphi}\rho_c^2 + I_c(\dot{\varphi} \pm \dot{\alpha}),$$

$$\frac{d(\partial T)}{dt(\partial \dot{\alpha})} = m\dot{\alpha}\rho_c^2 - m\dot{\varphi}R\rho_c \cos\alpha + m\dot{\varphi}\dot{\alpha}R\rho_c \sin\alpha - m\ddot{\varphi}\rho_c^2 + I_c(\ddot{\varphi} \pm \ddot{\alpha}),$$

$$\frac{\partial T}{\partial \alpha} = -m\dot{\varphi}^2R\rho_c \sin\alpha + m\dot{\varphi}\dot{\alpha}R\rho_c \sin\alpha,$$

To compose the second differential equation, one introduces increment $\Delta\alpha$ and, given that work is force multiplied by displacement one gets:

$$Q_2\Delta\alpha = P\rho_c \sin(\varphi - \alpha)\Delta\alpha + P_{cp}L\Delta\alpha,$$

where:

$$Q_2 = P\rho_c \sin(\varphi - \alpha) + P_{cp}L,$$

Having carried out the transformations, one writes the second equation of motion of the system:

$$B\ddot{\alpha} - \ddot{\varphi}(B + C\cos\alpha) + C\dot{\varphi}^2 \sin\alpha = P\sin(\varphi - \alpha) + P_{aver}L,$$

The first and second differential equations describe the motion of the rotor and the hinged cutting element.

Using high-speed photography, it was found that the rotor rotates evenly. Given the uniform rotation, the equation will become:

$$(B + C\cos\alpha)\ddot{\alpha} + (\dot{\alpha} - 2\omega)C\dot{\alpha}\sin\alpha = M_{\partial в} - M_c - PR\sin\varphi - P_{aver}R\cos\alpha,$$

$$B\ddot{\alpha} + C\omega^2 \sin\alpha = P\rho_c \sin(\varphi - \alpha) + P_{aver}L.$$

The last equation describes the movement of the cutting element. The contribution of term $P\rho_c \sin(\varphi - \alpha)$ to the differential equation of motion is extremely small. The moment of inertia of the cutting element is much less than centrifugal force. Based on the assumption made:

$$B\ddot{\alpha} + C\omega^2 \sin\alpha - P_{aver}L = 0,$$

or

$$IB\ddot{\alpha} + mR\rho_c\omega^2 \sin\alpha - P_{aver}L = 0.$$

This second-order differential equation relates centrifugal forces to knife parameters: mass, length, angular rotation speed, shear force.

System solution leads to the following expression:

$$B\ddot{\alpha} + C\omega^2 \sin\alpha - P_{aver}L = 0, \text{ or } IB\ddot{\alpha} + mR\rho_c\omega^2 \sin\alpha - P_{aver}L = 0,$$

which has the following solution:

$$\alpha_{vozv} = \alpha_{ost} - \frac{l}{m}\frac{tg^2\left(\frac{k\sqrt{m}\left(t_{rot} - t_{ost}^{total}\right)}{\sqrt{2}}\right)}{1 + tg^2\left(\frac{k\sqrt{m}\left(t_{rot} - t_{ost}^{total}\right)}{\sqrt{2}}\right)}.$$

3. Results and discussion

Passing to the period of oscillation of the knife in the field of centrifugal forces, one obtains the dependence of the period of oscillation on the angular frequency of rotation (Figure 3):

$$T = \frac{2\pi}{\sqrt{\frac{mR\rho_c\omega^2{}_б}{I_в}}}$$

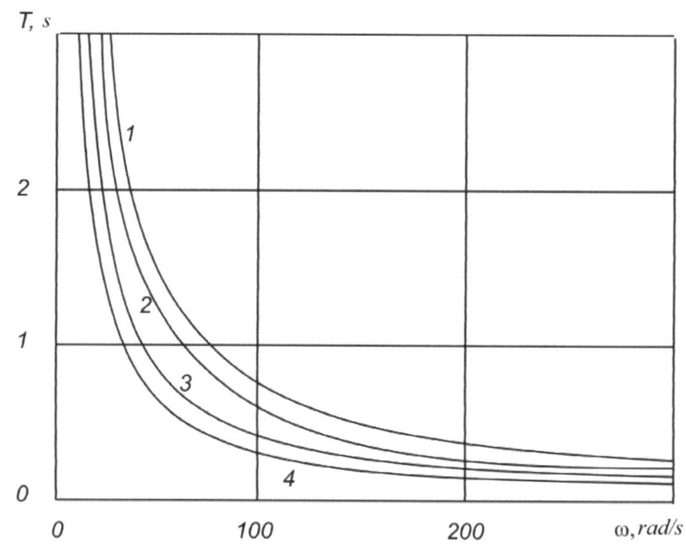

Figure 3. Dependence of the oscillation period of the knife on the rotor speed and the weight of the knife: *1* - 0.7 kg, *2* - 1.1 kg, *3* - 1.3 kg, *4* - 1.5 kg.

In the operating mode, at an angular speed of rotation of the rotor of more than 100 rad s^{-1} the dependence is almost linear in nature and can be replaced by the formula in the characteristic mass range of rotating knives 0.7-1.5 kg:

$$T \approx 0.6 - 0.001\omega.$$

Using mathematical modeling, the oscillation period was determined depending on the angle of deviation of the cutting tool (Figure 4).

Stable cutting occurs in the range of deviation of the knife from the radial position from 0° to 30°.

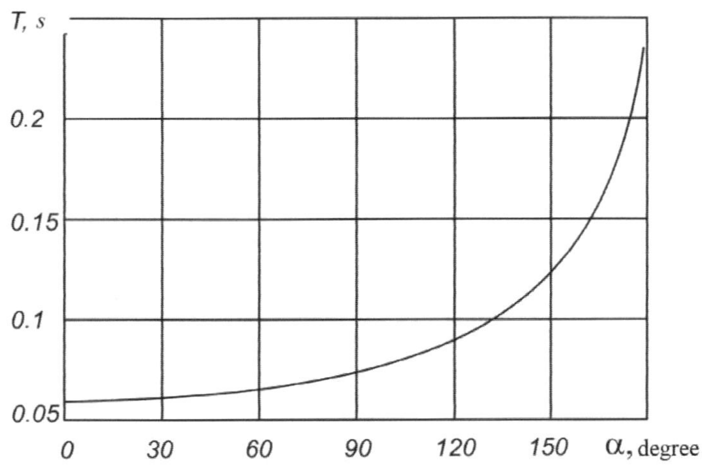

Figure 4. The dependence of the oscillation period on the angle of deviation of the cutting tool

4. Conclusion

The proposed methodology for modeling the operation of the rotor with cutting hinge elements made it possible to determine the relationship between the weight and size characteristics of the hinge knives and the quality of cutting the tops. The deflection angle of the knives should not exceed 30°. At these angles, optimal conditions are provided for cutting potato tops. At angles greater than 30°, the knife oscillation begins and the quality of harvesting decreases.

References

[1] Abramov Y N, Byshov N V, Uglanov M B and Yudaev Y A 2019 *Mater. Sci. Conf. RGATU* 8-12

[2] Yudaev Y A, Kozhanova T V and Yudaev M Y 2015 *Bulletin of the Ryazan State Agrotechnological University* **4(28)** 144-148

[3] Gorokhova M N, Yudaev Y A, Sannikova M L and Gorokhov A A 2013 Polythematic Network Electronic Scientific J. of the Kuban State Agrarian University **90** 397-407

[4] Byshov N V, Borychev S N, Uspensky I A, Yukhin I A, Golikov A A and Filyushin O V 2019 *IOP Conf. Series: Earth and Environmental Science* **341(1)** 145

[5] Byshov N V *et al* 2018 ARPN *J. of Engineering and Applied Sciences* **13(10)** 3502-3508

[6] Byshov N, Bogdanchikov I, Bachurin A, Oleinik D, Martyshov A and Kacharmin A 2019 *IOP Conf. Series: Earth and Environmental Science* **403** 1-8

[7] Ulyanov V, Utolin V, Luzgin N, Borontova M, Kiryanov A and Polunkin A 2019 *IOP Conf. Series: Earth and Environmental Science* **403** 8-11

[8] Byshov N, Uspensky I, Alekseev V and Fadeev V 2019 *Engineering Technologies and Systems* **29(2)** 295-305

[9] Byshov N, Simdyankin A and Uspensky I 2017 *Transportation Research Procedia* **20** 107-111

[10] Byshov N V, Borychev S N, Abramov Yu N, Uglanov M B, Kostenko M Yu and Rembalovich G K 2019 *Bioscience Biotechnology Research Communications* **12(5)** 312-322

IOP Conf. Series: Earth and Environmental Science **965** (2022) 012002 doi:10.1088/1755-1315/965/1/012002

Improving soil ecology when applying bacterial fertilizers

O G Chamurliev[1], I B Borisenko[1], A V Zelenev[1], G O Chamurliev[2], T V Konstantinova[1] and L A Feofilova[1]

[1] Volgograd State Agrarian University, 26 University Avenue, Volgograd, 400002, Russia
[2] RUDN University, 6 Miklukho-Maklaya Street, Moscow, 117198, Russia

E-mail: ignateva.l@bk.ru

Abstract. Increasing the yield of spring barley on light chestnut soils of the Volgograd region is possible through the use of microbiological fertilizers that activate the processes of plant growth and development. The influence of the methods of basic tillage and the application of mineral and bacterial fertilizers "Azotovit" and "Phosphatovit" on the content of nitrogen in the soil was studied. The degree of decomposition of the linen cloths was the maximum for deep flat-cutting processing. According to the binary interaction, the best results were obtained on the variant with deep flat-cutting processing with double application of bacterial fertilizers. The introduction of bacterial fertilizers sharply increases the content of azotobacter, and especially when they are applied twice. The lowest toxicity of the soil in barley crops compared to the control was noted on the variant with deep flat-cutting treatment, and the maximum on the variant of dump treatment. The analysis of data on the structure of the crop also indicates the advantage of the variant of deep flat-cutting tillage with double application of microbiological fertilizers. The most cost-effective variants were with double application of bacterial fertilizers on the background of deep flat-cutting tillage.

1. Introduction

Currently, there is an increased interest in bacterial preparations in agriculture. Their advantage over agrochemicals is the ability to transfer nutrients from fertilizers, root secretions and organic residues into a form accessible to plants. Bacterial preparations are able to absorb atmospheric nitrogen, produce physiologically active substances, and inhibit the development of pathogenic microflora. The use of bacterial fertilizers improves plant growth by reducing the stressful influence of the environment (unfavorable conditions), contributes to the fullest disclosure of the potential of the variety, which relates to both quantitative and qualitative indicators of agricultural products [1, 2, 3].

There are different opinions on the binary effect of mineral fertilizers and bacterial preparations. Thus, there is an increase in yield on sod-podzolic soils when sowing with inoculated seeds against the background of mineral nitrogen [4]. At the same time, it is widely believed that the nitrogen of mineral fertilizers slows down the process of associative nitrogen fixation. According to other data [5, 6, 7], against the background of nitrogen fertilizers, there is an increase in nitrogen fixation by 30-45%, which is probably due to better plant development when applying fertilizers.

In the complex of measures ensuring high and stable yields of agricultural crops and a steady increase in soil fertility, the use of the most rational methods of soil cultivation with the introduction of mineral and bacterial fertilizers that contribute to an increase in the yield of spring barley in the Volgograd region.

2. Materials and methods

The object of research is spring barley of the Prairie variety. The research was carried out on the experimental field of the educational farm "Gornaya Polyana" of the Volgograd State Agrarian University. The soil of the experimental site is light chestnut. Light chestnut soils traditionally have a low nitrogen and phosphorus content and a high supply of potassium. The content of mobile forms of nitrogen is low, the amount of total nitrogen does not exceed 20 mg/kg of soil. Phosphorus provision is average - 30-35 mg/kg of soil, exchangeable potassium is increased - 350-400 mg/kg of soil.

The assessment of weather conditions was carried out using data from the university's meteorological station, which provided the maximum characteristic for the years of research. The average annual precipitation varied from 300 to 450 mm. In dry years, it fell to 150 mm, and sometimes the difference was 3-4 times. In addition, precipitation fell unevenly according to the seasons and the growing season. The maximum is in the month of May, and the minimum is in September. Moreover, May precipitation, as a rule, was in the nature of a downpour, it happened that more than 25% of the average annual precipitation fell per day. In the summer, precipitation was insignificant.

The studies were carried out in a subzone of light chestnut, weakly solonetz soils with a heavy loamy granulometric composition. The presence of salt licks up to 25-30% with a humus content of 1.8%.

The agrotechnology of cultivating spring barley in the experiments was generally accepted for the Volgograd region.

The basic tillage was carried out according to three variants. For this purpose, a plow PN-4-35 was used for control, in the second and third variants a cultivator KPG 2.2 was used. The studies were carried out in a three-field crop rotation pure steam-winter wheat-barley. After harvesting winter wheat, stubble cultivation was not carried out, as the provided non-moldboard cultivation was carried out. The basic tillage was carried out in August according to the variants of the experiment.

Usually in the first decade of April, the site was harrowed with heavy tooth harrows BZTS-1.0. Then, pre-sowing cultivation of KPS-4 with lancet paws was carried out to a depth of 6-8 cm and sowing with SZ-3.6 seeders across the basic tillage.

Azotobacter accounting was carried out by the method of fouling of soil lumps (on Ashby medium). Biological activity was determined by the method of applications, by laying linen cloths according to the variants of the experiment in the phase of mass emergence of seedlings. Biological toxicity was determined by the method of plant biotests. Barley was used as a test culture in the experiments. Crop accounting was carried out by a continuous method. The grain yield is brought to the standard humidity of 14%. Statistical processing of the research results was carried out by the method of variance analysis according to Dospekhov.

3. Results and discussion

Currently, interest in the introduction of various plant growth stimulants in agricultural production is increasing. The appearance of new, more modern bacterial fertilizers makes it necessary to study them.

We have evaluated the binary effect of the methods of basic tillage and fertilizers on its microbiological parameters, crop structure and yield of spring barley.

Much less attention is paid to the study of biological factors of soil fertility than to physical ones. Meanwhile, the theoretical basis of tillage is the creation of favorable conditions for the development of biological processes in the soil, the accumulation of humus, the purification of the soil from pathogenic microorganisms, the strengthening of the biological activity of the soil.

Azotobacter is a biological indicator of soil fertility, as the most environment-sensitive microorganism (Table 1).

Table 1. Azotobacter content, % of soil clump fouling (on average for 2016-2017).

Experience variants		Indicator	
Factor A	Factor B		
A_1	B_1	59.17	23.3
	B_2		47.0
	B_3		70.7
	B_4		95.7
A_2	B_1	69.22	32.3
	B_2		55.0
	B_3		91.3
	B_4		98.3

2016 Lowest Significant Difference $(LSD)_{05}$ – 3.73, LSD_{05} A – 1.88, LSD_{05} B – 2.17, LSD_{05} AB – 1.87; 2017 LSD_{05} – 3.88, LSD_{05} A – 1.94, LSD_{05} B – 2.24, LSD_{05} AB – 1.96.

The maximum content of azotobacter in the studies was noted on the variant of flat-cutting processing to a depth of 0.20-0.22 m - 69.22%, which is 10.5% higher than the control - dump processing to the same depth. The introduction of bacterial fertilizers sharply increases the content of azotobacter, and especially when they are applied twice - 95.7% against 23.3% on the variant without fertilizers. According to the binary interaction of factors, the highest percentage of fouling of soil lumps was established on the variant of flat-cutting processing to a depth of 0.20-0.22 m with double application of bacterial fertilizers on the mineral background A_2B_4 - 98.3% [8, 9].

Soil fertility is determined by biological factors, mainly the activity of microorganisms. Plant life is closely related to microorganisms, the influence of which is multifaceted.

Under the influence of microorganisms, the mineralization of organic substances in the soil occurs, as a result of which nutrients pass from an inaccessible state to an accessible one.

The study of biological processes in the soil makes it possible to determine its fertility not only by the level of yield, but also by its biological activity (Table 2). Analysis of microbiological processes in the studied soil was determined by the method of disintegration of linen cloth.

Table 2. Biological activity of the soil in the 0-0.3 m layer, % (on average for 2016-2017).

Experience variants		Loss of fiber	
Factor A	Factor A		
A_1	B_1	29.50	16.8
	B_2		22.9
	B_3		37.4
	B_4		40.9
A_2	B_1	31.65	17.9
	B_2		24.5
	B_3		39.8
	B_4		44.4

It was found that the degree of decomposition of the linen cloth was the maximum for flat-cutting processing to a depth of 0.20-0.22 m. On average, for 2 years, the loss of fiber was 31.6% versus 29.5% in the control.

Fertilization increases the decomposition rate of linen cloth compared to the control (without fertilizers) by 8.3%. The maximum value is set when bacterial fertilizers are applied twice. Its value, on average for the studied methods of tillage, was 30.57%. According to the binary interaction, the best results were obtained on flat-cutting processing to a depth of 0.20-0.22 m with double application of bacterial fertilizers - 44.4%, which is 8.6% higher than the similar variant on the control [10]. The use of NHK did not increase the intensity of decomposition of linen cloth.

Microorganisms determine not only agronomically useful processes in the soil - biological activity, but also processes that negatively affect its fertility, growth and development of agricultural plants - biological toxicity of the soil.

The works of many authors have shown that the initial decomposition products of straw and root residues are toxic to plants. However, according to these and other scientists, the accumulation of such substances in significant quantities occurs only when plowing a large amount of straw, when it decomposes in relatively anaerobic conditions. If plant residues decompose under conditions of aerobiosis, then the resulting products quickly disintegrate or turn into humus substances that do not have an inhibitory effect on plants, and therefore, after non-dumping treatment, such negative consequences are less noticeable than after dumping (Table 3).

In the practice of agronomic research, the most common method for determining soil phytotoxicity is the method of plant biotests. Barley was used as a test culture in the experiments.

Table 3. Biological toxicity of soil, layer 0.0-0.3 m, % (average for 2016-2017).

Method of tillage	Length of seedlings, mm	Weight of seedlings, g	Germination, %	Toxicity to control			
				by the length of the seedlings	by weight of seedlings	by germination	average
Dump to a depth of 0.20-0.22 m	38	2.4	90	13.2	14.9	5.2	11.1
Flat-cutting to a depth of 0.20-0.22 m	50	3.0	97	9.8	9.2	0	6.3

In our studies, the lowest soil toxicity in barley crops compared to the control was noted on the variant with deep flat-cutting tillage at 0.20-0.22 m - 6.3%, and the largest on dump plowing with a depth of 0.20-0.22 m - 11.1%. Many scientists have proven a positive relationship between the size of the crop and the microbiological processes of the soil [11].

Analysis of data on the structure of the crop (Table 4) indicates the advantage of the option of flat-cutting tillage to a depth of 0.20-0.22 m with double application of bacterial fertilizers.

Table 4. Structure of spring barley yield (average for 2016-2017).

Experience variants		Elements of the crop structure				
Factor A	Factor B	Number of plants, million pcs/ha	Number of grains per ear, pcs	Weight of grain per ear, g	Weight of 1000 grains, g	Biological yield, t/ha
A_1	B_1	1.69	12.6	0.49	38.9	0.83
	B_2	2.10	14.6	0.57	39.0	1.20
	B_3	2.11	15.0	0.60	41.0	1.26
	B_4	2.13	17.3	0.75	43.3	1.60
A_2	B_1	1.78	14.3	0.56	39.1	1.00
	B_2	2.06	16.0	0.63	39.3	1.30

Experience variants		Elements of the crop structure				
Factor A	Factor B	Number of plants, million pcs/ha	Number of grains per ear, pcs	Weight of grain per ear, g	Weight of 1000 grains, g	Biological yield, t/ha
	B_3	2.07	16.9	0.68	40.5	1.40
	B_4	2.09	18.3	0.87	47.5	1.82

By the weight of 1000 grains, the number and weight of grains from the ear, this variant exceeds the control by 9.6, 5.7 and 16.0%, respectively. Here, the maximum biological yield of barley was obtained – 1.82 t/ha, which is 0.22 t/ha higher than the similar variant on the treatment taken as control, and 0.99 t/ha higher than the varint without fertilizers.

Crop accounting showed that the use of microbiological fertilizers twice in terms of mineral background contributed to an increase in the productivity of spring barley (Table 5). So, in 2016, against the background of flat-cutting processing to a depth of 0.20-0.22 m with double application of "Azotovit" and "Phosphatovit", the yield of barley was 1.75 t/ha, which is higher than the control for plowing to a similar depth by 1.0 t/ha. The same pattern was observed in 2017. But, in the wetter 2017, the yield of spring barley was 10-15% higher.

Table 5. Yield of barley grain, t/ha.

Experience variants		Years		
Factor A	Factor B	2016	2017	average
A_1	B_1	0.75	0.84	0.80
	B_2	1.09	1.13	1.11
	B_3	1.35	1.44	1.40
	B_4	1.49	1.55	1.52
A_2	B_1	0.90	0.94	0.92
	B_2	1.20	1.24	1.22
	B_3	1.50	1.65	1.58
	B_4	1.75	1.81	1.78
	LSD_{05} A	0.021	0.022	
	LSD_{05} B	0.032	0.027	
	LSD_{05} AB	0.033	0.027	
	LSD_{05} total	0.051	0.034	

On average, for two years, the maximum yield of barley was formed with the double application of "Azotovit" and "Phosphatovit" by flat-cutting tillage by 0.20-0.22 m and amounted to 1.78 t/ha, which exceeds the control by 44.9%.

The economic assessment was carried out on the basis of operational calculations of direct costs according to technological maps of cultivation of experimental crops and standard production standards and prevailing prices for consumed material and technical resources and sales prices of the received products [12] (Table 6).

From the data given in the table, it can be seen that the most cost-effective options are with double application of bacterial fertilizers against the background of flat-cut tillage at 0.20-0.22 m. The cost of 1 ton of grain on this variant was equal to 3607 rubles, which is lower than the control variant on plowing to a depth of 0.20-0.22 m by 3886 rubles.

From the data shown in the Table 6, it can be seen that the most cost-effective variants are with double application of bacterial fertilizers against the background of flat-cutting tillage at 0.20-0.22 m.

The cost of 1 ton of grain on this variant was equal to 3607 rubles, which is lower than the control variant on plowing to a depth of 0.20-0.22 m by 3886 rubles.

The level of estimated profit on this option is 2892.7 rubles, which is higher than the control by 617.0 rubles.

The maximum level of profitability 80.2% was marked on the A_2B_4 variant, which is 26.3% higher than the similar variant on the control.

Table 6. Economic efficiency of barley cultivation during various tillage and fertilization.

Indicators	Variants							
	A_1				A_2			
	B_1	B_2	B_3	B_4	B_1	B_2	B_3	B_4
Yield, t/ha	0.8	1.11	1.4	1.52	0.92	1.22	1.58	1.78
Cost of funds per 1 ha, RUB	5994	6345	6400	6421	6732	6322	6389	6421
Labor costs per 1 ha, person/hour	6.05	7.05	7.11	7.26	5.11	5.98	6.00	6.05
Sale price, RUB	6500	6500	6500	6500	6500	6500	6500	6500
Cost of 1 ton, RUB	7493	5716	4571	4224	7317	5182	4.044	3607
Cost recovery of 1 ton, RUB	0.9	1.1	1.4	1.5	0.9	1.3	1.6	1.8
Labor intensity of 1 ton, person/hour	7.6	6.4	5.1	4.8	5.6	4.9	3.8	3.4
Estimated profit on:								
1 ton, RUB	-1240.6	706.1	1377.6	1497.1	-888.5	1080.4	1554.6	1625.1
1 hectare, RUB	-992.5	783.8	1928.6	2275.7	-817.4	1318.0	2456.3	2892.7
1 person/hour, RUB	-164.0	111.2	271.2	313.5	-160.0	220.4	409.4	478.1
1 RUB of costs	-0.1	0.1	0.4	0.5	-0.1	0.3	0.6	0.8
Profitability level, %	-13.2	13.7	42.2	53.9	-11.2	25.4	60.7	80.2

4. Conclusion

The results of the conducted studies have shown that the use of bacterial fertilizers on a mineral background in combination with flat-cutting processing to a depth of 0.20-0.22 m increases the content of azotobacter to 98.3%.

The analysis of microbiological processes showed the advantage of binary interaction of flat-cutting processing to a depth of 0.20-0.22 m and double application of bacterial fertilizers on a mineral background. The loss of fiber in this variant was 44.4%, which is 8.6% higher than the same variant in the control.

By the weight of 1000 seeds, the number and weight of grains from the ear, this option exceeds the control by 9.6, 5.7 and 16.0%, respectively.

The use of microbiological fertilizers twice in terms of mineral background contributed to an increase in the productivity of spring barley to 1.78 t/ha.

Economic indicators make it possible to recommend the use of flat-cutting processing to a depth of 0.20-0.22 m with double application of bacterial fertilizers "Azotovit" and "Phosphatovit" in the phases of 1-2 leaves and tillering.

References

[1] Kozlova L M, Popov F A and Noskova E N 2016 The influence of tillage methods and the use of biological products on diseases and crop yields of the crop rotation link *Perm Agrarian Bulletin* **2(14)** 39-44

[2] Kolodziejczyk M, Kulig B, Oleksy A and Szmigiel A 2013 The effectiveness of N-fertilization and microbial preparation on spring wheat *Plant Soil Environ.* **59(8)** 335-341

[3] Kolodziejczyk M 2013 Effect of nitrogen fertilization and application of soil properties improving microbial prepara-tions on the content of mineral nitrogen in soil after spring wheat harvesting *J. of Central European Agriculture* **14(1)** 306-318

[4] Fomin D S, Yamaltdinova V R, Bessonova L V and Vyatkina P I 2018 Application of mineral fertilizers and microbiological preparations for spring wheat on sod-podzolic soil in Perm territory *Perm Agrarian Bulletin* **1(21)** 93-98

[5] Koryagin Yu V, Koryagina N V 2018 The effect of microbiological fertilizers on the productivity and sowing qualities of winter wheat seeds *Niva Povolzhya* **4(49)** 71-78

[6] Koryagin Yu V, Koryagina N V, Efremova S Yu and Koryagina E Yu 2017 Ecological modernization of the technology of biological bacterial preparations *XXI Century: The Results of the Past and the Problems of the Present Plus* **5-6(39-40)** 18-23

[7] Koryagin Yu V, Koryagina N V, Efremova S Yu and Koryagina E Yu 2016 Microbiological preparations as ensuring the environmental friendliness of agricultural *XXI Century: The Results of the Past and the Problems of the Present Plus* **2(30)** 29-34

[8] Zhidkov V M and Feofilova L A 2014 Improvement of methods of cultivation of spring barley in the Volgograd region *Int. Scientific-Practical Conf. "Youth in Science"* **1** 15-18

[9] Zhidkov V M and Feofilova L A 2015 Methods for increasing the efficiency of spring barley cultivation on light chestnut soils of the Volga-Don interfluve *Strategic Development of the Agro-Industrial Complex and Rural Areas of the Russian Federation in Modern Int. Conditions* **1** 95-100

[10] Zhidkov V M and Feofilova L A 2016 The use of bacterial fertilizers in the cultivation of spring barley against the background of moldboard and flat-cutting soil cultivation in the Volgograd region *Proc. of the Nizhnevolzhsky Agrouniversity Complex: Science and Higher Professional Education* **1(41)** 27-33

[11] Zhidkov V M, Chamurliev O G and Feofilova L A 2018 Cultivation of spring barley on light chestnut soils of the Volgograd region *Proc. of the Nizhnevolzhsky Agrouniversity Complex: Science and Higher Professional Education* **1(49)** 66-72

[12] Chamurliev O G, Chamurliev G O, Feofilova L A and Parpura D I 2018 Influence of soil cultivation o bacterial fertilizers on the productivity of spring barley *RUDN J. of Agronomy and Animal Industries* **13(2)** 93-103

Promising agricultural technologies for growing winter wheat for sustainable agricultural development

A V Zelenev, O G Chamurliev, I V Krivtsov, A A Kholod, A N Sidorov and E S Vorontsova

Volgograd State Agrarian University, 26 University Avenue, Volgograd, 400002, Russia

E-mail: Zelenev.A@bk.ru

Abstract. In the arid zone of light chestnut soils of the Lower Volga region, in order to increase the influx of organic matter and basic nutrients into the soil, winter rye and yellow two-year-old clover should be cultivated in occupied fallows as sideral crops. To increase the replenishment of potassium in the soil, it is also recommended to grow phacelia on siderate as a fallow crop. Plowing oats in the form of green fertilizer led to a decrease in the input of organic matter and mineral nutrition elements into the soil compared to the control. The highest content of nitrate nitrogen, mobile phosphorus and exchangeable potassium in the arable layer of the soil during the sowing of winter wheat, when its plants leave for winter and spring regrowth, as well as during harvesting, was provided by the predecessors of the occupied fallows by winter rye, two-year-old clover and phacelia for siderate. The highest yield of winter wheat was obtained by the predecessors the pure fallow and occupied fallows by winter rye, sweet clover and phacelia for siderate. According to the predecessors, the occupied fallows by spring ginger and oats for siderate showed a significant decrease in the yield of winter wheat.

1. Introduction

The preservation and improvement of the fertility of zonal soils is the main problem of modern agriculture. The return of matter and energy back to the soil, instead of crops alienated with harvests, acquires an important role in the system of reproduction of soil fertility. The decrease in the main indicators of soil fertility and field crop yields occurs against the background of intensive use of arable land, a decrease in the share of perennial grasses in the structure of sown areas, a reduction in the use of mineral and, most importantly, organic fertilizers [1, 2, 3, 4].

Siderates are one of the powerful biological resources that improve soil fertility and crop yields. Organic matter, which is obtained as a result of growing green fertilizers, is the main reserve of valuable nutrients for plants. When plowing into the arable soil layer, these substances turn into an easily digestible form for plants not immediately, but over time, during the entire growing season of the crop, ensuring constant plant growth [5, 6, 7].

The most qualitative selection of sideral crops is necessary to increase the efficiency of the use of the occupied fallows in field biologized crop rotations. The cultivated siderates must meet the following important requirements: have the lowest possible seeding rate to reduce the cost of buying seeds, a low transpiration coefficient in order to preserve and efficiently use the productive moisture in the soil, have

IOP Conf. Series: Earth and Environmental Science **965** (2022) 012003 doi:10.1088/1755-1315/965/1/012003

an accelerated growth of biological mass so that it can be plowed into the soil as early as possible, guarantee high yields with a better quality composition [8, 9, 10].

In the arid conditions of light chestnut soils of the Lower Volga region, very little research on the selection, cultivation technology, use of green fertilizers, their effect on the diet and yield of winter wheat has been conducted. This was the reason for the generalization of the obtained research results, which contributed to the scientific confirmation of the use of the occupied fallows and the sideral crops cultivated in them as predecessors of winter wheat and the main biological factor in increasing soil fertility in the Lower Volga region.

2. Materials and methods

Studies on the effectiveness of the occupied fallows and sideral crops grown in them in improving nutrition and stabilizing the yield of winter wheat in field biologized crop rotations were carried out in 2013-2017 and 2017-2020 agricultural years in the dry-steppe zone of chestnut subzone of light chestnut soils of the Lower Volga region in the experimental field of the Federal State Budgetary Research Center of Agroecology and Complex Reclamation of the Russian Academy of Sciences.

The soil of the site is light chestnut medium-sized low-humus heavy loamy with a content of 1.74-2.0% humus in the arable 0-0.3 m layer, easily hydrolyzable nitrogen - 32-36 mg/kg, mobile phosphorus - 18-21 mg/kg and exchangeable potassium - 330-370 mg/kg of absolutely dry soil. The pH of the soil solution in the arable layer was 8.1. The cation exchange capacity was 25.7 mg·eq/100 g of soil.

Two stationary field experiments were laid. In experiment No. 1, the predecessors of winter wheat were: 1) pure fallow (control); 2) occupied fallow (winter rye for siderate); 3) occupied fallow (spring ginger for siderate). In experiment No. 2, the predecessors of winter wheat were: 1) pure fallow (control); 2) occupied fallow (two-year-old sweet clover for siderate); 3) occupied fallow (oats for siderate); 4) occupied fallow (phacelia for siderate).

Agricultural technology of winter wheat cultivation was generally accepted for the research area, except for the studied predecessors and methods of biologization. In experiment No. 1, sideral crops winter rye and spring ginger were sown in the spring. The basic tillage included chiseling at 0.30-0.32 m and turnover of the surface layer to a depth of 0.20-0.22 m with the OCHO-5-40 tool with multifunctional working bodies of the "RANCHO" modular type (blade and wide chisel). In experiment No. 2, the basic tillage was similar in experiment No. 1. Sideral crops, two-year-old sweet clover were sown under the cover of mustard, oats and phacelia by pure sowing in spring in the occupied fallow. In both experiments, in the first decade of June, winter rye in the tillering phase, spring ginger, phacelia and sweet clover in the budding phase, oats in the sweeping phase were crushed and embedded in the soil as a green fertilizer with a heavy disk harrow BDT-3 to a depth of 0.10-0.12 m. Winter wheat of the Kamyshanka 5 variety was sown with a seeding rate of 3.0 million pieces of germinating seeds per 1 ha in the usual ordinary way with row spacing of 0.21 m with a grain seeder Don-114 to a depth of 0.05-0.07 m, which can work according to traditional, minimal and no tillage.

The variants in the experiment were placed in a randomized way in one tier with three repetitions. The area of the experimental plot was 900 m^2, the accounting area was 782 m^2. The mass of crop-root residues of sideral crops was taken into account by the method of monoliths according to Stankov in an eight-fold repetition after their harvesting. The wet and air-dry aboveground mass of sideral crops was determined by selecting sheaves from 0.25 m^2 sites in four-fold repetition and further weighing them in the laboratory. The content of nitrate nitrogen in the soil was determined by the Grandval-Lyaju method with the conversion of the content of nitrates NO_3 to nitrate nitrogen $N\text{-}NO_3$ using a conversion factor of 0.226. The content of mobile phosphorus and exchangeable potassium in the soil was determined by the Machigin method. The economic yield of winter wheat was determined by the weight method using a continuous combine harvesting of each plot. Then it was brought to the standard 14% humidity and 100% purity.

The amount of precipitation in 2013-2014, 2014-2015, 2015-2016, 2016-2017, 2017-2018, 2018-2019 and 2019-2020 agricultural years, respectively, were 435.5, 266.8, 554.8, 374.9, 391.0, 388.3 and 376.7 mm. The average annual precipitation was 339.7 mm.

3. Results and discussion

Previous studies have found that the cultivation and embedding of sideral crops on green fertilizer contributed to the compensation of part of traditional organic fertilizers and, in combination with nitrogen fixation of legumes, led to a reduction in the range of disruption of the organic matter and energy cycle in agrobiocenoses. Plowing 35-40 t/ha of green mass of siderates into the soil equated to 30-40 t of manure. When using sweet clover and phacelia for siderate, the soil received a total aboveground and crop-root absolutely dry mass, respectively, 5.9-9.5 and 6.7-8.8 t/ha, which is 1.8-2.4 times higher than that of oats, where this indicator was 2.8-4.9 t/ha [11, 12, 13]. In our experiments, the amount of aboveground and root air-dry mass of sideral crops that entered the arable soil layer was determined (Table 1).

Table 1. Aboveground and root air-dry mass of sideral crops in experiments, t/ha.

Variant No.	Crop	Air-dry mass		
		root	aboveground	total
	Experience No. 1 (average for 2014-2017)			
2	Winter rye	3.65	1.89	5.54
3	Spring ginger	1.00	1.14	2.14
	Experience No. 2 (average for 2017-2020)			
2	Two-year-old sweet clover	0.74	1.17	1.91
3	Oats	0.53	1.07	1.60
4	Phacelia	0.62	1.09	1.71

Analysis of the Table 1 data shows that in experiment No. 1, the highest air-dry root and aboveground mass, which was plowed into the soil, was provided for winter rye for siderate, respectively, 3.65 and 1.89 t/ha, the lowest for spring ginger for siderate, respectively, 1.00 and 1.14 t/ha. As a result, the total air-dry mass entering the soil in winter rye was 5.54 t/ha, and in spring ginger was 2.14 t/ha, which was lower than in winter rye by 3.4 t/ha or 61.4%.

In experiment No. 2, the highest amount of organic matter with air-dry root and aboveground mass in the soil was supplied with two-year-old sweet clover siderate, respectively, 0.74 and 1.17 t/ha, with air-dry sideral mass of phacelia, respectively, 0.62 and 1.09 t/ha, and the lowest amount - with air-dry sideral mass of oats, respectively, 0.53 and 1.07 t/ha. The largest total air-dry mass of sideral crops was formed in two-year-old sweet clover - 1.91 t/ha, the smallest in oats - 1.60 t/ha, which was lower than in sweet clover by 0.31 t/ha or 16.2%. Phacelia for siderate formed more organic matter than oats by 0.11 t/ha or 6.9% and less than sweet clover by 0.2 t/ha or 10.5%.

Studies have found that the currently existing natural resources of the supply of nutrients to the soil did not compensate for their losses with the cultivated crop yield. Cultivated sideral crops, due to their unique root system, transformed the elements of the mineral nutrition of plants into an accessible form, and also, getting into the soil solution during the period of constant and long mineralization of organic matter, they did not accumulate in excessive amounts in the soil [14, 15, 16]. In our experiments, the regulation of nutrients was carried out due to the entry of organic matter into the soil in the form of a sideral mass (Table 2).

Analysis of the Table 2 data shows that in experiment No. 1, the highest balance of basic nutrients was provided as a result of plowing into the soil of the sideral air-dry root and aboveground mass of winter rye, respectively, nitrogen, phosphorus and potassium - +47.8, +12.6, +26.7 kg/ha and +42.9, +7.1, +23.1 kg/ha. From the root and aboveground air-dry sideral mass of spring ginger, nitrogen, phosphorus and potassium were supplied to the arable soil layer, respectively, +13.2, +3.8, +7.6 kg/ha and +26.0, +4.3, +13.3 kg/ha. The largest total number of basic nutrients coming from the root and aboveground air-dry sideral mass was observed in winter rye: nitrogen - +90.7 kg/ha, phosphorus - +19.7 kg/ha and potassium - +49.8 kg/ha, which was higher than in spring ginger, respectively, nitrogen by 51.5 kg/ha, phosphorus - 11.6 kg/ha and potassium - 28.9 kg/ha.

Table 2. Balance of basic nutrition elements with organic matter of sideral crops, kg/ha.

Variant No.	Crop	Balance, ±								
		Root mass			Aboveground mass			Total		
		N	P_2O_5	K_2O	N	P_2O_5	K_2O	N	P_2O_5	K_2O
		Experience No. 1 (average for 2014-2017)								
2	Winter rye	+47.8	+12.6	+26.7	+42.9	+7.1	+23.1	+90.7	+19.7	+49.8
3	Spring ginger	+13.2	+3.8	+7.6	+26.0	+4.3	+13.3	+39.2	+8.1	+20.9
		Experience No. 2 (average for 2017-2020)								
2	Two-year-old sweet clover	+10.1	+2.0	+5.6	+24.6	+6.2	+12.7	+34.7	+8.2	+18.3
3	Oats	+5.0	+1.6	+4.0	+16.1	+3.0	+13.9	+21.1	+4.6	+17.9
4	Phacelia	+9.0	+1.9	+7.7	+18.2	+5.3	+15.1	+27.2	+7.2	+22.8

In experiment No. 2, the highest positive balance of nitrogen and phosphorus was provided with the root and aboveground air-dry mass of the sweet clover, respectively, +10.1, +2.0 kg/ha and +24.6, +6.2 kg/ha, potassium - in phacelia, respectively, +7.7 and +15.1 kg/ha. The highest total balance in nitrogen and phosphorus was also provided in the two-year-old sweet clover, respectively, +34.7 and +8.2 kg/ha, in potassium - in phacelia - +22.8 kg/ha. The lowest total balance of nutrients was provided in oats: nitrogen - +21.1 kg/ha, phosphorus - +4.6 kg/ha and potassium - +17.9 kg/ha.

It was previously established that the most effective way to increase the reserves of basic nutrients in the soil was considered to be the entry of organic matter in the form of green fertilizers into the arable layer. Monitoring of the dynamics of the basic elements of nutrition is necessary both for understanding the processes that occur in the soil, but also for the implementation of practical measures aimed at improving nitrogen, phosphorus and potassium regimes in grain crops [17, 18, 19, 20]. The result of our studies confirming the dependence of the nutrient content in winter wheat crops on the incoming elements with organic matter of sideral crops is presented in Table 3.

Table 3. The content of basic nutrients in the arable soil layer in winter wheat crops, mg/kg of absolutely dry soil.

Variant No.	Predecessor, biologization method	Sowing			Leave for winter			Spring regrowth			Harvesting		
		N-NO_3	P_2O_5	K_2O	N-NO_3	P_2O_5	K_2O	N-NO_3	P_2O_5	K_2O	N-NO_3	P_2O_5	K_2O
		Experience No. 1 (average for 2014-2017)											
1 (c)	Pure fallow	5.01	32.1	355.8	7.49	28.4	342.0	5.56	27.5	355.5	2.87	26.7	396.4
2	Occupied fallow (winter rye)	7.26	41.8	385.2	9.21	36.9	377.5	8.33	30.6	414.8	4.08	30.6	445.3
3	Occupied fallow (spring ginger)	6.39	31.1	366.2	8.55	26.9	353.8	7.34	25.3	384.1	3.66	29.9	419.6
		Experience No. 2 (average for 2017-2020)											
1 (c)	Pure fallow	6.11	21.5	324.2	7.48	17.6	314.3	7.49	27.1	299.2	4.60	30.8	292.3
2	Occupied fallow (sweet clover)	6.44	30.2	352.9	7.91	26.9	343.9	7.91	32.7	321.6	5.49	33.5	343.3
3	Occupied fallow (oats)	6.28	26.9	338.5	7.84	22.6	329.7	7.69	30.6	308.1	5.70	32.0	320.0
4	Occupied fallow (phacelia)	6.47	27.3	348.9	8.18	23.4	341.3	8.23	27.6	314.6	5.37	31.0	333.4

Analysis of the Table 3 data allows us to conclude that in experiment No. 1, the highest content of nitrate nitrogen in the arable soil layer was provided when winter wheat plants were left for winter by winter rye for siderate – 9.21 mg/kg, which is higher than in the control variant for pure fallow by 1.72

IOP Conf. Series: Earth and Environmental Science **965** (2022) 012003 doi:10.1088/1755-1315/965/1/012003

mg/kg of soil or 23.0%. During this period, the content of nitrate nitrogen in the soil according to the predecessor the occupied fallow by spring ginger for siderate was 8.55 mg/kg, which was also higher than in the control by 1.06 mg /kg of soil or 14.1%. The content of nitrate nitrogen for the spring regrowth of winter wheat plants was slightly lower than when leaving for winter due to low temperatures, which affected the activity of the nitrification process. By sowing winter wheat, the highest content of nitrate nitrogen was provided in the arable soil layer according to the predecessor the occupied fallows by winter rye for siderate – 7.26 mg/kg of soil, which is higher than the control by 2.25 mg/kg or 44.9%. By harvesting winter wheat, the content of nitrate nitrogen in the soil was the lowest - 2.87-4.08 mg/kg of soil. It should be noted that the content of nitrate nitrogen in the arable soil layer less than 10 mg/kg of absolutely dry soil is considered very low according to Grandval-Lyaju.

The highest content of mobile phosphorus in the arable soil layer was provided when sowing winter wheat on the occupied fallow by winter rye for siderate - 41.8 mg/kg of soil, which is higher than the control by 9.7 mg/kg or 30.2%. The content of this element when sowing winter wheat on the occupied fallow of spring ginger for siderate decreased compared to the control variant by 1.0 mg/kg of soil or 3.1%. By the time winter wheat plants left and during spring regrowth, the content of mobile phosphorus in the arable soil layer decreased. During the harvesting of winter wheat, the phosphorus content in the soil for the occupied fallows was 29.9-30.6 mg/kg of soil, which was higher than the control variant, pure fallow, by 3.2-3.9 mg/kg or 12.0-14.6%. It should be noted that according to Machigin's classification, the content of mobile phosphorus in the arable layer from 31 to 45 mg/kg is considered increased, and from 15 to 30 mg/kg of absolutely dry soil is considered average.

The highest content of exchangeable potassium was provided during harvesting of winter wheat in the variant according to the predecessor the occupied fallow by winter rye for siderate – 445.3 mg/kg, which was higher than in the control for pure fallow by 48.9 mg/kg of soil or 12.3%. For the occupied fallow of spring ginger for siderate, the content of this element in the harvesting was higher than in the control by 23.2 mg/kg of soil or 5.8%. Also, a high content of exchangeable potassium in the arable layer was noted during the spring regrowth of winter wheat plants – 414.8 mg/kg of soil, which is 59.3 mg/kg or 16.7% higher than the control variant. In other variants, when sowing winter wheat, leaving for winter and spring regrowth of its plants in the control and according to the occupied fallow of spring ginger for siderate, the content of exchangeable potassium in the arable layer was provided by Machigin at an elevated level - from 301 to 400 mg/kg of absolutely dry soil.

In experiment No. 2, the highest content of nitrate nitrogen in the arable layer was provided during the spring regrowth of winter wheat on the occupied fallow by phacelia for siderate - 8.23 mg/kg of soil, which is higher than the control, the predecessor the pure fallow, by 0.74 mg/kg or 9.9%. Slightly lower content of this nutrient was noted in the same variant, but when winter wheat plants left for winter - 8.18 mg/kg of soil. According to the predecessor the occupied fallow by sweet clover for siderate, the content of nitrate nitrogen during the leaving for winter of winter wheat plants and spring regrowth was the same, and amounted to 7.91 mg/kg of soil. When sowing and harvesting of winter wheat, the content of nitrate nitrogen in the arable layer was also very low and, accordingly, on the occupied fallows equaled 6.28-6.47 and 5.37-5.70 mg/kg of absolutely dry soil.

The highest content of mobile phosphorus in the arable soil layer was provided during the harvesting and spring regrowth of winter wheat for the occupied fallow by sweet clover for siderate, respectively, 33.5 and 32.7 mg/kg of soil, which is higher than the control variant, the predecessor the pure fallow, by 2.7 and 5.6 mg/kg or 8.8 and 20.7%. Also, an increased content of mobile phosphorus was noted in the harvesting of winter wheat on the occupied fallow by oat and phacelia for siderate, pure fallow, respectively, 32.0, 31.0 and 30.8 mg/kg, and during the spring regrowth of plants on the predecessor the occupied fallow by oat for siderate - 30.6 mg/kg of absolutely dry soil. When sowing and leaving for winter of winter wheat, the content of mobile phosphorus was average and, respectively, equaled 6.11-6.47 mg/kg and 17.6-26.9 mg/kg of soil. The average content of this element was also noted during the spring regrowth of winter wheat plants on pure fallow and the occupied fallow by phacelia for siderate, respectively, 27.1 and 27.6 mg/kg of absolutely dry soil.

The highest increased content of exchangeable potassium in the arable layer of the soil was provided when sowing winter wheat according to the predecessor the occupied fallow by sweet clover for siderate - 352.9 mg/kg, which is higher than the control variant, the predecessor the pure fallow, by 28.7 mg/kg of soil or 8.8%. Variants with the occupied fallow by phacelia and oats for siderate also exceeded the control for the content of this element, respectively, by 24.7 and 14.3 mg/kg soil or 7.6 and 4.4%. When winter wheat plants leave for winter, spring regrowth and for harvesting, the best option was also the one where this crop was cultivated according to the occupied fallow by sweet clover for siderate. It should be noted that in terms of the content of exchangeable potassium in the arable layer, all variants corresponded to an increased level and only in the control variant, where the predecessor of winter wheat was pure fallow, the content of this element during spring regrowth and harvesting was average and, respectively, amounted to 299.2 and 292.3 mg/kg of absolutely dry soil.

Previous studies have established that sideral crops for fertilizer in the occupied fallows are an important biological factor in the reproduction of soil fertility and increasing the yield of grain crops in agrocenoses [21, 22, 23, 24, 25]. In our experiments, the yield of winter wheat depended on the input of organic matter into the soil in the form of root and aboveground air-dry mass of sideral crops (Table 4).

Table 4. Influence of predecessors and sideral crops on the yield of winter wheat in experiments, t/ha.

Variant No.	Predecessor, biologization method	Yield
	Experience No. 1 (average for 2014-2017)	
1 (control)	Pure fallow	2.69
2	Occupied fallow (winter rye)	2.68
3	Occupied fallow (spring ginger)	2.29
	LSD_{05}	0.06
	Experience No. 2 (average for 2017-2020)	
1 (control)	Pure fallow	1.80
2	Occupied fallow (sweet clover)	1.80
3	Occupied fallow (oats)	1.68
4	Occupied fallow (phacelia)	1.78
	LSD_{05}	0.09

The analysis of the Table 4 data allows us to conclude that in experiment No. 1, the highest yield of winter wheat was obtained according to the predecessors the pure fallow (control) and the occupied fallow by winter rye for siderate, respectively, 2.69 and 2.68 t/ha. According to the predecessor the occupied fallow by spring ginger for siderate, the yield of winter wheat was significantly inferior to the control variant by 0.4 t/ha or 14.9%.

In experiment No. 2, the yield of winter wheat on predecessors the occupied fallows by sweet clover and phacelia for siderate was at the level with the control variant and amounted to 1.80 and 1.78 t/ha, respectively. In the control, this indicator was 1.80 t/ha. And only according to the predecessor on the occupied fallow by oats for siderate, the yield of winter wheat was significantly inferior to the control by 0.10 t/ha or 6.7%.

4. Conclusion

In arid conditions of light chestnut soils of the Lower Volga region, to improve the nutritional regime and stabilize the yield of winter wheat, it is necessary to grow this grain crop on the occupied fallow by winter rye, sweet clover and phacelia for siderate.

References

[1] Kislov A V, Glinushkin A P, Kashcheev A V and Sudarenkov G V 2018 Ecologization of crop rotations and the biological system of soil fertility reproduction in the steppe zone of the Southern Urals *Agriculture* **6** 6-10

[2] Morozov V I, Toygildin A L and Podsevalov M I 2018 Biologization of crop rotations and productivity of links with winter wheat in the forest-steppe zone of the Volga region *Bulletin of the Ulyanovsk State Agricultural Academy* **3(43)** 84-90
[3] Toygildin A L, Podsevalov M I, Toygildina I A and Dozorov A V 2017 Comparative productivity of crop rotation links with winter wheat during their biologization in the forest-steppe conditions of the Volga region International *Agricultural J.* **6** 53-55
[4] Saidasheva G, Kulikova A, Nikitin S and Lashchenkov A 2020 Influence of fertilizers, including biomodified ones, on the balance of nutrients in the soil and yield capacity in crop rotation *BIO Web of Conf.* **27** 00041
[5] Akhtyamova A A and Eremin D I 2018 Destruction of plant residues on leached chernozem *Proc. of the Orenburg State Agrarian University* **4(72)** 30-33
[6] Borisova E E 2015 The use of green manure in the world *Bulletin NGIEI* **6(49)** 24-33
[7] Chadaev I M and Gurin A G 2020 Accumulation of nutrients in leguminous crops used as a predecessor *Legumes and Cereals* **1(33)** 59-63
[8] Zezin N N, Namyatov M A, Postnikov P A and Zubarev Y N 2019 Assessment of the effectiveness of biologization factors in agriculture of the Ural region *Perm Agrarian Bulletin* **1(25)** 34-41
[9] Zelenev A V, Pleskachev Y N and Seminchenko E B 2018 Field crop rotations providing the highest yield of crop production in the dry steppe zone of the Lower Volga region *Bulletin of the Peoples' Friendship University of Russia. Agronomy and Animal Husbandry* **13** 216-223
[10] Shatokhin A A, Chamurliev O G, Zelenev A V, Chamurliev G O and Vorontsova E S 2020 Field crop rotations in organic agriculture of the Volgograd region *BIO Web of Conf.* **27** 00152
[11] Zelenev A V and Seminchenko E V 2019 Biological methods of increasing soil fertility in organic farming in the Lower Volga region *Bulletin of Kurgan State Agricultural Academy* **1** 4-6
[12] Smurov S I and Popova T V 2015 Evaluation of various types of crops and their combinations as steaming green manure *Achievements of Science and Technology of the Agro-Industrial Complex* **11** 74-77
[13] Shramko N V and Vikhoreva G V 2016 The role of leguminous grasses in the change in the humus content of sod-podzolic soils of the Upper Volga region *Legumes and Cereals* **3(19)** 125-132
[14] Zelenev A V and Seminchenko E V 2019 Influence of predecessors, biologized techniques on the productivity and quality of winter wheat grain in the Lower Volga region *Proc. of the Nizhnevolzhsky Agrouniversity Complex: Science and Higher Professional Education* **3(55)** 65-72
[15] Kaminsky V F, Litvinov D V and Sharonova N L 2014 Biological cycle of organic matter and nutrients in short-rotation crop rotations *Achievements of Science and Technology of the Agro-Industrial Complex* **3** 11-14
[16] Turusov V I, Novichikhin A M, Bogatykh O A and Bocharnikova E G 2017 Biological methods of increasing soil fertility and increasing the productivity of agricultural crops *Achievements of Science and Technology of the Agro-Industrial Complex* **10** 27-31
[17] Goryanin O I, Chichkin A P, Dzhangabaev B Zh and Shcherbinina E V 2017 Content of main nutrients in ordinary chernozem and productivity of field crops in crop rotations of the Middle Volga region *Agrochemical Bulletin* **6** 62-66
[18] Dedov A A, Nesmeyanova M A, Dedov A V and Voronin V I 2016 Influence of biologization techniques and various methods of tillage on fertility and productivity of crop rotation *Voronezh State Agrarian University Bulletin* **3(50)** 47-56
[19] Tyutyunov S I, Navolneva E V, Solovichenko V D and Stupakov A G 2016 Influence of dietary regime and organic matter on soil fertility and yield of winter wheat *Agrochemical Bulletin* **5** 23-27
[20] Kharitonov V V and Shakhov S S 2016 Assessment of the influence of green manure on the agrochemical parameters of the soil *Agrochemical Bulletin* **3** 39-43
[21] Bogomazov S V, Grishin G E, Tikhonov N N and Kochmin A G 2015 The role of agricultural techniques and abiotic factors in the formation of winter wheat yield *Niva Volga region* **2(35)** 2-8

[22] Boluchevsky D A 2014 Yield and quality of winter wheat, depending on the methods of biologization and soil cultivation *Agrochemical Bulletin* **2** 39-40
[23] Turusov V I, Garmashov V M, Abanina O A and Mikhina TI 2016 Siderate fallow as a method of increasing soil fertility and productivity of winter wheat *Int. Research J.* **3-3(45)** 125-126
[24] Khakimov R A, Nikiforova S A and Khakimova N V 2021 Formation of the yield of winter wheat on the occupied fallow, depending on the level of mineral nutrition *Achievements of Science and Technology of the Agro-Industrial Complex* **2** 33-40
[25] Chichkin A P and Goryanin O I 2015 Influence of biologization of soil fertility reproduction systems and technologies on the availability of exchangeable potassium in ordinary chernozem and the productivity of field crops in the Middle Volga region *Bulletin of the Orenburg State Agrarian University* **2(52)** 18-20

Expert system for determining the technical condition of long-term operational structures of water management facilities of the agro-industrial complex

D P Arkov[1], **A A Yastrebov**[2], **A S Matveev**[2] **and A A Sukhov**[1]

[1] Federal Research Center of Agroecology of the Russian Academy of Sciences, 97 University Avenue, Volgograd, 400062, Russia
[2] Volgograd State Agrarian University, 26 University Avenue, Volgograd, 400002, Russia

E-mail: arkov-dmitriy@rambler.ru

Abstract. This paper presents a promising solution to the issue of determining the category of technical condition of hydraulic structures in operation, namely, the collaboration of IT solutions and a set of current building codes and rules in the developed expert system. Most of the reclamation and hydraulic structures operating in the Volgograd region have exceeded their operational life of 25 years laid down in the design. Therefore, it is necessary to control and maintain such structures with the establishment of the technical condition. The developed expert system would contribute to the solution of the listed problems by the operational services of the organizations on the balance of which the structures are located, and the result should comply with all relevant norms and rules.

1. Introduction

At the moment, one of the most urgent problems for the Russian Federation is the problem of maintaining in normal working condition hydraulic structures belonging to different classes.

In Russia, among other things, the new regulation of January 1, 2021 [1] "Regulation on the operation of a hydraulic structure and ensuring the safety of a hydraulic structure whose construction and operation permit has been canceled, as well as a hydraulic structure that does not have an owner" came into force. In short, each hydraulic structure whose construction or operation permit has been revoked is placed under special control, and each owner of such a hydraulic structure or operating organization, in accordance with the instructions of the state supervisory authority, is obliged to develop and implement measures to eliminate the reasons that led to the cancellation of the permit for the construction and operation of the hydraulic structure.

Everywhere on the territory of Russia, the company expires the validity of the permit for the operation of the structure. Most have been in operation for more than 30 years, and some have been in operation for more than 70 years without repair. 3388 hydraulic structures remain unattended [2], most of them are not suitable for operation, for one reason or another, according to Figure 1, 313 more structures were identified compared to the previous year, 5 structures were liquidated during the year, 201 structures were excluded from the list of ownerless hydraulic structures after the survey by the decisions of the emergency commissions, as they lost the signs of hydraulic structures and do not pose a danger. And today there is an acute question of how to profitably and effectively put into operation deplorable structures. Designing a hydraulic structure is one of the most challenging engineering tasks, as it requires extensive knowledge in structural engineering, fluid mechanics, geotechnics, and materials science among other fields, as well as innovation skills, since every hydraulic structure is somehow a prototype. While the general purpose of hydraulic structures has not changed since early times, design methods and analysis tools have evolved significantly in recent years. Currently, computational fluid dynamics and artificial intelligence modeling methods are common additions to more classical physical modeling and direct measurement tools that help analyze the complex features of fluid flow around hydraulic structures [3].

Gavrilova and Khoroshevsky define expert systems as follow: "These are complex software systems that accumulate the knowledge of specialists in specific subject areas and replicate this empirical experience for consulting less qualified user" [4].

These tools can help better address the new challenges posed by social and climate change. The underlying uncertainties require greater design flexibility to achieve the current project objectives, while allowing for adaptation to new conditions. Existing hydraulic structures, in turn, are often older than their original intended service life. This is a common problem all over the world, and hydraulic engineers are increasingly faced with the task of finding innovative solutions to ensure the safe operation of the structure in the future, taking into account social, economic and environmental problems. These circumstances indicate the need to create an expert system that allows you to solve these problems.

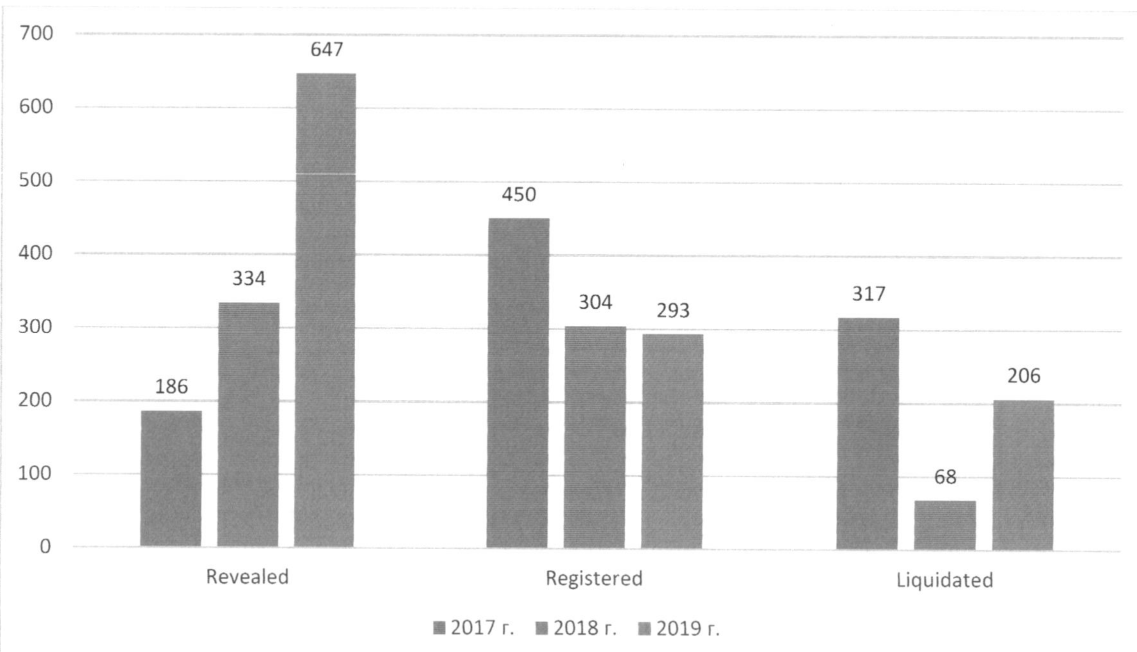

Figure 1. Ensuring the safety of ownerless hydraulic structures.

2. Materials and methods

During the examination of structures, the technical condition of structures, components and the building as a whole is evaluated for a specific period of time. At the same time, the main thing is to identify defects, damage, assess the causes of occurrence, the degree of danger and the forecast of the remaining strength resource.

The result of the expert examination is a written opinion that competently assesses the causes that led to an accident, an accident of building structures or structures. It can be continuous (for example, design and estimate documentation), one-time, departmental, selective in the order of control of a higher-level organization, etc. However, the results of the classical examination can be obtained after months of work, but everything can be simplified thanks to the expert system and the competent layout of requests.

Expert systems - automated systems focused on solving problems that are difficult to unambiguously and formalized description and are usually solved on the basis of experience and informal logic (heuristic methods), usually with the involvement of highly qualified experts. Systems are based on the concentration of the maximum possible number of forms and heuristic knowledge from the most qualified specialists in a particular application area, and then use this knowledge to solve the same problems that specialists usually face in this field. A significant difference between developed expert systems and conventional computer systems used to support decision-making is such an element of artificial intelligence as the ability to self-learn.

The use of a combination of formal and heuristic knowledge and the rejection of formal algorithms, which very conditionally reflect the variety of ways to solve specific problems, increase the adequacy of the actions of the systems under consideration to real conditions in comparison with ordinary ones. At the same time, it becomes possible to quickly process a large amount of information, which is difficult (sometimes impossible) for a human expert. The practical use of such systems looks like an active human-machine dialogue (interactive mode), during which not only a person asks questions to the machine, but also vice versa. In addition, the user can, if desired, find out the reason for making a particular decision, i.e., without delving into the essence of the software, get an explanation of the actions of the machine when justifying the result of the decision. Some of the developments themselves recognize when the user needs help. In practice, a corrective procedure is used to determine the degree of confidence of each potential solution in order to cut off the wrong solutions and leave the acceptable ones.

However, the expert still needs to examine the design to obtain an initial conclusion, but further, the extensive knowledge base of the expert system can quickly make a final conclusion and determine the necessary resources for repairs.

Any expert system should have a data output and a sequence of "thinking" of the system, its structure is shown in Figure 2.

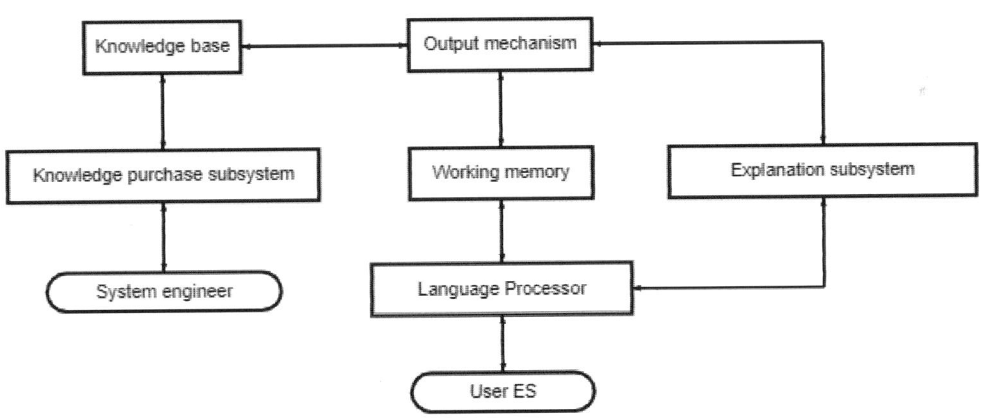

Figure 2. Structure of the expert system.

The basis for the functioning of this system is the knowledge base. Unlike databases, which are the information support of traditional systems, it contains two groups of knowledge: declarative (facts about a specific application area) and procedural (heuristic methods or rules for solving problems, including the development of hypotheses, information processing, and logic for obtaining conclusions). In addition to the knowledge base, the expert system is introduced: for communication of the user in a language that is understandable to him, working memory for storing preliminary hypotheses and results that the system

IOP Conf. Series: Earth and Environmental Science **965** (2022) 012004 doi:10.1088/1755-1315/965/1/012004

comes to during the solution of the problem, an acquisition component that performs the correction procedure of knowledge and a component that explains the actions to the user.

3. Results and discussion

The development is based on an expert system [5], it is designed to consult with the user in any application area (for which the downloaded knowledge base is configured) in order to determine probabilities.

To work with the expert system, it is necessary to download the knowledge base, then at the start of the consultation, it offers the specialist to select the necessary categories of structural defects (Figure 3).

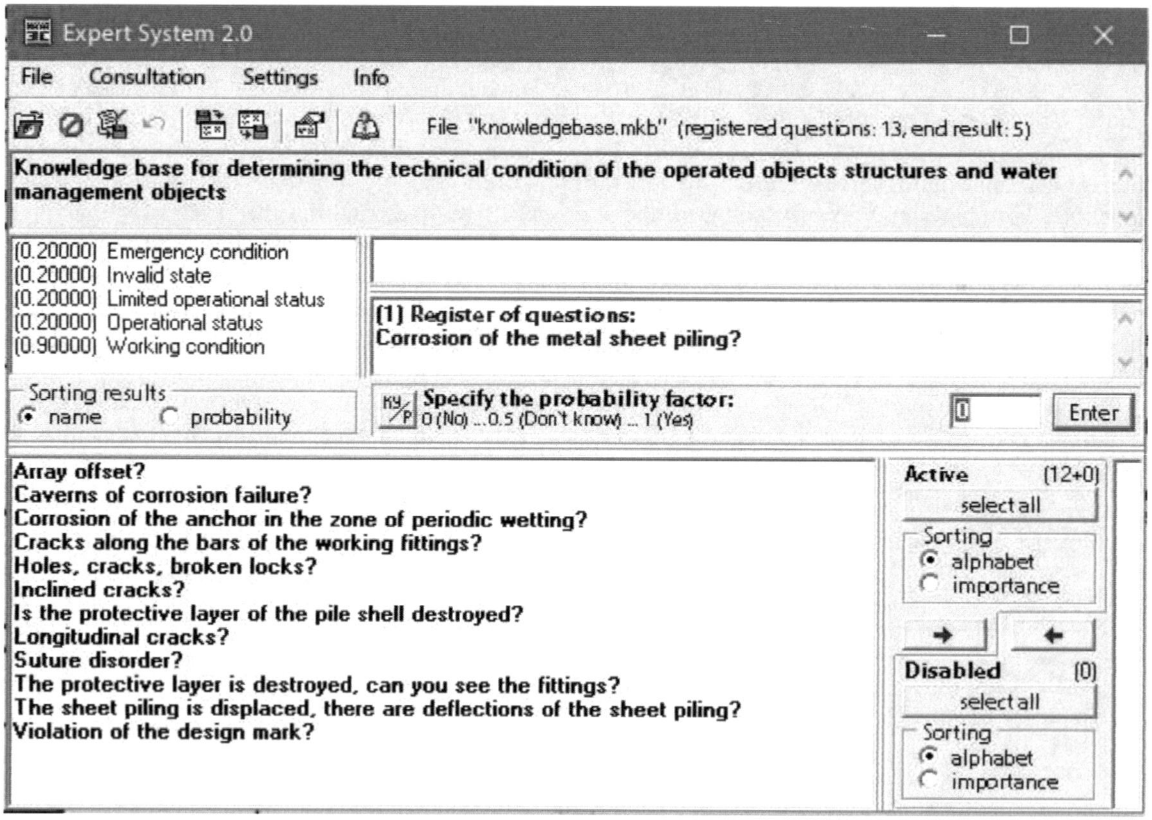

Figure 3. The main form of the expert system.

After the end of the consultation, the program will highlight the status according to the completed list of questions. To determine further work with the structure, it is necessary to familiarize yourself with the state assessments according to the code of rules for design and construction 13-102-2003 [6]. For each condition of the structures, there are necessary recommendations for maintenance or measures to restore and strengthen them. Additional methods of non-destructive quality control of the structure can also be applied, which is the ultrasonic method [7, 8, 9].

With the completion of the software, the addition of a neural network structure capable of analyzing the source data, it is possible to obtain an advanced product in the diagnosis of structures. The essence is simple, the user uploads images of the object, neural networks recognize defects, visually assess the state of the object, only on the recognized defects begins a consultation with the user, and then issues a decision on the state of the object. This will help to reduce the list of consulting issues, thereby coming to a solution in the shortest possible time.

The principle of recognition will work on a ready-made sample, which contains all classes of damage and defects in building structures of buildings and structures. For each image, there is a set of features that have been highlighted by some feature, whether it is Haar, HOG, SURF, or some kind of wavelet. The learning algorithm must build a model that allows it to analyze the new image and decide which of the defects is present in the image. How is this done? Each of the images is a point in the feature space. Its coordinates are the weight of each of the features in the image. Let our signs be: "The presence of bare reinforcement", "The destruction of a concrete structure", examples in Figure 4.

Figure 4. Signs of structural defects.

All these signs will be identified by our existing detectors, which are trained to detect rebar, or the presence of cracks and peeling, all that characterizes a design defect. For a structure with a serious defect in such a space, the correct point will be [1; 1; 1; 1…]. For a less serious defect, the point [1; 0; 1; 0...]. For a serviceable structure [1; 0; 0; 0...]. The classifier is trained from a sample of examples. But not all of the images can be seen as cracks, others are peeling, and on the third, a serviceable structure may have bare reinforcement due to a classifier error. The trained defect classifier automatically splits the feature space in such a way as to say: if the first feature lies in the range $0.5 < x < 1$, the second $0.7 < y < 1$, etc., then this is definitely a serious design defect.

In essence, the purpose of the classifier is to draw areas in the feature space that are characteristic of classification objects. Figure 5 shows a sequential approximation to the answer for one of the AdaBoost classifiers. AdaBoost combines weak classifiers with specific strategies to produce a strong classifier [10].

IOP Conf. Series: Earth and Environmental Science **965** (2022) 012004 doi:10.1088/1755-1315/965/1/012004

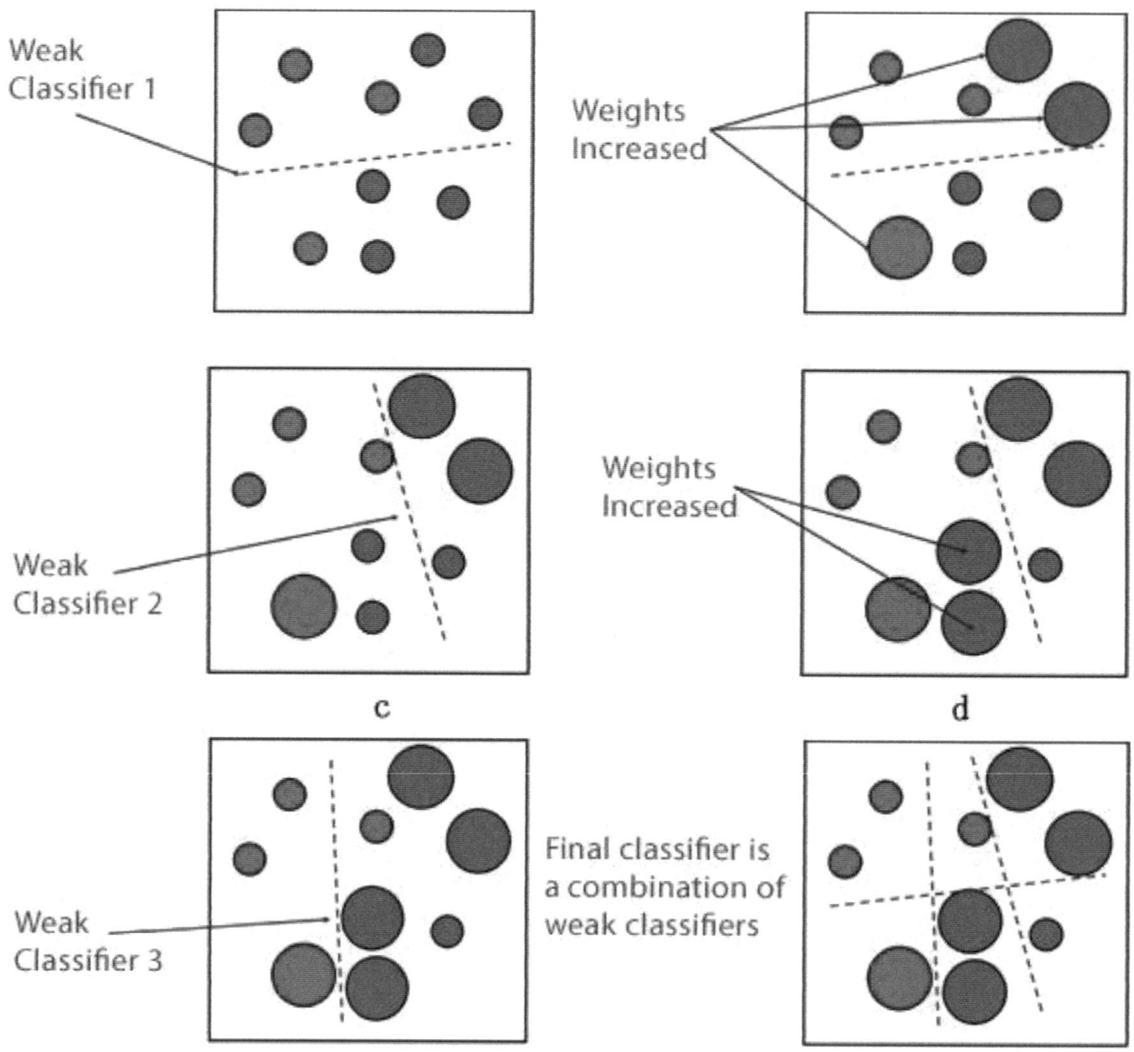

Figure 5. Combining AdaBoost classifiers.

4. Conclusion

Expert systems are widely used in our life, design and survey activities should not be an exception when examining structures of water management facilities of the agro-industrial complex. The developed expert system allows solving the problems of long-term diagnostics of the structure, making a practical assessment of the state of the structure immediately after a detailed examination, applying the necessary technological processes in time and constantly predicting design defects at the initial stage for successful control of the structure. Reducing the time of diagnosis will significantly help to solve the problem of the state of hydraulic structures.

References

[1] On the government commission for administrative reform on approval of the regulations on the operation of a hydraulic structure and ensuring the safety of a hydraulic structure whose construction and operation permit has been revoked (including a hydraulic structure in an emergency condition), a hydraulic structure that does not have an owner or whose owner is

unknown, or whose owner has renounced the right of ownership Resolution of the Government of the Russian Federation No. 1606 of October 05 2020 *Sobranie Zakonodatelstva* **41** 6446

[2] Annual report on the activities of the Federal Service for Environmental, Technological and Nuclear Supervision in 2019 2020 (Moscow: Scientific and Technical Center For Industrial Safety Research) p 389

[3] Gondarev V V 2021 Automation of monitoring of the hydraulic structures-development of the RPC "phase" *Hydraulic Engineering* **1(62)** 34-35

[4] Gavrilova T A and Khoroshevsky V F 2000 *Knowledge Bases of Intellectual Systems* (St. Petersburg: Piter) p 384

[5] *Small Expert System 2.0 Knowledge Base Editor 1.0, User Manual* 2000 (Kiev: National Technical University of Ukraine Kyiv Polytechnic Institute)

[6] Rules for Design and Construction *13-102-2003 Rules for the Inspection of Load-Bearing Building Structures of Buildings and Structures* 2003 (Moscow: Center of Design Products in Construction)

[7] Semenenko S Ya, Arkov D P and Marchenko S S 2018 Verification and practical implementation of the method of brand diagnostics for waterproofness of concrete structures of meliorative structures *Proc. of the Nizhnevolzhsky Agrouniversity Complex: Science and Higher Professional Education* **1(49)** 254-262

[8] Sambursky G A, Lukyanov O V and Khrapov I V 2011 An approach to the construction of hybrid forecasting systems based on neural networks *Transactions of the TSTU* **4** 932-935

[9] Ali M A 2006 Neural networks time series forecasting *Mat. Int. Scientific Conf. Mathematical methods in Engineering and Technology* vol 6 (Voronezh) pp 46-48

[10] Stadnik A V, Kravchuk A V and Gulina K I 2014 Building classifiers using artificial neural networks and the Ada boost principle *Discrete and Continuous Models and Applied Computational Science* **2** 431-436

IOP Conf. Series: Earth and Environmental Science **965** (2022) 012005 doi:10.1088/1755-1315/965/1/012005

Calculation of the strength of objects of water management of the environment

N A Gureeva[1], R Z Kiseleva[2], A P Nikolaev[2], A P Kiselev[2] and V V Ryabukha[2]

[1] Federal State Educational Budgetary Institution of Higher Education "Financial University under the Government of the Russian Federation", 49 Leningradsky Avenue, Moscow, 125993 (City service post office-3), Russia
[2] Federal State Educational Institution of Higher Education "Volgograd State Agrarian University" 26 University Avenue, Volgograd, 400002, Russia

E-mail: nagureeve@fa.ru

Abstract. The article discusses the objects of water management of the environment. An algorithm for forming a matrix of the stress-strain state of a prismatic finite element with a triangular cross-section in a mixed formulation is proposed to take into account the physical nonlinearity of water management objects in the form of rotation shells. The basic relations of the flow theory for arbitrary loading are used. The desired displacements and stresses of the inner point of the finite element were approximated by linear functions in terms of their nodal values. The stress-strain state matrix was formed using a functional expressing the equality of the actual work of external and internal work of external and internal forces, in which the actual work of internal forces was represented by the difference between the total and additional work of these forces.

1. Introduction

Calculations for the strength of nonlinearly deformable shells of rotation are performed using the finite element method in various formulations [1, 2, 3, 4, 5]. Finite elements in the mixed formulation have a number of advantages due to the continuity of the desired values not only at the nodes of the finite element, but also at its borders [6, 7, 8]. To take into account the physical nonlinearity of rotation shells, the relations of the deformation theory of plasticity or flow theory are used [9]. With the step loading method, the total increments of deformations are the sum of the increments of elastic and plastic deformations.

In this paper, a finite element of a prismatic shape with a triangular cross-section is developed in a mixed formulation.

The dependences of increments of elastic deformations on increments of stresses in the plastic region are expressed by Hooke's law [9].

The increments of plastic deformations were determined on the basis of the hypothesis that the increments of plastic deformations are proportional to the components of the total stress deviator. The proportionality coefficient is a function of the ratio of the increment of stress intensity to stress intensity. After representing the increment of stress intensity in terms of stress increments, matrix dependences of strain increments on stress increments are formed.

The matrix of the stress-strain state of the finite element at the loading step is formed using a functional based on the equality of possible and actual work of external and internal forces with the replacement of the actual work of stress increments at the loading step by the difference between the total and additional work [10, 11].

2. Materials and methods

2.1. Flow theory relations

At the loading step, the increments of deformations are assumed to be the sum of the increments of elastic and plastic deformations [9]:

$$\Delta\varepsilon_{ij} = \Delta\varepsilon_{ij}^{e} + \Delta\varepsilon_{ij}^{p}, \tag{1}$$

where $\Delta\varepsilon_{ij}$ - total increments of deformations; $\Delta\varepsilon_{ij}^{e}$ - increment of elastic deformations; $\Delta\varepsilon_{ij}^{p}$ - increments of plastic deformations.

Increments of elastic deformations are related to stress increments by Hooke's law [9]:

$$\begin{gathered}
\Delta\varepsilon_{rr}^{e} = \frac{1}{E}\Delta\sigma_{rr} - \frac{\vartheta}{E}(\Delta\sigma_{\theta\theta} + \Delta\sigma_{zz}); \\
\Delta\varepsilon_{\theta\theta}^{e} = \frac{1}{E}\Delta\sigma_{\theta\theta} - \frac{\vartheta}{E}(\Delta\sigma_{rr} + \Delta\sigma_{zz}); \\
\Delta\varepsilon_{zz}^{e} = \frac{1}{E}\Delta\sigma_{zz} - \frac{\vartheta}{E}\left(\Delta\sigma_{rr} + \Delta\sigma_{\theta\theta}\right); \\
\Delta\varepsilon_{r\theta}^{e} = \frac{1+\vartheta}{E}\Delta\varepsilon_{r\theta};\ \Delta\varepsilon_{rz}^{e} = \frac{1+\vartheta}{E};\ \Delta\varepsilon_{\theta z}^{e} = \frac{1+\vartheta}{E},
\end{gathered} \tag{2}$$

where E - is the elastic modulus of the material; ϑ - is the coefficient of transverse deformation; $\Delta\sigma_{ij}$ - and is the stress increment.

Increments of plastic deformations are determined by the relations [9]:

$$\begin{gathered}
\Delta\varepsilon_{rr}^{p} = \frac{3}{2}\frac{\varepsilon_i^{\Delta p}}{\sigma_i}(\sigma_{rr} - \sigma_c);\ \Delta\varepsilon_{\theta\theta}^{p} = \frac{3}{2}\frac{\varepsilon_i^{\Delta p}}{\sigma_i}(\sigma_{\theta\theta} - \sigma_c); \\
\Delta\varepsilon_{zz}^{p} = \frac{3}{2}\frac{\varepsilon_i^{\Delta p}}{\sigma_i}(\sigma_{zz} - \sigma_c);\ \Delta\varepsilon_{r\theta}^{p} = \frac{3}{2}\frac{\varepsilon_i^{\Delta p}}{\sigma_i}\Delta\sigma_{r\theta}; \\
\Delta\varepsilon_{rz}^{p} = \frac{3}{2}\frac{\varepsilon_i^{\Delta p}}{\sigma_i}\Delta\sigma_{rz};\ \Delta\varepsilon_{\theta z}^{p} = \frac{3}{2}\frac{\varepsilon_i^{\Delta p}}{\sigma_i}\Delta\sigma_{\theta z},
\end{gathered} \tag{3}$$

where $\Delta\varepsilon_i^{\Delta p}$ –intensity of increments of plastic deformations ; σ_i - intensity of stresses; σ_{ij} - stresses; $\sigma_c = \dfrac{\sigma_{rr} + \sigma_{\theta\theta} + \sigma_{zz}}{3}$ - average stress.

For small deformations at the loading step, the following equalities are assumed:

$$\varepsilon_i^{\Delta p} = \Delta\varepsilon_i^{p};\ \Delta\varepsilon_i = \Delta\varepsilon_i^{e} + \Delta\varepsilon_i^{p}, \tag{4}$$

where $\Delta\varepsilon_i^{p}$ – increment of the intensity of plastic deformations; $\Delta\varepsilon_i$ - increment of the intensity of deformations; $\Delta\varepsilon_i^{e}$ – increment of the intensity of elastic deformations.

At the loading step, the relations take place between increments of strain intensity and increments of stress intensity:

$$\Delta\varepsilon_i^e = \frac{\Delta\sigma_i}{E_1};\ \Delta\sigma_i = E_k \Delta\varepsilon_i, \tag{5}$$

where E_1 is the elastic modulus of the initial section of the material deformation diagram; E_k is the tangent modulus of the deformation diagram at the point under consideration; $\Delta\sigma_i$ is the increment of stress intensity at the loading step.

The increment in the intensity of plastic deformations based on the relations (4,5) is written by the expression:

$$\Delta\varepsilon_i^p = \Delta\varepsilon_i - \Delta\varepsilon_i^e = \frac{\Delta\sigma_i}{E_k} - \frac{\Delta\sigma_i}{E_1} = \Delta\sigma_i\left(\frac{1}{E_k} - \frac{1}{E_1}\right). \tag{6}$$

The increment of stress intensity can be determined by the following general formula

$$\Delta\sigma_i = \frac{\partial\sigma_i}{\partial\sigma_{rr}}\Delta\sigma_{rr} + \frac{\partial\sigma_i}{\partial\sigma_{\theta\theta}}\Delta\sigma_{\theta\theta} + \frac{\partial\sigma_i}{\partial\sigma_{zz}}\Delta\sigma_{zz} + \\ + \frac{\partial\sigma_i}{\partial\sigma_{r\theta}}\Delta\sigma_{r\theta} + \frac{\partial\sigma_i}{\partial\sigma_{\theta z}}\Delta\sigma_{\theta z} + \frac{\partial\sigma_i}{\partial\sigma_{rz}}, \tag{7}$$

where

$$\sigma_i = (\sigma_{rr}^2 + \sigma_{\theta\theta}^2 + \sigma_{zz}^2 - \sigma_{rr}\sigma_{\theta\theta} - \sigma_{rr}\sigma_{zz} - \sigma_{\theta\theta}\sigma_{zz} \\ + 3\sigma_{r\theta}^2 + 3\sigma_{\theta z}^2 + 3\sigma_{rz}^2)^{\frac{1}{2}}$$ - stress intensity;

$$\frac{\partial\sigma_i}{\partial\sigma_{rr}} = \frac{\sigma_{rr} - \frac{1}{2}(\sigma_{\theta\theta} + \sigma_{zz})}{\sigma_i} = s_{11};$$

$$\frac{\partial\sigma_i}{\partial\sigma_{\theta\theta}} = \frac{\sigma_{\theta\theta} - \frac{1}{2}(\sigma_{rr} + \sigma_{zz})}{\sigma_i} = s_{22};$$

$$\frac{\partial\sigma_i}{\partial\sigma_{zz}} = \frac{\sigma_{zz} - \frac{1}{2}(\sigma_{rr} + \sigma_{\theta\theta})}{\sigma_i} = s_{33};$$

$$\frac{\partial\sigma_i}{\partial\sigma_{r\theta}} = \frac{1}{2\sigma_i}6\sigma_{r\theta} = 3s_{12};\ \frac{\partial\sigma_i}{\partial\sigma_{rz}} = \frac{1}{2\sigma_i}6\sigma_{rz} = 3s_{13};$$

$$\frac{\partial\sigma_i}{\partial\sigma_{\theta z}} = \frac{1}{2\sigma_i}6\sigma_{\theta z} = 3s_{23}.\ \psi_1 = \frac{3}{2}\frac{\Delta\varepsilon_i^p}{\Delta\sigma_i}.$$

Increments of plastic deformations (3) at the loading step, taking into account (6) and (7), can be written as:

$$
\begin{aligned}
&\Delta\varepsilon_{rr}^{p} = \Delta\sigma_{rr}\frac{3}{2}\frac{s_{11}}{\sigma_i}\varphi(\sigma_{rr}-\sigma_c)+\Delta\sigma_{\theta\theta}\frac{3}{2}\frac{s_{22}}{\sigma_i}\varphi(\sigma_{rr}-\sigma_c)+\\
&+\Delta\sigma_{zz}\frac{3}{2}\frac{s_{33}}{\sigma_i}\varphi(\sigma_{rr}-\sigma_c)+\Delta\sigma_{r\theta}\frac{3}{2}\frac{3s_{12}}{\sigma_i}\varphi(\sigma_{rr}-\sigma_c)+\\
&+\Delta\sigma_{rz}\frac{3}{2}\frac{3s_{13}}{\sigma_i}\varphi(\sigma_{rr}-\sigma_c)+\Delta\sigma_{\theta z}\frac{3}{2}\frac{3s_{23}}{\sigma_i}\varphi(\sigma_{rr}-\sigma_c);\\
&\Delta\varepsilon_{\theta\theta}^{p} = \Delta\sigma_{rr}\frac{3}{2}\frac{s_{11}}{\sigma_i}\varphi(\sigma_{\theta\theta}-\sigma_c)+\Delta\sigma_{\theta\theta}\frac{3}{2}\frac{s_{22}}{\sigma_i}\varphi(\sigma_{\theta\theta}-\sigma_c)+\\
&+\Delta\sigma_{zz}\frac{3}{2}\frac{s_{33}}{\sigma_i}\varphi(\sigma_{\theta\theta}-\sigma_c)+\Delta\sigma_{r\theta}\frac{3}{2}\frac{3s_{12}}{\sigma_i}\varphi(\sigma_{\theta\theta}-\sigma_c)+\\
&+\Delta\sigma_{rz}\frac{3}{2}\frac{3s_{13}}{\sigma_i}\varphi(\sigma_{\theta\theta}-\sigma_c)+\Delta\sigma_{\theta z}\frac{3}{2}\frac{3s_{23}}{\sigma_i}\varphi(\sigma_{\theta\theta}-\sigma_c);\\
&\Delta\varepsilon_{zz}^{p} = \Delta\sigma_{rr}\frac{3}{2}\frac{s_{11}}{\sigma_i}\varphi(\sigma_{zz}-\sigma_c)+\Delta\sigma_{\theta\theta}\frac{3}{2}\frac{s_{22}}{\sigma_i}\varphi(\sigma_{zz}-\sigma_c)+\\
&+\Delta\sigma_{zz}\frac{3}{2}\frac{s_{33}}{\sigma_i}\varphi(\sigma_{zz}-\sigma_c)+\Delta\sigma_{r\theta}\frac{3}{2}\frac{3s_{12}}{\sigma_i}\varphi(\sigma_{zz}-\sigma_c)+\\
&+\Delta\sigma_{rz}\frac{3}{2}\frac{3s_{13}}{\sigma_i}\varphi(\sigma_{zz}-\sigma_c)+\Delta\sigma_{\theta z}\frac{3}{2}\frac{3s_{23}}{\sigma_i}\varphi(\sigma_{zz}-\sigma_c);\\
&\Delta\varepsilon_{r\theta}^{p} = \Delta\sigma_{ss}\frac{3}{2}\frac{s_{11}}{\sigma_i}\varphi\,\sigma_{r\theta}+\Delta\sigma_{\theta\theta}\frac{3}{2}\frac{s_{22}}{\sigma_i}\varphi\,\sigma_{r\theta}+\\
&+\Delta\sigma_{zz}\frac{3}{2}\frac{s_{33}}{\sigma_i}\varphi\,\sigma_{r\theta}+\Delta\sigma_{r\theta}\frac{3}{2}\frac{3s_{12}}{\sigma_i}\varphi\,\sigma_{r\theta}+\\
&+\Delta\sigma_{rz}\frac{3}{2}\frac{3s_{13}}{\sigma_i}\varphi\,\sigma_{r\theta}+\Delta\sigma_{\theta z}\frac{3}{2}\frac{3s_{23}}{\sigma_i}\varphi\,\sigma_{r\theta};\\
&\Delta\varepsilon_{rz}^{p} = \Delta\sigma_{ss}\frac{3}{2}\frac{s_{11}}{\sigma_i}\varphi\,\sigma_{rz}+\Delta\sigma_{\theta\theta}\frac{3}{2}\frac{s_{22}}{\sigma_i}\varphi\,\sigma_{rz}+\\
&+\Delta\sigma_{zz}\frac{3}{2}\frac{s_{33}}{\sigma_i}\varphi\,\sigma_{rz}+\Delta\sigma_{r\theta}\frac{3}{2}\frac{3s_{12}}{\sigma_i}\varphi\,\sigma_{rz}+\\
&+\Delta\sigma_{rz}\frac{3}{2}\frac{3s_{13}}{\sigma_i}\varphi\,\sigma_{rz}+\Delta\sigma_{\theta z}\frac{3}{2}\frac{3s_{23}}{\sigma_i}\varphi\,\sigma_{rz};
\end{aligned}
\tag{8}
$$

$$\Delta\varepsilon^{p}_{\theta z} = \Delta\sigma_{ss}\,\frac{3}{2}\frac{s_{11}}{\sigma_i}\,\varphi\,\sigma_{\theta z} + \Delta\sigma_{\theta\theta}\,\frac{3}{2}\frac{s_{22}}{\sigma_i}\,\varphi\,\sigma_{\theta z} +$$

$$+\Delta\sigma_{zz}\,\frac{3}{2}\frac{s_{33}}{\sigma_i}\,\varphi\,\sigma_{\theta z} + \Delta\sigma_{r\theta}\,\frac{3}{2}\frac{3s_{12}}{\sigma_i}\,\varphi\,\sigma_{\theta z} +$$

$$+\Delta\sigma_{rz}\,\frac{3}{2}\frac{3s_{13}}{\sigma_i}\,\varphi\,\sigma_{\theta z} + \Delta\sigma_{\theta z}\,\frac{3}{2}\frac{3s_{23}}{\sigma_i}\,\varphi\,\sigma_{rz},$$

where $\varphi = \frac{1}{E_\kappa} - \frac{1}{E_1}$.

Relations (8) can be written in matrix form:

$$\underset{6x1}{\{\Delta\varepsilon\}} = \underset{6x6}{[C]}\underset{6x1}{\{\Delta\sigma\}}, \tag{9}$$

where $\underset{1x6}{\{\Delta\varepsilon\}}^T = \{\Delta\varepsilon_{rr}\ \Delta\varepsilon_{\theta\theta}\ \Delta\varepsilon_{zz}\ 2\Delta\varepsilon_{r\theta}\ 2\Delta\varepsilon_{rz}\ 2\Delta\varepsilon_{\theta z}\}$ -matrix-string of strain increments; $\underset{6x6}{[C]}$ - matrix of material compliance; $\underset{1x6}{\{\Delta\sigma\}}^T = \{\Delta\sigma_{rr}\ \Delta\sigma_{\theta\theta}\ \Delta\sigma_{zz}\ \Delta\sigma_{r\theta}\ \Delta\sigma_{rz}\ \Delta\sigma_{\theta z}\}$ - matrix-a string of stress increments.

2.2. Displacements and deformations

Under the action of a load, the shell deforms and its arbitrary point receives a displacement described by the vector:

$$\vec{V} = \upsilon^1\vec{g}^{\,0}_1 + \upsilon^2\vec{g}^{\,0}_2 + \upsilon^3\vec{g}^{\,0}_3 = \underset{1x3}{\left\{\vec{g}^{\,0}\right\}}\underset{3x1}{\{\upsilon\}}, \tag{10}$$

where $\underset{1x3}{\left\{\vec{g}^{\,0}\right\}}^T = \left\{\vec{g}^{\,0}_1\ \vec{g}^{\,0}_2\ \vec{g}^{\,0}_3\right\}$ - matrix is a string of basis vectors of an arbitrary point of the shell; $\underset{1x3}{\{\upsilon\}}^T = \left\{\upsilon^1\ \upsilon^2\ \upsilon^3\right\}$ - matrix is a string of point displacements.

Increments of deformations at an arbitrary point of the shell are expressed in terms of its displacements by the Cauchy relations:

$$\Delta\varepsilon_{rr} = \frac{\partial\upsilon^1}{\partial r};\ \Delta\varepsilon_{\theta\theta} = \frac{\partial\upsilon^2}{\partial\theta};\ \Delta\varepsilon_{zz} = \frac{\partial\upsilon^3}{\partial z};$$

$$\Delta\varepsilon_{r\theta} = \frac{1}{2}\left(\frac{\partial\upsilon^1}{\partial\theta} + \frac{\partial\upsilon^2}{\partial r}\right);\ \Delta\varepsilon_{rz} = \frac{1}{2}\left(\frac{\partial\upsilon^1}{\partial z} + \frac{\partial\upsilon^3}{\partial r}\right);\ \Delta\varepsilon_{\theta z} = \frac{1}{2}\left(\frac{\partial\upsilon^2}{\partial z} + \frac{\partial\upsilon^3}{\partial\theta}\right), \tag{11}$$

which are represented in matrix form:

$$\underset{6x1}{\{\Delta\varepsilon\}} = \underset{6x3}{[L]}\underset{3x1}{\{\upsilon\}}, \tag{12}$$

where $\underset{6\times3}{[L]}$ is the matrix of differential operators.

3. Results and discussion

The finite element is a prism with a cross-section in the form of arbitrary triangles with nodes *i, j, k, l*. To perform numerical integration, the finite element is mapped to a local prism with a base in the form of a right triangle $0 \le \xi, \eta \le 1$ and with a height coordinate that varies within $-1 \le \zeta \le 1$. The global coordinates of the inner point of the final element are determined in terms of nodal coordinates by linear relations of local coordinates ξ, η, ζ

$$r = \underset{1x6}{\left\{ f(\xi,\eta,\zeta) \right\}}^T \underset{6x1}{\{r_y\}};\ \theta = \underset{1x6}{\left\{ f(\xi,\eta,\zeta) \right\}}^T \underset{6x1}{\{\theta_y\}};\ z = \underset{1x6}{\left\{ f(\xi,\eta,\zeta) \right\}}^T \underset{6x1}{\{z_y\}}, \tag{13}$$

where $\underset{1x6}{\left\{ f(\xi,\eta,\zeta) \right\}}^T = \left\{ \begin{matrix} (1-\xi-\eta)\frac{1-\zeta}{2} : \xi\frac{1-\zeta}{2} : \eta\frac{1-\zeta}{2} : \\ (1-\xi-\eta)\frac{1+\zeta}{2} : \xi\frac{1+\zeta}{2} : \eta\frac{1+\zeta}{2} \end{matrix} \right\}$ - shape functions;

$\underset{1x6}{\{r_y\}}^T = \left\{ r^i\ r^j\ r^k\ r^m\ r^n\ r^p \right\};\ \underset{1x6}{\{\theta_y\}}^T = \left\{ \theta^i \dots \theta^p \right\};\ \underset{1x6}{\{z_y\}}^T = \left\{ z^i \dots z^p \right\}$ -matrices-strings of nodal values of global coordinates r, θ, z.

By differentiating (13) the derivatives of global coordinates in the local system $r,_\xi, r,_\eta, r,_\zeta, \theta,_\xi, \theta,_\eta, \theta,_\zeta, z,_\xi, z,_\eta\ z,_\zeta$ and the derivatives of local coordinates in the global system are determined $\xi,_r, \eta,_r, \zeta,_r, \xi,_\theta, \eta,_\theta, \zeta,_\theta, \xi,_z, \eta,_z, \zeta,_z$.

Increments of displacements of the inner point of a finite element are expressed in terms of increments of displacements of its nodal values at the loading step in the same way as (13) and written as

$$\Delta\upsilon^1 = \underset{1x6}{\left\{ f(\xi,\eta,\zeta) \right\}}^T \underset{6x1}{\{\Delta\upsilon_y^1\}};\ \Delta\upsilon^2 = \underset{1x6}{\left\{ f(\xi,\eta,\zeta) \right\}}^T \underset{6x1}{\{\Delta\upsilon_y^2\}};$$

$$\Delta\upsilon^3 = \underset{1x6}{\left\{ f(\xi,\eta,\zeta) \right\}}^T \underset{6x1}{\{\Delta\upsilon_y^3\}}, \tag{14}$$

where $\underset{1x6}{\{\Delta\upsilon_y^1\}}^T = \left\{ \Delta\upsilon^{1i}\ \Delta\upsilon^{1j}\ \Delta\upsilon^{1k}\ \Delta\upsilon^{1m}\ \Delta\upsilon^{1n}\ \Delta\upsilon^{1p} \right\};$

$\underset{1x6}{\{\Delta\upsilon_y^2\}}^T = \left\{ \Delta\upsilon^{2i}\ \Delta\upsilon^{2j}\ \Delta\upsilon^{2k}\ \Delta\upsilon^{2m}\ \Delta\upsilon^{2n}\ \Delta\upsilon^{2p} \right\};$

$\underset{1x6}{\{\Delta\upsilon_y^3\}}^T = \left\{ \Delta\upsilon^{3i}\ \Delta\upsilon^{3j}\ \Delta\upsilon^{3k}\ \Delta\upsilon^{3n}\ \Delta\upsilon^{3n}\ \Delta\upsilon^{3p} \right\}$ -matrices-rows of nodal values of displacements.

Stress increments at the inner point of the finite element at the loading step are expressed in terms of stress increments at the nodal points by relations similar to (13) and are written as:

$$\Delta\sigma_{rr} = \left\{\underset{1x6}{f(\xi,\eta,\zeta)}\right\}^T \left\{\underset{6x1}{\Delta\sigma_{rr\,y}}\right\};\ \Delta\sigma_{\theta\theta} = \left\{\underset{1x6}{f(\xi,\eta,\zeta)}\right\}^T \left\{\underset{6x1}{\Delta\sigma_{\theta\theta\,y}}\right\};$$
$$\Delta\sigma_{zz} = \left\{\underset{1x6}{f(\xi,\eta,\zeta)}\right\}^T \left\{\underset{6x1}{\Delta\sigma_{zz\,y}}\right\};\ \Delta\sigma_{r\theta} = \left\{\underset{1x6}{f(\xi,\eta,\zeta)}\right\}^T \left\{\underset{6x1}{\Delta\sigma_{r\theta\,y}}\right\}; \quad (15)$$
$$\Delta\sigma_{rz} = \left\{\underset{1x6}{f(\xi,\eta,\zeta)}\right\}^T \left\{\underset{6x1}{\Delta\sigma_{rz\,y}}\right\};\ \Delta\sigma_{\theta z} = \left\{\underset{1x6}{f(\xi,\eta,\zeta)}\right\}^T \left\{\underset{6x1}{\Delta\sigma_{\theta z\,y}}\right\},$$

where $\underset{1x6}{\left\{\Delta\sigma_{rr\,y}\right\}^T} = \left\{\Delta\sigma^i_{rr}\ \Delta\sigma^j_{rr}\ \Delta\sigma^k_{rr}\ \Delta\sigma^m_{rr}\ \Delta\sigma^n_{rr}\ \Delta\sigma^p_{rr}\right\};$

$$\underset{1x6}{\left\{\Delta\sigma_{\theta\theta\,y}\right\}^T} = \left\{\Delta\sigma^i_{\theta\theta}\ \Delta\sigma^j_{\theta\theta}\ \Delta\sigma^k_{\theta\theta}\ \Delta\sigma^m_{\theta\theta}\ \Delta\sigma^n_{\theta\theta}\ \Delta\sigma^p_{\theta\theta}\right\};$$

$$\underset{1x6}{\left\{\Delta\sigma_{zz\,y}\right\}^T} = \left\{\Delta\sigma^i_{zz}\ \Delta\sigma^j_{zz}\ \Delta\sigma^k_{zz}\ \Delta\sigma^m_{zz}\ \Delta\sigma^n_{zz}\ \Delta\sigma^p_{zz}\right\};$$

$$\underset{1x6}{\left\{\Delta\sigma_{r\theta\,y}\right\}^T} = \left\{\Delta\sigma^i_{r\theta}\ \Delta\sigma^j_{r\theta}\ \Delta\sigma^k_{r\theta}\ \Delta\sigma^m_{r\theta}\ \Delta\sigma^n_{r\theta}\ \Delta\sigma^p_{r\theta}\right\};$$

$$\underset{1x6}{\left\{\Delta\sigma_{rz\,y}\right\}^T} = \left\{\Delta\sigma^i_{rz}\ \Delta\sigma^j_{rz}\ \Delta\sigma^k_{rz}\ \Delta\sigma^m_{rz}\ \Delta\sigma^n_{rz}\ \Delta\sigma^p_{rz}\right\};$$

$\underset{1x6}{\left\{\Delta\sigma_{\theta z\,y}\right\}^T} = \left\{\Delta\sigma^i_{\theta z}\ \Delta\sigma^j_{\theta z}\ \Delta\sigma^k_{\theta z}\ \Delta\sigma^m_{\theta z}\ \Delta\sigma^n_{\theta z}\ \Delta\sigma^p_{\theta z}\right\}$ - matrices-rows of nodal stress values.

On the basis of expressions (14) and (15), matrix relations are formed for the inner point of a finite element:

$$\underset{3x1}{\{\Delta\upsilon\}} = \underset{3x18}{[A]}\underset{18x1}{\{\Delta\upsilon_y\}};\ \underset{6x1}{\{\Delta\sigma\}} = \underset{6x36}{[S]}\underset{36x1}{\{\Delta\sigma_y\}}, \quad (16)$$

where $\underset{1x18}{\left\{\Delta\upsilon_y\right\}^T} = \left\{\underbrace{\Delta\upsilon^{2i}\ldots\Delta\upsilon^{1p}}_{6}\ \underbrace{\Delta\upsilon^{2i}\ldots\Delta\upsilon^{2p}}_{6}\ \underbrace{\Delta\upsilon^{3i}\ldots\Delta\upsilon^{3p}}_{6}\right\};$

$\underset{1x36}{\left\{\Delta\sigma_y\right\}^T} = \left\{\begin{matrix}\Delta\sigma^i_{rr}\ldots\Delta\sigma^p_{rr}\ \Delta\sigma^i_{\theta\theta}\ldots\Delta\sigma^p_{\theta\theta}\ \Delta\sigma^i_{zz}\ldots\Delta\sigma^p_{zz}\ \Delta\sigma^i_{r\theta}\ldots\Delta\sigma^p_{r\theta}\\ \Delta\sigma^i_{rz}\ldots\Delta\sigma^p_{rz}\ \Delta\sigma^i_{\theta z}\ldots\Delta\sigma^p_{\theta z}\end{matrix}\right\}$ - matrices-strings of nodal unknowns of a finite element.

Taking into account (16), the increments of deformations (12) are represented in matrix form:

$$\underset{6x1}{\{\Delta\varepsilon\}} = \underset{6x3}{[L]}\underset{3x1}{\{\Delta\upsilon\}} = \underset{6x3}{[L]}\underset{3x18}{[A]}\underset{18x1}{\{\Delta\upsilon_y\}} = \underset{6x18}{[B]}\underset{18x1}{\{\Delta\upsilon_y\}}. \quad (17)$$

To form the matrix of the stress-strain state of a finite element at the loading step, the equality of possible and actual work of external and internal forces is used in the form [12]:

$$\int_V \left[\underset{1x6}{\{\sigma\}^T} + \frac{1}{2}\underset{1x6}{\{\sigma\}^T} \right] \underset{6x1}{\{\Delta\varepsilon\}} dV = \int_S \underset{1x3}{\{\Delta u\}^T} \left[\underset{3x1}{\{\Delta q\}} + \frac{1}{2}\underset{3x1}{\{\Delta q\}} \right] dS, \tag{18}$$

where $\underset{1x6}{\{\sigma\}^T} = \{\sigma_{rr}\ \sigma_{\theta\theta}\ \sigma_{zz}\ \sigma_{r\theta}\ \sigma_{rz}\ \sigma_{\theta z}\}$ -matrix-string of stresses per j loading steps; $\underset{1x3}{\{q\}^T} = \{q_1\ q_2\ q_3\}$ - vector of increments of loads per loading steps; V - volume of the finite element; S - surface area of the load application.

The actual work of the stress increments in (18) is replaced by the difference between the total and additional energy at the loading step in the form [12]

$$\frac{1}{2}\underset{1x6}{\{\Delta\sigma\}^T}\underset{6x1}{\{\Delta\varepsilon\}} = \underset{1x6}{\{\Delta\sigma\}^T}\underset{6x1}{\{\Delta\varepsilon\}} - \frac{1}{2}\underset{1x6}{\{\Delta\sigma\}^T}\underset{6x6}{[C]}\underset{6x1}{\{\Delta\sigma\}} =$$
$$= \underset{1x6}{\{\Delta\sigma\}^T}\underset{6x3}{[L]}\underset{3x1}{\{\Delta\upsilon\}} = \frac{1}{2}\underset{1x6}{\{\Delta\sigma\}}\underset{6x6}{[C]}\underset{6x1}{\{\Delta\sigma\}}. \tag{19}$$

Equality (18), taking into account (19), can be written as a functional [12]:

$$\Phi \equiv \int_V \underset{1x6}{\{\Delta\sigma\}^T}\underset{6x3}{[L]}\underset{3x1}{\{\Delta\upsilon\}} dV - \frac{1}{2}\int_V \underset{1x6}{\{\Delta\sigma\}^T}\underset{6x6}{[C]}\underset{6x1}{\{\Delta\sigma\}} dV -$$
$$-\frac{1}{2}\int_S \underset{1x3}{\{\Delta\upsilon\}^T}\underset{3x1}{\{\Delta q\}} dS - \int_S \underset{1x3}{\{\Delta\upsilon\}^T}\underset{3x1}{\{q\}} dS + \int_V \underset{1x6}{\{\sigma\}^T}\underset{6x1}{\{\Delta\varepsilon\}} dV. \tag{20}$$

Taking into account expressions (9) and (17), functional (20) for a single finite element at the loading step will take the form:

$$\Phi \equiv \underset{1x36}{\{\Delta\sigma_y\}^T} \int_V \underset{36x6}{[S]^T}\underset{6x18}{[B]} dV \underset{18x1}{\{\Delta\upsilon_y\}} -$$
$$-\frac{1}{2}\underset{1x36}{\{\Delta\sigma_y\}^T} \int_V \underset{36x6}{[S]^T}\underset{6x6}{[C]}\underset{6x36}{[S]} dV \underset{36x1}{\{\Delta\sigma_y\}} -$$
$$-\frac{1}{2}\underset{1x18}{\{\Delta\upsilon_y\}^T} \int_S \underset{18x3}{[A]^T}\underset{3x1}{\{\Delta q\}} dS -$$
$$-\underset{1x18}{\{\Delta\upsilon_y\}^T} \int_S \underset{18x3}{[A]^T}\underset{3x1}{\{q\}} dS - \underset{1x18}{\{\Delta\upsilon_y\}^T} \int_V \underset{18x6}{[B]^T}\underset{6x1}{\{\sigma\}} dV. \tag{21}$$

By varying the functional (21) in $\{\Delta\sigma_y\}^T$ and $\{\Delta\upsilon_y\}^T$, systems of equations are obtained:

$$\frac{\partial\Phi}{\partial\{\Delta\sigma_y\}^T} \equiv \underset{36x18}{[Q]}\underset{18x1}{\{\Delta\upsilon_y\}} - \underset{36x36}{[H]}\underset{36x1}{\{\Delta\sigma_y\}} = 0;$$
$$\frac{\partial\Phi}{\partial\{\Delta\upsilon_y\}^T} \equiv \underset{18x36}{[Q]^T}\underset{36x1}{\{\Delta\sigma\}} - \underset{18x1}{\{\Delta f_q\}} - \underset{18x1}{\{R\}} = 0, \tag{22}$$

where $\underset{36x18}{[\mathrm{Q}]} = \int_V \underset{36x6}{[S]}^T \underset{6x18}{[B]} dV$; $\underset{36x36}{[\mathrm{H}]} = \int_V \underset{36x6}{[S]}^T \underset{6x6}{[C]} \underset{6x36}{[S]} dV$;

$\underset{18x1}{\{\Delta f_q\}} = \int_S \underset{18x3}{[A]}^T \underset{3x1}{\{\Delta q\}} dS$; $\underset{18x1}{\{R\}} = \int_S \underset{18x3}{[A]}^T \underset{3x1}{\{q\}} dS - \int_V \underset{18x6}{[B]}^T \underset{6x1}{\{\sigma\}} dV$ -Raphson's discrepancy.

Systems (22) can be represented in the traditional finite element formulation [13]:

$$\underset{54x54}{[k]} \underset{54x1}{\{z_y\}} = \underset{54x1}{\{F_y\}}, \tag{23}$$

where $\underset{54x54}{[k]} = \begin{bmatrix} -\underset{36x36}{[\mathrm{H}]} & \underset{36x18}{[\mathrm{Q}]} \\ \underset{18x36}{[\mathrm{Q}]^T} & \underset{18x18}{[\mathrm{O}]} \end{bmatrix}$ is the matrix of the stress-strain state of the finite element at the loading step;

$\underset{1x54}{\{z_y\}^T} = \left\{ \underset{1x36}{\{\Delta\sigma_y\}^T} \quad \underset{1x18}{\{\Delta\upsilon_y\}^T} \right\}$ - vector of nodal unknowns;

$\underset{1x27}{\{F_y\}^T} = \left\{ \underset{1x18}{\{O\}} \quad \underset{1x9}{\{\Delta f_q\}^T} + \underset{1x9}{\{R\}^T} \right\}$ - vector of nodal loads.

The stress-strain state matrix of the entire structure is formed using the traditional FEM procedure [6, 11].

4. Conclusion

The desired unknown displacement increments and stress increments obtained using the developed finite element in the mixed FEM formulation when using flow theory to take into account the physical nonlinearity of rotation shells are continuous not only in the nodes of the finite element, but also on its faces and boundaries.

5. Acknowledgments

The research was carried out with the financial support of the Russian Foundation for Basic Research and the Administration of the Volgograd Region in the framework of the scientific project No. 19-41-340004 r_a.

References

[1] Rikards R B 1988 *The Finite Element Method in the Theory of Shells and Plates* (Riga: Zinatne) p 284
[2] Golovanov A I, Tyuleneva O N and Shigabutdinov A F 2006 *Finite Element Method in the Statics and Dynamics of Thin-Walled Structures* (Moscow: Fizmatlit) p 392
[3] Bate K 2010 *Finite Element Methods* (Moscow: Fizmatlit) p 1024
[4] Oden J T 1976 *Mechanics (*Moscow: Mir*)* p 464
[5] Gureeva N A, Klochkov Yu V and Nikolaev A P 2009 *Proc. of the Nizhnevolzhsky Agrouniversitetskiy complex: science and higher professional education.* **2(14)** 123-129
[6] Postnov V A, Kharkhurim Ya 1974 *Finite Element Method in Calculations of Ship Structures* (Leningrad: Shipbuilding) p 344
[7] Kayumov R A, Kupriyanov V N, Mukhamedova I Z, Suleymanov A M, Shakirova A M 2007 Method of analysis of the deformation process of a film-fabric composite taking into account geometric and physical nonlinearity *Modeler. and Edges. Tasks* vol 1 119-121

[8] Sedov L I 1976 *Continuum Mechanics* vol 1 (Moscow: Science) p 535
[9] Malinin N N 1975 Applied Theory of Plasticity and Creep (Moscow: Mashinostroenie) p 400
[10] Gureeva N A, Klochkov Yu V and Nikolaev A P 2008 *Computing Technologies* **4(13**) 51-59
[11] Gureeva N A, Klochkov Yu V and Nikolaev A P 2010. *Izv. VUZ. Aviatsionnaya Tekhnika* **3** 7-10
[12] Gureeva N A, Klochkov Yu V and Nikolaev A P 2015 *Scientific Notes of Kazan University. Physical and Mathematical Sciences* **2(157)** 28-39
[13] Samul V I 1970 (Moscow: Vysshaya shkola) p 288

IOP Conf. Series: Earth and Environmental Science **965** (2022) 012006 doi:10.1088/1755-1315/965/1/012006

Theoretical substantiation of the principle of operation of the rod flow meter of irrigation canals

A A Pakhomov[1], E A Khodyakov[1], V S Bocharnikov[1], V G Grin[2], N A Kolobanova[1], E S Vorontsova[1, 3]

[1] Volgograd State Agrarian University, 26 University Avenue, Volgograd, 400002, Russia
[2] Kuban State Agrarian University named after I T Trubilin, 13 Kalinin Street, Krasnodar, 350044, Russia
[3] All-Russian Research Institute of Irrigated Agriculture, 9 Timiryazev Street, Volgograd, 400002, Russia

E-mail: kolobanova.nina@yandex.ru

Abstract. The paper presents a theoretical substantiation of the developed design of a rod flow meter for open canals of irrigation systems. The design features of the lattice elements are revealed, the methods of their hydraulic calculation are given. It was found that the lattices were not used as sensitive elements for measuring water flow rates. A feature of the hydraulic calculation of lattice elements is that it is necessary to take into account the following pressure losses: from the degree of restriction of the living section of the flow, from the ratio of the rod width to the size of the gap, from the rod width, from the rod thickness, from the shape of the cross section of the rod. The operability of the proposed flow meter is theoretically substantiated. It is found that with a significant change in the angle α of the lattice deviation, the degree of tension of the elastic element (angle β) increases by 1-3% from the nominal position. An analytical dependence for determining the value of the hydrodynamic pressure is also obtained.

1. Introduction

Currently, a large number of different designs of flow meters and liquid quantity meters are known, which are recommended for use in industry and reclamation systems [1, 2, 3], as well as methods and means of water metering in open canals [4, 5, 6].

The analysis of these tools showed that the most effective designs are those that use acoustic (ultrasonic) flow sensing. However, devices and instruments for measuring the flow rate and water levels of domestic and foreign manufacturers have limited use in open canals of reclamation systems, either due to the lack of acceptable measurement accuracy, or due to the lack of electricity supply at water accounting posts, or due to the high cost. Taking into account the opinion of the authors dealing with the problems of water accounting on land reclamation systems [7, 8, 9, 10, 11], it can be concluded that it is necessary to develop simple in design and reliable in operation, non-volatile and providing high accuracy of measurement of water accounting tools that would not worsen the canal capacity and would not require the reconstruction of existing hydraulic structures.

In this paper, a study on the mathematical description of the operation process of the proposed flow meter is carried out.

2. Materials and methods

The research methodology is based on complex theoretical, laboratory and field studies. The research was carried out using industry-specific methods, factorial planning of the experiment, computer mathematics tools with verification of the results using statistical criteria based on the theory of errors, variance and regression analysis.

3. Mathematical description of the rod flow meter design

The main element of the proposed design of the flow meter [12] is its sensitive element, which senses the hydrodynamic pressure of the flow. Below is the rationale for choosing a sensitive element in the form of a lattice.

A feature of the hydraulic calculation of lattice elements [13] is that it is necessary to take into account the following pressure losses: from the degree of restriction of the live flow section, from the ratio of the rod width to the size of the gap, from the rod width, from the rod thickness, from the shape of the cross section of the rod. The main issue in this paper is the mathematical description of the process of operation of the proposed flow meter [14].

The design contains a rotary sensitive element (SE) 1, a two-arm lever 2 with shoulders l_2 and l_1, a cylindrical hinge at the point O, an elastic element 3, an L-shaped stand 4 fixed in the support 5, a limiter 6 for turning the upper arm l_1 of the lever 2, an indicator head 7, a limiter 8 of rotation of the two-arm lever 2 in the cylindrical hinge. One end of the elastic element 3 is placed on the upper arm (point B) of the two-armed lever 2, and its other end on the stand 4 (point A). Support 5 allows to move the stand 4 vertically, and, consequently, the hinge in the point O and SE 1 to any depth H_{gate} of the canal.

With an increase in the speed of the water flow v, the force P on the surface of SE 1 increases. Under the influence of this force P, the two-arm lever 2 rotates in the cylindrical hinge in the point O, stretching the turns of the elastic element 3. Since the upper arm l_1 of the lever 2 rotates in the point O, then the point B when turning SE 1 through an angle α, it will move along the arc of a circle with a size l_1 in the point B`. To identify the regularity of the force R_{BX} – the tension force of the turns of the elastic element 3, we establish an analytical expression of the change in the angle β with an increase in the value of the angle α – rotation of the lever 2 SE 1.

During the calculation, the following dependencies and expressions were obtained.

Analytical dependence of the change in the value of the angle β on the angle α:

$$\beta = arctg\frac{l_1}{d} - \arccos\left\{\frac{b^2 + (l_1 \sin\alpha)^2 + l_1^2(1-\cos\alpha)^2 - l_1^2}{2\cdot b\sqrt{(d + l_1\cdot\sin\alpha)^2 + l_1^2(1-\cos\alpha)^2}}\right\} \quad (1)$$

From the analytical dependence (1), it is established that with a significant change in the angle α, the value of the angle β increases by 1-3% from the nominal position, therefore, the force R_{BX} remains constant.

Functional relationship between angles β and α:

$$\angle\beta = arctg[l_1(1-\cos\alpha)/(d + l_1\sin\alpha)]. \quad (2)$$

Equations of the projection of forces on the axis OX and OY:

$$\sum F_{kx} = R_B\cdot\cos\alpha + R_{XO} + P; \quad (3)$$

$$\sum F_{ky} = 0; \quad (4)$$

$$\sum F_{kz} = R_B\cdot\sin\beta + R_{ZO} - G. \quad (5)$$

Equation of moments of forces acting on the system relative to the point O:

$$\sum M_{(O)} = R_{BZ}\cdot l_1\cdot\sin\alpha + R_{BX}\cdot l_1\cdot\cos\alpha + \\ + G\cdot(l_2 + h/2)\cdot\sin\alpha - P\cdot(l_2 + h/2)\cdot\cos\alpha, \quad (6)$$

where $R_{BZ} = R_B\cdot\sin\beta$, $R_{BX} = R_B\cdot\cos\beta$.

IOP Conf. Series: Earth and Environmental Science **965** (2022) 012006 doi:10.1088/1755-1315/965/1/012006

Reaction R_B in the point B from the elastic forces of tension of the spring:

$$R_{BX} = c \cdot \overline{AB} = c \cdot d, \tag{7}$$

where c is caliber (stiffness) of the elastic element, N/mm; d is distance between points AB, mm.

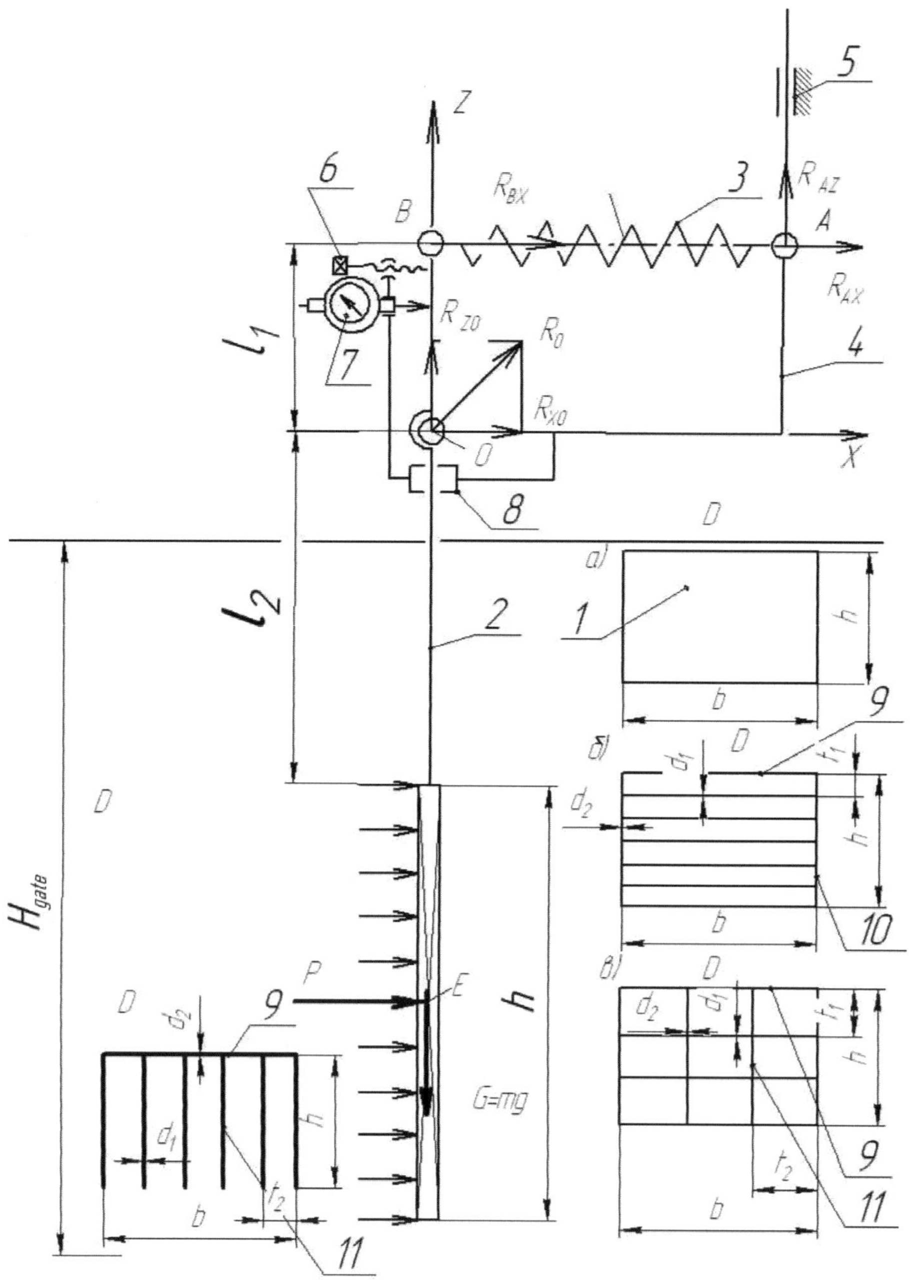

Figure 1 – Flow meter design scheme: 1 - rotary SE, 2 - two-arm lever, 3 - elastic element, 4 -stand, 5 - support, 6 - upper arm rotation limiter, 7 - indicator head, 8 - rotation limiter, 9 - rods, 10 -stands, 11 - vertical rods, 12 – rod.

Reaction value during flow meter operation:

$$R_B = c \cdot \sqrt{(d + l_1 \sin\alpha)^2 + [l_1(1-\cos\alpha)]^2}, \text{ N} \tag{8}$$

Hydrodynamic flow force:

$$P = \zeta_p \cdot \rho \cdot v^2 \cdot F_p = \zeta_p \cdot \rho \cdot Q^2 \cdot F_p / \omega_{l.s.}, \tag{9}$$

where ζ_p is coefficient of hydrodynamic resistance of the lattice; ρ is liquid density; v is speed of liquid movement, m/s; F_p is active lattice area, m^2; Q is flow rate, m^3/s; $\omega_{l.s.}$ is area of live section of flow, m^2.

From the equation of moments, the hydrodynamic force of the flow:

$$P = \frac{R_B \cdot \sin\beta \cdot l_1 \cdot \sin\alpha + R_B \cdot \cos\beta \cdot l_1 \cdot \cos\alpha + mg(l_2 + h/2) \cdot \sin\alpha}{(l_2 + h/2) \cdot \cos\alpha}. \tag{10}$$

Analytical expression of the speed of the water flow v:

$$v = \sqrt{\frac{R_B \cdot \sin\beta \cdot l_1 \cdot \sin\alpha + R_B \cdot \cos\beta \cdot l_1 \cdot \cos\alpha + mg(l_2 + h/2) \cdot \sin\alpha}{(l_2 + h/2) \cdot \cos\alpha \cdot |\zeta_p \cdot \rho \cdot F_p|}}. \tag{11}$$

Expression (11) takes into account both the design parameters of the flow meter and the dynamic parameters of the water flow in the canal.

4. Conclusion

In the course of the study, the possibility of using a sensitive element in the form of a movable lattice, the hydraulic resistance of which depends on the value of the hydrodynamic pressure, the geometric parameters of the lattice cell, the deflection angle, the degree of overlap of the live cross section of the flow and its turbulence, are substantiated. The operability of the proposed flow meter is theoretically substantiated. It is found that with a significant change in the angle α of the lattice deviation, the degree of tension of the elastic element (angle β) increases by 1-3% from the nominal position. An analytical dependence for determining the value of the hydrodynamic pressure is also obtained.

References

[1] Fatxulloyev A, Gafarova A and Hamroqulov J 2021 Improvement of water accounting for irrigation systems *IOP Conf. Ser.: Mater. Sci. Eng.* **1030** 012145

[2] Ghosh A, Kumar B and Sarkar R 2020 Real time PC-based flow measurement technique using optical channel as a signal transmitting media *IEEE Sensors Journal* **20(21)** 12827-12836

[3] Mulik S S, Patange A D, Jegadeeshwaran R, Pardeshi S S and Rahegaonkar A 2021 Development and experimental assessment of a fluid flow monitoring system using flow sensor and arduino interface Innovative *Design, Analysis and Development Practices in Aerospace and Automotive Engineering* (Singapore: Springer) pp 115-122

[4] Belavtseva T M and Altunin V S 1995 Present state of and outlook for the development of water metering means *Gidrotekhnicheskoe Stroitel'stvo* **10** 21-26

[5] Le Coz J, Blanquart B, Pobanz K, Dramais G, Pierrefeu G, Hauet A and Despax A 2016 Estimating the uncertainty of streamgauging techniques using in situ collaborative interlaboratory experiments *J. of Hydraulic Engineering* **142(7)**

[6] Olgarenko V I, Olgarenko I V, Kisarov O P and Selyukov V I 2012 Information technologies of water use planning in farms *Polythematic Network Electronic Scientific Journal of the Kuban State Agrarian University* **78** 279-290

[7] Olgarenko V I, Olgarenko I V, Zakharchenko N S and Kisarov O P 2012 Program complex of information support for operational water distribution management in irrigation network canals *Land Reclamation and Water Management the Current State and Prospects for the Development of Land Reclamation, Forest Reclamation and Water Management Complexes in the South of Russia* (Novocherkassk) pp 294-301

[8] Bochkarev V Ya 2011 *Water Accounting and Water Distribution on Meliorative Systems* (Novocherkassk: Russian Research Institute of Land Reclamation Problems) p 43

[9] Bochkarev V Ya 2012 *New Technologies and Measuring Instruments, Methods of Organizing Water Accounting on Irrigation Systems* (Novocherkassk: Russian Research Institute of Land Reclamation Problems) p 227
[10] Shchedrin V N, Vasiliev S M and Slabunov V V 2013 *Basic Rules and Regulations for the Operation of Meliorative Systems and Structures, Water Accounting and Production of Operational Works* (Novocherkassk: Gelikon LLC) p 262
[11] Shchedrin V N, Pakhomov A A and Kolobanova N A 2015 The choice of the water distribution scheme for regulating water supply in irrigation systems *Scientific Foundations of Nature Management in Russia: Problems, Current State, Steps to the Future* (Volgograd: Volgograd State University) pp 167-172
[12] Ovchinnikov A S, Pakhomov A A, Melikhov K M and Kolobanova N A 2009 Flowmeter of liquid media in open canals utility model patent 86300 of the Russian Federation MPK G 01 F 1/00 applicant and patent holder Volgograd State Agricultural University 2009109482/22
[13] Pakhomov A A, Bolshakov I A and Kolobanova N A 2010 Features of hydraulic calculation of lattice elements in hydraulic engineering construction *Proc. of the Nizhnevolzhsky Agrouniversity Complex: Science and Higher Professional Education* **2(18)** 167-172
[14] Pakhomov A A, Kolobanova N A and Saldaev A M 2010 Perspective means of water accounting New *Directions in Solving Problems in the Agro-Industrial Complex on the Basis of Modern Resource-Saving, Innovative Technologies* vol 1 (Volgograd) pp 362-366

IOP Conf. Series: Earth and Environmental Science **965** (2022) 012007 doi:10.1088/1755-1315/965/1/012007

Development of new ways to preserve the water content of lake basins in the natural park of the Volga-Akhtuba floodplain

O A Solovyeva, V F Loboiko, A Y Ovcharova and A V Loboiko

Volgograd State Agrarian University, 26 University Avenue, Volgograd, 400002 Russia

E-mail: olqamatveeva@inbox.ru

Abstract. The article presents new modern ways to increase the receiving capacity and preserve the water content of lake basins on the territory of the Volga-Akhtuba floodplain natural Park. The main factor that caused the need for rehabilitation of the rivers and lakes of the Volga-Akhtuba floodplain is the change in the hydrological and ecological situation that arose as a result of the change in the hydrological regime by regulating the flow of the Volga River after the construction of the Volga hydroelectric power station. The rehabilitation of water objects is aimed at preserving and restoring the water objects to a state that provides environmentally favorable living conditions for the population; ensuring the protection of the population from low water (water scarcity), for the restoration and environmental rehabilitation of the water objects and the adjacent territory.

1. Introduction

The Volga-Akhtuba floodplain within the borders of the Volgograd region has a high biological and landscape diversity. A large number of water bodies are located on the territory of the floodplain: the Akhtuba river, 90 km long within the Volgograd region; the Krasnoslobodsky tract, 33 km long; the Kashirinsky tract, 99.3 km long; backwaters - 8 units; large channels - 6 units; shallow channels - 173 units; more than 200 large lakes representing unique accumulators of liquid and solid river flow; marshes - 2 units with a total area of 206.8 hectares [1].

The main factor that caused the need for rehabilitation of the rivers and lakes of the Volga-Akhtuba floodplain is the change in the hydrological and ecological situation that arose as a result of the change in the hydrological regime by regulating the flow of the Volga River after the construction of the Volga hydroelectric power station.

The fundamental element of the landscape complex of the Volga-Akhtuba floodplain is a complex hydrographic network of 6 shallow channels, channels, lakes, estuaries and other water objects. Currently, there are processes of deterioration of the water supply of the Volga-Akhtuba floodplain territory associated with the regime of releases of the Volga-Kama cascade of hydroelectric power plants and a set of local anthropogenic changes in the landscape. Under the influence of artificial regulation of the floodplain hydro regime, as well as the expansion of residential territories, the ecological parameters of reservoirs are transformed and they lose many ecosystem properties. Intensive and unregulated anthropogenic loads lead to violations of the natural physical and chemical parameters of reservoirs, loss of biodiversity, degradation of the structure and functions of biological communities and, as a result, to a reduction in the range of potential ecosystem services [2]. In this regard, the work aimed at improving the state of water supply of degraded water objects is relevant and problematic. Such objects in the system of the Krasnoslobodsky water management tract include Lake Zapornoye and the Sudomoyka and the Sakharny shallow channels feeding it, located on a specially protected natural territory "The Volga-Akhtuba Floodplain Natural Park" on the lands of the Frunzensky rural settlement in the Sredneakhtubinsky municipal district of the Volgograd region.

2. Materials and methods

According to the data, Lake Zapornoye is a typical example of degradation due to high anthropogenic load: the reservoir is located near a busy highway, surrounded by the outer perimeter of the settlements of Gospitomnik, Sugar and dacha societies, it is collapsed, the land adjacent to the reservoir is used for agriculture (arable land). The pace and density of construction is constantly growing, which was facilitated by the commissioning of a bridge crossing over the Volga in 2010. The lake, as a natural object, is experiencing a powerful load that is destructive to the ecosystem. With the increase in the area of development and the construction of the highway, the lake lost its flowability. In the conditions of reducing the duration and scale of 7 floods, which accelerate the processes of siltation, the lake is quickly overgrown with air-water plants. Overgrowth accelerates the processes of silting of shallow waters and increases waterlogging, slows down the flow of water during high water, contributes to the decline in the diversity of flora and fauna both in shallow water and in adjacent meadows. Due to repeated depressions of the bottom of the northern part of the reservoir, the fauna of hydrobionts and aquatic flora, which should maintain water quality, are depleted in it. Fish productivity is steadily decreasing. The ground vegetation cover is strongly transformed as a result of high anthropogenic load. The animal population corresponds to the parameters of an anthropogenically disturbed landscape. The habitats of aquatic plants, benthic invertebrates and feeding grounds for juvenile fish have been lost. Thus, Lake Zapornoye has almost completely lost most of its ecosystem functions and needs to be restored through the implementation of special rehabilitation measures.

3. Results and discussion

Lake Zapornoye (Figure 1) is a depression in the relief, oriented in a north-easterly direction. The shoreline marks in the flood period of 2016 were minus 4.6 m, the depth of the lake is from 2 m in the southern 17 part to 4.5-5.0 m in the northern part. The territory adjacent to the lake has a mark of minus 4-minus 3 m, which indicates the drainless nature of the lake [3].

Carrying out geoplastics of the relief in the shallow part of the lake with a deepening in certain places up to 2.0-3.0 m will increase the water content of the object, ensure year-round preservation of water in it, create conditions for the development of aquatic organisms. During engineering and geological surveys, the groundwater level in the territory adjacent to the lake was fixed at about minus 6.0 m. Increase of the normal retaining level in the lake to the level of -4.5....-5.0 m will provide an increase in the groundwater level in the territory adjacent to the lake [4] (Figure 1).

When organizing works on the rehabilitation of water objects, the following features of the Volga-Akhtuba floodplain are taken into account:

- work is being carried out on a specially protected area - the Volga-Akhtuba Floodplain Natural Park;
- from April to mid-July, the floodplain may be flooded by flood waters.

These features impose restrictions on the terms of work (July-March of the next year) and impose special requirements for environmental protection measures.

The main objectives of the project are:

- development of measures to restore the ecosystem of lakes Zapornoye and of the Sudomoyka and the Sakharny shallow channels during the discharge costs of the Volgograd hydroelectric complex provided for by the operating rules (at the level of 25,000-27,000 cubic meters at a second, lasting 5-7 days) and artificial pumping of water by the Head pumping station into the Krasnoslobodsky water tract during the low streamflow period;

- formation of a complex lake relief to create habitats for typical groups of biodiversity;
- restoration of typical groups of biodiversity in the formed habitats (tree-shrub and herbaceous vegetation);
- water protection measures on the formed terrain (optimization of tree and shrub vegetation on coastal slopes);
- development of a program of local (project-based) environmental monitoring to assess the effectiveness of measures taken for the environmental rehabilitation of Lakes Zapornoye and the Sudomoyka and the Sakharny shallow channels;
- determination of the permissible withdrawal of water for economic needs without compromising the ecosystem of the reservoir.

Figure 1. Top view of Lake Zapornoye.

In the order of technical rehabilitation for the Sudomoyka shallow channel the following activities are planned:

- repair of three hydraulic structures (No. 55, 56, 46), which consists in restoring corroded steel pipes by sanitizing and installing valves;
- cleaning of the riverbed of the shallow channel in the head part of the hydraulic structures from trees - 220 pcs;
- clearing of bushes of 2.01 ha;
- cleaning of the riverbed at the hydraulic structures from sediment - 800 m^3;
- cleaning of the riverbed from foreign objects - 2 railway supports with a volume of 0.94 m^3;
- cleaning the Sudomoyka shallow channel from sunken logs during rafting -700/178 pcs/t;
- cleaning the Sudomoyka shallow channel from floating garbage 36 t.

In the order of technical rehabilitation for the Sakharny shallow channel the following activities are planned:

- cleaning of the riverbed in the head part of the shallow channel from bushes of 0.3 ha;
- installation of gates of the hydraulic structures (release into Lake Zapornoye).

In the order of technical rehabilitation for Lake Zapornoye the following activities are planned:

- formation of the littoral and pelagial by geoplastics of the bottom relief with a profile volume of earthworks -140,960 m^3 (performed by a floating excavator and a dredger);
- related earthworks related to the construction of the base from geotubes, the collapse of the alluvial maps by an excavator, the alluvial maps by a dredger;

IOP Conf. Series: Earth and Environmental Science **965** (2022) 012007 doi:10.1088/1755-1315/965/1/012007

- strengthening of slopes formed in the water area of embankments by planting perennial grasses and woody and shrubby plants.

This will allow for biological rehabilitation, which provides for the creation of a balanced ecosystem in which self-purification mechanisms operate and consist in the formation of littoral and pelagial, reducing the area of coverage with air-water plants (reeds, cattails), and restoring submerged vegetation of the littoral. In the coastal protective strip (width of 30-40 m), it is possible to restore tree and shrub communities [5] (Figure 2).

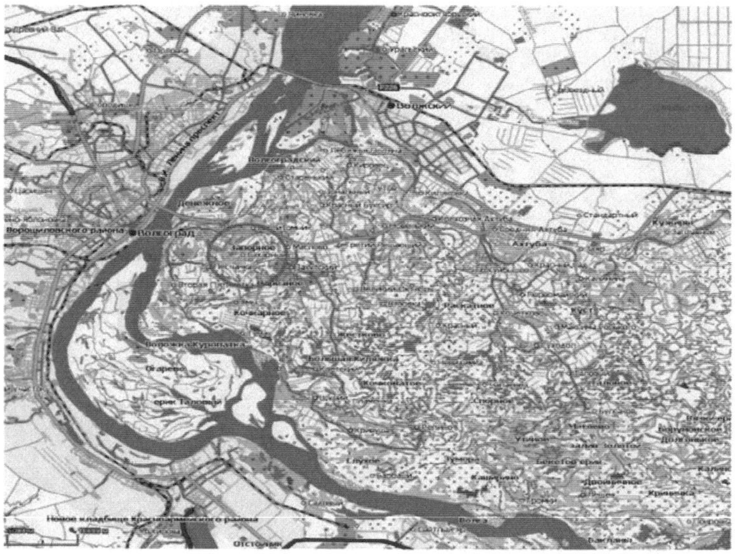

Figure 2. Map-scheme of the area of planned work on the rehabilitation of lakes.

According to the design solutions, the rehabilitation objects are simple (profile recesses and embankments). According to construction and installation processes, they refer to objects with construction and installation processes that are repeated in the same volumes and technological alignment on most grippers. According to the construction conditions, they are considered complex objects due to the hydrological and hydrogeological situation. The production of works can be started after the passage of the flood and the runoff of water from the lakes or the preliminary drainage. The work on the rehabilitation of Lake Zapornoye is carried out in the water area at a water depth of up to 1.5 m (with the discharge mode of the Volga hydroelectric power station according to the type of 2016, with the discharge modes of previous years, the water depth can be 0.21....0.5 m).

The peculiarities of carrying out work in the conditions of the Volga-Akhtuba floodplain leads to the adoption of the following organizational and technological decisions:

- taking into account the limited construction time between the flood period and the limitations of the requirements of the fisheries, special attention should be paid to the preparatory work that must be performed before the start of the main work;
- the production of basic works begins after the passage of the flood, the end of the fish spawning period, the decrease in the water level in the lake below 1.5 m.

The organizational and technological scheme of the lake rehabilitation includes:

- spatial division of the complex of works into sections – excesses and depressions;
- the sequence of the device of recesses and the construction of embankments;
- characteristics of the main methods of construction of objects – filling dams with alluvial maps with an excavator, making depressions with a dredger with pumping into alluvial maps, planting trees, shrubs and aquatic plants [6].

The work on the biological rehabilitation of the Sudomoyka shallow channel, which provides for cleaning the riverbed from floating debris, foreign objects, from bushes (in the head part), is performed at any time of the inter-flood period [7]. In the riverbed of the Sakharny shallow channel, it is planned to remove shrubs on an area of 0.3 hectares. Works on Lake Zapornoye are carried out during the inter-flood period on the territory flooded with water, after the level drops below 1.5 m [8] (Figure 3).

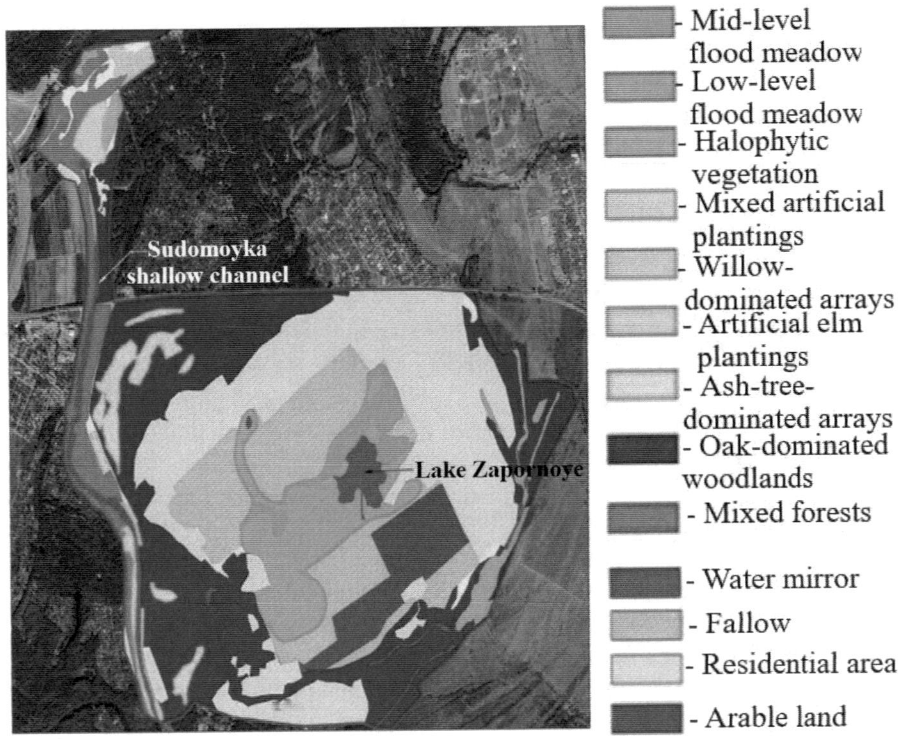

Figure 3. Map of the flora of the studied area.

Taking into account the above provisions, the ecological rehabilitation of water objects is carried out in two stages:

- technical stage;
- the biological stage.

4. Conclusion

The rehabilitation of water objects is aimed at preserving and restoring the water objects to a state that provides environmentally favorable living conditions for the population; ensuring the protection of the population from low water (water scarcity), for the restoration and environmental rehabilitation of the water objects and the adjacent territory. Rehabilitation will improve the state of aquatic ecosystems, as a necessary factor for restoring species diversity and providing conditions for the reproduction of aquatic biological resources. The quality of surface water in the water body is ensured by the annual flooding of the water area by spring floods and corresponds to the quality of the Volga water. The functioning of aquatic ecosystems to a certain extent will have a positive impact on the quality of surface water in reservoirs. Maintaining the water horizon in the lake at the levels of -5.5-5.0 m (the mark minus 5.0 m according to the Baltic Altitude System was taken as the normal retaining level) immediately after the flood will contribute to the supply of groundwater to the territory adjacent to the lake. The planned measures are not the object of capital construction, do not involve the implementation of any economic or other activities. Ecological rehabilitation of water objects is an environmental protection measure aimed at optimizing the hydro regime and restoring the normal functioning of the main links of the ecological systems of reservoirs. The indicator of the effectiveness of rehabilitation measures is high and amounts to 16,185 rubles per 1 ruble of investments in rehabilitation. Minimization of the negative impact on the environment during the short-term period of construction works on rehabilitation is provided by organizational and technical measures. In the project of rehabilitation of water objects, the costs are planned for: compensation for damage to aquatic biological resources from the production of works; compensation for environmental damage from the production of construction works. Recommendations for monitoring the state of the environment. Environmental monitoring is carried out in order to ensure the environmental safety of the object during the implementation of an investment and construction project. Environmental monitoring includes monitoring of atmospheric air, land, water bodies, objects of flora and fauna. When implementing the project, you should be guided by: recommendations on environmental support for investment and construction projects; instructions for the implementation of state control over the protection of atmospheric air (approved by order of the Federal Service for Supervision of Natural Resources Management dated March 1, 2011 No. 112); regulations on the Unified State Environmental Monitoring System.

References

[1] Loboyko V F, Ovcharova A Y and Nikitina N S 2018 Features of the water regime of the Lower Volga and its influence on the state of the northwestern part of the Volga-Akhtuba floodplain Proc. *of the Nizhnevolzhsky Agrouniversity Complex: Science and Higher Professional Education* **4(52)** 89-96

[2] Hoagland B, Schmidt C, Russo T A, Adams R and Kaye J 2019 Controls the total environment *Sci. Total Environ.* **649** 979–994

[3] Kalinin V V, Loboyko V F, Solovyeva O A, Likhomanova M A, Ovcharova A Y and Loboyko A V 2020 Environmental rehabilitation of Volga-Akhtuba floodplain *IOP Conf. Series: Materials Science and Engineering* **828** 012012

[4] Chen J, Zhu W, Zheng Y, Tian Y Q and Yu Q 2019 Monitoring seasonal variations of colored dissolved organic matter for the Saginaw River based on Landsat-8 data *Water Science & Technology: Water Supply* **19(1)** 274-281

[5] Brylev V A, Korinets V V and Revebtsov V V 1999 Problems of scientific substantiation and practical implementation of the project of the Volga-Akhtuba Floodplain roadside Park *Povolzhskiy J. of Ecology* **6**

[6] Loboiko V F and Kuznetsov P I 2009 Conservation of biodiversity in the main wetlands of the Lower Volga in conditions of technogenic load Proc. *of the Nizhnevolzhsky Agrouniversity Complex: Science and Higher Professional Education* **3(15)** 25-40

[7] Wilk-Woźniak E, Walusiak E, Burchardt L, Cerbin S, Chmura D, Gąbka M and Żbikowski J 2019 Effects of the environs of waterbodies on aquatic plants in oxbow lakes (habitat 3150) *Ecological Indicators* **98** 736-742

[8] Barmin A N and Golub V B 2000 Instructive lesson of the results of the exploitation of reed beds in the Volga River delta *Izvestia of Samara Scientific Center of the Russian Academy of Sciences* **2** 295-299

IOP Conf. Series: Earth and Environmental Science **965** (2022) 012008 doi:10.1088/1755-1315/965/1/012008

Effect of the structure of artificial rain on the soil

V S Bocharnikov, O V Kozinskaya, M A Denisova, O V Bocharnikova, T V Repenko and E V Pustovalov

Volgograd State Agrarian University, 26 University Avenue, Volgograd, 400002, Russia

E-mail: kozinska1977@mail.ru

Abstract. The quality of irrigation by sprinklers is determined not only by the uniformity of the distribution of the rain layer over the irrigated area. If the water permeability exceeds the intensity of precipitation, then surface runoff does not occur. With an increase in soil moisture, water permeability decreases, and surface runoff increases, contributing to the washout of the fertile layer. The sprinkling machine "Kuban-LS" is installed 18...20 km west of Volgograd, located on the western slope of the Volga Upland. The soil cover has a complex character and is represented by carbonate light chestnut varieties of varying degrees of washout. The data on the effect of artificial rain created by a sprinkler machine on the soil structure and absorption rate are presented. Due to the partial destruction of agronomically valuable soil aggregates, the density of the arable layer from sowing to harvesting the grass mixture after each irrigation increased from 1.10...1.28 to 1.26...1.35 t/m^3. The quality of irrigation by sprinklers is determined not only by the uniformity of rain distribution, but also by the conditions for the absorption of irrigation water into the soil. Based on these studies, we offer certain recommendations to reduce the negative impact on the soil.

1. Introduction

A prerequisite for the use of reclaimed land is not only obtaining high and sustainable crop yields, but also respect for the land as the main means of agricultural production.

The quality of irrigation by sprinklers is determined not only by the uniformity of the distribution of the rain layer over the irrigated area [1, 2, 3, 4]. Equally important is the determination of the optimal rainfall rate depending on the soil absorbency. If the water permeability exceeds the intensity of precipitation, then surface runoff does not occur [5-10]. Soil moisture and the nature of precipitation are of great importance. With an increase in soil moisture, water permeability decreases, and surface runoff increases, contributing to the washout of the fertile layer. Given the insufficient experimental development of water absorption into the soil and methods of its intensification during irrigation with frontal sprinklers [11-16], it is necessary to conduct experimental studies to develop measures to prevent irrigation soil erosion [17, 18, 19, 20].

2. Materials and methods

The object of research is the frontal sprinkler machine "Kuban-LS", which irrigates in motion. The movement is provided by electric motors installed on each carts. The irrigation rate can vary within 60...900 m^3/ha. The presence of an electrical automatic control and protection system allows irrigation in a given mode at any time of the day without operator intervention. The articulated connection of the spans of the machine gives it a certain flexibility in the vertical and horizontal planes, which allows you to overcome the unevenness of the field when driving. The joints between the spans are connected by a coupling with sealing sleeves. The cantilever part of the pipeline is rigidly attached to the ends of the wings, supported by cable ties. The power unit consists of a diesel pump unit connected by a torsion shaft to a generator. The machine has a short circuit protection system. The direction and speed of movement can be changed using a position switch located on the control panel.

Figure 1. The sprinkler machine "Kuban-LS".

To maintain the straightness of the movement of the main water pipeline, all support carts, except for the extreme ones, are equipped with a movement control mechanism consisting of a cable-lever system connected to the pipeline. This mechanism allows you to control the switching on and off of the electric motor of the drive of the carts when it lags or runs ahead of the adjacent carts.

The use of frontal sprinklers «Kuban» allows you to fully automate the irrigation process, ensure high quality rain distribution, water savings by 10...15% compared to machines equipped with medium-jet devices. The presence of the machine of systems for stabilizing movement along the course, emergency protection of the water pipeline from unacceptable bends and the engine from overloads, the ability to automatically stop at any place, a high degree of reliability (0.95) allows one operator to service 4...5 such machines per shift.

The sprinkler machine "Kuban-LS" is installed 18...20 km west of Volgograd, located on the western slope of the Volga Upland. The soil cover has a complex character and is represented by carbonate light-chestnut varieties of varying degrees of erosion, sometimes solonetzic, with spots of solonetz soils has a complex character. The humus content in the arable layer ranges from 1.30...2.10%, the reaction of the soil medium is slightly alkaline (pH within 7.2...8.1).

The studies were carried out in accordance with the test method for sprinkling technology, according to which the irrigation rate before the runoff was determined at the beginning, middle and end of the wing using drain pads made of sheet steel in the form of a square frame with an internal dimension of 316 × 316 mm. Frame height 75 mm, wall thickness 2 mm. The onset of 10% runoff was determined by the ratio of the volumes of water in the vessel and rain gauges, expressed in the following relationship:

$$\frac{0.1F}{F_a} = \frac{V_{\text{runoff}}}{V_{\text{average}}} \tag{1}$$

where F is receiving area of the drain platform, 100 cm^2; F_a – rain gauge receiving area, cm^2; V_{runoff} – volume of water flow in the receiving vessel, cm^3; V_{average} – average water volume for four rain gauges, cm^3.

The irrigation rate before the runoff was calculated:

IOP Conf. Series: Earth and Environmental Science **965** (2022) 012008 doi:10.1088/1755-1315/965/1/012008

$$m_C = 100\frac{V_{\text{average}}}{F_a},\ \text{m}^3/\text{ha} \tag{2}$$

During the experiment, the wind speed and water flow pressure on the hydrant were measured in triplicate.

3. Results and discussion

The main types of soils differ significantly in free-flow water permeability, the structure of the soils and the degree of salinity are affected (Table 1).

The permissible rain rate is shown (*P*-value) depending on the rate of absorption (Table 2).

Table 1. Environmental restrictions on rain intensity depending on soil water permeability.

Soil water permeability	Absorption rate in 1 hour, cm/hour (mm/min)	Intensity, mm/min
High	More 15 (2.5)	0.5…0.8
Average	15-5 (2.5-0.83)	0.2…0.3
Low	Less than 5 (0.83)	0.1…0.2

Table 2. Dependence of rain intensity on the size of raindrops.

Soils	Permissible intensity (mm/min) for droplet diameter, mm				
	0.5	1.0	1.5	2.0	2.5
Sandy beaches	1.19	0.46	0.30	0.19	0.13
Sandy loam	0.93	0.36	0.25	0.17	0.12
Medium Loamy	0.72	0.28	0.19	0.12	0.09
Heavy loams and clays	0.48	0.21	0.12	0.08	0.06

There is a concept of «critical» slope steepness, at which intensive soil erosion begins. The intensity of artificial rain, at which there is no surface runoff, depending on the type of soil and the steepness of the slope (i), the absence or presence of vegetation cover for different types of irrigated soil is presented in Table 3.

Table 3. The intensity of artificial rain (*P*-value) depending on the steepness of the slope (i) and the absence or presence of vegetation cover

Soils	Intensity (mm/min)							
	i ≤ 0.05		i - 0.05…0.08		i - 0.08…0.12		i > 0.12	
	Steam	Seeding	Steam	Seeding	Steam	Seeding	Steam	Seeding
Medium loamy	0.21	0.42	0.17	0.34	0.13	0.25	0.09	0.17
Medium-loamy, underlain by denser rock	0.13	0.25	0.11	0.21	0.07	0.17	0.04	0.13
Heavy loams and clays	0.07	0.09	0.04	0.07	0.03	0.05	0.02	0.04

After observing the dynamics of soil structure from irrigation to irrigation on crops of continuous sowing, grass mixtures (rump + pea-oat mixture), a decrease in the structure coefficient by 30% was observed (Figure 2).

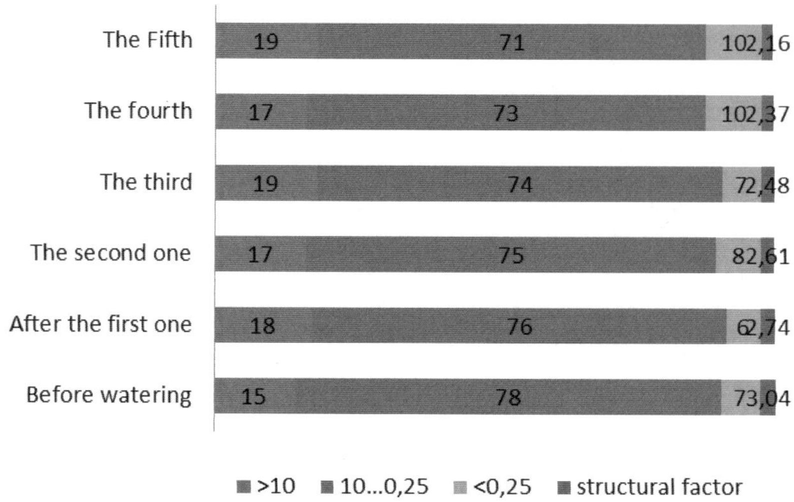

Figure 2. Dynamics of the soil structure in a layer of 0...0.30 m under the influence of irrigation of the grass mixture (rump + pea mixture) by sprinkling.

Due to the partial destruction of agronomically valuable soil aggregates, the density of the arable layer from sowing to harvesting the grass mixture after each irrigation increased from 1.10...1.28 to 1.26...1.35 t/m^3 (Figure 3).

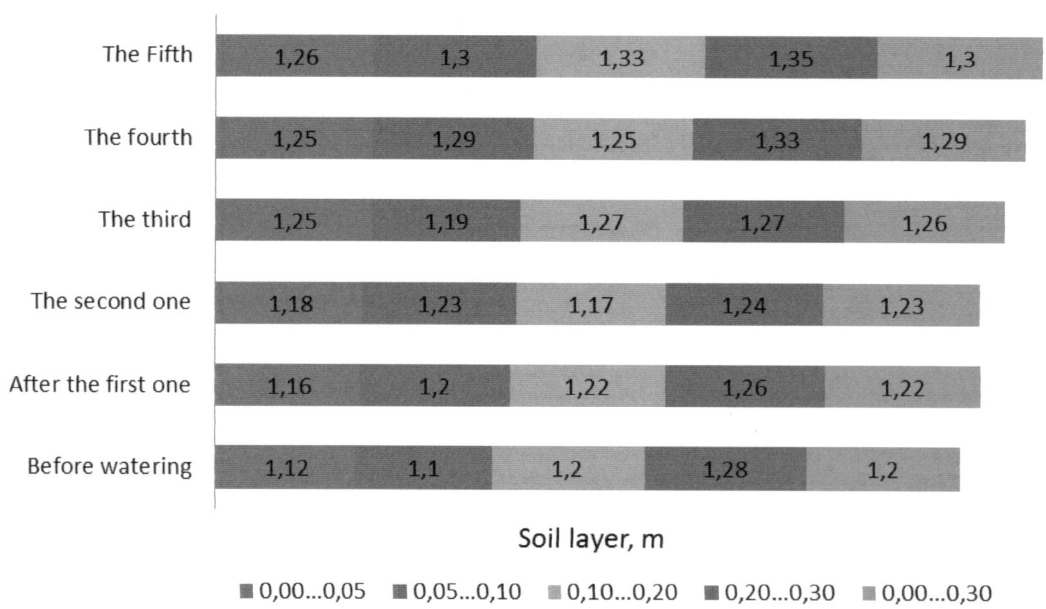

Figure 3. Changes in soil density under the influence of irrigation.

Our studies during the entire growing season showed that the percentage of unabsorbed water increased from the first irrigation to the next.

So, during the first irrigation, with a 60.0 mm layer of precipitation, no runoff was observed, with the second irrigation of 50.0 mm, the appearance of the first signs of surface water redistribution was noted (up to 4.7%, of the irrigation norm). The amount of runoff in all accounting plots during the third irrigation with an irrigation rate of 600 m^3/ha was 20.3%, the average rainfall rate was 0.5 mm/min; during the fourth irrigation, this value increased to 14% of the irrigation rate of 59.9 mm. Rain intensity, the size of the drops, as well as its uniform distribution over the width of the sprinkler's grip, in different ways affect soil properties such as water permeability, moisture capacity, water capacity and water vaporization ability when watering. Under the end of the machine, soil compaction is faster, destruction of soil aggregates is observed. The duration of the unpressurized stage of infiltration depends on the intensity of the rain and the size of the droplets. The duration of the drainless period decreases with increasing rain intensity. The free movement of water is impeded by the density of the vegetation cover and the roughness of the soil. On poorly watertight, heavy loamy and clayey soils, with an irrigation rate of more than 400 m^3/ha, runoff outside the irrigated contour may begin, which is usually accompanied by active erosion.

Stock platforms were installed at the beginning, middle and end of the sprinkler wing. With dense grass, no irrigation water losses outside the irrigated contour were observed. The formation of a runoff was noted only under the terminal apparatus.

The platforms are located at a distance of 1.5 m from the second support. The sites numbered 1 and 2 received 50.6 and 52.2 mm of precipitation with an average rainfall rate of 0.3 mm/min. The size of the droplets was the most optimal - 0.8 mm in diameter. In these areas of the irrigated contour, all the water had time to be absorbed by the soil during sprinkling. At site number 3, the main irrigation rate, about 70%, is formed by drops with a diameter of 1.6 mm with an average rainfall rate of 0.55 mm/min. With a higher irrigation rate on this site, the water did not have time to be absorbed into the soil, out of 20.3% of precipitation falling during the sprinkling process, only 14.33% was absorbed.

The pads located in the middle of the sprinkler were located under the third sprinkler support. The average intensity on site 1 was 0.45 mm/min with a droplet size of 0.6-1.2 mm and on site 2 - 0.5 mm/min with a droplet size of ~ 0.7 mm. On site 3, the average rain rate is 0.48 mm/min with a droplet diameter: on the area – 0.8-0.9 mm. The amount of not absorbed water in this case was insignificant and amounted to 4.7% of the irrigation rate. The increase in the irrigation rate caused an increase in the runoff.

An insignificant layer of water was formed mainly in the first half of irrigation on plots 2 and 3 located at the end of the sprinkler. In this experiment, no significant puddling was noted. In another experiment, with the fourth vegetation irrigation and a wind speed of 4-6 m/s, plots 3 and 7 were watered at a rate of 83.3 mm. The bulk of water, approximately 97 and 92% of the norm, to the survey sites from the second position with an average intensity of 0.45...0.6 mm/min.

4. Conclusion

Our studies during the entire growing season showed that the percentage of unabsorbed water increased from the first irrigation to the next. During the first irrigation, with a 60.0 mm layer of precipitation falling under the sprinkler machine «Kuban –LS», no runoff was observed, and already with the second irrigation of 50.0 mm, the first signs of surface water redistribution were noted. The duration of the unpressurized stage of infiltration depends on the intensity of the rain and the size of the droplets. With an increase in the rain intensity, the duration of the non-runoff period decreases. The free movement of water is impeded by the density of the vegetation cover and the roughness of the soil. On the basis of these studies, we offer certain recommendations to reduce the negative impact on the soil, for example, the use of such measures as ditching, ridging, field crevice. Empirical dependencies have become widespread in irrigation practice, which, with a sufficient degree of accuracy, make it possible to assess the influence of certain factors in specific conditions.

After observing the dynamics of soil structure from irrigation to irrigation on crops of continuous sowing, a decrease in the structure factor by 30% was observed. When carrying out counter-drainage methods, the absorbent capacity of light - chestnut heavy - loamy soil increased by 55...85%.

The quality of irrigation by sprinklers is determined not only by the uniformity of rain distribution, but also by the conditions for the absorption of irrigation water into the soil. We recommend, when carrying out irrigation, to assess the erosion and select the most optimal methods and scheme for irrigation of the hazard of irrigated lands. Thus, after assessing the erosion hazard of an irrigated field, it is necessary to choose the optimal irrigation scheme and measures to restrain their negative impact, corresponding to specific conditions.

5. Acknowledgments

The research was carried out within the State Assignment of Ministry of Agriculture of the Russian Federation (theme No. 13/2673).

References

[1] Slyusarenko V V Machines and equipment for irrigation of agricultural crops https://2lib.org/book/3001739/8a47a5?id=3001739&secret=8a47a5

[2] Kuznetsova N V, Kuznetsov Yu V, Kozinskaya O V and Denisova M A 2020 Influence of hydraulic parameters on irrigation quality *Proc. of the Nizhnevolzhsky Agrouniversity Complex: Science and Higher Professional Education* **2(58)** 73-83

[3] Kozinskaya O V 2011 The influence of wind speed and direction on the quality of irrigation by small-sized frontal sprinklers *Proc. of the Nizhnevolzhsky Agrouniversity Complex: Science and Higher Professional Education* **2(22)** 231-236

[4] Bocharnikov V S, Kozinskaya O V, Denisova M A and Bocharnikova O V 2020 Research of the hydraulic characteristics of the pivot sprinkler *Proc. of the Nizhnevolzhsky Agrouniversity Complex: Science and Higher Professional Education* **2(58)** 319-327

[5] Bocharnikov V S, Kozinskaya O V, Denisova M A and Bocharnikova O V 2020 Technology for preparation of contaminated water from poultry farms for irrigation *IOP Conf. Series: Earth and Environmental Science* **579** 012034

[6] Bocharnikova O V, Denisova M A and Bocharnikov V S 2020 Technology of preparation of natural waters for irrigation *IOP Conf. Series: Earth and Environmental Science* **579** 012033

[7] Ovchinnikov A S, Loboyko V F, Bocharnikov V S, Ovcharova A Yu and Fomin S D 2019 State of the small rivers of the Volga basin within the lower Volga *IOP Conf. Series: Earth and Environmental Science* **341** 012107

[8] Ovchinnikov A S, Borodychev V V, Lytov M N, Bocharnikov V S, Fomin S D, Bocharnikova O V and Vorontsova E S 2018 Optimum control model of soil water regime under irrigation *Bulgarian J. of Agricultural Science* **24** 909-913

[9] Mo J, Huang X, Li W and Li Y 2020 Research and optimization of hydraulic characteristics of large-scale variable sprinkler irrigation machine based on PWM technology *Nongye Gongcheng Xuebao/Transactions of the Chinese Society of Agricultural Engineering* **36(19)** 76-85

[10] Bakirov S M, Logacheva O V and Shlyupikov S V 2020 Justification of dependence of the sprinkler machine power supply system efficiency on the irrigation process parameters *IOP Conf. Series: Earth and Environmental Science* **488** 012005

[11] Junping L, Tao L, Jien X, Jipeng L and Nan J 2020 Effect of different fertilizer concentration on hydraulic performance of rotary sprinkler *ASABE 2020 Annual International Meeting*

[12] Zverkov M and Olgarenko G 2020 Estimating raindrops sizes for research of irrigation equipment *Int. Multidisciplinary Scientific Geo Conf. Surveying Geology and Mining Ecology Management SGEM 2020* pp 107-112

[13] Salah H M, Ayoub G, Abdelmalek A and Khaled M 2019 Theoretical sprinkler-spacing configurations in center pivot irrigation system *Water and Energy Int.* **62(9)** 54-59

[14] Jobbágy J, Michlian N, Dačanin P and Rigó I 2019 Application and evaluation of performance quality of hose-reel irrigation machine *Acta Technological Agriculture* **22(4)** 109-114

[15] Li D, Zhu D, Liu K, Wang F and Zhu P 2019 Determination and application verification for driving resistance of lateral move sprinkling machine *Nongye Gongcheng Xuebao/Transactions of the Chinese Society of Agricultural Engineering* **35(17)** 19-27
[16] Run Y 2019 The irrigation uniformity analysis based on translation variable-rate sprinkler *IOP Conf. Series: Materials Science and Engineering* **576** 012028
[17] Rovelo C O R, Ruiz N Z, Tolosa J B, Félix J R F and Latorre B 2019 Characterization and simulation of a low-pressure rotator spray plate sprinkler used in center pivot irrigation systems *Water* **11(8)** 1684
[18] Al Al-Baaj A A and Lewis A 2019 Variable pulsed irrigation algorithm (VPIA) to reduce runoff losses under a low-pressure lateral move irrigation machine *Horticulturae* **5(1)** 10
[19] Wang Y, Sun P, Sun W, Wang T and Chong B 2015 Key components design and experiment of roll wheel line move sprinkling irrigation machine *Paiguan Jixie Gongcheng Xuebao / J. of Drainage and Irrigation Machinery Engineering* **33(10)** 915-920
[20] Kruzhilin I P, Ganiev M A, Kuznetsova N V and Rodin K A 2018 Dynamics of total water consumption and yield of periodically moistened rice during sprinkling and drip irrigation in the Volgograd region *Proc. of the Nizhnevolzhsky Agrouniversity Complex: Science and Higher Professional Education* **3(51)** 34-42

IOP Conf. Series: Earth and Environmental Science **965** (2022) 012009 doi:10.1088/1755-1315/965/1/012009

Calculation of the strength of water management facilities for the sustainable development of rural areas

N A Gureeva[1], R Z Kiseleva[2], Yu V Klochkov[2] and A P Nikolaev[2]

[1] Federal State Educational Budgetary Institution of Higher Education "Financial University under the Government of the Russian Federation", 49 Leningradsky Avenue, Moscow, 125993 (City service post office-3), Russia
[2] Federal State Educational Institution of Higher Education "Volgograd State Agrarian University" 26 University Avenue, Volgograd, 400002, Russia

E-mail: nagureeve@fa.ru

Abstract. The physical relations are presented at the loading step of a bulk body in two variants. In the first version, the theory of plastic flow is used. In the second variant, to determine the increments of plastic deformations, a hypothesis is proposed about the proportionality of the increments of plastic deformations to the components of the deviator of stress increments, which made it more correct and easier to obtain the defining equations at the loading step. Variants of physically nonlinear defining equations in the calculation of three-dimensional elements of water management objects are implemented in the developed tetrahedral finite element with nodal unknowns in the form of displacement increments and stress increments.

1. Introduction

When determining the stress-strain state of engineering objects, the finite element method is widely used [1, 2, 3, 4, 5].

The calculation of the elements of water management facilities for strength, taking into account the physical nonlinearity, is relevant and is carried out using the theory of plasticity – deformation or the theory of plastic flow [6, 7, 8, 9]. In the step-by-step loading process, the flow theory is particularly widely used, in which, at the loading step, the increments of deformations are divided into elastic and plastic parts. Elastic deformations are determined by Hooke's law. The increments of plastic deformations are determined on the basis of the hypothesis of proportionality of the components of the tensor of the increments of plastic deformations and the components of the deviator of total stresses. The proportionality coefficient in the implementation of this hypothesis turns out to be a function of the increment of the stress intensity, represented by the differential of the components of the stress increment tensor.

By summing the components of the increments of elastic and plastic deformations, the defining equations of the flow theory at the loading step are obtained.

In this paper, a variant of the flow theory for plastic strain increments is developed, based on the proposed hypothesis of the proportionality of the components of the tensor of plastic strain increments and the components of the deviator of stress increments. The defining equations of the proposed version of the flow theory turned out to be more correct. To implement physical nonlinearity, a tetrahedral finite element in a mixed formulation with nodal unknowns in the form of displacement increments and stress increments is developed.

2. Materials and methods

2.1. Variants of the defining equations of the theory of plastic flow

In the first variant, when constructing the defining relations of the theory of plastic flow at the loading step, the hypothesis of the separation of the increments of deformations into elastic and plastic parts is used [9]. The increments of elastic deformations $\Delta\varepsilon_{ij}^{e}$ are determined by the equations of Hooke's law [10]:

$$\begin{aligned}
\Delta\varepsilon_{11}^{e} &= \frac{1}{E}(\Delta\sigma_{11} - \nu\Delta\sigma_{22} - \nu\Delta\sigma_{33});\\
\Delta\varepsilon_{22}^{e} &= \frac{1}{E}(\Delta\sigma_{22} - \nu\Delta\sigma_{11} - \nu\Delta\sigma_{33});\\
\Delta\varepsilon_{33}^{e} &= \frac{1}{E}(\Delta\sigma_{33} - \nu\Delta\sigma_{11} - \nu\Delta\sigma_{22});\\
\Delta\varepsilon_{12}^{e} &= \frac{1+\nu}{E}\Delta\sigma_{12};\quad \Delta\varepsilon_{13}^{e} = \frac{1+\nu}{E}\Delta\sigma_{13};\quad \Delta\varepsilon_{23}^{e} = \frac{1+\nu}{E}\Delta\sigma_{23},
\end{aligned} \tag{1}$$

where $\Delta\sigma_{ij}$ - is the increments of normal stresses; E- is the elastic modulus of the material; ν - coefficient of transverse deformation.

To determine the increments of plastic deformations, the hypothesis of proportionality of the components of the increments of plastic deformations to the components of the total stress deviator is used, on the basis of which the relations are written in the form [9]:

$$\Delta\varepsilon_{ij}^{p} = \frac{3}{2}\frac{\Delta\varepsilon_i^p}{\sigma_i}\left(\sigma_{ij} - \frac{1}{3}\delta_{ij}\sigma_c\right), \tag{2}$$

where $\Delta\varepsilon_i^p$ - is the increment of the intensity of plastic deformations; σ_{ij} - the total stresses achieved during the previous loading steps; $\sigma_c = \sigma_{11} + \sigma_{22} + \sigma_{33}$ - the average stress; δ_{ij} - Kronecker symbol; σ_i - stress intensity.

The value $\Delta\varepsilon_i^p$ is determined by the expression [9]:

$$\Delta\varepsilon_i^p = \Delta\varepsilon_i - \Delta\varepsilon_i^e = \frac{\Delta\sigma_i}{E_K} - \frac{\Delta\sigma_i}{E_1} = \Delta\sigma_i\left(\frac{1}{E_K} - \frac{1}{E_1}\right), \tag{3}$$

where $\Delta\varepsilon_i$ is the increment of the intensity of the total deformations; $\Delta\varepsilon_i^e$ - the increment of the intensity of elastic deformations; E_K - the tangent modulus of the deformation diagram; E_1 - the modulus of the initial section of the deformation diagram; $\Delta\sigma_i$ - the increment of the intensity of total stresses.

The value $\Delta\sigma_i$ is defined in this general form:

$$\begin{aligned}
\Delta\sigma_i = {} & \frac{\partial\sigma_i}{\partial\sigma_{11}}\Delta\sigma_{11} + \frac{\partial\sigma_i}{\partial\sigma_{22}}\Delta\sigma_{22} + \frac{\partial\sigma_i}{\partial\sigma_{33}}\Delta\sigma_{33} + \frac{\partial\sigma_i}{\partial\sigma_{12}}\Delta\sigma_{12} + \\
& + \frac{\partial\sigma_i}{\partial\sigma_{13}}\Delta\sigma_{13} + \frac{\partial\sigma_i}{\partial\sigma_{23}}\Delta\sigma_{23}.
\end{aligned} \tag{4}$$

Taking into account (3) and (4), the increments of plastic deformations (2) are written as:

$$\Delta\varepsilon_{ij}^{p}=\frac{3}{2}\frac{\left(\sigma_{ij}-\frac{1}{3}\delta_{ij}\sigma_c\right)}{\sigma_i}\left(\frac{1}{E_K}-\frac{1}{E_1}\right)\frac{\partial\sigma_i}{\partial\sigma_{mn}}\Delta\sigma_{mn}. \quad (5)$$

By summing (5) and (1), the matrix relation is formed:

$$\underset{6\times1}{\{\Delta\varepsilon\}}=\underset{6\times6}{\left[C_1^{\Pi}\right]}\underset{6\times1}{\{\Delta\sigma\}}, \quad (6)$$

where $\underset{1\times6}{\{\Delta\varepsilon\}^T}=\{\Delta\varepsilon_{11}\Delta\varepsilon_{22}\Delta\varepsilon_{33}2\Delta\varepsilon_{12}2\Delta\varepsilon_{13}2\Delta\varepsilon_{23}\}$;

$\underset{1\times6}{\{\Delta\sigma\}^T}=\{\Delta\sigma_{11}\Delta\sigma_{22}\Delta\sigma_{33}\Delta\sigma_{12}\Delta\sigma_{13}\Delta\sigma_{23}\}$;

$\underset{6\times6}{\left[C_1^{\Pi}\right]}$ - is the elastic-plasticity matrix of the first variant.

In the second version of the defining equations for determining the increments of plastic deformations, a hypothesis is proposed about the proportionality of the increments of plastic deformations to the components of the deviator of the stress increments:

$$\Delta\varepsilon_{ij}^{p}=\psi_1\left(\Delta\sigma_{ij}-\frac{1}{3}\delta_{ij}\Delta\sigma_c\right), \quad (7)$$

where the proportionality coefficient is determined ψ_1 by the expression:

$$\psi_1=\frac{3}{2}\frac{\Delta\varepsilon_i^{p}}{\Delta\sigma_i}. \quad (8)$$

Taking into account (8) and (3) the relations (7) will take the form:

$$\Delta\varepsilon_{ij}^{p}=\frac{3}{2}\left(\frac{1}{E_K}-\frac{1}{E_1}\right)\left(\Delta\sigma_{ij}-\frac{1}{3}\delta_{ij}\Delta\sigma_c\right). \quad (9)$$

By summing (9) and (1), the matrix relation of the second variant of the relationship between strain increments and stress increments is formed:

$$\underset{6\times1}{\{\Delta\varepsilon\}}=\underset{6\times6}{\left[C_2^{\Pi}\right]}\underset{6\times1}{\{\Delta\sigma\}}. \quad (10)$$

2.2. Displacements and deformations

Under the action of the load an arbitrary point of the body at the loading step receives an displacement, which is determined by the vector:

$$\vec{W}=w_1\vec{i}+w_2\vec{j}+w_3\vec{k}. \quad (11)$$

The increments of deformations at the point under consideration are determined by the Cauchy relations:

$$\Delta\varepsilon_{11}=\frac{\partial w_1}{\partial x};\quad \Delta\varepsilon_{22}=\frac{\partial w_2}{\partial y};\quad \Delta\varepsilon_{33}=\frac{\partial w_3}{\partial z};$$
$$\Delta\varepsilon_{12}=\frac{1}{2}\left(\frac{\partial w_1}{\partial y}+\frac{\partial w_2}{\partial x}\right);\quad \Delta\varepsilon_{13}=\frac{1}{2}\left(\frac{\partial w_1}{\partial z}+\frac{\partial w_3}{\partial x}\right);\quad \Delta\varepsilon_{23}=\frac{1}{2}\left(\frac{\partial w_2}{\partial z}+\frac{\partial w_3}{\partial y}\right), \quad (12)$$

which are represented in matrix form:

$$\underset{6\times1}{\{\Delta\varepsilon\}}=\underset{6\times3}{[L]}\underset{3\times1}{\{v\}}, \quad (13)$$

where $\underset{3\times1}{\{v\}^T}=\{w_1\ w_2\ w_3\}$; $\underset{6\times3}{[L]}$ -matrix of differential operators.

3. Results and discussion

An arbitrary tetrahedron with nodes *i, j, k, l* is accepted as a finite element in the cartesian coordinate system. To perform numerical integration, an arbitrary tetrahedron is mapped to a local rectangular tetrahedron, the local coordinates of which vary within $0 \le \xi, \eta, \zeta \le 1$.

The cartesian coordinates of the inner point of the tetrahedron are determined through their nodal values using linear relations:

$$\lambda = \xi\lambda^i + \eta\lambda^j + \zeta\lambda^k + (1-\xi-\eta-\zeta)\lambda^l = \underset{1\times4}{\{\varphi(\xi,\eta,\zeta)\}^T} \underset{4\times1}{\{\lambda_y\}}, \tag{14}$$

where the symbol λ refers to the cartesian coordinates *x, y,* and *z*, and the symbol $\{\lambda_y\}$ - refers to the column of values of the coordinates of the nodes of the tetrahedron.

By differentiating (14) the derivatives of cartesian coordinates $x,_\xi, x,_\eta, x,_\zeta, y,_\xi\ y,_\eta, y,_\zeta, z,_\xi\ z,_\eta, z,_\zeta$ in the local coordinate system and the derivatives of local coordinates in cartesian coordinates are determined $\xi,_x, \xi,_y, \xi,_z, \eta,_x, \eta,_y, \eta,_z, \zeta,_x, \zeta,_y, \zeta,_z$. The components of the displacement vector of the inner point of the finite element $\{w_i\}$ are approximated by their nodal values also by expressions (14):

$$w_1 = \underset{1\times4}{\{\varphi\}^T} \underset{4\times1}{\{w_{1y}\}}, \quad w_2 = \underset{1\times4}{\{\varphi\}^T} \underset{4\times1}{\{w_{2y}\}}, \quad w_3 = \underset{1\times4}{\{\varphi\}^T} \underset{4\times1}{\{w_{3y}\}}. \tag{15}$$

On the basis of (15), a matrix relation is formed:

$$\underset{3\times1}{\{v\}} = \underset{3\times12}{[A]} \underset{12\times1}{\{u_y\}}, \tag{16}$$

where $\underset{1\times12}{\{u_y\}^T} = \left\{ \underset{1\times4}{\{w_{1y}\}^T} \underset{1\times4}{\{w_{2y}\}^T} \underset{1\times4}{\{w_{3y}\}^T} \right\}$ - a string of nodal unknown displacements.

Taking into account (16), the relation (13) for the increments of deformations gets the expression:

$$\underset{6\times1}{\{\Delta\varepsilon\}} = \underset{6\times3}{[L]} \underset{3\times12}{[A]} \underset{12\times1}{\{u_y\}} = \underset{6\times12}{[B]} \underset{12\times1}{\{u_y\}}. \tag{17}$$

The approximation of the individual components of the stress tensor was also performed by the relations (14). The string of nodal unknown stresses is taken as:

$$\underset{1\times24}{\{\Delta\sigma_y\}^T} = \left\{\Delta\sigma^i_{xx}\, \Delta\sigma^j_{xx}\, \Delta\sigma^k_{xx}\, \Delta\sigma^l_{xx}\, \Delta\sigma^i_{yy}\Delta\sigma^j_{yy}\Delta\sigma^k_{yy}\Delta\sigma^l_{yy} .. \Delta\sigma^i_{yz}\Delta\sigma^j_{yz}\Delta\sigma^k_{yz}\Delta\sigma^l_{yz}\right\}.$$

For the increments of the stresses of the inner point of the finite element, a matrix expression is formed:

$$\underset{6\times1}{\{\Delta\sigma\}} = \underset{6\times24}{[S]} \underset{24\times1}{\{\Delta\sigma_y\}}, \tag{18}$$

where $\underset{1\times6}{\{\Delta\sigma\}^T} = \{\Delta\sigma_{xx}\, \Delta\sigma_{yy}\Delta\sigma_{zz}\Delta\sigma_{xy}\, \Delta\sigma_{xz}\Delta\sigma_{yz}\}$

To form the matrix of elastic-plastic deformation of the tetrahedron at the loading step, a mixed functional in the form of [5]:

$$\Phi = \int_V \left[\underset{1\times6}{\{\sigma\}^T} + \frac{1}{2} \underset{1\times6}{\{\Delta\sigma\}^T} \right] \underset{6\times1}{\{\Delta\varepsilon\}} dV - \frac{1}{2}\int_V \underset{1\times6}{\{\Delta\sigma\}^T} \underset{6\times6}{\left[C^{\Pi}_{\rho}\right]} \underset{6\times1}{\{\Delta\sigma\}} dV - \int_w \underset{1\times3}{\{v\}^T} \left[\underset{3\times1}{\{q\}} + \frac{1}{2}\{\Delta q\} \right] dw, \tag{19}$$

where $\{\sigma\}$ - is the stress column for the previous loading steps; $\rho = 1,2$ - for the plasticity matrix by (6) or by (10); $\{q\}$ - the load for the previous loading steps; $\{\Delta q\}$ - the load at the considered loading step; w – the area of application of a given load.

Taking into account (16), (17) and (18), the functional (19) can be represented as:

$$\Phi = \int\limits_{V} \underset{1\times6}{\{\sigma\}^T} \underset{6\times2}{[B]} dV \underset{12\times1}{\{u_y\}} + \frac{1}{2} \underset{1\times24}{\{\Delta\sigma_y\}^T} \int\limits_{V} \underset{24\times6}{[S]^T} \underset{6\times12}{[B]} dV \underset{12\times1}{\{u_y\}} - \frac{1}{2} \underset{1\times24}{\{\Delta\sigma_y\}^T} \int\limits_{V} \underset{24\times6}{[S]} \underset{6\times6}{\left[C_\rho^{\Pi}\right]} \underset{6\times24}{[S]} dV \underset{24\times1}{\{\Delta\sigma_y\}} - \underset{1\times12}{\{u_y\}^T} \int\limits_{w} \underset{12\times3}{[A]^T} \underset{3\times1}{\{q\}} dw - \frac{1}{2} \underset{1\times12}{\{u_y\}^T} \int\limits_{w} \underset{12\times3}{[A]^T} \underset{3\times1}{\{\Delta q\}} dw. \quad (20)$$

After minimizing the functional (20) with respect to the nodal unknowns, $\{\Delta\sigma_y\}^T$ a $\{u_y\}^T$ system of equations is obtained:

$$\frac{\partial \phi}{\partial \{\Delta\sigma_y\}^T} \equiv \underset{24\times12}{[Q]} \underset{12\times1}{\{u_y\}} - \underset{24\times24}{[H]} \underset{24\times1}{\{\Delta\sigma_y\}} = 0; \quad (21)$$

$$\frac{\partial \phi}{\partial \{u_y\}^T} \equiv \underset{12\times24}{[Q]^T} \underset{24\times1}{\{\Delta\sigma_y\}} - \underset{12\times1}{\{f_{\Delta q}\}} + \underset{12\times1}{\{f\}_\sigma} = 0, \quad (22)$$

where $\underset{24\times12}{[Q]} = \int\limits_{V} \underset{24\times6}{[S]^T} \underset{6\times12}{[B]} dV$; $\underset{24\times24}{[H]} = -\int\limits_{V} \underset{24\times6}{[S]^T} \underset{6\times6}{\left[C_\rho^{\Pi}\right]} \underset{6\times24}{[S]} dV$; $\underset{12\times1}{\{f_q\}} = \int\limits_{w} \underset{12\times3}{[A]^T} \underset{3\times1}{\{q\}} dw$;

$\underset{12\times1}{\{f_{\Delta q}\}} = \int\limits_{w} \underset{12\times3}{[A]^T} \underset{3\times1}{\{\Delta q\}} dw$; $\underset{12\times1}{\{f_\sigma\}} = \int\limits_{V} \underset{12\times6}{[B]^T} \underset{6\times1}{\{\sigma\}} dV$.

Systems (21) and (22) are represented in the finite element form:

$$\underset{36\times36}{[R]} \underset{36\times1}{\{Z_y\}} = \underset{36\times1}{\{F_y\}}, \quad (23)$$

where $\underset{36\times36}{[R]} = \begin{bmatrix} \underset{24\times24}{[H]} & \underset{24\times12}{[Q]} \\ \underset{12\times24}{[Q]} & \underset{12\times12}{[0]} \end{bmatrix}$ - is the matrix of elastic-plastic deformation of the tetrahedral element;

$\underset{1\times36}{\{z_y\}^T} = \left\{ \underset{1\times24}{\{\Delta\sigma_y\}^T} \ \underset{1\times12}{\{u_y\}^T} \right\}$ -a string of nodal unknowns of the finite element;

$\underset{1\times36}{\{F_y\}^T} = \left\{ \underset{1\times24}{\{0\}^T} \ \underset{1\times12}{\{f\}^T} \right\}$; $\underset{1\times12}{\{f\}^T} = \left\{ \underset{1\times12}{\{f_{\Delta q}\}^T} + \underset{1\times12}{\{R\}^T} \right\}$; $\underset{1\times12}{\{R\}^T} = \underset{1\times12}{\{f_q\}^T} - \underset{1\times12}{\{f_\sigma\}^T}$ - the Newton Raphson discrepancy.

The formation of the matrix of the stress-strain state of the calculated structure is carried out by the traditional finite element method [7].

4. Conclusion

In the first version of the theory of plastic flow, the stress increments appeared on the basis of the differential representation of the stress intensity increment.

The developed variant of obtaining the defining equations (10) differs in that in it the stress increments appear directly on the basis of the proposed hypothesis.

The defining equations of the second option are more correct.

5. Acknowledgments

The research was carried out with the financial support of the Russian Foundation for Basic Research and the Administration of the Volgograd Region in the framework of the scientific project No. 19-41-340004 r_a.

References

[1] Rikards R B 1988 *The Finite Element Method in the Theory of Shells and Plates* (Riga: Zinatne)

p 284
[2] Golovanov A I, Tyuleneva O N and Shigabutdinov A F 2006 *Finite Element Method in the Statics and Dynamics of Thin-Walled Structures* (Moscow: Fizmatlit) p 392
[3] Bate K 2010 *Finite Element Methods* (Moscow: Fizmatlit) p 1024
[4] Oden J T 1976 *Mechanics (*Moscow: Mir*)* p 464
[5] Gureeva N A, Klochkov Yu V and Nikolaev A P 2009 *Proc. of the Nizhnevolzhsky Agrouniversitetskiy complex: Science and Higher Professional Education.* **2(14)** 123-129
[6] Ovchinnikov A S, Kiselev A P and Kiseleva R Z 2018 *Proc. of the Nizhnevolzhsky Agrouniversitetskiy complex: Science and Higher Professional Education.* **4(52)** 108-115
[7] Postnov V A, Kharkhurim Ya 1974 *Finite Element Method in Calculations of Ship Structures* (Leningrad: Shipbuilding) p 344
[8] Kayumov R A, Kupriyanov V N, Mukhamedova I Z, Suleymanov A M, Shakirova A M 2007 Method of analysis of the deformation process of a film-fabric composite taking into account geometric and physical nonlinearity *Modeler. and Edges. Tasks* vol 1 119-121
[9] Malinin N N 1975 Applied Theory of Plasticity and Creep (Moscow: Mashinostroenie) p 400
[10] Sedov L I 1976 *Continuum Mechanics* vol 1 (Moscow: Science) p 535

IOP Conf. Series: Earth and Environmental Science **965** (2022) 012010 doi:10.1088/1755-1315/965/1/012010

Efficiency of using the Kazakhstan zeolites for the purification of water from iron ions

Zh B Khamzina[1] and S L Belopukhov[2]

[1] Almaty Technological University, 100 Tole Bi Street Almaty, 050012, Kazakhstan
[2] Russian State Agrarian University - Moscow Timiryazev Agricultural Academy, 49 Timiryazevskaya Street, Moscow, 127434, Russia

E-mail: Zhuldyz_hamzina@mail.ru

Abstract. The problem of provision of the population with high-quality drinking water is currently one of the most topical in connection with the unsatisfactory technical condition of the existing water supply systems and pollution of the water sources. Natural zeolites, which have a sufficiently high sorption capacity, cation-exchange properties, relatively low cost and availability, are increasingly being used in recent times for water purification. The mineralogical properties of natural zeolites of the Tayzhuzgen and Shankanai deposits have been investigated. A comparative study of the sorption properties of the natural and modified zeolites of the Kazakhstan deposits has been carried out on model solutions, containing iron ions (Fe^{2+}). The exchange mechanism and the degree of extraction of ions from the solutions have been established. The efficiency of the municipal water purification in the Auezovskiy district of Almaty with zeolites of the Shankanai field has been equal to 85.94%. This suggests a possibility of using these zeolites to purify water from iron ions with the purpose of water treatment in the dairy industry.

1. Introduction

For the provision of the population and industry with milk, both full and powdered milk is used. The seasonality of milk production in agriculture is the main reason for the partial replacement of the full milk for powdered milk. To restore the milk powder, water is used, which should meet the requirements of the Technical Regulations of the Customs Union “On food safety” [1]. Apart from the general requirements to drinking water, in the production of dairy products, the water, used for technological needs, should meet a number of specific requirements regarding hardness, the presence of nitrate and nitrithions, chlorine ions, heavy metals, etc. [2]. It has been established that the chemical composition of water affects the solubility of milk powder [3]. The main methods of water treatment in the food production are filtration, coagulation, deferrization, disinfection, ion exchange, electrodialysis, and reverse osmosis. These methods, along with the positive aspects, have also disadvantages. Besides, along with the hardness ions it is required to extract from the water, entering the production processes of the dairy products, more hazardous components, such as heavy metal cations, radioactive isotopes, organochlorine and surfactants, pesticides, etc. Accounting for the properties of these pollutants, those filter materials are promising, which have the pronounced sorption and ion-exchange properties [4]. At the same time, these materials should be chemically resistant and mechanically strong. Natural zeolites meet almost all of these requirements. Extensive studies have

shown that zeolites meet all the requirements and are unique in their high content of exchangeable iron ions, fine dispersion and swelling in water [5].
This work aims at studying the sorption of metal ions Fe_{total} by Kazakhstani deposit zeolites.

2. Materials and methods

In the experiments, a pre-sieved fraction of zeolites from the Tayzhuzgen and Shankanai deposits of Kazakhstan with a size of 0.5 and 1 mm were used. To increase the exchange capacity of natural zeolite and improve its sorption properties, natural zeolites from the Tayzhuzgen and Shankanai deposits were subjected to "hard" modification. For modification, a 4% solution of 1 N hydrochloric acid was used, which was prepared from a 37% concentrated hydrochloric acid solution with a density of 1.198 g/cm^3.

X-ray diffraction analysis was carried out using DRON-3 Automated Diffractometer with $Cu_{K\alpha}$ radiation, β-filter. Diffractogram scanning environment: U = 35 kV; I = 20 mA; scanning – θ-2θ; detector – 2 DPM. The semiquantitative X-ray phase analysis was carried out using diffractograms of powdered samples by the method of equal weighted quantities and artificial mixtures. Quantitative ratios of crystalline phases were determined. Diffractograms were interpreted using data from the ICDD and diffractograms of impurity-free minerals. The component content was calculated for the major phases.

In order to achieve the goal, analyses were conducted with the use of standardized test solutions. Standardized test solutions were prepared from state standard reference samples of Fe^{2+} using distilled water with 2.5 mg/l concentration of C(Fe^{2+}). Once the standardized test solution was filtered at a rate of 4 ml/min, each filtrate was selected and the Fe^{2+} concentration was determined using KFK-2 photocolorimeter. After positive results, further similar experiments were carried out using tap water. The resulting filtrate was examined on the spectrometer "KVANT-Z.ETA".

3. Results and discussion

The quality of water is assessed not only in terms of the impact on the technological processes, the quality of the finished products, but also on the hygienic food safety indicators [6]. Currently, much attention is paid to the environment and the internal environment of the human body from the ever-increasing action of chemicals of natural and anthropogenic origin, in particular heavy metals.

Heavy metals are highly toxic substances and pose a danger to human health [7]. At the same time, macro- and microelements are indispensable factors of nutrition, which, on the one hand, makes it possible to normalize their content in the diet and food products, on the other hand, dictates the need for such rationing under the conditions of environmental pollution [8].

The sources of heavy metals, which enter the human body, can be air, food, and water [9]. All of these factors take place especially in the industrialized regions.

It has been of interest to study the sorption of a number of heavy metals from the aqueous solutions of their salts using materials, based on zeolites from the Tayzhuzgen and Shankanai deposits.

At this stage of the work, both natural and hydrochloric acid-modified zeolites of the Tayzhuzgen (the Tarbagatai district of East Kazakhstan region) and Shankanai (the Kerbulak district of Almaty region) deposits have been used, the main mineral composition of which has been determined by the X-ray diffraction method. It has been found that clinoptillolite is the basis of the mineral composition of zeolites. The X-ray phase analysis data are presented in Table 1.

In general, the chemical composition of zeolites is described by the formula: $Me_{2/n}O \cdot Al_2O_3 \cdot xSiO_2 \cdot yH_2O$, where n is the valence of the metal cation, x is the molar ratio of SiO_2/Al_2O_3, y is the number of moles of water. The total volume of cavities and channels in zeolites is about 50% of the volume.

Table 1. X-ray phase analysis of zeolites.

Natural zeolite of the Tayzhuzgen deposit			Natural zeolite of the Shankanai deposit		
d, Å	*I*, %	mineral	*d*, Å	*I*, %	mineral
1	2	3	4	5	6
14.02855	19.6	smectite	14.28128	23.7	montmorrilonite (smectite)
9.03618	24.7	clinoptilolite	9.46528	34.4	lamontite
7.94808	14.8	clinoptilolite	8.98021	52.4	clinoptilolite
6.65278	14.9	clinoptilolite	7.90552	24.4	clinoptilolite
6.37924	4.2	PS (albite)	6.83373	33.7	clinoptilolite
5.27303	15.9	clinoptilolite	5.60729	27.8	PS (albite)
5.11657	16.5	clinoptilolite	5.12042	29.2	clinoptilolite
4.67201	18.7	clinoptilolite	4.67339	29.0	clinoptilolite
4.47112	20.8	smectite	4.49995	27.6	montmorrilonite (smectite)
4.25442	32.5	quartz	4.25775	38.5	quartz
3.97849	36.3	clinoptilolite	4.15598	31.3	lamontite
3.92433	27.1	clinoptilolite	4.03244	41.9	PS (albite)
3.78425	24.9	PF (orthoclase)	3.97388	48.7	clinoptilolite
3.56673	16.5	PF (orthoclase)	3.91763	38.5	clinoptilolite
3.34444	100.0	quartz	3.76496	35.4	PF (orthoclase)
3.23349	28.8	PF (orthoclase)	3.67024	38.6	PF (albite)
3.18953	66.5	PS (albite)	3.50194	34.9	lamontite, PF (orthoclase)
2.98814	28.8	PF (orthoclase)	3.34337	100.0	quartz
2.80169	14.2	clinoptilolite	3.19007	70.3	PS (albite)
2.57073	14.7	PF (orthoclase), PS (albite)	2.98939	43.2	clinoptilolite
2.45667	16.8	quartz	2.93668	32.0	PS (albite)
2.28263	12.7	quartz	2.79857	32.2	clinoptilolite
2.23723	10.3	quartz	2.70356	32.4	hematite, lamontite
2.16957	9.8	PF (orthoclase)	2.52047	31.9	clinoptilolite
2.12730	15.4	quartz	1.79911	18.2	PF (orthoclase)
1.81744	13.4	quartz			

The frame of zeolites is negatively charged, as a result of which charge-compensating counterions can be replaced by protons or other cations (Figure 1). The negative charge on the surface of zeolite promotes the adsorption of the polar particles [10].

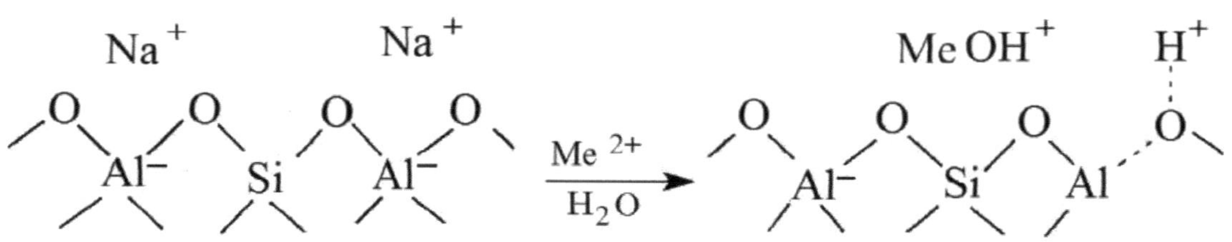

Figure 1. Adsorption of metal ions on the zeolite frame.

To analyze the sorption properties, the samples of the zeolite materials have beee crushed in a jaw crusher and separated into fractions, using a mechanical sieve. The fractions of 0.5 and 1 mm have been used in the experiments. The mineral fractions have been selected on the basis of the previous studies, which have shown that the use of particles less than 1.0 mm increases the hydrodynamic resistance of the mineral layer when filtering the solutions. The modification of minerals has been carried out in the following way: a weighed portion of a mineral weighing 5 grams has been placed in a glass, poured with a 4% solution of hydrochloric acid in a ratio of 2:5 and stirred for 1 hour at the room temperature, then zeolite has been separated from the acid, washed with water and dried at 60 °C in a drying cabinet. To determine the conditions for purifying water from iron ions in a dynamic mode, water has been passed through a column filled with a sorbent at a rate of 4 ml/min. The concentration of Fe^{2+} ions has been determined by the photocolorimetric method (photocolorimeter KFK-2). The results are presented in Table 2, Table 3, Table 4 and Table 5.

Table 2. The results of reducing the content of iron ions in the model solutions in a dynamic mode, when using natural zeolite of the Tayzhuzgen deposit.

No.	Fraction, mm	Sample weight, g	The volume of passed water, ml	Fe^{2+} concentration in the model solution, mg/l	After passing through a zeolite filter, mg/l	Purification degree α, %
1	0.5	4	20	2.5	1.1	56
2	0.5	4	30	2.5	0.95	62
3	0.5	4	30	2.5	0.75	70
4	0.5	4	30	2.5	0.6	76
1	1	5.0	20	2.5	1.0	60
2	1	5.0	30	2.5	0.9	64
3	1	5.0	30	2.5	0.9	64
4	1	5.0	30	2.5	1.05	58

As can be seen from the presented data, in the dynamic mode at the tested rate of water passing through the column with the load of 4 g, with the fraction of 0.5 mm, the degree of purification increases up to 76%, and when used with the load of 5 g, with the fraction of 1 mm, the degree of purification has eached 64%.

Table 3. The results of reducing the content of iron ions in the model solutions in a dynamic mode, when using modified zeolite from the Tayzhuzgen deposit.

No.	Fraction, mm	Sample weight, g	The volume of passed water, ml	Fe^{2+} concentration in the model solution, mg/l	After passing through a zeolite filter, mg/l	Purification degree α, %
1	0.5	3.7	20	2.5	0.95	62
2	0.5	3.7	30	2.5	0.92	63.2
3	0.5	3.7	30	2.5	1.05	58
4	0.5	3.7	30	2.5	1.1	56
1	1	5.0	20	2.5	0.95	62
2	1	5.0	30	2.5	0.95	62
3	1	5.0	30	2.5	0.9	64
4	1	5.0	30	2.5	1.2	52

It is shown that as a result of the modification, the sorption capacity of zeolite in the dynamic mode at the tested rate of water passing through the column with the load of 3.7 g, with the fraction of 0.5 mm, increases up to 63.2%, and when used with the load of 5 g, with the fraction 1 mm up to 64%.

Table 4. Results of reducing the content of iron ions in the model solutions in a dynamic mode, using natural zeolite from the Shankanai deposit.

No.	Fraction, mm	Sample weight, g	The volume of passed water, ml	Fe^{2+} concentration in the model solution, mg/l	After passing through a zeolite filter, mg/l	Purification degree α, %
1	0.5	15	30	2.5	1.15	54
2	0.5	15	40	2.5	0.9	64
3	0.5	15	40	2.5	0.85	66
4	0.5	15	50	2.5	0.45	82
1	1	15	30	2.5	1.1	56
2	1	15	40	2.5	0.9	64
3	1	15	40	2.5	0.7	72
4	1	15	50	2.5	0.5	80

Table 5. Results of a decrease in the content of iron ions in the model solutions in a dynamic mode, when using the modified zeolite of the Shankanai deposit.

No.	Fraction, mm	Sample weight, g	The volume of passed water, ml	Fe^{2+} concentration in the model solution, mg/l	After passing through a zeolite filter, mg/l	Purification degree α, %
1	0.5	10	30	2.5	0.9	64
2	0.5	10	40	2.5	0.82	67.2
3	0.5	10	40	2.5	0.8	68
4	0.5	10	50	2.5	0.5	80
1	1	15	30	2.5	0.92	63.2
2	1	15	40	2.5	0.92	63.2
3	1	15	40	2.5	0.7	72
4	1	15	50	2.5	0.65	74

Table 4 presents the results before and after passing water through a column with natural zeolite with the filtering mass of 15 g, with the fraction of 0.5 mm, the purification is achieved by 82%, and with the load of 5 g, with the fraction of 1 mm, by 80%. Table 5 shows that when using the modified zeolite with the mass of 10 g, with the fraction of 0.5 mm, the degree of purification increases up to 80%, and with the load of 15 g with the fraction of 1 mm, up to 74%.

As can be seen from the data obtained, the extraction of iron ions by zeolites is ambiguous. Based on all these data, we can conclude that it is possible to improve the cleaning performance in a dynamic mode with an increase in the mass of zeolite.

The analysis of iron content in drinking water of Almaty before and after its processing with natural zeolites by atomic-asorption spectrometry method on the spectrometer "KVANT-Z.ETA" (Table 6, Table 7) has been carried out.

Table 6. Results of the study of the content of iron ions before and after passing through zeolite of the Tayzhuzgen deposit.

Sample weight, g	Fe^{2+} concentration in the model solution, mg/l		Purification degree α, %
	In the non-purified municipal water	After passing through zeolite	
3.2	23.076	3.37	85.94

Table 7. Results of the study of the content of iron ions before and after passing through zeolite of the Shankanai deposit.

Sample weight, g	Fe^{2+} concentration in the model solution, mg/l Fe^{2}_{total} concentration in the model solution, mg/l		Purification degree α, %
	In the non-purified municipal water	After passing through zeolite	
3.2	23.076	3.486	84.89

Table 6 and Table 7 shows that before and after the purification of municipal water with zeolites of the Tayzhuzgen deposit from Fe_{total} has been 85.94%, and with zeolites of the Shankanai deposit - 84.89%.

4. Conclusion

The mineralogical composition of natural zeolites of the Tayzhuzgen and Shankanai deposits has been studied. The main component of the zeolite-containing rock is clinoptilolite. At the initial concentration of iron ions Fe^{2+} (2.5 mg/L), when passing the model solution through a filter on the basis of natural zeolite of the Tayzhuzgen deposit, the removal efficiency has been 76%, and that of the Shankanai deposit - 82%, and when the model solution has been passed through a modified zeolite filter, the removal efficiency of Fe^{2+} has been 80%. The results of the study before and after purification of the municipal water with zeolites of the Tayzhuzgen deposit from Fe_{total} has been 85.94%, and with zeolites of the Shankanai deposit - 84.89%. It has been found that iron ions interact with zeolites by the ion exchange mechanism, i.e. its extraction by zeolite is possible. The recovery increases depending on the weight of zeolite.

Thus, the studies carried out allow us to speak of the efficiency of using zeolites of the Tayzhuzgen and Shankanai deposits in the water treatment system in the dairy industry in order to ensure its technological indicators.

References

[1] Technical Regulations of the Customs Union 021/2011 "On food safety" 2011 Decision of the Customs Union Commission (CU) 874 http://www.tsouz.ru/db/techreglam/Documents/

[2] Babkina T A 2017 Requirements for water quality in the dairy industry *Dairy Industry* **5** 60-63

[3] Popova N V and Potoroko I Yu 2014 Ensuring the quality and preservation of the reconstituted milk processing products *Bulletin of South Ural State University* **3** 37-46

[4] Scaglioni P T and Badiale-Furlong E 2016 Rice husk as an adsorbent: A new analytical approach to determine aflotoxins in milk *Talanta* **152** 423-431

[5] Wang S and Peng Y 2010 Natural zeolites as effective adsorbents in water and wastewater treatment *Chemical Engineering J.* **156** 11-24

[6] Boguniewicz-Zablocka J, Klosok-Bazan I and Naddeo V 2019 Water quality and resource management in the dairy industry *Environ. SciPollut. Res. Int.* **26(2)** 1208-1216

[7] Plotnikov E, Martemianova I, Mertemianov D, Zhuravkov S, Voronova O, Korotkova E and Slonikov V 2015 Water purification with natural sorbents: effect of surface modification with nano-structured particles *16th International Scientific Conf. "Chemistry and Chemical Engineering in XXI century" dedicated to professor L.P. Kulyov* (Tomsk)

[8] Garcia S N, Osbum B I and Cullor J S 2019 A one health perspective on the dairy production and dairy food safety. *One Health* **7** 100086

[9] Hassan A H 2021 Study of certain heavy metals with potential human health effects *IOP Conf. Series: Earth and Environmental Science* **722** 012047

[10] Dyachkova T Yu, Makarova I A, Vaganova E S, Davydova O A and Mosunova T V 2019 Sorption purification of the solutions from heavy metal ions, using zeolite, modified with carbon nanotubes *Bulletin of South Ural State University. Chemistry* **11(2)** 16-27

IOP Conf. Series: Earth and Environmental Science **965** (2022) 012011 doi:10.1088/1755-1315/965/1/012011

Dynamics of various forms of manganese in the pond waters

S Ch Kazanchev[1], O O Getokov[1], Sh B Khashegulgov[2], M M Dolov[2] and L U Yusupova[2]

[1] Kabardino-Balkarian State Agrarian University named after V M Kokov 1V Lenin Avenue, Nalchik, 360030, Russia
[2] Ingush State University, 7 Zyazikov Avenue, Magas, Republic of Ingushetia, 386001, Russia

E-mail: getokov777@mail.ru

Abstract. The forms of migration of manganese in pond waters are suspended matter and truly dissolved forms, including free manganese cations, as well as complex compounds formed by manganese with organic substances. The aim of the work was to study the dynamics of the content of various forms of manganese in the water of fish ponds. The content of individual forms of manganese was established according to the method developed by Linnik and Nabivanets. According to the methodology of Alekin, Semenov and Skopintsev determined hydrochemical parameters. It has been established that the dynamics of the content of manganese forms in different seasons depends mainly on the concentration of dissolved oxygen, pH value, inflow from sludge waters and the content of dissolved organic complexing substances in the water. In winter, free uncomplexed ions are the dominant form of migration. In other seasons, the content of complex forms exceeds the concentration of uncomplexed manganese. The dominant form of manganese migration in spring, summer, and autumn is Mn suspended, while in some periods of the same seasons, Mn dissolved predominates, which is explained by the influence of the intensity of water exchange, where suspended particles settle under conditions of slow runoff.

1. Introduction

The main forms of manganese migration in pond waters are suspensions and truly dissolved forms, including free manganese cations (Mn^{2+}) and its complex compounds with dissolved organic substances (molecular weight 0.5-1.5 thousand and 1.5-5 thousand). The manganese content in the water of fish ponds, especially in phytoplankton, depends on the pH of the water, soil and the nature of fertilizers [1, 2]. Only with fertilization with ammonium sulfate is it possible to increase the manganese content in water, soil and in organisms of hydrobionts, i.e. the amount of manganese received by hydrobionts is determined by its content in individual ingredients of the trophic chain of water bodies. Literary sources indicate that the lack of manganese in the habitat of ingredients and hydrobionts causes disturbances in growth and development, changes in the skeleton, the coefficient of mobility and fish death [3, 4, 5].

The aim of this work was to find out the dynamics of the content of various forms of manganese in the water of fish-breeding ponds and compare the results with previously obtained similar data, which are influenced by the biological resource potential of water bodies.

2. Materials and methods

The studies were conducted in 2016 and again in 2017 in the pond farms of LLC "Khamovykh" (V ecological and phenological zone) of the Tersk district of the Kabardino-Balkar Republic [3]. Water samples to determine morphometric parameters were taken in the morning and evening every week.

The content of individual forms of manganese was established according to the method developed by Linnik and Nabivanets [4]. Water samples taken at the surface and bottom were mixed and, in addition to the content of certain forms of manganese, some hydrochemical parameters (the amount of dissolved oxygen, pH, dichromate oxidizability) were determined in them according to the method of Alekin, Semenov and Skopintsev [1].

3. Results and discussion

The content of Mn_{total} and its individual forms in the water of fish-breeding ponds (V fish-breeding zone, Figure 1, 2) varies over a fairly wide range throughout the year. The maximum concentrations of Mn_{total} and $Mn_{dissolved}$ fall mainly in the winter and early spring (pre-flood) periods, 1.7 and 1.65 mg/l, respectively. In winter $Mn_{dissolved}$ dominates, and the $Mn_{suspended}$ content is minimal (39-79 µg/L). The smallest amount of $Mn_{dissolved}$ (33-42 µg/l) was recorded in the spring-summer period (end of April-August). In addition to the well-defined peak of $Mn_{dissolved}$, in the winter-spring time the two largest maxima of the concentration of this form were noted at the end of August and at the end of October.

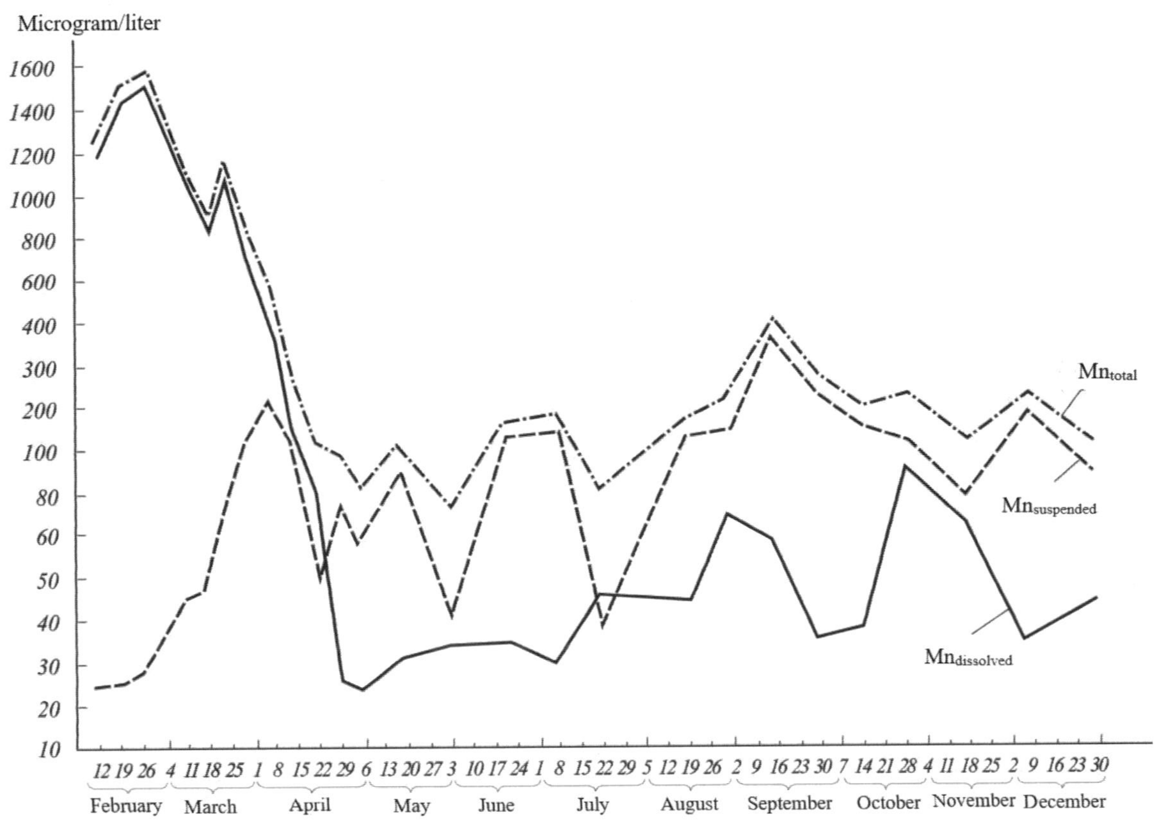

Figure 1. Dynamics of total, dissolved and suspended manganese in the water of fish ponds (2016).

In spring, summer and autumn the predominant form of manganese migration is a fine mechanical suspension. This is consistent with the conclusions that the main amount of manganese is transported by river waters in a suspended state [6, 7, 8]. The largest amount of $Mn_{suspended}$ (about 415 µg/l) was found in August.

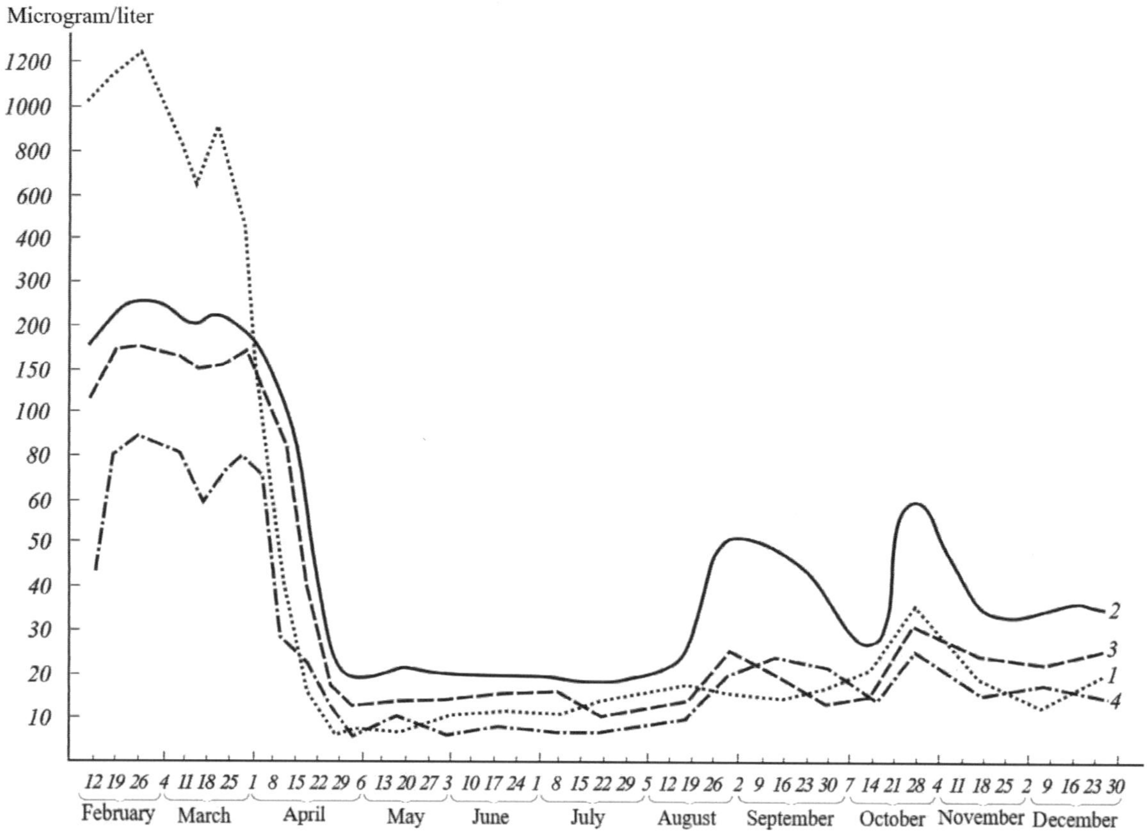

Figure 2. Dynamics of dissolved forms of manganese: 1 - uncomplexed ions (Mn^{+2}); 2 - sum of complex compounds; 3, 4 - complex compounds with a molecular weight of, respectively, 0.6-1.6 thousand and 1.6-5.1 thousand.

The manganese content in natural waters is subjected to significant fluctuations depending on the values of the redox potential and the pH conditions [9].

High concentrations of $Mn_{dissolved}$ in winter can be explained by the occurrence of reducing conditions in the water supply, which contribute to an increase in mobility, and, consequently, in the content of manganese. At the same time the concentration of dissolved oxygen in February decreased to a minimum (about 0.51 mg/l), and the pH was equal to 7.46 (Figure 3). During the ice period the content of $Mn_{suspended}$ decreases due to the recovery of its oxidized forms, and the amount of suspended particles, on which it depends, is minimal. In winter uncomplexed ions predominate among the dissolved forms of manganese (Fig. 2); their concentration in February is maximum (1.3 mg/l, or 80% of the total amount of $Mn_{dissolved}$). Complexed forms are represented mainly by compounds with a molecular weight of about 0.5-1.5 thousand and 1.5-5.0 thousand: concentrations are equal to 180 and 80 μg/l, or 11.6 and 5.0% of $Mn_{dissolved}$. In addition to them complexes with a molecular weight of about 70 thousand and more than 120-150 thousand were found, but their share in the total amount of dissolved forms of manganese does not exceed 3-5% (only 50-70 μg/l). Since in other seasons of the year the concentrations of these complexes were even lower or were not detected at all, their dynamics is not considered.

Figure 3. Dynamics of dissolved oxygen (1), PH (2), and bichromatic oxidability (3).

The increase in the concentration of free uncomplexed manganese in the winter-spring period is explained by its influx from the silt waters. The latter are characterized by relatively low pH values (6.8-7.2) and redox potential (-80: -140 mV) as a result of which the accumulation of divalent manganese is possible in them. In the course of special experiments, it was found that water-soluble forms of manganese in silt waters contain uncomplexed ions in an amount sometimes more than 12-17 mg/l. Part of the manganese coming from the silt water is bound into complex compounds by dissolved organic substances. However, the water in ponds is characterized by a low content of dissolved organic compounds, especially in winter (the concentration of organic matter was judged by the value of dichromate oxidizability (Figure 3), and, therefore, for complete binding of the incoming manganese, there are apparently lack of complexing compounds ... The proportion of complex forms of manganese in the winter-spring period did not exceed 19% of $Mn_{dissolved}$. The facts of detecting manganese complexes with a molecular weight of about 70 thousand and more than 110-160 thousand in river water indicate the possibility of their supply from silt waters, where concentrations reach 260-320 μg/l.

With the onset of snow melting and the disappearance of the ice cover aeration of the river water occurs which can be seen from the rise in the dissolved oxygen concentration curve (Figure 3). The onset of oxidizing conditions leads to a sharp decrease in the content of dissolved forms and, above all, in the concentration of uncomplexed manganese as a result of its oxidation to tetravalent and precipitation in the form of hydrated MnO_2 oxide (IV).

In summer the dynamics of $Mn_{dissolved}$ is characterized by the predominance of not complexed forms over noncomplexed ones (Figure 2). Similar data were obtained in the autumn period. At relatively high pH values (8.7-9.2) noted in these seasons, the accumulation of free uncomplexed manganese ions in river water is practically impossible, since at pH ~ 8.5 it already undergoes hydrolysis and oxidation.

At the end of August, the concentrations of $Mn_{dissolved}$ and especially of its complex forms, slightly increase (Figure 2), which, apparently, can be explained by an increase of the role of intra-water body biological processes in the manganese cycle. A similar peak, but more clearly expressed, was noted at about the same time for ponds in 2017, the bioproductivity of which is undoubtedly higher than in rivers. This phenomenon can be explained by the decomposition of phytoplankton (mainly blue-green algae), which is capable of accumulating manganese during development.

About 60-70% of the $Mn_{dissolved}$ released during such decomposition is free uncomplexed manganese, which can later bind into complex compounds by dissolved organic substances contained in natural water [10]. As you can see (Figure 3), the concentration of dissolved organic matter in summer sufficient to bind Mn^{2+} free ions. This can explain the increase in the concentrations of complex forms of manganese in August and their dominance in the total amount of dissolved forms.

The appearance of a peak in dissolved forms in October is possible either as a result of the decomposition of the remains of aquatic organisms that consumed manganese in the process of development or in connection with an increase in the role of groundwater supply. For a final conclusion many years of research are required. It is characteristic that in this period, too, complex forms prevailed over free manganese, accounting for about 62% of $Mn_{dissolved}$.

Comparison of data on the dynamics of manganese migration forms in ponds in 2016-2017 allows us to note that its content and distribution between different forms in different seasons depends mainly on the concentration of dissolved oxygen, pH value, input from sludge waters and the content of dissolved organic complexing substances in the water. In winter when recovery conditions arise in the waters the dominant form of migration is free uncomplexed ions. Their concentration at this time reaches 80% of $Mn_{dissolved}$ in 2016 and 60% in 2017. In other seasons the content of complex forms exceeds the content of uncomplexed manganese.

It is also important that in the dynamics of manganese migration forms characteristic stages are observed that practically coincide in time but more clearly expressed in 2017.

In 2017 the diversity of manganese complex compounds in ponds is wider than in 2016.In winter complexes with a molecular weight of ~ 70 thousand and ~ 0.5-1.5 thousand prevail, while in summer during the period of intensive vegetation of aquatic vegetation, the share of the latter is minimal, apparently, due to their consumption by hydrobionts. Throughout the year other complexes with a molecular weight of 120-150 thousand and ~ 1.5-5 thousand were discovered. In 2016 complex manganese compounds with a molecular weight of 120-150 thousand and about 70 thousand were presented in a noticeable quantity only in winter. Apparently, in this case, a different nature of the feeding of both ponds is manifested, and therefore in 2017 there are more high-molecular-weight humic compounds in ponds than in ponds in 2016, and their role in the complexation of heavy metals, including manganese, is well known.

In 2016, the dominant form of manganese migration in spring, summer, and autumn is $Mn_{suspended}$, where as in 2017 in some periods of the same seasons $Mn_{dissolved}$ prevails. The latter is explained by the influence of the intensity of water exchange where suspended particles settle under conditions of slow runoff.

Conclusion

Analysis of data on the dynamics of various forms of manganese in water in 2016 showed that the content of uncomplexed forms sharply increases and dominates in the under-ice period, while in spring, summer, and autumn, complexed forms predominate.

In all seasons with the exception of the winter-spring (February-March), the relative concentrations of dissolved forms of manganese are lower than $Mn_{suspended}$.

Compounds with a molecular weight of ~ 0.5-5 thousand dominate among the complex forms. Complexes with a molecular weight of more than 120-150 thousand and about 70 thousand are present in noticeable quantities only in the winter-spring period.

References

[1] Alekin A A, Semenova A D and Skopintseva B A 2003 *Manual on Chemical Analysis of Waters* (Leningrad: Hydrometeoizdat) p 115

[2] Kazanchev S Ch and Kazancheva L A 2003 *Ecological and Hydrochemical Characteristics of Fisheries Reservoirs* (Nalchik) p 150

[3] Mirzoeva A A, Kazancheva L A and Kushcheterov A V 2013 Morphohydroecological characteristics of ichthyic reservoirs *Izvestia of Kabardino-Balkarian State Agrarian University named after V M Kokov* **2** 42-46
[4] Kazancheva L A, Kazanchev S Ch, Mirzoeva A A and Kumysheva Yu A 2015 Some information about the possibilities of accumulation of Mp (V11) in natural waters *Modern Problems of Science and Education* **6** 56-59
[5] Buchtová H, Svobodová Z, Kocour M and Velisek J 2008 Chemical composition of edible parts of three-year-old experimental scaly crossbreds of common carp (Cyprinus carpio, Linnaeus 1758) *Acta Alimentaria* **37(3)** 311-322
[6] Wong J H *et al* 2009 Digestibility of protein and starch from sorghum (Sorghum bicolor) is linked to biochemical and structural features of grain endosperm *J. of Cereal Science* **49(1)** 73-82
[7] Khabzhokov A B, Kazanchev S Ch, Shakhmurov M M, Getokov O O and Kazancheva L A 2018 Ecological and biological features and prospects of industrial use of a non-ton carp community *The South of Russia: Ecology, Development* **13(3)** 167-175
[8] Khabzhokov A B, Kazanchev S Ch, Shakhmurov M M, Getokov O O and Kazancheva L A 2019 Ecological parameters of outbreeding during intraspecific carp mating *The South of Russia: Ecology, Development* **14(4)** 25-34
[9] Dolov M M, Khabzhokov A B, Getokov O O and Kazanchev S. Ch. 2019 Ecological and biological characteristics of brook trout when grown in the factory *Fish Farming and Fisheries* **10(165)** 66-73
[10] Dzagurov B, Getokov O, Gukezhev V, Kokov T and Vorokov V 2021 Biotechnological aspect of transformation of heavy metal salts and their detoxication in milk cows E3S Web of Conferences **262** 02010

IOP Conf. Series: Earth and Environmental Science **965** (2022) 012012 doi:10.1088/1755-1315/965/1/012012

Promising technology of wastewater treatment with natural ameliorants

V S Bocharnikov, M A Denisova, A S Ovchinnikov, O V Bocharnikova, O V Kozinskaya and M P Meshcheryakov

Volgograd State Agrarian University, 26 University Avenue, Volgograd, 400002, Russia

E-mail: masha2008-1988@mail.ru

Abstract. The article presents the results of testing filter media with various natural sorbents. When studying the absorbing properties of zeolite, charcoal and gravel sand, their properties of removing heavy metals such as Zn, Ca, Cu, Fe, as well as ammonium nitrogen were studied. During experimental studies, all studied ion exchangers showed absorption properties to varying degrees. After calculating the static capacities of each of the studied sorbents, he showed that the best absorption properties are aluminosilicate (zeolite). After completing experimental studies within 48 hours, the best moisture-absorbing effect was observed with zeolite, after 12 hours of contact with poultry wastewater, it removed 64.7 percent of the total iron, and activated carbon and gravel sand only 15.4 percent. The result of research in the study of the absorption properties of heavy metals, ammonium nitrogen ions, showed that zeolite has the highest quality wastewater treatment. The removal of ions of chemical elements was more than 60 percent within 12 hours of being in the liquid phase, in contrast to charcoal and gravel sand.

1. Introduction

At present, the state of surface and underground waters is of concern, since their quality does not meet the standards of sanitary supervision. The reason for this is not conscientious purification of wastewater [1, 2, 3] and their discharge into rivers, lakes, etc. When studying the qualitative indicators of water analysis, it shows an increased content of toxic substances in the form of metals, sulfates, nitrates, ammonium, oil products [4, 5, 6, 7].

Wastewater is purified by various methods, such as reagent, without reagent, filtration, cascade filtration, multistage, sorption [8, 9, 10, 11].

The best from the point of technical and economic predominance are the processes of sedimentation and filtration, since in these methods of loading exhibit better properties of sorption in relation to toxic chemical elements.

Water absorption occurs when the sorbent absorbs a liquid or solid substance in this, and this is the process of sorption itself. This method makes it possible to produce a high degree of wastewater treatment [12, 13, 14, 15] from toxic impurities, which is the main advantage of this process. The absorption process largely depends on the correctly selected sorption material, on its granule size, crystal lattice, porosity, and resistance to a chemical environment.

The water absorption capacity of the selected sorption ion exchangers is influenced by factors such as residence time in an aqueous medium, pH, volumetric capacity, etc. There are several types of volumetric containers - static, absolute and dynamic.

When selecting a filtration material and its quality of poultry wastewater treatment in relation to toxic heavy metals and ammonium ion, all of the above parameters are necessary. Zeolite, also called aluminosilicate, has good static ion exchange capacity. Aluminosilicate has a porous structure and a well-developed crystal lattice, which allows it to conduct high selectivity to large ions. For the complete absorption of harmful substances from wastewater, an optimal contact time of aluminosilicate with soluble substances is required. There are many sorbents in the modern world, both natural and synthetic. The most high-quality and durable are natural sorbents, while studying the absorbing properties of zeolite, charcoal and gravel sand, their properties of removing heavy metals such as Zn, Ca, Cu, Fe, as well as ammonium nitrogen were studied [16, 17, 18, 19, 20].

To study the most effective sorbent for the absorption of chemical impurities were determined under static conditions [21, 22, 23].

2. Materials and methods

When studying the moisture-absorbing properties of zeolite, charcoal, gravel sand, the method for determining static conditions was used, developed at the Scientific Research Institute of Mineral Raw Materials – Russian Geological Society Standard «STO RosGeo 08-002-98».

Before performing the experiment, it is necessary to prepare containers for injecting the studied sorbent, add poultry wastewater, stir for several seconds to develop the reaction of the sorbent with a liquid medium, then separate the liquid phase from the solid by settling, filter the solution and perform photocolorimetric analysis of the test filtrate.

The volume of wastewater was taken in an amount of 250 ml, for this, volumetric flasks were used to which each sorbent was added in equal weights of 60 grams, the experiment was carried out under laboratory conditions.

3. Results and discussion

During experimental studies, all studied ion exchangers showed absorption properties to varying degrees. After calculating the static capacities of each of the studied sorbents, he showed that the best absorption properties are aluminosilicate (zeolite). After completing experimental studies for 48 hours, the zeolite showed the best moisture-absorbing effect after 12 hours of contact with poultry wastewater, it removed 64.7 percent of the total iron, and activated carbon and gravel sand only 15.4 percent. The combined composition of sorbents absorbed only 19.7 percent of heavy metals. Contact within 48 hours showed that the aluminosilicate (zeolite) reduced its sorption capacity to 54.6 percent, as the crystal lattice saturation occurred, and in the complex of activated carbon, gravel sand and zeolite up to 3.1%, respectively, in the activated coal and gravel sand sorption was not observed. Therefore, the best contact time of the sorbent with the liquid phase is observed within 12 hours (Figure 1).

Figure 1. The ability of various sorbents to absorb heavy metals and ammonium ions.

All studied ion exchangers showed their sorption properties in various percentages, the results of which are summarized in Table 1.

Table 1. Determination of sorption of various sorbents in waste water in relation to heavy metals and ammonium ions.

Name of substance	S_{ish}, mg/l	Zeolite Fraction 0.75, 1.0, 4.0 mm			Activated carbon and gravel sand			Activated carbon, gravel sand and zeolite		
		Comparison, mg/l	Siskh-Comparison	SOE, mg/g	Comparison, mg/l	Siskh-Comparison	SOE, mg/g	Comparison, mg/l	Siskh-Comparison	SOE, mg/g
Time of contact of wastewater with sorbent 20 minutes (0.3 hours)										
Iron total	35.5	25.5	10	0.04	35.0	0.5	0.44	32.2	3.3	0.01
Iron II	15.6	10.6	5	0.021	15.3	0.3	0.001	14.6	1.0	0.004
Iron III	25.3	16.3	9	0.037	24.3	1.2	0.004	22.3	3.0	0.008
Chromium	1.1	0.55	0.55	0.002	1.0	0.1	0.0004	1.0	0.1	0.0004
Zinc	1.3	0.63	0.67	0.003	1.3	0	No sorption	1.0	0.3	0.001
Copper	3.5	1.62	1.88	0.007	3.5	0	No sorption	2.5	1.0	0.004
Cadmium	1.6	1.3	0.3	0.001	1.6	0	No sorption	1.3	0.3	0.001

Name of substance	S_{ish}, mg/l	Zeolite Fraction 0.75, 1.0, 4.0 mm			Activated carbon and gravel sand			Activated carbon, gravel sand and zeolite		
		Comparison, mg/l	Siskh-Comparison	SOE, mg/g	Comparison, mg/l	Siskh-Comparison	SOE, mg/g	Comparison, mg/l	Siskh-Comparison	SOE, mg/g
Ammonium	10.5	6.4	4.1	0.017	9.5	1.0	0.004	8.5	2.0	0.008
Time of contact of wastewater with sorbent 1 hour										
Iron total	35.5	15.5	20	0.08	32.0	3.5	0.45	30.1	5.4	0.02
Iron II	15.6	5.25	10.35	0.043	14.2	1.4	0.005	13.5	2.1	0.008
Iron III	25.3	8.35	16.95	0.07	24.0	1.3	0.005	21.4	3.9	0.016
Chromium	1.1	0.35	0.75	0.003	1.0	0.1	0.0004	0.95	0.15	0.0006
Zinc	1.3	0.26	1.04	0.004	1.3	0	0.0008	0.93	0.37	0.0007
Copper	3.5	1.1	2.4	0.01	3.5	0	0.002	2.0	1.5	0.006
Cadmium	1.6	0.8	0.8	0.003	1.3	0.3	0.001	1.0	0.6	0.0025
Ammonium	10.5	3.4	7.1	0.029	8.5	2.0	0.008	7.3	3.2	0.012
Time of contact of wastewater with sorbent 12 hours										
Iron total	35.5	12.5	23	0.09	30	5.5	0.46	28.5	7	0.03
Iron II	15.6	4.8	10.8	0.045	13.7	1.9	0.007	11.6	4.0	0.01
Iron III	25.3	6.3	19	0.08	23.6	1.7	0.007	19.3	6.0	0.025
Chromium	1.1	0.14	0.96	0.004	0.9	0.2	0.0008	0.75	0.35	0.001
Zinc	1.3	0.16	1.14	0.004	1.0	0.3	0.001	0.83	0.47	0.002
Copper	3.5	1.0	2.5	0.01	2.8	0.7	0.003	1.82	1.68	0.007
Cadmium	1.6	0.7	0.9	0.003	1.1	0.5	0.002	0.9	0.7	0.003
Ammonium	10.5	1.9	8.6	0.035	6.5	4.0	0.016	5.5	5.0	0.02
Time of contact of wastewater with sorbent 24 hours										
Iron total	35.5	14.3	21.2	0.09	30	5.5	0.46	29.7	5.8	0.02
Iron II	15.6	5.1	10.5	0.043	14.0	1.6	0.006	11.9	3.7	0.015
Iron III	25.3	6.9	18.4	0.07	23.9	1.1	0.006	19.9	5.4	0.022
Chromium	1.1	0.17	0.93	0.003	1.0	0.1	0.0004	0.78	0.32	0.0013
Zinc	1.3	0.17	1.13	0.005	1.2	0.1	0.0004	0.85	0.45	0.0018
Copper	3.5	1.3	2.2	0.009	3.1	0.4	0.001	1.85	1.65	0.006
Cadmium	1.6	0.75	0.85	0.0035	1.2	0.4	0.001	0.93	0.67	0.0027
Ammonium	10.5	2.0	8.5	0.035	6.9	3.6	0.015	5.9	4.6	0.019
Time of contact of wastewater with sorbent 48 hours										
Iron total	35.5	16.1	19.4	0.08	30	5.5	0.46	31.4	4.1	0.017
Iron II	15.6	5.6	10	0.041	14.9	0.7	0.003	12.2	3.4	0.014
Iron III	25.3	7.5	17.8	0.074	24.2	1.1	0.004	20.5	4.8	0.02
Chromium	1.1	0.23	0.87	0.003	1.0	0.1	0.0004	0.82	0.28	0.0011
Zinc	1.3	0.29	1.01	0.004	1.3	0	No sorption	0.91	0.39	0.0016
Copper	3.5	1.5	2.0	0.008	3.4	0.1	0.0004	1.93	1.57	0.0065
Cadmium	1.6	0.85	0.75	0.0031	1.35	0.25	0.001	0.98	0.62	0.0025
Ammonium	10.5	2.7	7.8	0.032	7.2	3.3	0.013	6.3	4.2	0.017

Having completed the experimental study on the sorption properties under static conditions of ion exchangers and having performed the calculation of the volumetric capacity, it can be seen that aluminosilicate (zeolite) showed the best result than charcoal and gravel sand.

4. Conclusion

The result of research in the study of the absorption properties of heavy metals, ammonium nitrogen ions, showed that zeolite has the highest quality wastewater treatment. The removal of ions of chemical elements was more than 60 percent within 12 hours of being in the liquid phase, in contrast to charcoal and gravel sand.

Experimental data are focused on reducing the environmental load of water bodies into which wastewater is discharged. According to these results, it can be seen that the wastewater treatment process showed a decrease in the toxicity of chemical elements to the maximum permissible concentration, which corresponds to sanitary standards. Therefore, it is possible to discharge wastewater into water bodies, as well as use them as irrigation water for irrigation of industrial crops.

5. Acknowledgments

The research was carried out within the State Assignment of Ministry of Agriculture of the Russian Federation (theme No. 13/2673).

References

[1] Ovchinnikov A S, Loboyko V F, Bocharnikov V S, Ovcharova A Yu and Fomin S D 2019 State of the small rivers of the Volga basin within the Lower Volga *IOP Conf. Series: Earth and Environmental Science* **341** 012107

[2] Kruzhilin I P, Ovchinnikov A S, Kuznetsova N V, Kozinskaya O V, Fomin S D, Bocharnikov V S and Vorontsova E S 2018 Water pressure monitoring in irrigation piping as quality management tools of sprinkler irrigation *ARPN J. of Engineering and Applied Sciences* **13** 4181-4184

[3] Ovchinnikov A S and Pustovalov E V 2017 The results of studies of the use of zeolite-containing rocks as a filter material in the treatment of livestock effluents *Proc. of the Nizhnevolzhsky Agrouniversity Complex: Science and Higher Professional Education* **4(48)** 71-77

[4] Kolesnikov V A, Nistratov A V, Kolesnikova O Y and Kandelaki G I 2019 Integrated approach to neutralization of wastewater containing copper ions and EDTA ligand *News of Higher Education Institutions. Chemistry and Chemical Engineering Series* **62** 108-114

[5] Kolesnikov V A, Il'in V I, Kapustin Y I, Varaksin S O, Kisilenko P N and Kokarev G A 2007 *Flotation Wastewater Treatment of Industrial Enterprises* (Moscow: Khimiya)

[6] Kolesnikov V A, Menshutina N V and Desyatov A V 2016 *Equipment, Technology and Design of Sewage Treatment Systems* (Moscow: new Delhi and NCR plus)

[7] Ksenofontov B S 2013 Intensification of wastewater treatment of engineering industries using ion flotation *Sanitary Engineering* **5** 30-33

[8] Rybalova O, Artemiev S, Sarapina M, Tsymbal B, Bakhareva A, Shestopalov O and Filenko O 2018 Development of methods for estimating the environmental risk of degradation of the surface water state *Eastern-European J. of Enterprise Technologies* **2** 4-17

[9] Brasoveanu F, Petru A and Brezeanu L 2012 European policy concerning the protection of the quality of the environmental factor-water *Challenges of the Knowledge Society* **2** 1058-1063

[10] Feng X, Wang X, Chen Z and Chen J 2019 Nitrogen removal from iron oxide red wastewater via partial nitritation-Anammox based on two-stage zeolite biological aerated filter *Bioresource Technology* **279(8-9)** 17-24

[11] Matsak A and Tsytlishvili K 2018 Using different filter media of stormwater treatment performance *Norwegian J. of development of the International Science* **1(20)** 19–22

[12] Kazeminejadfard F and Hojjati M R 2019 Preparation of superabsorbent composite based on acrylic acid-hydroxypropyldistarch phosphate and clinoptilolite for agricultural applications *J. of Applied Polymer Science* **136(16)** 4736

[13] Bochkarev G R and Pushkareva G I 1988 On a new natural sorbent for the extraction of metals from aqueous media *Physical and Technical Problems of Mining* **4** 46-51

[14] Kochetov G, Prikhna T, Kovalchuk O and Samchenko D 2018 Research of the treatment of depleted nickel-plating electrolytes by the ferritization method *Eastern-European J. of Enterprise Technologies* **3** 52-60

[15] Ovchinnikov A S, Bocharnikov V S and Denisova M A 2019 Technology of wastewater treatment of poultry enterprises using natural sorbents with the addition of ferrite suspension *Proc. of the Nizhnevolzhsky Agrouniversity Complex: Science and Higher Professional Education* **1(1)** 15-22

[16] Bocharnikov V S, Kozinskaya O V, Denisova M A and Bocharnikova O V 2020 Study of the modes of sedimentation of the loading using a hydraulic installation *Proc. of the Nizhnevolzhsky Agrouniversity Complex: Science and Higher Professional Education* **1(57)** 260-267

[17] Yurchenko I F 2020 Development of innovative management systems for agricultural production on reclamed lands *Business. Education. Right* **1(50)** 42-49

[18] Loiko A V, Shibanov I V, Kagramanov G G and Blanco-Pedrejon A M 2018 Experience in the implementation of membrane technology for purification of artesian waters with a high content of iron and manganese *Water Purification. Water Treatment. Water Supply* **4(124)** 58-62

[19] Novikova I V, Luneva E N and Gritsay A V 2019 Means and technologies of water treatment for drip irrigation of agricultural land *Scientific J. of the Russian Research Institute of Melioration Problems* **3(35)** 1-17

[20] Domracheva V A and Shiyrav G 2013 Adsorption extraction of heavy metal ions by carbon sorbents under static conditions *Non-Ferrous Metals* **1** 43-48

[21] Pustovalov E V 2018 Influence of irrigation by livestock runs on the technological qualities of mustard oil seeds *Proc. of the Nizhnevolzhsky Agrouniversity Complex: Science and Higher Professional Education* **2(50)** 181-186

[22] Kruzhilin I P, Ganiev M A, Kuznetsova N V and Rodin K A 2018 Dynamics of total water consumption and yield of periodically moistened rice during sprinkling and drip irrigation in the Volgograd region *Proc. of the Nizhnevolzhsky Agrouniversity Complex: Science and Higher Professional Education* **3(51)** 34-42

[23] Kuznetsova N V, Kuznetsov Yu V, Kozinskaya O V and Denisova M A 2020 Influence of hydraulic parameters on irrigation quality *Proc. of the Nizhnevolzhsky Agrouniversity Complex: Science and Higher Professional Education* **2(58)** 73-83

Stress state of hydrotechnical facilities resistant to natural disasters

A P Nikolaev[1], N A Gureeva[2], V N Yushkin[1], S S Marchenko[1], R Z Kiseleva[1] and S D Fomin[1]

[1] Federal State Educational Institution of Higher Education "Volgograd State Agrarian University", 26 University Avenue, Volgograd, 400002, Russia
[2] Federal State Educational Budgetary Institution of Higher Education "Financial University under the Government of the Russian Federation", 49 Leningradsky Avenue, Moscow, 125993 (City service post office-3), Russia

E-mail: aup-volgau@yandex.ru

Abstract. The equations of the theory of the stress-strain state of an elastic body for practical engineering structures of the agro-industrial complex do not have analytical solutions. Therefore, the development of numerical methods for determining the strength parameters of agricultural facilities is an urgent task. Among the numerical methods of strength calculations, the finite element method is currently widely used. In a Cartesian coordinate system, a finite element of a hexahedral shape is used to determine the stress-strain state of an elastic body under bulk loading in two formulations: in the formulation of the method of displacements with nodal unknowns in the form of displacements and their derivatives, and in a mixed formulation with nodal unknowns in the form of displacements and stresses. Approximation of displacements through nodal unknowns in obtaining the finite element stiffness matrix was performed using a form function, the elements of which were Hermite polynomials of the third degree. When obtaining the deformation matrix, displacements and stresses of the internal points of the finite element were approximated in terms of nodal unknowns using bilinear functions. The calculation example shows a significant advantage of using a finite element in a mixed formulation.

1. Introduction

With a sufficiently complete development of the theory of deformation of loaded solids [1, 2, 3, 4, 5], specific analytical results can only be obtained in some cases that are far from the practice of engineering calculations. To obtain the results of calculations for determining the stress-strain state (SSS) of practical engineering structures, it is necessary to use numerical methods [6, 7]. Among numerical methods, the finite element method in the formulation of the displacement method is widely used [8, 9, 10, 11, 12]. Significant disadvantages of this method include the lack of continuity of the derived displacements on the contours and faces of finite elements while maintaining continuity at the nodal points. The use of finite elements in the mixed formulation [13, 14, 15, 16] leads to the fulfillment of the conditions for continuity of stresses and deformations not only at the nodal points, but also on the contours and faces of finite elements.

2. Materials and methods

In this paper, finite element algorithms are presented for a hexahedral element in the formulation of the displacement method and in a mixed formulation for determining the SSS of arbitrarily loaded elastic bodies. The example of calculating the stress state of a cantilever beam shows the advantage of using a finite element in a mixed formulation.

The relations of elasticity theory used. When loading an elastic body in the coordinate system $0xyz$, deformations and displacements are related by Cauchy dependences [1]

$$\varepsilon_{xx}=\frac{\partial u}{\partial x};\ \varepsilon_{yy}=\frac{\partial v}{\partial y};\ \varepsilon_{zz}=\frac{\partial w}{\partial z};\ 2\varepsilon_{xy}=\frac{\partial v}{\partial x}+\frac{\partial u}{\partial y};$$
$$2\varepsilon_{xz}=\frac{\partial w}{\partial x}+\frac{\partial u}{\partial z};\ 2\varepsilon_{yz}=\frac{\partial w}{\partial y}+\frac{\partial v}{\partial z}, \quad (1)$$

or in matrix form

$$\underset{6\times1}{\{\varepsilon\}}=\underset{6\times3}{[L]}\underset{3\times1}{\{V\}}, \quad (2)$$

where $\underset{1\times6}{\{\varepsilon\}^T}=\{\varepsilon_{xx},\varepsilon_{yy},\varepsilon_{zz},2\varepsilon_{xy},2\varepsilon_{xz},2\varepsilon_{yz}\}$ – is a string of deformations;

$\underset{1\times3}{\{V\}^T}=\{u,\ v,\ w\}$ – string of point displacements;

$\underset{6\times3}{[L]}$ – is a matrix of differential operators.

In elastic deformation, stresses and deformations are related by Hooke's law

$$\underset{6\times1}{\{\sigma\}}=\underset{6\times6}{[C]}\underset{6\times1}{\{\varepsilon\}}, \quad (3)$$

where $\underset{1\times6}{\{\sigma\}^T}=\{\sigma_{xx},\sigma_{yy},\sigma_{zz},\sigma_{xy},\sigma_{xz},\sigma_{yz}\}$ – is the stress string.

Hexagon finite element. In the Cartesian coordinate system $0xyz$, a hexagon with nodes i, j, k, l, m, n, p, h is assumed. To perform numerical integration over the volume of an element, it is mapped to a local cube in the coordinate system ξ, η, ζ, which vary from -1 to 1. Cartesian coordinates of the inner point of a hexagon are determined using their nodal values using bilinear functions

$$\lambda=\underset{1\times8}{\{\varphi(\xi,\eta,\zeta)\}^T}\underset{8\times1}{\{\lambda_y\}}, \quad (4)$$

where the symbol λ means x, y and z coordinates; $\underset{1\times8}{\{\varphi(\xi,\eta,\zeta)\}^T}$ – is a string of trilinear functions of the form; $\underset{1\times8}{\{\lambda_y\}^T}$ – is a string of nodal values of the coordinate λ.

Differentiation of (4) are determined by derivatives of the Cartesian coordinates in the local system $(x_{,\xi}, x_{,\eta}, x_{,\zeta}, y_{,\xi}, y_{,\eta}, y_{,\zeta}, z_{,\xi}, z_{,\eta}, z_{,\zeta})$ and local coordinates in the Cartesian system $(\xi_{,x}, \xi_{,y}, \xi_{,z}, \eta_{,x}, \eta_{,y}, \eta_{,z}, \zeta_{,x}, \zeta_{,y}, \zeta_{,z})$.

The rigidity matrix of a hexahedral finite element in the formulation of the displacement method. The displacements and their first derivatives with respect to the x, y and z coordinates are taken as nodal unknowns. Each component u, v, w of the displacement vector of the inner point of a finite element is approximated in terms of nodal unknowns by the expressions

$$\lambda=\underset{1\times32}{\{\psi(\xi,\eta,\zeta)\}^T}\underset{32\times1}{\{\lambda_y^l\}}, \quad (5)$$

where $\underset{1\times32}{\{\lambda_y^l\}^T} = \{\lambda^i, \lambda^j, \lambda^k, \lambda^l, \lambda^m, \lambda^n, \lambda^p, \lambda^h, \lambda^i_{,\xi}, \lambda^j_{,\xi}, \lambda^k_{,\xi}, \lambda^l_{,\xi}, \lambda^m_{,\xi}, \lambda^n_{,\xi}, \lambda^p_{,\xi}, \lambda^h_{,\xi},$
$\lambda^i_{,\eta}, \lambda^j_{,\eta}, \lambda^k_{,\eta}, \lambda^l_{,\eta}, \lambda^m_{,\eta}, \lambda^n_{,\eta}, \lambda^p_{,\eta}, \lambda^h_{,\eta}\lambda^i_{,\eta}, \lambda^j_{,\eta}, \lambda^k_{,\eta}, \lambda^l_{,\eta}, \lambda^m_{,\eta}, \lambda^n_{,\eta}, \lambda^p_{,\eta}, \lambda^h_{,\eta}, \lambda^i_{,\zeta}, \lambda^j_{,\zeta}, \lambda^k_{,\zeta}, \lambda^l_{,\zeta}, \lambda^m_{,\zeta}, \lambda^n_{,\zeta}, \lambda^p_{,\zeta}, \lambda^h_{,\zeta}\}$ – a string of nodal unknowns in the local coordinate system;

$\underset{1\times32}{\{\psi(\xi, \eta, \zeta)\}^T}$ – is an approximating function whose elements are products of Hermite polynomials in the coordinates ξ, η, ζ in the third power.

Derivatives of displacements in Cartesian coordinates are defined by the expressions

$$\begin{aligned}
\lambda_{,x} &= \left[\underset{1\times32}{\{\psi_{,\xi}\}^T}\xi_{,x} + \underset{1\times32}{\{\psi_{,\eta}\}^T}\eta_{,x} + \underset{1\times32}{\{\psi_{,\zeta}\}^T}\zeta_{,x}\right]\underset{32\times1}{\{\lambda_y^l\}};\\
\lambda_{,y} &= \left[\underset{1\times32}{\{\psi_{,\xi}\}^T}\xi_{,y} + \underset{1\times32}{\{\psi_{,\eta}\}^T}\eta_{,y} + \underset{1\times32}{\{\psi_{,\zeta}\}^T}\zeta_{,y}\right]\underset{32\times1}{\{\lambda_y^l\}};\\
\lambda_{,z} &= \left[\underset{1\times32}{\{\psi_{,\xi}\}^T}\xi_{,z} + \underset{1\times32}{\{\psi_{,\eta}\}^T}\eta_{,z} + \underset{1\times32}{\{\psi_{,\zeta}\}^T}\zeta_{,z}\right]\underset{32\times1}{\{\lambda_y^l\}}.
\end{aligned} \tag{6}$$

Based on (6), a matrix expression is formed

$$\underset{3\times1}{\{V\}} = \underset{3\times96}{[A]}\underset{96\times1}{\{V_y^l\}}, \tag{7}$$

where $\underset{1\times96}{\{V_y^l\}^T} = \left\{\underset{1\times32}{\{u_y^l\}^T}, \underset{1\times32}{\{v_y^l\}^T}, \underset{1\times32}{\{w_y^l\}^T}\right\}$.

Taking into account (6) and (7), the matrix expression for deformations (1) can be represented as

$$\underset{6\times1}{\{\varepsilon\}} = \underset{6\times3}{[L]}\underset{3\times96}{[A]}\underset{96\times1}{\{V_y\}} = \underset{6\times96}{[B]}\underset{96\times1}{\{V_y\}}. \tag{8}$$

To form the stiffness matrix of a finite element, a functional is used that reflects the equality of the actual work of external and internal forces

$$P_L = \frac{1}{2}\int_V \sigma_{ij}\varepsilon_{ij}dV - \frac{1}{2}\int_S (q_{xx}u + q_{yy}v + q_{zz}w)dS, \tag{9}$$

where V – is the volume of the finite element; S – is the area of application of the load.

Taking into account relations (3) and (8), the functional (9) can be represented in matrix form

$$P_L = \frac{1}{2}\underset{1\times96}{\{V_y^l\}^T}\int_V \underset{96\times6}{[B]^T}\underset{6\times6}{[C]}\underset{6\times96}{[B]}dV\underset{96\times1}{\{V_y^l\}} - \frac{1}{2}\underset{1\times96}{\{V_y^l\}^T}\int_S \underset{96\times3}{[A]^T}\underset{3\times1}{\{q\}}dS. \tag{10}$$

Derivatives of displacements in the local coordinate system are defined through derivatives in the global system by the relations

$$\begin{aligned}
\lambda_{,\xi} &= \lambda_{,x}x_{,\xi} + \lambda_{,y}y_{,\xi} + \lambda_{,z}z_{,\xi};\\
\lambda_{,\eta} &= \lambda_{,x}x_{,\eta} + \lambda_{,y}y_{,\eta} + \lambda_{,z}z_{,\eta};\\
\lambda_{,\zeta} &= \lambda_{,x}x_{,\zeta} + \lambda_{,y}y_{,\zeta} + \lambda_{,z}z_{,\zeta}.
\end{aligned} \tag{11}$$

Based on (11), a vector of nodal unknowns in the local system is transformed into a vector of nodal unknowns in the Cartesian system in matrix form

$$\underset{96\times1}{\{V_y^l\}} = \underset{96\times96}{[T]}\underset{96\times1}{\{V_y^g\}}. \tag{12}$$

After minimizing the transformed functional (10) on the basis of (12) with respect to the nodal unknowns $\{V_y^g\}^T$, the finite element stiffness matrix is obtained

$$\underset{96\times96}{[K_1]}\underset{96\times1}{\{V_y^g\}}=\underset{96\times1}{\{f_y\}}, \tag{13}$$

where $\underset{96\times96}{[K_1]}=\underset{96\times96}{[T]^T}\int_V\underset{96\times6}{[B]^T}\underset{6\times6}{[C]}\underset{6\times96}{[B]}dV\,\underset{96\times96}{[T]}$ – finite element stiffness matrix;

$\underset{96\times1}{\{f_y\}}=\underset{96\times96}{[T]^T}\int_S\underset{96\times3}{[A]^T}\underset{3\times1}{\{q\}}dS$ – vector of nodal forces of the finite element.

When minimizing the functional (10), it is taken into account that the load and displacement are related by a linear relationship, so when differentiating the coordinate function, the coefficient $1/2$ in front of the force vector was reduced.

Finite element deformation matrix in a mixed formulation. Displacements and stresses are taken as nodal unknowns of the hexahedral finite element.

The displacements of the inner point of a finite element u, v, w are approximated in terms of nodal unknowns by expressions (2).

The matrix expression for deformations is written as

$$\underset{6\times1}{\{\varepsilon\}}=\underset{6\times3}{[L]}\underset{3\times1}{\{V\}}=\underset{6\times3}{[L]}\underset{3\times24}{[A]}\underset{24\times1}{\{v_y\}}=\underset{6\times24}{[B]}\underset{24\times1}{\{v_y\}}, \tag{14}$$

where $\underset{1\times24}{\{v_y\}^T}=\{u^i,\ldots,u^h,v^i,\ldots,v^h,w^i,\ldots,w^h\}.$

Each component of the stress tensor is also approximated in terms of nodal unknowns by bilinear expressions

$$\sigma_{mn}=\underset{1\times8}{\{\varphi(\xi,\eta,\zeta)\}^T}\underset{8\times1}{\{\sigma_{mn\,y}\}};\ (m,n=1,3), \tag{15}$$

where $\underset{1\times8}{\{\sigma_{mn\,y}\}^T}=\{\sigma_{mn}^i,\ldots,\sigma_{mn}^h\}$ – a string of nodal unknown stresses.

On the basis of (15), the matrix relation is formed

$$\underset{6\times1}{\{\sigma\}}=\underset{6\times48}{[G]}\underset{48\times1}{\{\sigma_y\}}, \tag{16}$$

where $\underset{1\times48}{\{\sigma_y\}^T}=\left\{\underset{1\times8}{\{\sigma_{xx\,y}\}^T},\underset{1\times8}{\{\sigma_{yy\,y}\}^T},\underset{1\times8}{\{\sigma_{xx\,y}\}^T},\ldots,\underset{1\times8}{\{\sigma_{yz\,y}\}^T}\right\}.$

To form the deformation matrix of a finite element, functional (9) is transformed by replacing the actual work of internal forces with the difference between the total and additional work under loading of the deformable body

$$\frac{1}{2}\sigma_{mn}\varepsilon_{mn}=\underset{1\times6}{\{\sigma\}^T}\underset{6\times3}{[L]}\underset{3\times1}{\{V\}}-\frac{1}{2}\underset{1\times6}{\{\sigma\}^T}\underset{6\times6}{[C]^{-1}}\underset{6\times1}{\{\sigma\}}. \tag{17}$$

Taking into account (17), the functional (9) is written by the matrix expression

$$P=\underset{1\times48}{\{\sigma_y\}^T}\int_V\underset{48\times6}{[G]^T}\underset{6\times24}{[B]}dV\underset{24\times1}{\{v_y\}}-\frac{1}{2}\underset{1\times48}{\{\sigma_y\}^T}\int_V\underset{48\times6}{[G]^T}\underset{6\times6}{[C]^{-1}}\underset{6\times48}{[G]}dV\underset{48\times1}{\{\sigma_y\}}-$$
$$-\frac{1}{2}\underset{1\times24}{\{v_y\}^T}\int_S\underset{24\times3}{[A]^T}\underset{3\times1}{\{q\}}dS. \tag{18}$$

After performing a variation of the functional (18) over the nodal unknowns $\{\sigma_y\}^T$ and $\{v_y\}^T$ of the finite element, a system of equations is obtained

$$\frac{\partial P}{\partial\{\sigma_y\}^T} \equiv -\underset{48\times48}{[H]}\underset{48\times1}{\{\sigma_y\}} + \underset{48\times24}{[Q]}\underset{24\times1}{\{v_y\}} = 0;$$
$$\frac{\partial P}{\partial\{v_y\}^T} \equiv \underset{24\times48}{[Q]^T}\underset{48\times1}{\{\sigma_y\}} - \underset{24\times1}{\{f_y\}} = 0, \quad (19)$$

where $\underset{48\times24}{[Q]} = \int_V \underset{48\times6}{[G]^T}\underset{6\times24}{[B]}dV$; $\underset{48\times48}{[H]} = \int_V \underset{48\times6}{[G]^T}\underset{6\times6}{[C]^{-1}}\underset{6\times48}{[G]}dV$; $\underset{24\times1}{\{f_y\}} = \int_S \underset{24\times3}{[A]^T}\underset{3\times1}{\{q\}}dS$.

The systems of equations (19) are presented in the traditional FEM form

$$\underset{72\times72}{[K_2]}\underset{72\times1}{\{Z_y\}} = \underset{72\times1}{\{F_y\}}, \quad (20)$$

where $\underset{72\times72}{[K_2]} = \begin{bmatrix} -\underset{48\times48}{[H]} & \underset{48\times24}{[Q]} \\ \underset{24\times48}{[Q]^T} & \underset{24\times24}{[0]} \end{bmatrix}$ – finite element deformation matrix;

$\underset{1\times72}{\{F_y\}^T} = \left\{ \underset{1\times48}{\{0\}^T}\ \underset{1\times24}{\{f_y\}^T} \right\}$ – is the vector of nodal forces of the finite element;

$\underset{1\times72}{\{Z_y\}^T} = \left\{ \underset{1\times48}{\{\sigma_y\}^T}\ \underset{1\times24}{\{S_y\}^T} \right\}$ – is a vector of nodal unknowns of a finite element.

3. Results and discussion

We considered a cantilever beam loaded with a uniformly distributed load, with the following initial data: length $L = 0.5$ m, cross-sectional height $h = 0.05$ m, width $t = 0.01$ m, material elastic modulus $E = 2.0 \times 10^5$ MPa, Poisson's ratio $\nu = 0.3$, linear load intensity $q = 20.0$ kN/m.

The results of the computational process are shown in Tables 1 (based on the functional based on the difference between the actual work of external and internal forces) and Table 2 (based on a mixed functional), which shows the values of normal stresses in the extreme fibers of the cross-section of the seal (let's call point a) and the section located at a distance of 5 cm from the seal (let's call point b), depending on the number of finite elements when discretizing the cantilever beam. The last columns of the tables show the displacements at the free end of the beam (let's call point c) depending on the number of finite elements.

The number of finite elements in the beam thickness was assumed to be 1, 2, 3, 4, and 5. Along the length of the beam, the number of finite elements in each case was 10, 20, 30, 50, and 100.

Table 1. Numerical values of stress-strain state parameters when using elements in the formulation of the displacement method.

Sampling rate	96 × 96 element		
	Stress σ_{xx}, kPa (point a)	Stress σ_{xx}, kPa (point b)	Movement w, cm (point c)
10 × 1	4660	4026	0.754
20 × 1	4908	4038	0.756
30 × 1	4956	4046	0.756
50 × 1	4974	4048	0.756
100 × 1	4980	4048	0.756
10 × 2	5470	4496	0.752

Sampling rate	96 × 96 element		
	Stress σ_{xx}, kPa (point a)	Stress σ_{xx}, kPa (point b)	Movement w, cm (point c)
20 × 2	5604	4564	0.754
30 × 2	5630	4580	0.754
50 × 2	5642	4588	0.754
100 × 2	5646	4590	0.756
10 × 3	5610	4556	0.752
20 × 3	5772	4638	0.754
30 × 3	5808	4654	0.756
50 × 3	5826	4662	0.756
100 × 3	5828	4664	0.756
10 × 4	5678	4586	0.752
20 × 4	5870	4680	0.754
30 × 4	5914	4700	0.756
50 × 4	5938	4710	0.756
100 × 4	5946	4714	0.756

Table 2. The numerical values of the parameters of the stress-strain state when using elements in a mixed formulation.

Sampling rate	72 × 72 element		
	Stress σ_{xx}, kPa (point a)	Stress σ_{xx}, kPa (point b)	Movement w, cm (point c)
10 × 1	4660	4026	0.754
20 × 1	4908	4038	0.756
30 × 1	4956	4046	0.756
50 × 1	4974	4048	0.756
100 × 1	4980	**4048**	0.756
10 × 2	5470	4496	0.752
20 × 2	5604	4564	0.754
30 × 2	5630	4580	0.754
50 × 2	5642	4588	0.754
100 × 2	5646	**4590**	0.756
10 × 3	5610	4556	0.752
20 × 3	5772	4638	0.754
30 × 3	5808	4654	0.756
50 × 3	5826	4662	0.756
100 × 3	5828	**4664**	0.756
10 × 4	5678	4586	0.752
20 × 4	5870	4680	0.754
30 × 4	5914	4700	0.756
50 × 4	5938	4710	0.756
100 × 4	5946	**4714**	0.756

4. Conclusion

Analysis of the results shows that the convergence of the computational process when using a finite element in a mixed formulation is much faster. For comparison, approximate results of the beam calculation according to technical theory (taking into account the hypothesis of a straight normal) were taken: σ_{xx} = 6000 kPa (point a), σ_{xx} = 4860 kPa (point b), w = 0.75 cm (point c).

At point b, which is located at a distance *h* from the seal, the convergence of the computational process with respect to the normal stresses σ_{xx} (marked in bold in Table 2) is much better in the mixed finite element method.

At point a, the convergence of the computational process is affected by boundary conditions in the form of linear relationships. Therefore, it is more appropriate to compare the methods based on the numerical values of normal stresses at a point located at a distance *h* from the seal.

At each discretization of the beam by finite elements, the displacements turned out to be almost identical in both formulations of the finite element method.

5. Acknowledgments

The research was carried out with the financial support of the Russian Foundation for Basic Research and the Administration of the Volgograd Region in the framework of the scientific project No. 19-41-340004 r_a.

References

[1] Novozhilov V V 1962 *Theory of Thin Shells* (Leningrad: Sudpromgiz) p 432
[2] Galimov K Z and Paimushin V N 1985 *The Theory of Shells of Complex Geometry* (Kazan: Kazan State University) p 164
[3] Sedov L I 1994 *Continuum Mechanics* vol 1 (Moscow: Science) p 528
[4] Sedov L I 1994 *Continuum Mechanics* vol 2 (Moscow: Science) p 560
[5] Petrov V V 2014 *Nonlinear Incremental Structural Mechanics* (Moscow: Infra Engineering) p 479
[6] Zenkevich O and Morgan K 1986 *Finite Elements and Approximation* (Moscow: World) p 318
[7] Oden J 1976 Finite Elements in the Nonlinear Mechanics of Continuous Media (Moscow) p 464.
[8] Postnov V A 1977 Numerical Methods for Calculating Ship Structures (Leningrad: Shipbuilding) p 280
[9] Rikards R B 1988 The Finite Element Method in the Theory of Shells and Plates (Riga: Zinatne) p 284
[10] Bate K U 2010 *Finite Element Methods* ed L I Turchak (Moscow: Fizmatlit) p 1024
[11] Golovanov A I, Tyuleneva O N and Shigabutdinov A F 2006 *The Finite Element Method in the Statics and Dynamics of Thin-Walled Structures* (Moscow: Fizmatlit) p 392
[12] Kayumov R A, Kupriyanov V N, Mukhamedova I Z, Suleymanov A M, Shakirova A M 2007 Method of analysis of the deformation process of a film-fabric composite taking into account geometric and physical nonlinearity *Modeler. and Edges. Tasks* vol 1 119-121
[13] Ignatiev V A and Ignatiev A V 2013 The solution of the flat problem of the theory of elasticity by the finite element method in the form of the classical mixed method *Bulletin of the Volgograd State University of Architecture and Civil Engineering. Construction and Architecture* **31-2(50)** p 337-343
[14] Gureeva N A, Nikolaev A P and Yushkin V N 2020 Comparative analysis of finite element formulations at plane loading of an elastic body *Structural Mechanics of Engineering Constructions and Buildings* **16(2)** 139-145
[15] Gureeva N A, Klochkov Yu V, Nikolaev A P and Yushkin V N 2020 Stress-strain state of shell of revolution analysis by using various formulations of three-dimensional finite elements *Structural Mechanics of Engineering Constructions and Buildings* **16(5)** 361-379
[16] Gureeva N A, Nikolaev A P and Yushkin V N 2020 Elastic deformation of a solid based on various variants of finite element formulations *Bulletin of Volgograd State University of Architecture and Civil Engineering. Civil Engineering and Architecture* **2** 5-13

Disinfection of wastewater with ultraviolet light on devices of the cyclic principle of action

A S Ovchinnikov, V S Bocharnikov, O V Bocharnikova, E V Pustovalov, M A Denisova and T V Repenko

Volgograd State Agrarian University, 26 University Avenue, Volgograd, 400002, Russia

E-mail: pustovalov-evgeniy@mail.ru

Abstract. The article presents a diagram of a device for ultraviolet disinfection of wastewater of a cyclic principle of action, which allows you to control the time of liquid processing and, accordingly, the dose of ultraviolet radiation. The device consists of the following structural elements: round-section body; ultraviolet lamps attached to the inner surface of the housing walls; inlet and outlet pipelines with shut-off and control valves; level gauge made in the form of a float. One cycle of operation of the device consists of filling the body, processing (disinfecting) the original liquid, and emptying the body. Control tests of this device were carried out in the process of purifying wastewater from a cattle breeding complex. During the tests of this device, it was revealed that the effluents treated with ultraviolet light with a generally accepted radiation dose of 30 mJ/cm^2 do not meet the requirements of sanitary and hygienic standards, since the number of viable helminth eggs at the outlet is 54.2% of their initial value, and the effect of complete deworming can be achieved only at a dose of ultraviolet radiation of at least 45 mJ/cm^2.

1. Introduction

Maintaining an environmentally friendly environment is one of the key points in the development and improvement of crop irrigation technologies [1, 2, 3]. Reclamation measures should be aimed not only at increasing soil fertility and productivity of cultivated crops, but also at ensuring a stable ecological environment that does not violate the balance between man and the environment [4, 5, 6, 7].

Particular attention in ensuring environmental safety is paid to the issues of wastewater disposal. Both when discharged into surface water sources and when disposed of in irrigation fields, wastewater in most cases does not meet sanitary standards: the content of oil products, nitrites, sulfates, phenols, heavy metals, etc. is exceeded. [8, 9, 10, 11]. Based on this, there is a need to improve the known, as well as the development of new technologies for wastewater treatment, where special attention should be paid to disinfection [12-21].

Currently, the most widespread are disinfection with chemically active components (chlorination and ozonation) and treatment with ultraviolet radiation. Ultraviolet irradiation is considered by most scientists as the most promising way to disinfect wastewater. Firstly, this method of disinfection does not require specially designated areas for storing reagents. Secondly, ultraviolet rays do not change the chemical composition of the processed liquid.

Most of the ultraviolet disinfection devices are based on a flow-through principle of action. The main disadvantage of this constructive solution is the lack of the possibility of regulating the time of bactericidal irradiation and, as a consequence, the dose of ultraviolet irradiation remains constant. In addition, it should be noted that when using such devices for the disinfection of wastewater, the doses of ultraviolet radiation are 30 mJ/cm^2, which does not satisfy the necessary inactivation of the entire species composition of pathogenic microflora.

2. Materials and methods

To eliminate the disadvantages mentioned above, we have developed an ultraviolet disinfection device, which is based on a cyclic principle of operation, which will ensure control of ultraviolet radiation doses by regulating the contact time of ultraviolet lamps with the processed liquid.

The device consists of the following structural elements: round-section body 1; ultraviolet lamps 2, attached to the inner surface of the walls of the housing; inlet 3 and outlet 4 pipelines with shut-off and control valves 4; a level gauge made in the form of a float (Figure 1). One cycle of operation of the device consists of filling the body, processing (disinfecting) the original liquid, and emptying the body.

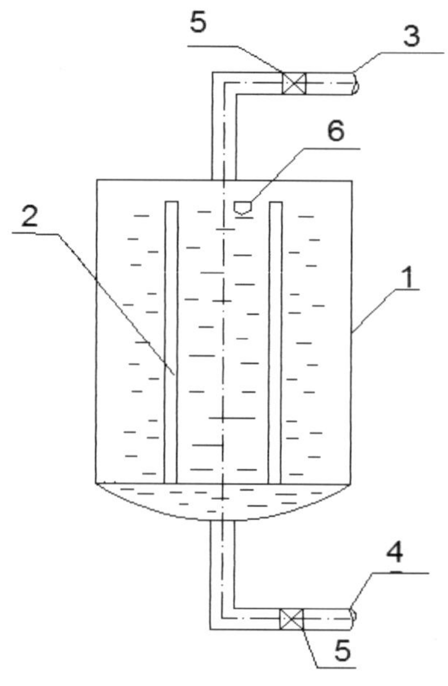

Figure 1. Scheme of the device for ultraviolet disinfection of waste water of the cyclic principle of operation: 1 - sealed housing; 2- ultraviolet lamps; 3 - liquid supply pipe; 4 - liquid discharge pipe; 5 - shut-off and control valves; 6 - liquid level sensor.

Control tests of this device were carried out in the process of purifying wastewater from a cattle breeding complex. Evaluation of the efficiency of wastewater disinfection was carried out according to the coli-index of viable helminth eggs. For this, liquid samples were taken with a step of 5 mJ/cm^2, which in time equivalent is equal to 1 second.

3. Results and discussion

Figure 2 shows the dynamics of inactivation of viable helminth eggs.

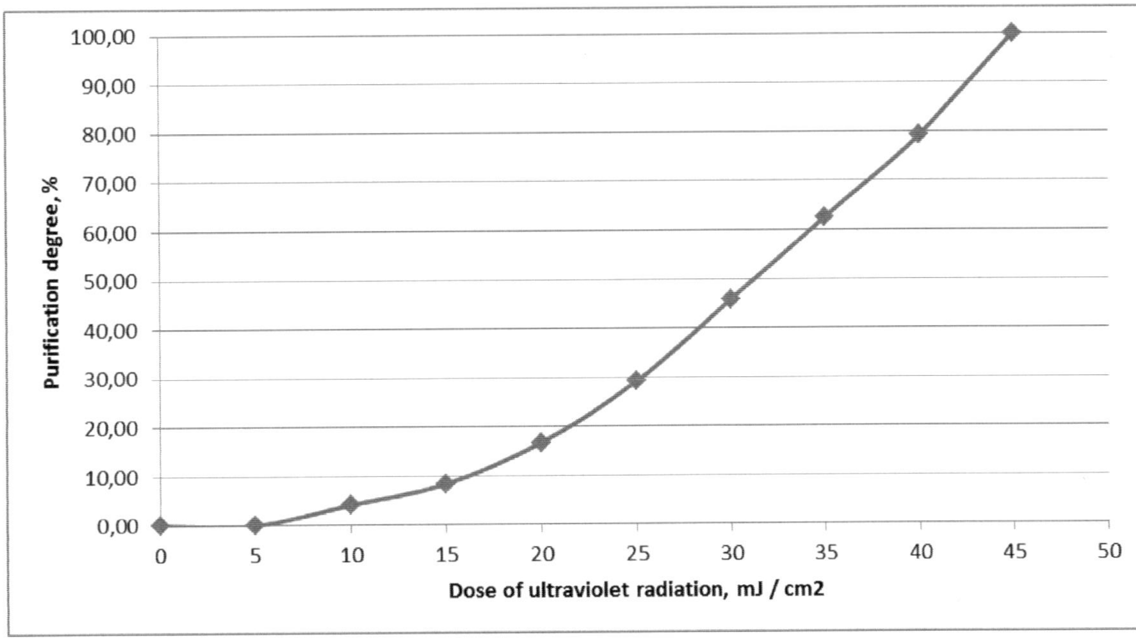

Figure 2. Dynamics of inactivation of viable eggs of helminths depending on the dose of ultraviolet radiation.

This graph shows that in the first seconds of disinfection, when the ultraviolet radiation dose is small (5-25 mJ/cm^2), the degree of death is 4.2-8.3%. At a dose of 30 mJ/cm^2, which corresponds to flow-through devices, only 45.8% of helminth eggs died. Most of the eggs of helminths (79.2-100.0%) died off at doses of ultraviolet irradiation of 40-45 mJ/cm^2. The effect of complete deworming is only at 9 seconds of liquid treatment, which corresponds to an irradiation dose of 45 mJ/cm^2.

4. Conclusion

The effluents treated with ultraviolet light with a radiation dose of 30 mJ/cm^2 do not meet the requirements of sanitary and hygienic standards, since the number of viable helminth eggs at the outlet is 54.2% of their initial value. Discharge of such effluents into open water sources, as well as their use for irrigation, is contraindicated. The effect of complete deworming can be achieved only by increasing the treatment time of the waste liquid and controlling the doses of ultraviolet radiation. Only those ultraviolet disinfection devices based on a cyclic principle of action are able to provide these conditions.

5. Acknowledgments

The research was carried out within the State Assignment of Ministry of Agriculture of the Russian Federation (theme No. 13/2673)

References

[1] Ovchinnikov A S, Loboyko V F, Bocharnikov V S, Ovcharova A Yu and Fomin S D 2019 State of the small rivers of the Volga basin within the lower Volga *IOP Conf. Series: Earth and Environmental Science* **341** 012107

[2] Ovchinnikov A S, Borodychev V V, Lytov M N, Bocharnikov V S, Fomin S D, Bocharnikova O V and Vorontsova E S 2018 Optimum control model of soil water regime under irrigation *Bulgarian J. of Agricultural Science* **24** 909-913

[3] Kuznetsova N V, Kuznetsov Yu V, Kozinskaya O V and Denisova M A 2020 Influence of hydraulic parameters on irrigation quality *Proc. of the Nizhnevolzhsky Agrouniversity Complex: Science and Higher Professional Education* **2(58)** 73-83
[4] Bocharnikov V S, Kozinskaya O V, Denisova M A and Bocharnikova O V 2020 Studies of hydraulic characteristics of a circular sprinkler machine *Proc. of the Nizhnevolzhsky Agrouniversity Complex: Science and Higher Professional Education* **2(58)** 319-327
[5] Snipich Yu F, Chelakhov V C and Kozinskaya O V 2019 Quality study of a circular irrigation sprinkler machine *Scientific J. of the Russian Research Institute of Melioration Problems* **4(36)** 43-54
[6] Ovchinnikov A S and Pustovalov E V 2018 Water balance and structure of total water consumption of sarepta mustard when watered with livestock runoff *Proc. of the Nizhnevolzhsky Agrouniversity Complex: Science and Higher Professional Education* **2(50)** 118-124
[7] Pustovalov E V 2018 Influence of irrigation by livestock runs on the technological qualities of mustard oil seeds *Proc. of the Nizhnevolzhsky Agrouniversity Complex: Science and Higher Professional Education* **2(50)** 181-186
[8] Kolesnikov V A, Nistratov A V, Kolesnikova O Y and Kandelaki G I 2019 Integrated approach to neutralization of wastewater containing copper ions and EDTA ligand *News of Higher Education Institutions. Chemistry and Chemical Engineering Series* **62** 108-114
[9] Kolesnikov V A, Il'in V I, Kapustin Y I, Varaksin S O, Kisilenko P N and Kokarev G A 2007 *Flotation Wastewater Treatment of Industrial Enterprises* (Moscow: Khimiya)
[10] Kolesnikov V A, Menshutina N V and Desyatov A V 2016 *Equipment, Technology and Design of Sewage Treatment Systems* (Moscow: new Delhi and NCR plus)
[11] Ksenofontov B S 2013 Intensification of wastewater treatment of engineering industries using ion flotation *Sanitary Engineering* **5** 30-33
[12] Ovchinnikov A S and Pustovalov E V 2017 The results of studies of the use of zeolite-containing rocks as a filter material in the treatment of livestock effluents *Proc. of the Nizhnevolzhsky Agrouniversity Complex: Science and Higher Professional Education* **4(48)** 71-77
[13] Rybalova O, Artemiev S, Sarapina M, Tsymbal B, Bakhareva A, Shestopalov O and Filenko O 2018 Development of methods for estimating the environmental risk of degradation of the surface water state *Eastern-European J. of Enterprise Technologies* **2** 4-17
[14] Brasoveanu F, Petru A and Brezeanu L 2012 European policy concerning the protection of the quality of the environmental factor-water *Challenges of the Knowledge Society* **2** 1058-1063
[15] Feng X, Wang X, Chen Z and Chen J 2019 Nitrogen removal from iron oxide red wastewater via partial nitritation-Anammox based on two-stage zeolite biological aerated filter *Bioresource Technology* 17-24
[16] Matsak A and Tsytlishvili K 2018 Using different filter media of stormwater treatment performance *Norwegian J. of Development of the Int. Science* **1(20)** 19-22
[17] Kochetov G, Prikhna T, Kovalchuk O and Samchenko D 2018 Research of the treatment of depleted nickel-plating electrolytes by the ferritization method *Eastern-European J. of Enterprise Technologies* **3** 52-60
[18] Bocharnikov V S, Kozinskaya O V, Denisova M A and Bocharnikova O V 2020 Study of the modes of sedimentation of the loading using a hydraulic installation *Proc. of the Nizhnevolzhsky Agrouniversity Complex: Science and Higher Professional Education* **1(57)** 260-267
[19] Yurchenko I. F. 2020 Development of innovative management systems for agricultural production on reclamed lands *Business. Education. Right* **1(50)** 42-49
[20] Loiko A V, Shibanov I V, Kagramanov G G and Blanco-Pedrejon A M 2018 Experience in the implementation of membrane technology for purification of artesian waters with a high

content of iron and manganese *Water Purification. Water Treatment. Water Supply* **4(124)** 58-62

[21] Novikova I V, Luneva E N and Gritsay A V 2019 Means and technologies of water treatment for drip irrigation of agricultural land *Scientific J. of the Russian Research Institute of Melioration Problems* **3(35)** 1-17

IOP Conf. Series: Earth and Environmental Science **965** (2022) 012015 doi:10.1088/1755-1315/965/1/012015

Influence of the water and food regime of the soil on the yield of rice varieties during sprinkling in the Lower Volga region

K A Rodin and A B Nevezhina

All-Russian Research Institute of Irrigated Agriculture, 9 Timiryazev Street, Volgograd, 400002, Russia

E-mail: rodin.ka@yandex.ru

Abstract. In the Russian Federation, as in most countries of the world, rice is cultivated under artificial irrigation by flooding the field with a layer of water, which is maintained during almost the entire growing season. Unfortunately, within the rice-growing zone in the Russian Federation, water resources are limited and this limits the expansion of sown areas under flooded rice. At the same time, judging by the biological characteristics of this plant, a fundamentally different method of rice cultivation is also possible, when the field occupied by it is not flooded. The rice's water demand is met by using the soil moisture reserves, which are maintained at a certain level by irrigation.

1. Introduction

Rice is considered to be the main grain crop of irrigated agriculture. It is cultivated in 125 countries on all agricultural continents [1]. In the Russian Federation, as in most countries of the world, rice is cultivated with artificial irrigation by flooding the field with a layer of water, which is maintained during almost the entire growing season [2, 3, 4, 5, 6]. Irrigation water consumption in this case reaches 12-25 thousand m^3 per hectare and more [7, 8].

Unfortunately, within the rice cultivation zone in the Russian Federation, water resources are limited and this limits the expansion of the cultivated areas under flooded rice. At the same time, judging by the biological characteristics of this plant, a fundamentally different method of rice cultivation is possible, when the field occupied by it is not heated. At the same time, the need for rice in water is satisfied through the use of soil moisture reserves, maintained not lower than a certain level of irrigation. As a result of using this irrigation technology, unproductive water losses are significantly reduced, and the total water consumption of rice approaches biologically necessary [9].

Our research was aimed at confirming the stated scientific hypothesis, some of the results of which are presented in this article.

The purpose of the research is to substantiate the parameters of the water regime of the soil and the irrigation regulations that provide it, which, in combination with the doses of mineral fertilizers, contribute to the yield at the level of 4 and 5 t/ha of grain with a significant decrease in the irrigation rate and negative influence on soil fertility.

2. Materials and methods

Experimental studies were carried out in the northern part of the Sarpinskaya lowland.

The soils of the site are light chestnut, heavy loamy with a thin humus horizon (0.00-0.28 m) and a low humus content (1.29-1.80%) in the arable layer. The supply of soils in the experimental area with mineral nitrogen is low, mobile phosphorus is medium and exchangeable potassium is high. In the calculated layers of 0.0-0.4 and 0.0-0.6 m, the soil density is 1.31 and 1.35 t/m^3, respectively, the lowest moisture capacity is 25.2 and 23.8% of the dry sample mass. The water permeability of heavy loamy soils is weak. The filtration rate for the first hour is about 0.08 m, and by the 7th hour it decreased to 0.024 m/h. The reaction of the soil solution is slightly alkaline, the pH of the aqueous extract is 7.5-7.8.

Field experiments were carried out according to a three-factor scheme, including: the first factor - rice varieties, two of which, "Razdolny" and "Virage", are mid-ripening, and one, "Volgogradsky", belongs to the group of early ripening ones. The second factor (water regime of the soil) - vegetation irrigation in the first variant was carried out with a pre-irrigation moisture threshold of the calculated (0.6 m) soil layer of 70% of the smallest moisture capacity; the second - maintaining in the calculated layer (0.6 m) a differentiated water regime (not lower than 70-80-70% of the smallest moisture capacity) according to the scheme: during the period from sowing to tillering in a soil layer of 0.4 m and from waxy to full ripeness of grain below 70% of the smallest moisture capacity in the soil layer 0.6, and from tillering to waxy ripeness in the same layer - 80% of the smallest moisture capacity; the third - the water regime of the soil was maintained at least 80% of the smallest moisture capacity, during the period from sowing to tillering in the layer - 0.4 m, and from tillering to full ripeness of the grain - 0.6 m; the fourth - the water regime in the calculated soil layer of 0.6 m during the entire growing season of rice was maintained by irrigation of at least 80% of the smallest moisture capacity. The third factor of the experiments included the study of the effect of fertilization doses on the productivity of rice varieties for all variants of the water regime of the soil. They were calculated to obtain the planned yield of 4 and 5 t/ha and amounted to $N_{90}P_{50}K_{60}$ and $N_{110}P_{60}K_{70}$, respectively [10]. The option without fertilizers was adopted as a control.

Experiments were laid with a single-tier systematic arrangement of options according to irrigation regimes and rice varieties, and randomly - according to mineral nutrition. The experiment was repeated in irrigation regimes once with a plot area of 15436 m^2, and for rice varieties and mineral nutrition - three times with an accounting plot area of 1230 m^2 and 615 m^2, respectively. Irrigation method is sprinkling.

The sowing time was determined by the warming up of the soil at the seeding depth of up to 14 °C. Rice was sown with a SN-16 seeder in a narrow-row method in the third decade of April or the first decade of May with a rate of 5 million germinating seeds per hectare. Post-sowing soil consolidation was carried out with ZKK-6 ring rollers. Harvesting was carried out separately by varieties by direct combining.

Phenological observations, measurements and accounting were carried out according to the methods of experimental work [5, 11, 12].

3. Results and discussion

In different years during the growing season of rice, the maintenance of the water regime of soils in the variants of the same name was ensured by a different number of irrigations. In 2013, the first vegetation irrigation fell on May 13, in 2014 and 2015 - on May 20 and 28, respectively. The number of irrigations and irrigation norms for the variants of the soil water regime are shown in Table 1.

Maintaining soil moisture not lower than 70% of the smallest moisture capacity was ensured by carrying out in 2014 and 2015. 7 irrigations with a rate of 600 m^3/ha, and in 2013 with the same rate - 5 irrigations. At the same time, the irrigation rate varied within 3000-4200 m^3/ha.

In the variant with a differentiated pre-irrigation moisture threshold, 70-80-70% of the smallest moisture capacity, and a soil wetting layer increasing from the tillering phase from 0.4 to 0.6 m in 2013 and 2015. 10 irrigations were given at a rate of 400 m^3/ha, in 2014 - 10 irrigations at 400 m^3/ha and 1 irrigation at a rate of 600 m^3/ha. The volume of water supplied for irrigation of rice according to the variants of the water regime, on average over 3 years, varied in the range of 4000-4600 m^3/ha.

Table 1. The number and rates of rice irrigation by options for the water regime of the soil.

Pre-irrigation soil moisture, % of the smallest moisture capacity	Number of waterings, pcs / Irrigation rate, m^3/ha			Irrigation rate, m^3/ha
	2013	2014	2015	
70, h = 0.6 m	$\frac{5}{600}$	$\frac{7}{600}$	$\frac{7}{600}$	3000-4200
70-80-70, h = 0.4 и 0.6 m	$\frac{10}{400}$	$\frac{10}{400}$ and $\frac{1}{600}$	$\frac{10}{400}$	4000-4600
80, h = 0.4 и 0.6 m	$\frac{2}{300}$ and $\frac{11}{400}$	$\frac{2}{300}$ and $\frac{12}{400}$	$\frac{12}{400}$	4800-5400
80, h = 0.6 m	$\frac{12}{400}$	$\frac{13}{400}$	$\frac{12}{400}$	4800-5200

The most favorable effect on the growth, development and formation of the rice yield was the maintenance of soil moisture not lower than 80% of the smallest moisture capacity, until the tillering phase at a depth of 0.4 m, and from the tillering phase in the 0.6 m layer. Maintaining such a water regime in 2013 was ensured by 2 irrigations with a rate of 300 m^3/ha and 11 irrigations at a rate of 400 m^3/ha, in 2014 with 2 irrigations of 300 m^3/ha and 12 irrigations with a rate of 400 m^3/ha and in 2015 - 12 irrigations with a rate of 400 m^3/ha.

The data obtained in the studies show that the application of doses of mineral fertilizers ($N_{88}P_{50}K_{60}$ and $N_{109}P_{62}K_{75}$) calculated for different levels of yield, in combination with the options for the water regime of the soil, provided yields for all rice varieties of 4 and 5 t/ha of grain. Thus, the yield of 4 t/ha with a deviation from - 0.23 to +0.32 t/ha on crops of all three varieties was obtained by combining the water regime of the soil with a pre-irrigation moisture threshold of 80% of the smallest moisture capacity and a differentiated layer of moisture against the background of the application of $N_{88}P_{50}K_{60}$.

Rice yield close to 5 t/ha of grain was also formed on all varieties in the same version of the soil water regime as for obtaining 4 t/ha, but against a higher background of fertilization, $N_{109}P_{62}K_{75}$ (Table 2).

Table 2. Combination of regulated factors to obtain the planned yield of rice, t / ha of grain (average over the years of experiments).

Yield, t/ha		Deviation of the yield from the planned one		Combination of factors		
Planned activity	Actual price	t/ha	%	Variety	Pre-irrigation soil moisture, % of the smallest moisture capacity	Doses of mineral fertilizers, kg of active substance per hectare
4.00	3.60	-0.40	-10.0	Volgogradsky	70-80-70 (h = 0.4 and 0.6 m)	$N_{88}P_{50}K_{60}$
	3.77	-0.23	-5.75	Virage	80 (h = 0.4 and 0.6 m)	$N_{88}P_{50}K_{60}$
	4.01	+0.01	+0.25	Razdolny	80 (h = 0.4 and 0.6 m)	$N_{88}P_{50}K_{60}$
	4.28	+0.28	+7.00	Volgogradsky	80 (h = 0.4 and 0.6 m)	$N_{88}P_{50}K_{60}$
	4.09	+0.09	+2.25	Volgogradsky	80	$N_{88}P_{50}K_{60}$

Yield, t/ha		Deviation of the yield from the planned one		Combination of factors		
Planned activity	Actual price	t/ha	%	Variety	Pre-irrigation soil moisture, % of the smallest moisture capacity	Doses of mineral fertilizers, kg of active substance per hectare
5.00					(h = 0.6 m)	
	4.32	+32	+8.00	Volgogradsky	70-80-70 (h=0.4 and 0.6 m)	$N_{109}P_{62}K_{75}$
	4.63	-0.37	-7.40	Virage	80 (h = 0.4 and 0.6 m)	$N_{109}P_{62}K_{75}$
	4.89	-0.11	-2.20	Razdolny	80 (h = 0.4 and 0.6 m)	$N_{109}P_{62}K_{75}$
	5.25	+0.25	+5.00	Volgogradsky	80 (h = 0.4 and 0.6 m)	$N_{109}P_{62}K_{75}$
	5.09	+0.09	+1.80	Volgogradsky	80 (h = 0.6 m)	$N_{109}P_{62}K_{75}$

The total water consumption during the growing season of periodically watered rice varied depending on the value of the formed yield and the meteorological conditions of the growing season. The greatest amount of water, on average for three years 6663 m^3/ha, was consumed by rice plants in the variant where the moisture content was maintained by irrigation of at least 80% of the smallest moisture capacity in soil layers of 0.4 and 0.6 m with a yield close to 5 t/ha of grain (Table 3).

The share of irrigation water participation in the structure of water consumption in the third variant of the soil water regime was 76.1%, atmospheric precipitation - 20.9 and soil moisture - 3.0%. In the variant with a pre-irrigation moisture content of 80% of the smallest moisture capacity in a layer of 0.6 m, the water consumption by plants decreased to 5067 m^3/ha. Maintaining a differentiated threshold of pre-irrigation moisture content, 70-80-70% of the smallest moisture capacity, with an increasing depth of soil wetting by irrigation from 0.4 to 0.6 m was accompanied by a further decrease in yield and water consumption by plants to 4200 m^3/ha.

Table 3. Water balance by options for periodically watered rice (average over years of experiments).

Pre-irrigation soil moisture, % of the smallest moisture capacity	Irrigation rate		The arrival of moisture from precipitation		Soil moisture residue		Total water consumption, m^3/ha
	m^3/ha	%	m^3/ha	%	m^3/ha	%	
70, h = 0.6 m	3600	67.1	1361	25.4	401	7.5	5362
70-80-70, h = 0.4 and 0.6 m	4200	70.5	1385	23.2	373	6.3	5958
80, h = 0.4 and 0.6 m	5067	76.1	1393	20.9	203	3.0	6663
80, h = 0.6 m	4933	75.6	1393	21.3	151	3.1	6477

The maximum water consumption by rice plants was noted in the interfacial period "booting-sweeping" and coincides with the period of the most intensive growth of the vegetative mass. On average, over three years, depending on the maintained water regime of the soil, it varied from 2543 to

IOP Conf. Series: Earth and Environmental Science **965** (2022) 012015 doi:10.1088/1755-1315/965/1/012015

2602 m^3/ha. The smallest amount of water, 124 m^3/ha, was consumed during the seeding-emergence period.

The processing of experimental data made it possible to establish a relationship between the total water consumption (E, m^3/ha) and the yield (Y, t/ha) of rice, which is described by the straight-line regression equation $E = 333.01\ Y + 4520$ at $R = 0.88$. It follows that the formation of a yield at the level of 4 t/ha is ensured with a total water consumption on average of 5852.0 ± 100 m^3/ha, with a yield of 5 t/ha - 6185.1 ± 100 m^3/ha.

4. Conclusion

The research results indicate that rice plants have the ability to adapt to cultivation with periodic watering and provide a yield of 4-5 t/ha of grain with the consumption of irrigation water per unit area 2.4-6.9 times less compared to traditional flood irrigation technology. Along with significant savings in irrigation water with this irrigation technology, rice does not require the construction of more expensive specialized irrigation systems. It seems possible to place it in field crop rotations on general-purpose irrigation systems. And this means that in addition to the high economic efficiency of rice cultivation using innovative irrigation technology, a number of environmental problems associated with the risk of rising groundwater levels, salinization, waterlogging and soil acidification, disposal of drainage and waste waters contaminated with herbicides and fertilizers, and often mineral compounds, are removed.

References

[1] Rice Market Monitor 2018 *FAO* **21(1)** 1-3

[2] Balakai G T and Dokuchaeva L M 2018 On the issue of development of standards for rice water demand and water consumption from rice irrigation systems *Scientific J. of the Russian Research Institute of Melioration Problems* **3(31)** 1-22

[3] Borodychev V V, Dedova E B and Shabanov R M 2017 Technology of rice cultivation on general-purpose reclamation systems with sprinkler irrigation *Proc. of the Nizhnevolzhsky Agrouniversity Complex: Science and Higher Professional Education* **1(45)** 20-29

[4] Kruzhilin I P, Ganiev M A, Rodin K A, Dubenok N N and Abdu N M 2017 The water regime of the soil and the dose of macrofertilizers during the cultivation of rice on drip irrigation systems *Bulletin of the Russian Agricultural Science* **2** 12-15

[5] Dospekhov B A 1985 *Field Experiment Methodology with the Basics of Statistical Processing of Research Results* (Moscow: Agropromizdat) p 351

[6] Ionova L P and Arykbaev R K 2017 Agrobiological and economic aspects of growing Russian and Iranian rice varieties by seedling method with intermittent irrigation in the Volga delta *Bulletin of Michurin State Agrarian University* **3** 43-56

[7] Kruzhilin I P, Dubenok N N, Ganiev M A, Melikhov V V, Abdu N M and Rodin K A 2016 Combination of natural and anthropogenically controlled conditions for obtaining different yields of rice using drip irrigation systems *Bulletin of the Russian Agricultural Science* **5** 41-44

[8] Kruzhilin I P, Doubenok N N, Ganiev M A, Ovchinnikov A S, Melikhov V V, Abdou N M, Rodin K A and Fomin S D 2017 Mode of rice drip irrigation *J. of Engineering and Applied Sciences* **12(24)** 7118-7123

[9] Kruzhilin I P, Ganiev M A, Rodin K A and Kuznetsova N V 2019 Less water-consuming and ecologically preferable technology of irrigation of rice with periodic irrigation *Proc. of the Nizhnevolzhsky Agrouniversity Complex: Science and Higher Professional Education* **2(54)** 49-55

[10] Filin V I 1994 *Reference Book on Crop Production with the Basics of Crop Programming* (Volgograd: Volgograd State Agricultural Academy) p 274

[11] Pleshakov V N 1983 *Methodology of Field Experiment in Irrigation Conditions* (Volgograd: All-Russian Research Institute of Irrigated Agriculture) p 149

[12] Nikitenko G F 1982 *Experienced Business in Field Cultivation* (Moscow: Rosselkhozizdat) p 190

Surface flow of melted snow water on the central Russian upland

A I Petelko[1], A T Barabanov[2] and V V Melikhov[3]

[1] Novosilskaya Zonal Agroforestry Experimental Station - branch of the Federal Scientific Center of Agroecology, Complex Reclamation and Protective Afforestation of the Russian Academy of Sciences, 2A Semashko Street, Mtsensk, Oryol region, 303035, Russia
[2] Federal Scientific Center of Agroecology, Complex Reclamation and Protective Afforestation of the Russian Academy of Sciences, 97 University Avenue, Volgograd, 400062, Russia
[3] All-Russian Research Institute of Irrigated Agriculture, 9 Timiryazev Street, Volgograd, 400002, Russia

E-mail: vniioz@yandex.ru

Abstract. In areas with a pronounced relief, water erosion causes great harm to agriculture and other sectors of the national economy. It leads to the destruction of the soil cover, a reduction in the area of arable land, a decrease in fertility, the removal of erosion products into floodplains, the siltation of ponds and rivers and the destruction of the road network. The scientific goal was to identify the influence of the main natural factors (snow reserves, freezing and soil moisture) on the formation of the spring melt water runoff on the plow. To study the questions raised, the water-balance method was used with the use of runoff sites. Observations showed that the formation of the melt water runoff depended on the depth of soil freezing. With shallow soil freezing up to 50 cm, there was no runoff. All the melt water was absorbed by the soil. As a result of 30 years of research (1991-2020), it was found that the average annual runoff on the plow was very weak - 3.8 mm on the scale of Surmach, the runoff coefficient is 0.05, shallow soil freezing - 19-28 cm, duration of snow melting - 16 days.

1. Introduction

For reliable soil protection from water erosion, it is necessary to apply a set of anti-erosion measures, which includes: anti-erosion organization of the territory, forest reclamation, agrotechnical, meadow reclamation and hydrotechnical measures [1, 2, 3, 4].

The basis for the implementation of these measures aimed at protecting the soil from washout and erosion is the anti-erosion organization of the territory. It includes land classification, crop rotation system. Agrotechnical methods of cultivation of any agricultural crops in our conditions should contain soil-protective elements that prevent the flow of melt and storm water and help to protect the soil from destruction. Protection should be carried out not only in areas prone to flushing, but also in fields where there is no flushing, but runoff is formed [5, 6]. The generalization of long-term research materials showed that the arsenal of agrotechnical anti-erosion methods is large, but their effectiveness is low. The most effective of these is deep winter plowing. Effective retention of melt water on the plow is achieved by creating a deep arable layer rich in organic matter by introducing large doses of organic fertilizers and reducing the number of treatments due to the cultivation of perennial grasses.

2. Materials and methods

Field experiments were carried out at the Novosilskaya Zonal Agroforestry Experimental Station in 1991-2020. The soil in the experimental plots is gray forest, washed away. Every year, drainage plots with a size of 20 × 100 m were created on the plowing fields. Snow deposition, freezing and soil moisture, melt water runoff, and soil thawing were studied on them. Before the spring snowmelt, the snow height was measured at runoff sites using two snow-measuring runs every 2-4 m in 3-5 times. Snow density was determined with a weight snow gauge at each runoff site at 6 points in 2-fold repetition. The depth of soil freezing was determined by the presence of ice crystals during drilling. The thermostat-weight method was used to determine soil moisture. Soil samples for moisture were taken from the following depths: 0-3, 10, 20, 30, 40, 50, 75-100 cm. Runoff measurements at the thresholds of weirs with a triangular cut were carried out with a ruler with millimeter divisions every hour [7, 8]. Observations of soil thawing were carried out using a metal rod in the upper, middle, and lower parts of the runoff plots in 5 replicates. The study studied the regularities of the influence of natural and anthropogenic factors on the indicators of the runoff of thawed waves on the plow [9].

3. Results and discussion

Water erosion is a big problem for agriculture. For the successful protection of soils from washout and erosion, it is necessary to understand the patterns of erosion-hydrological processes. Barabanov previously found that the relationship between the runoff layer and the duration of snow melting is very weak. This is explained by the fact that the period of snow melting at the Novosilskaya Zonal Agroforestry Experimental Station is long (10-27 days) and even its minimum duration (10 days) is enough for the topsoil (0-30 cm) to be moistened to the lowest and even full moisture capacity. The depth of freezing affects the formation of runoff, which is confirmed by our research. If, before the snowmelt, the soil is thawed or frozen up to 50 cm, then the spring runoff is not formed or is weak, regardless of soil moisture, snow reserves and the duration of snow melting. When the depth of soil freezing is more than 50 cm, the runoff of melt water is formed depending on the level of moisture (ice content) of the soil and snow reserves.

On the basis of generalization and analysis of long-term data characterizing the relationship between the layer of melt water runoff from the plow and compacted arable land with natural factors, the law of limiting factors of melt water runoff was discovered earlier [2]. The essence of the law is that at a certain (limiting) value of one of the factors (freezing depth, soil moisture and snow reserves), runoff is not formed regardless of the level of others. If the soil is thawed or frozen to a depth of no more than 50 cm, runoff is not formed, regardless of the level of soil moisture and the amount of snow reserves. An increase in the depth of soil freezing above the limiting level has practically no effect on the amount of runoff, i.e. at any depth of soil freezing above the limiting one, the runoff is formed the same. In this case, soil moisture and water reserves in the snow have a decisive influence on it. When the upper (0-50 cm) layer is moistened to a level of 120-130 mm, runoff is also not formed, regardless of the depth of soil freezing and the amount of water in the snow, i.e. the limiting factor is soil moisture.

Earlier it was found [2] that the soil is capable of absorbing and retaining a certain amount of water (the maximum in the frozen state can reach the value of the total moisture capacity of the upper layer). Moisture deficiency (the difference between the total moisture capacity and the actual moisture content) determines the amount of water absorption. The runoff layer depends on the moisture deficit in the soil and snow reserves before snow melting.

Over the long-term observation period, the prevailing hydrometeorological conditions of the winter period with weak soil freezing and other factors determined the absorption of all melt water into the soil for 23 years (Table 1). Water absorption ranged from 40 to 151 mm. In 1994, the largest spring runoff was formed - 37 mm with a runoff coefficient of 0.31, soil freezing of 75 cm.

Table 1. Slope runoff of melt water and water absorption on the plow.

Years	Runoff, mm	Water supply, mm		The depth of soil freezing, cm	Duration of snow melting, days	Precipitation during the period of snow melting, mm	Runoff coefficient	Water absorption, mm
		in soil 0-50, cm	in the snow					
1991	34	172	58	10	5	17	0,05	71
1992	0	169	54	0-17	14	16	0	70
1993	17	139	40	45-83	10	22	0.22	45
1994	37	263	119	75	12	17	0.31	99
1995	0	228	107	25	21	10	0	117
1996	29	166	73	70-80	16	8	0.40	52
1997	1	227	56	38-53, in places thawed	2	5	0.02	60
1998	0	188	48	18-20	18	12	0	60
1999	0	166	144	20-25, in places thawed	11	7	0	151
2000	0	209	57	7-25	12	0	0	57
2001	0	198	81	0-5	12	21	0	102
2002	0	184	81	8-10	10	40	0	121
2003	26	212	97	41-70	13	25	0.25	96
2004	0	189	58	10-18	6	14	0	72
2005	0	203	71	5-12	12	14	0	85
2006	0	212	109	0-3	5	21	0	130
2007	0	225	51	0	13	30	0	81
2008	0	208	67	20-31	8	14	0	81
2009	0	202	105	12-25	21	33	0	138
2010	0	262	89	30-40	15	22	0	111
2011	0	156	119	thawed	15	14	0	133
2012	0	168	61	18-40	18	25	0	86
2013	0	171	108	0-15	12	18	0	126

Years	Runoff, mm	Water supply, mm		The depth of soil freezing, cm	Duration of snow melting, days	Precipitation during the period of snow melting, mm	Runoff coefficient	Water absorption, mm
		in soil 0-50, cm	in the snow					
2014	0	203	26	14-17 in places thawed	15	14	0	40
2015	0	264	43	15-20	13	20	0	63
2016	0	235	54	0-5	15	25	0	79
2017	0	212	80	10-21	15	15	0	95
2018	8	221	84	30-45	22	29	0.07	105
2019	0	142	99	15-35	20	18	0	117
2020	0	135	48	3-10	9	5	0	53
Average	4	198	76	19-28	16	18	0,05	90

On the scale of Surmach [10] it was a moderate runoff. In the years with runoff (1991, 1993, 1995, 1996, 1997, 2003, 2008), it varied from 1 to 34.0 mm, i.e. was very weak, weak and moderate. For all the years of field observations, the average annual runoff was 4 mm, the runoff coefficient was 0.05 (very weak), and water seeped into the soil was 90 mm.

Analysis of the data showed that only three years (1994, 1996, 2003) there was a moderate runoff of melt water. In 1996, it was 29 mm with a runoff coefficient of 0.40, the depth of soil freezing was 70-80 cm, water reserves in the snow were 73 mm, and the duration of snow melting was 16 days. In 1994, the runoff increased to 37 mm. In 2003, the runoff was 24 mm, snow reserves were 97 mm, soil freezing was 41-70 cm, the runoff coefficient was 0.25, and the duration of snow melting was 13 days. A weak runoff of melt water (17 and 8 mm), with a runoff coefficient of 0.25 and 0.07, respectively, and soil freezing of 10 and 30-45 cm, was formed in 1993 and 2018. A very weak runoff was only in 1997 and was 1 mm, the runoff coefficient was 0.02. For 30 years of research, the surface runoff of melt water of various sizes was formed for 7 years and there was no runoff for 23 years (1992, 1998, 1999, 2001, 2002, 2004, 2005, 2006, 2007, 2008, 2009, 2010, 2011, 2012, 2013, 2014, 2015, 2016, 2017, 2019, 2020). Our data confirm the law of flow limiting factors.

4. Conclusion

Long-term studies (1991-2020) established the influence of the main natural factors on the slope spring runoff. It depends on the depth of soil freezing, snow storage and soil moisture before spring snow melting. With shallow soil freezing, there is no runoff of melt water, and when freezing deeper than 50 cm, spring runoff is formed.

The prevailing hydrometeorological conditions of the winter period for 30 years (shallow freezing of the soil before snow melting and its weak moisture in some years) caused the total water absorption of melt water into the soil (40-151 mm) for 23 years and no runoff was formed during these years.

The system of measures to protect soil from erosion should be based on knowledge of the regularities of erosion-hydrological processes under the influence of natural and anthropogenic factors. The analysis of long-term studies of the regularities of the formation of surface melt water runoff over a 30-year period made it possible to establish the leading natural factors, influencing which it is possible to control the erosion-hydrological process.

References

[1] Drums A T, Petelko A I and Levshin A O 2006 Influence of runoff-regulating forest belts on snow deposition *Nature Management in Agrarian Regions of Russia* (Moscow) pp 86-89

[2] Drums A T 2017 *Erosion-Hydrological Assessment of the Interaction of Natural and Anthropogenic Factors in the Formation of Surface Melt Water Runoff and Adaptive Landscape Agriculture* (Volgograd) p 188

[3] Barabanov A T, Dolgov S V, Koronkevich N I, Panov V I and Petelko A I 2018 Surface runoff and infiltration in soils of melt water on arable land in the forest-steppe and steppe zones of the East European Plain *Pochvovedenie* **11** 62-69.

[4] Petelko A I 2018 Influence of landscape on the formation of surface melt water runoff in the forest-steppe of the Central Non-Black Earth Region *New Methods and Results of Landscape Studies in Europe, Central Asia and Siberia. Optimization of Agricultural Landscapes* ed V G Sychev and L Müller vol 4 (Moscow: All-Russian Research Institute of Agrochemistry) pp 298-303

[5] Garshinev E A 2002 *Erosion-Hydrological Process and Forest Melioration. Experimental Evaluation, Calculation, Design* (Volgograd: All-Russian Research Institute of Agroforestry) pp 220

[6] Petelko A I and Panov V I 2019 Influence of the main natural factors on the formation of surface melt water runoff in forest reclamation agrolandscapes in the forest-steppe and steppe zones of the European part of Russia *Scientific Bulletin of Belgorod State University. Natural Sciences* **43(1)** 16-29

[7] Surmach G P 1967 *Methods for Studying the Water-Regulating and Anti-Erosion Role of Forest Belts and Agricultural Practices* (Volgograd) p 39

[8] Mastinskaya S B 1959 *To Clarify the Method for Determining Soil Moisture in Different Soil and Climatic Zones* (Moscow: Gidrometeoizdat) pp 57-75

[9] Lvovich M I 1954 Influence of tillage on runoff *News of the USSR Academy of Sciences* **5** 40-48

[10] Surmach G P 1975 *Water Erosion and its Control* (Leningrad: Gidrometeoizdat) p 254

IOP Conf. Series: Earth and Environmental Science **965** (2022) 012017 doi:10.1088/1755-1315/965/1/012017

Modernization of structural elements of a hydrocyclone to improve the efficiency of irrigation water treatment

A S Ovchinnikov, M A Denisova, E V Pustovalov, V S Bocharnikov, O V Bocharnikova and O V Kozinskaya

Volgograd State Agrarian University, 26 University Avenue, Volgograd, 400002, Russia

E-mail: masha2008-1988@mail.ru

Abstract. There are many methods of water treatment and wastewater treatment, this includes sorption filters with different loads, the use of reagents, but mechanical cleaning filters carry a heavy load. The correct selection of equipment assumes effective water purification and high-quality irrigation of crops. The article presents the results of modernization of structural elements of a hydrocyclone to improve the efficiency of irrigation water treatment. The modernization consists in the fact that an impeller with a bearing was additionally installed on the outlet of the clarified product, which increases the centrifugal force of the swirling liquid flow, thereby increasing the efficiency of cleaning irrigation water. After the cleaning process, a chemical analysis of the irrigation water was carried out, which showed that the proposed design of a hydrocyclone with an impeller effectively reduces the concentration of impurities in the irrigation water. The amount of impurities in the purified water decreased due to, increasing the centrifugal force of the swirling fluid flow, suspended solids decreased by 3 times, turbidity by 2.5 times, color by more than 29%.

1. Introduction

Cultivation of agricultural crops is not possible without high-quality irrigation, especially in arid places where there is little rain [1, 2, 3, 4]. However, water for irrigation needs can come from any water source with a different composition of chemical elements. When irrigating large areas, surface water sources are mainly used [5, 6], that is, canals, lakes, as it is economically profitable. This water is considered the most polluted and therefore must be mechanically cleaned from coarse impurities. If you do not carry out preliminary cleaning, then the pump and other equipment will break down. The process of water treatment consists in the fact that water is purified from coarse and fine particles, as well as metal-containing [7, 8, 9] elements of various types of salts, which at one time have properties to accumulate in irrigation systems.

There are many methods of water treatment and wastewater treatment [10, 11, 12], these include sorption filters with different loads [13, 14, 15, 16, 17], the use of reagents, but mechanical cleaning filters carry a heavy load. The correct selection of equipment assumes effective water purification and high-quality irrigation of crops [18-23].

2. Materials and methods

For the study, the design of the hydrocyclone was modernized (Figure 1). The hydrocyclone consists of a body that includes mating cylindrical and conical shells, a tangential inlet for supplying the initial product, has a drain pipe for withdrawing the clarified product, equipped with a filter element with a filtering side surface, and a sand pipe for withdrawing a heavy product, it is equipped with a cylindrical sorption filter, the lower edge of which is at the level of the upper edge of the tangential inlet, installed coaxially with a drain pipe on a disk that forms an additional upper chamber in the cylindrical shell for collecting the clarified product with an additional drain pipe for its output [20].

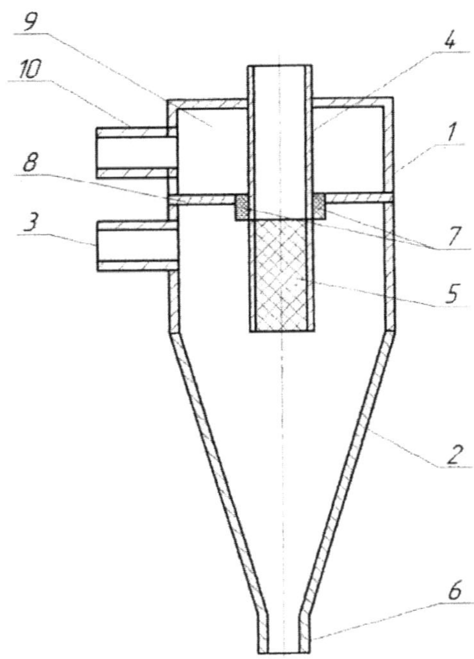

Figure 1. Hydrocyclone: 1 - Cylindrical shell, 2 - conical shell, 3 - tangential inlet, 4 - drain pipe, 5 - filter element, 6 - sand pipe, 7 - cylindrical sorption filter, 8 - disk, 9 - additional chamber for collecting the clarified product, 10 - additional drain pipe for the output of the clarified product.

The disadvantage of the above-described hydrocyclone design is a decrease in the kinetic energy inside the flow, due to the resistance arising from the collision of the flow with the wall located perpendicular to the axis of the cylinder, as a result of which the quality indicators of irrigation water are reduced.

The modernization consists in the fact that an impeller with a bearing was additionally installed on the outlet of the clarified product, which increases the centrifugal force of the swirling fluid flow, thereby increasing the efficiency of cleaning irrigation water (Figure 2).

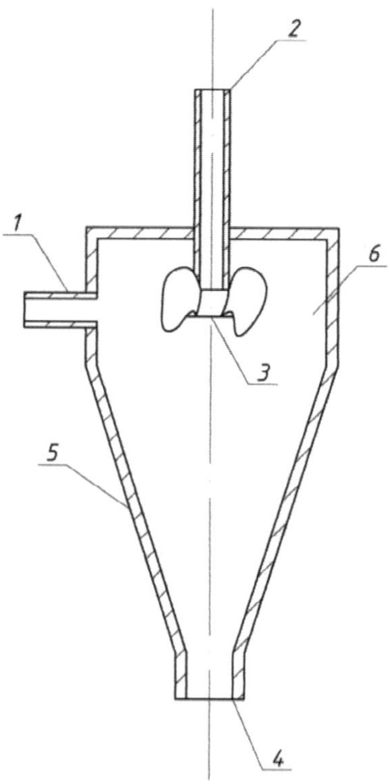

Figure 2. Hydrocyclone with impeller: 1. - tangential inlet, 2 - drain pipe, 3 - impeller with bearing, 4 - sand pipe, 5 - body with cylindrical and conical shells.

3. Results and discussion

The principle of operation of the hydrocyclone is that the initial product, not clarified irrigation water, enters through the tangential inlet into the opposite wall of the upper shell of the hydrocyclone body hitting the impeller fixed on the outlet of the clarified product. Due to which the water flow swirls without significant losses of kinetic energy inside the flow, while increasing the centrifugal force, the heavy product hits the walls of the body, goes down along the conical shell and is discharged through the sand pipe to remove the heavy product. At the end of the cleaning process, an analysis of the irrigation water was carried out, the results of which are shown in Table 1.

Table 1. Results of chemical analysis of treated irrigation water.

No.	Impurities, mg/l	Initial water, mg/l	Purified water, mg/l	The results of the analysis of the above described construction Figure 1	State Standard GOST 2874-82
1	Suspended substances	14.3	3.7	4.1	-
2	Turbidity of water, EMF	9.5	2.73	3.4	2.6-3.5
3	Color, hail.	31.5	21.2	22.0	20.0-35.0
4	$Fe(OH)_2$	0.65	0.212	0.286	0.3
5	Fe_2O_3	0.251	0.115	0.125	0.3

According to the analysis results, it can be seen that the proposed design of a hydrocyclone with an impeller effectively reduces the concentration of impurities in the irrigation water than the above-described Figure 1. The amount of impurities in the treated water decreased due to, increasing the centrifugal force of the swirling fluid flow, suspended solids decreased by 3 times, turbidity 2.5 times, color more than 29%.

4. Conclusion

Thus, using the proposed hydrocyclone design, effective cleaning of irrigation water from coarse impurities is achieved by increasing the centrifugal force while reducing the loss of kinetic energy inside the fluid flow, which makes it possible to reduce suspended solids by 3 times, turbidity by 2.5 times, color more than by 29%.

5. Acknowledgments

The research was carried out within the State Assignment of Ministry of Agriculture of the Russian Federation (theme No.13/2673).

References

[1] Ovchinnikov A S, Loboyko V F, Bocharnikov V S, Ovcharova A Yu and Fomin S D 2019 State of the small rivers of the Volga basin within the lower Volga *IOP Conf. Series: Earth and Environmental Science* **341** 012107

[2] Kruzhilin I P, Ovchinnikov A S, Kuznetsova N V, Kozinskaya O V, Fomin S D, Bocharnikov V S and Vorontsova E S 2018 Water pressure monitoring in irrigation piping as quality management tools of sprinkler irrigation *ARPN J. of Engineering and Applied Sciences* **13** 4181-4184

[3] Ovchinnikov A S, Borodychev V V, Lytov M N, Bocharnikov V S, Fomin S D, Bocharnikova O V and Vorontsova E S 2018 Optimum control model of soil water regime under irrigation *Bulgarian J. of Agricultural Science* **24** 909-913

[4] Yurchenko I. F. 2020 Development of innovative management systems for agricultural production on reclamed lands *Business. Education. Right* **1(50)** 42-49

[5] Rybalova O, Artemiev S, Sarapina M, Tsymbal B, Bakhareva A, Shestopalov O and Filenko O 2018 Development of methods for estimating the environmental risk of degradation of the surface water state *Eastern-European J. of Enterprise Technologies* **2** 4-17

[6] Brasoveanu F, Petru A and Brezeanu L 2012 European policy concerning the protection of the quality of the environmental factor-water *Challenges of the Knowledge Society* **2** 1058-1063

[7] Kolesnikov V A, Nistratov A V, Kolesnikova O Y and Kandelaki G I 2019 Integrated approach to neutralization of wastewater containing copper ions and EDTA ligand *News of Higher Education Institutions. Chemistry and Chemical Engineering Series* **62** 108-114

[8] Feng X, Wang X, Chen Z and Chen J 2019 Nitrogen removal from iron oxide red wastewater via partial nitritation-Anammox based on two-stage zeolite biological aerated filter *Bioresource Technology* **279(8-9)** 17-24

[9] Kochetov G, Prikhna T, Kovalchuk O and Samchenko D 2018 Research of the treatment of depleted nickel-plating electrolytes by the ferritization method *Eastern-European J. of Enterprise Technologies* **3** 52-60

[10] Kolesnikov V A, Il'in V I, Kapustin Y I, Varaksin S O, Kisilenko P N and Kokarev G A 2007 *Flotation Wastewater Treatment of Industrial Enterprises* (Moscow: Khimiya)

[11] Kolesnikov V A, Menshutina N V and Desyatov A V 2016 *Equipment, Technology and Design of Sewage Treatment Systems* (Moscow: new Delhi and NCR plus)

[12] Ksenofontov B S 2013 Intensification of wastewater treatment of engineering industries using ion flotation *Sanitary Engineering* **5** 30-33

[13] Matsak A and Tsytlishvili K 2018 Using different filter media of stormwater treatment performance *Norwegian J. of Development of the Int. Science* **1(20)** 19-22

[14] Kazeminejadfard F and Hojjati M R 2019 Preparation of superabsorbent composite based on acrylic acid-hydroxypropyldistarch phosphate and clinoptilolite for agricultural applications *J. of Applied Polymer Science* **136(16)** 4736
[15] Bochkarev G R and Pushkareva G I 1988 On a new natural sorbent for the extraction of metals from aqueous media *Physical and Technical Problems of Mining* **4** 46-51
[16] Ovchinnikov A S, Bocharnikov V S and Denisova M A 2019 Technology of wastewater treatment of poultry enterprises using natural sorbents with the addition of ferrite suspension *Proc. of the Nizhnevolzhsky Agrouniversity Complex: Science and Higher Professional Education* **1(1)** 15-22
[17] Bocharnikov V S, Kozinskaya O V, Denisova M A and Bocharnikova O V 2020 Study of the modes of sedimentation of the loading using a hydraulic installation *Proc. of the Nizhnevolzhsky Agrouniversity Complex: Science and Higher Professional Education* **1(57)** 260-267
[18] Loiko A V, Shibanov I V, Kagramanov G G and Blanco-Pedrejon A M 2018 Experience in the implementation of membrane technology for purification of artesian waters with a high content of iron and manganese *Water Purification. Water Treatment. Water Supply* **4(124)** 58-62
[19] Novikova I V, Luneva E N and Gritsay A V 2019 Means and technologies of water treatment for drip irrigation of agricultural land *Scientific J. of the Russian Research Institute of Melioration Problems* **3(35)** 1-17
[20] Lamskova M I, Filimonov M I and Novikov A E 2016 The use of swirling flows and sorption effects in water treatment in low-pressure irrigation systems with local supply *Scientific J. of the Russian Research Institute of Melioration Problems* **4(24)** 189-201
[21] Pustovalov E V 2018 Influence of irrigation by livestock runs on the technological qualities of mustard oil seeds *Proc. of the Nizhnevolzhsky Agrouniversity Complex: Science and Higher Professional Education* **2(50)** 181-186
[22] Kruzhilin I P, Ganiev M A, Kuznetsova N V and Rodin K A 2018 Dynamics of total water consumption and yield of periodically moistened rice during sprinkling and drip irrigation in the Volgograd region *Proc. of the Nizhnevolzhsky Agrouniversity Complex: Science and Higher Professional Education* **3(51)** 34-42
[23] Kuznetsova N V, Kuznetsov Yu V, Kozinskaya O V and Denisova M A 2020 Influence of hydraulic parameters on irrigation quality *Proc. of the Nizhnevolzhsky Agrouniversity Complex: Science and Higher Professional Education* **2(58)** 73-83

IOP Conf. Series: Earth and Environmental Science **965** (2022) 012018 doi:10.1088/1755-1315/965/1/012018

The use of wastes from the fat-and-oil industry in feeding calves in order to improve environmental safety

O Yu Bryukhno, S I Nikolaev, S Yu Agapov, E A Lipova, V N Agapova and D N Spiridonov

Volgograd State Agrarian University, 26 University Avenue, Volgograd, 400002, Russia

E-mail: o.bruxno@yandex.ru

Abstract. The present article provides the results of the commercial farm-based scientific trial conducted in calves up to six months of the age. In the rearing period of young animals, a promising technology would be the way, when premixes are included in the diet in the dairy period. The conducted study showed that the use of fat and oil industry waste as the basis for premixes helps to solve several problems, namely, to expand the ingredient list and provide alternative materials for farm animals, which allows to increase the protein level in animal diet, as well as to solve the problem of disposal the waste from the fat-and-oil industry. The experiment carried out on the Holstein dairy cows positively estimated the studied premix ZP61-2S. The results show that giving it to the calves had a beneficial effect on the digestibility of dry matter, which was better in the animals of the experimental group by 2.1% than in the control and amounted to 75.2%. Nitrogen utilization by animals in both control and experimental groups were not reduced, and the balance was positive. Thus, the ZP61-2S premix in the compound feed for young cattle increases the average daily gain of animals by 6.85%.

1. Introduction

Present time, the most important challenge of cattle breeding in the country is to obtain fast-growing animals, which are resistant to diseases and with a well-developed constitution [1, 2, 3].

For the improving of the production processes in animal rearing, it is necessary to combine of the biological characteristics and individual development of animals with modern technological practices [4]. A highly productive adult animal can be obtained only if since the beginning of growth, the ingredient base was well controlled and all nutritional requirements of the animals were strictly covered.

For many years, scientists around the world have been studying the impact of feed and feeding principles on the productivity of farm animals [5]. The results of a number of scientific research works testify, that animal productivity and production indicators are determined by growth with an improvement in the feed base, reasonable use of feed resources and an increase in the quality of diets [6].

The production of livestock products largely depends on the condition and diversity of the forage base. Improvement of the fodder base should go along the ways of organizing an intensive fodder production system and a modern fodder industry. The inclusion of various balancing materials and biologically active substances in the diet allows to provide to the animals of all types with the adequate nutrition, which is currently far behind of the needs of animal husbandry [4, 7].

To ensure high productivity of modern cattle breeds, it is necessary to increase the request for the quality of their nutrition [5]. This can be done by using nutritionally balanced diets in animal feeding with the use of biologically active substances and highly effective feed additives [8, 9].

Plant food is an essential source of minerals and vitamins for all types of animals. The quality of feed and its mineral composition greatly depends on many parameters. When animals lack any macro- and microelements, many farms use mineral supplements to balance the diets [10, 11].

In order to produce livestock products, a large amount of vegetable protein is required. An animal needs to use about 7 kg of vegetable protein to get one kg of animal protein. Thus, it is advantageous to use oilcakes, meal, and industrial waste in animal feeding [12].

The search for fundamentally new ways and economic justification of technological solutions in the production of new types of feed based on food waste is of current importance [13].

In this connection, the purpose of the experiment was to investigate the effect of the ZP61-2S premix on the growth and digestibility of feed nutrients by dairy calves.

2. Materials and methods

Current experiment was conducted in the facilities of the commercial farm in Volgograd region. The studies were carried out on young cattle. The experiment consisted of two stages. First of all, the raw materials for the base of the ZP61-2S premix were studied and the chemical and amino acid analysis of feed were carried out. In the second stage, the effect of the ZP61-2S premix on the growth of calves was determined in the trial.

The Volgograd region occupies one of the leading places in the production of mustard cake. At the end of the 20th century our colleagues of Volgograd State Agricultural Academy invented a technology for the neutralization of antinutritional factors in mustard cake. The final product of this neutralization was named the feed concentrate "Sarepta" [9]. It appears as a well-flowing substance. Its particle size is 0.98 mm in average. Its pH is close to the neutral, about 6.7. It contains 39.0% of crude protein, 8.90% of crude fat, the sum of amino acids in the Sarepta concentrate is 25.54%. According to the results of the study, it was determined as feed material, which has no disadvantages compare to the traditional kinds of materials used as the premix base, and could be used for this purpose.

The largest nutritional company in our region LLC "Megamix" developed formula of the experimental premix ZP61-2S for dairy calves based on feed concentrate from vegetable raw materials "Sarepta".

When developing a new premix, all modern research on the nutritional requirements of animals for energy, protein, fiber, vitamins, macro- and microelements, and other nutrients, and also parameters like gender and targeted productivity, were taken into account.

One kg of the ZP61-2S premix contains: vitamin A (92000 IU), vitamin D3 (71000 IU), vitamin E (77.0 mg), copper (69.0 mg), zinc (1074.0 mg), manganese (767.0 mg), cobalt (15.0 mg), iodine (22.0 mg), selenium (3.0 mg), lysine (36.0 g), methionine (14.6 g), toxin binder "Toxfin" (500.0 mg), probiotic "Bacell" (5000.0 mg), antioxidant "Luktanox" (45.0 mg), Limstone (20.0 g).

Premix for calves of the control group included the same components, but the filler in it was regular sunflower cake.

Studies were conducted on young cattle 30 heads in total. Calves were distributed into two groups using the method of analogue-groups. When distributing animals into groups, age, physiological state, body weight and body condition were taken into account. The conditions of their feeding and keeping were identical.

The ration of both groups consisted of: alfalfa and Sudanese hay, corn silage, motley grass haylage and compound feed (Table 1). In order to provide animals of all groups with macro- and microelements, vitamins, amino acids, a premix was introduced into the feed. In the diet of the control group - the standard premix ZP61-2S, and in the experimental group - the premix based on the concentrate "Sarepta".

Table 1. Average daily ration for experimental calves.

Indicator	Control group	Experimental group
Alfalfa hay, kg	1.0	1.0
Sudanese hay, kg	1.3	1.3
Corn silage, kg	4.0	4.0
Mixed grass haylage, kg	4.0	4.0
Compound feed, kg	3.0	3.0
	The diet contains:	
Fodder Unit	3.6	3.6
Metabolic energy, MJ	38.3	37.9
Dry matter, g	3.5	3.5
Crude protein, g	591.7	593.7
Digest. protein, g	402.0	404.0
Crude fiber, g	691.3	647.3
Sugar, g	353.7	354.7
Crude fat, g	174.0	164.0
Calcium, g	34.5	35.0
Phosphorus, g	21.5	21.6
Magnesium, g	6,6	6.7
Potassium, g	52.7	53.0
Sulfur, g	8.2	8.2
Iron, mg	893.7	912.3
Zinc, mg	214.7	215.7
Cobalt, mg	4.8	4.8
Manganese, mg	442.0	444.3
Iodine, mg	0.8	0.8
Carotene, mg	545.0	546.0
Vitamin D, IU	0.8	0.8
Vitamin E, mg	618.8	618.7

3. Results and discussion

The experimental premix supported to improve metabolic processes in the body of the calves of the experimental group. This was reflected in the increased digestibility coefficients of all studied feed components (Table 2).

Table 2. Coefficients of digestibility of nutrients in the diet, %.

Group	Matter		Protein	Fiber	Fat	Nitrogen-free extractive substances
	Dry	Organic				
Control	73.1 ± 1.3	75.2 ± 1.6	72.2 ± 1.1	64.0 ± 2.1	73.9 ± 2.4	78.1 ± 2.3
Experimental	75.2 ± 1.5	76.5 ± 1.3	75.1 ± 2.6	65.2 ± 2.3	74.3 ± 8.4	79.2 ± 0.8

From the second table we see that the digestibility of nutrients in the diet of calves in the experimental group were higher to the analogs from the control group. The digested dry matter in the body in the experimental group was 2.11% higher than in the control group. Digestibility of organic matter in the experimental group was 1.36% higher than in the control group. Inclusion of ZP61-2S premix in the diet given better digestibility of the most important nutrient – crude protein. Digestibility of crude protein in the control group was 72.22%, which is 2.28% lower than in the experimental group. In the experimental group of calves digestibility of crude fat and crude fiber were 0.4% and 1.21% higher than in the control group respectively. Digestibility of nitrogen-free extractive substances was 1.05% higher in the experimental group, compare to the control one.

The results of the trial allow us to see that the inclusion of the ZP61-2S premix in the diet of calves has a positive effect on the digestion processes, especially of the nitrogenous part of the diet (Table 3).

Table 3. Nitrogen utilization balance, g/head (M ± m).

Parameter	Control group	Experimental group
Consumed with feed	94.7 ± 5.8	94.9 ± 6.1
Excreted	28.5 ± 2.3	28.1 ± 2.4
Digested	66.2 ± 3.7	66.7 ± 4.0
Digestibility, %	68.9 ± 4.1	70.2 ± 4.4
Excreted with the urine	32.0 ± 2.9	32.0 ± 3.0
Balance	34.2 ± 4.0	34.9 ± 4.6
Utilized, %		
of the consumed	36.1 ± 4.4	36.7 ± 4.7
of the digested	51.6 ± 4.9	52.3 ± 4.6

The nitrogen balance was positive in all experimental calves. Young animals of the experimental group retained more nitrogen in the body than animals of the control group. Nitrogen in the body of young animals that received a premix based on the "Sarepta" concentrate in the diet was increased due to its decrease in feces. The amount of nitrogen utilized from the consumed was slightly higher in the calves of the experimental group (by 1.47%) (Figure 1).

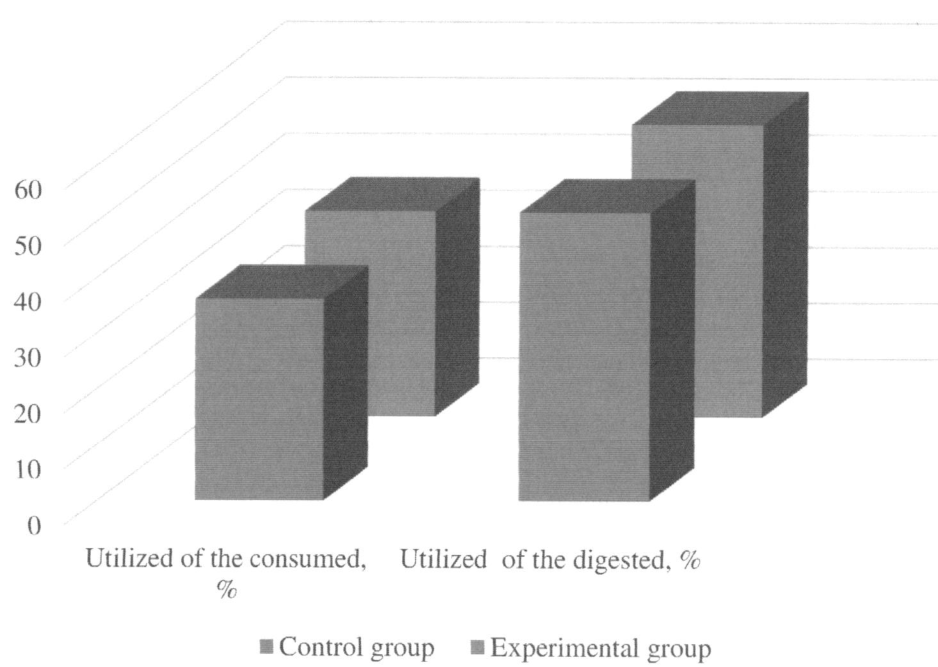

Figure 1. Nitrogen utilization.

Due to the fact that the use of nitrogen by a living organism is interconnected with the metabolism of mineral substances, in our experiment balance and utilization of calcium and phosphorus in the body of experimental calves were also studied (Table 4).

Table 4. Calcium and phosphorus balance of experimental calves, g/head (M ± m).

Group	Consumed with feed	Excreted with faeces	Excreted with urine	Balance	Utilized of consumed, %
	Calcium				
Control	34.50 ± 1.16	10.28 ± 2.41	0.12 ± 0.71	24.10 ± 0.27	69.90 ± 0.92
Experimental	35.00 ± 1.14	10.30 ± 2.21	0.13 ± 0.62	24.57 ± 0.72	70.20 ± 2.56
	Phosphorus				
Control	21.50 ± 1.16	3.97 ± 2.41	0.78 ± 0.71	16.75 ± 0.27	77.91 ± 1.92
Experimental	21.60 ± 1.14	3.77 ± 2.21	0.89 ± 0.62	16.94 ± 0.14	78.43 ± 0.85

Calcium balance in the experimental and control groups of calves was positive. Experimental animals of the control group utilized calcium by 0.4% less than their counterparts from the experimental group. The control group utilized 69.89%.

Phosphorus with feed was consumed by calves of the experimental group by 0.8% more than analogs from the control group. Utilized phosphorus from the consumed in the control group was 77.9%, while in the experimental group it was 0.51% higher.

To analyze the growth of calves according to the research methodology, control of live body weights was carried out, the data of which are presented in Table 5.

Table 5. Live weight and average daily growth of young cattle.

Parameters	Control group	Experimental group
Live body weight: kg		
at the beginning of the experiment, kg	45.90 ± 3.11	45.60 ± 2.94
at the end of the experiment, kg	144.10 ± .13	151.61 ± 5.86
Body weight gain:		
total, kg	98.20	106.01
average daily increase, g	680 ± 47.61	730 ± 45.55

According to the results of the study, it was found that the use of the ZP61-2S premix in calves feeding had a positive effect on their weight gain and height. The average daily gain of the experimental calves in the experimental group was higher than that of their counterparts in the control group by 50 grams. Young cattle weighed at the end of the experiment in the control group 144.10 kg, and the animals in the experimental group 151.61 kg.

The inclusion of the ZP61-2S premix in the compound feed for the calves of the experimental group provided a good average daily gain.

When examined by a veterinarian, no deviations from the norm conditions were observed. The use of a premix based on wastes of the fat and oil industry did not give a negative effect on the physiological state of animals.

4. Conclusions

The experiment carried out on the Holstein dairy cows positively estimated the studied premix ZP61-2S based on the "Sarepta" concentrate. The results show that giving it to the calves had a beneficial effect on the digestibility of dry matter, which was better in the animals of the experimental group by 2.1% than in the control and amounted to 75.2%. Nitrogen utilization by animals in both control and experimental groups were not reduced, and the balance was positive.

Thus, the ZP61-2S premix in the compound feed for young cattle increases the average daily gain of animals by 6.85%. And this waste of the oil and fat industry can be used as a filler for premixes.

References

[1] Mednova V V and Moshkina S V 2018 Efficiency of using an enzymatic probiotic in feeding calves *Modern Aspects of Production and Processing of Agricultural Products* pp 237-241

[2] Nikolaenko E I, Lukina D V and Glebova I V 2019 Features of feeding calves during the colostrum period *Science and Society in the Context of Gobalization* **1(6)** 30-33

[3] Esaulenko N N 2014 Feeding a probiotic improves the digestibility of nutrients in calf feed *Collection of Scientific Papers of the Stavropol Research Institute of Livestock and Forage Production* **3(7)** 200-203

[4] Abuov S K 2017 Use of feed concentrate in feeding calves *Scientific and Technical Progress in Agricultural Production* (Velikie Luki: Velikie Luki State Agricultural Academy) pp 88-92

[5] Frolov A I, Ivanova G V, Malaev D V and Lobkov V Yu 2013 The use of the Pekmelin premix in feeding cattle *Bulletin of the Agro-Industrial Complex of the Upper Volga Region* **1(21)** 93-100

[6] Stepurina M A, Varakin A T, Filatov A S, Khakimov I N, Vorontsova E S and Kokhanov A P 2020 Productive qualities of dairy cows while optimizing the mineral supply of rations *Proc. of the Nizhnevolzhsky Agrouniversity Complex: Science and Higher Professional Education* **4(60)** 269-280

[7] Agapov S Yu, Bryukhno O Yu and Shevchenko P A 2019 Adaptive feeding technology for dairy cattle Development of the Agro-Industrial Complex Based on the Principles of Rational Nature Management and the Use of Convergent Technologies pp 582-587

[8] Bryukhno O Yu, Agapova V N 2015 The use of chickpea in feeding dairy calves *VII All-Russian Scientific and Practical Correspondence Conf. of Young Scientists* (Lesnikovo: Kurgan State Agricultural Academy named after T.S. Maltseva) pp 127-129

[9] Nikolaev S I, Karapetyan A K and Chekhranova S V 2016 Comparative analysis of the chemical composition of oilseeds processing products *Polythematic Network Electronic Scientific J. of the Kuban State Agrarian University* **118** 1293-1303

[10] Nikolaev S I, Volkolupov G V, Chekhranova S V and Akmaliev T A 2015 Influence of feeding premixes on physiological indicators of cows *Proc. of the Nizhnevolzhsky Agrouniversity Complex: Science and Higher Professional Education* **3(39)** 137-141

[11] Chamurliev N G and Petrukhina E A 2015 Comparative efficiency of using urea concentrate and urea concentrate granules for feeding dairy cows *Agricultural Science: Search, Problems, Solutions* pp 174-177.

[12] Bryukhno O Yu, Karapetyan A K, Sherstyugina M A and Agapova V N 2016 Efficiency of using chickpea in feeding calves *Proc. of the Nizhnevolzhsky Agrouniversity Complex: Science and Higher Professional Education* **2(42)** 197-204

[13] Chamurliev N G, Filatov A S, Melnikov A G, Melnikova E A and Vorontsova E S 2020 Milk productivity of cows, the quality of milk and its processing products in the normalization of protein nutrition *Proc. of the Nizhnevolzhsky Agrouniversity Complex: Science and Higher Professional Education* **1(57)** 202-212

Effectiveness of using the Feed-Food Magic Antistress Mix additive in broiler chicken diets for sustainable agricultural development

O V Budtuev, S P Pereryadkina, M N Mishurova, O D Budtueva, Yu G Bukaeva and A A Baksarova

Volgograd State Agrarian University, 26 University Avenue, Volgograd, 400002, Russia

E-mail: olegbudtuev@yandex.ru, pereryadkina.svetlana@mail.ru, mishurova.mari@mail.ru, budtuevaolga@mail.ru, shabasheva.yuliya@mail.ru, aa.alseitova@gmail.com

Abstract. The article presents the results of using the anti-stress additive when introducing it into the mixed fodder for broiler chickens. Studies were carried out at the research base of Volgograd State Agrarian University. The birds of the control group were given a basic diet. The birds of the experimental group I received an anti-stress additive "Feed Food Magic Antistress Mix" 0.05% per ton of mixed fodder in addition to the basic diet, and experimental group II - 0.02% of anti-stress additive per ton of mixed fodder. Good results were obtained in terms of poultry survival, and good dynamics of growth and development of chickens were observed in all groups and sections. At all ages, the body weight of broiler chickens was within the standard values of the broiler breed. When comparing the indicators of live weight of a bird at the age of 35 days, it was found that the experimental group I prevailed over the control group by 2.96%, and over the experimental group II by 0.76%.

1. Introduction

Livestock breeding is a whole complex of industries. One of the fastest-developing against the background of its other sub-sectors is poultry farming. It allows you to get at the lowest cost of labor a very high percentage of the output of products - meat, by-products and eggs [1, 2, 3].

One of the main problems of this segment of livestock breeding, which can significantly reduce the volume and quality of the products are stresses [4, 5, 6].

Stresses affect the body at the cellular level, worsening the blood parameters of poultry and its condition in general. In order to reduce the occurrence of stress and its negative effects, anti-stress preparations, which are biologically active additives, are introduced into the diet of farm poultry [7, 8].

The aim of the experiment was to study the effect of the anti-stress preparation "Feed-Food Magic Antistress Mix" on the productive qualities of experimental poultry.

2. Materials and methods

The object of the study to study the impact of this additive on the body of farm poultry were chickens meat cross "Ross - 308".

At the enterprise JSC "Poultry farm Krasnodonskaya" were bought 360 heads of chickens at the age of 1 day. After that, they were randomly divided into three groups, one of which became the control and the other two - experimental under the order numbers I and II. Each group had 120 chicks. The duration of the experiment was 35 days.

In accordance with the requirements of All-Russian Scientific Research and Technological Institute of Poultry Farming, the birds of the control group were given a basic diet. In addition to the basic diet, the birds of the experimental group I received an anti-stress additive "Feed Food Magic Antistress Mix" 0.05% per ton of mixed fodder, and experimental group II - 0.02% of anti-stress additive per ton of mixed fodder (Table 1).

Table 1. Experiment execution scheme.

Group	Nutritional conditions for experimental chickens
Control	Basic Diet (BD)
I experienced	BD + Feed Food Magic Antistress Mix (500 g/t feed)
II experienced	BD + Feed Food Magic Antistress Mix (200 g/t feed)

At the same time, it should be noted that, in accordance with the requirements of All-Russian Scientific Research and Technological Institute of Poultry Farming, the content and feeding of poultry of all groups was identical.

3. Results and discussion

Individual weighing of daily broiler chickens was performed before planting (Figure 1).

Figure 1. Weighing day old chicks before planting.

After seating, the chicks adapted well to the conditions created. Chicks felt comfortable, quickly found nipples with water, began to actively consume pelleted feed.

The results of control weekly weighing of poultry are clearly presented in Table 2.

Table 2. Live weight of a bird in dynamics (n = 120).

Group	Age, days					
	1 day	7 7 days	14 days	21 days	28 days	35 days
Control	40.81± 0.43	200.09± 0.99	456.16± 4.04	880.53± 7.84	1479.25± 10.21	2201.93± 20.4
I experienced	41.35± 0.48	214.11± 2. 04	458.31± 3.98	896.54± 7.12	1493.25± 9.83	2267.12± 21.78*
II experienced	40.88± 0.41	208.00± 2.17**	460.46± 4.25	892.5± 7.91	1491.74± 10.45	2238.84± 22.1

* P > 0,95, ** P > 0,99, ***P > 0,999.

Analysis of growth and development of experimental broiler chickens showed good growth dynamics in all groups and sections. At all ages, the live weight of broiler chickens was within or slightly exceeded the normative indicators of the cross. The excess of the live weight indicator at the age of 7 days in the I experimental group was 7.91 g (3.95%), and in the II experimental group - 14.02 g (7.01%) relative to control.

The indicator "live weight" at day 35 of the experiment in poultry of group I experimental prevailed over this indicator of group II by 1.26 percent. However, it should be noted that the difference in this indicator between the control group and groups I, II experimental group was respectively 65.19 and 36.91 g higher (Figure 2).

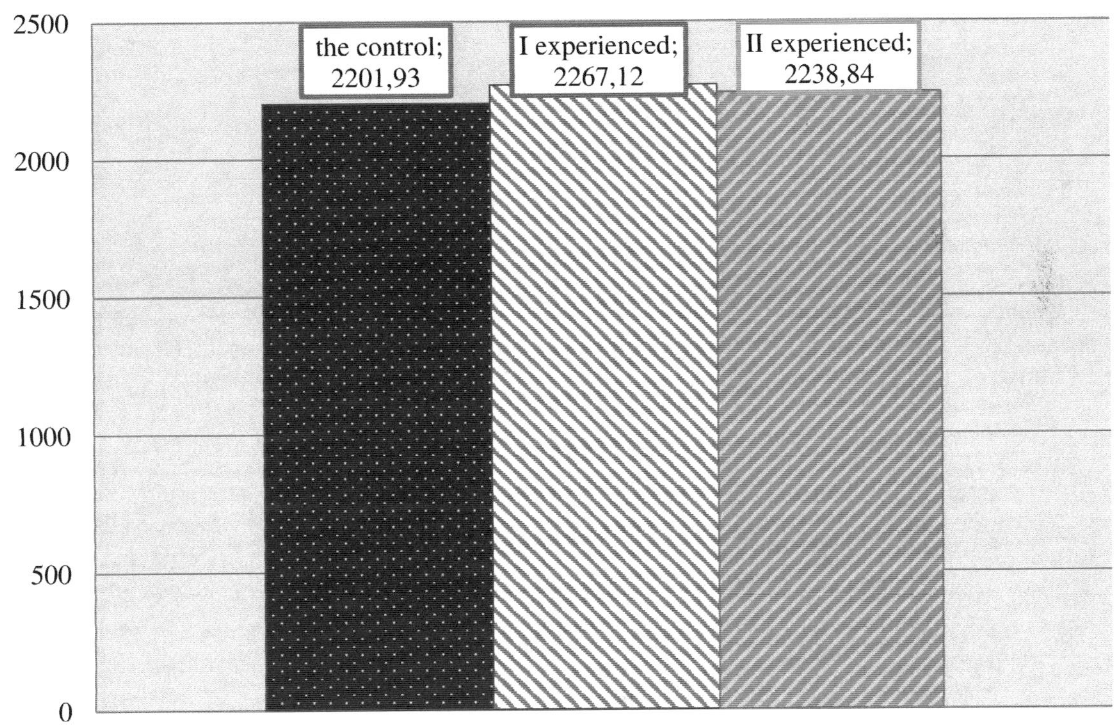

Figure 2. Live weight of the experimental bird on the day of slaughter, g.

The safety of poultry is presented in Table 3.

Table 3. Poultry safety at the time of slaughter (n = 120).

Group	Number of heads
Control	116
I experienced	120
II experienced	119

According to the Table 3, the safety of birds in the I experimental group was 100%, in the II experimental group this indicator is 0.84% lower. The control group showed 96.66% safety, which is 3.33% lower than the indicator of the I experimental group.

Fodder costs per 1 kg of live weight gain in all groups were within the standards throughout the entire period of fattening and by the end of rearing were: in the control group 1.52 kg, and in I and II experimental groups 1.53 kg, which is slightly higher than in the control (Figure 3).

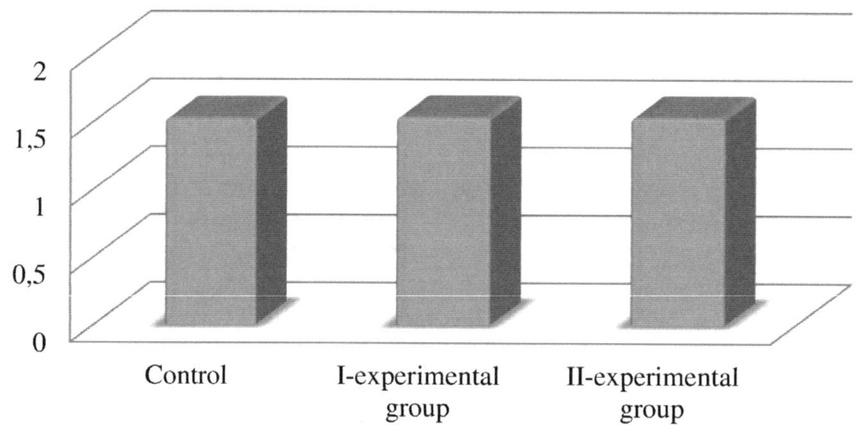

Figure 3. Feed consumption per 1 kg of live weight, kg.

Taking into account the obtained results of the research in the course of the experiment, a conditional calculation of economic efficiency was made. Due to the fact that all production costs, except for feed, were the same in all experimental groups, we calculated the efficiency of using the feed additive on this basis (Table 4).

Table 4. Economic efficiency of use feed additive of the studied groups.

Indicators	Group		
	Control	I experienced	II experienced
Absolute gain, kg	87.118	85.730	86.374
Feed consumed, kg	132.379	131.198	132.090
Feed consumption, kg	1.52	1.53	1.53
Feed cost, RUB	4032.87	4072.51	4053.56
Cost of 1 kg of growth, RUB	46.29	47.50	46.93
Selling price, RUB	85.00	85.00	85.00
Sales proceeds, RUB	7405.06	7287.62	7341.78
Profit, RUB	3372.19	3215.11	3288.23
Profitability level, %	83.62	78.95	81.12

Profitability of poultry meat production was the highest in the control group and was 83.62%, and in the experimental groups 78.95 and 81.12% respectively.

4. Conclusion

In order to reduce the negative effects of stress, an anti-stress drug was included in the diet of meat chickens, which had a positive effect on their performance. Good results were obtained in terms of poultry survival, and good dynamics of growth and development of chickens were observed in all groups and sections. At all ages, the body weight of broiler chickens was within the standard values of the broiler breed. However, it was noted that a significant difference between the control and experimental groups was observed only in the first week of life. Analyzing the data of growth indicator in all the groups it can be noted that the bird of the group I experimental, received in the composition of the diet we studied the feed additive at the dose of 500 g/t mixed feed, showed greater gains than the bird of the group II experimental. When comparing the indicators of live weight of a bird at the age of 35 days, it was found that the experimental group I prevailed over the control group by 2.96%, and over the experimental group II by 0.76%.

References

[1] Budtueva O D, Struk M V, Pleshakova I G and Pleshakov D V 2018 The use of nutovit feed additive in the diets of laying hens *Proc. of the Nizhnevolzhsky Agrouniversity Complex: Science and Higher Professional Education* **1(49)** 237-243

[2] Egorov I A, Vertiprakhov V G and Lenkova T N 2020 Gene expression, gut microbiome composition and biochemical blood parameters when using white lupine in feed for broilers *Poultry* **12** 15-20

[3] Sabyrzhanov U A, Mullakaev O T and Kushaliev K J 2017 Morphology of blood of young and laying hens receiving feed additives "Vilomix" and "Suvar" *Scientific Notes of the Kazan State Academy of Veterinary Medicine named after N.E. Bauman* **4(232)** 123-127

[4] Manukyan V A, Bajkovskaya E Yu and Silaeva A V 2020 Effect of electrolyte balance in the diet on some indicators of mineral metabolism in laying hens *Poultry* **10** 35-39

[5] Sabyrzhanov U A, Mullakaev O T and Kushaliev K J 2016 Relevance of the use of feed additives in industrial and private poultry *Scientific Notes of the Kazan State Academy of Veterinary Medicine named after N.E. Bauman* **2(226)** 138-141

[6] Chekhranova S V, Dyuzhev N A, Ponomarchenko I A and Kornilova V A 2018 Increasing the productivity of breeding poultry cross Haisex brown when using premixes in feeding *Proc. of the Nizhnevolzhsky Agrouniversity Complex: Science and Higher Professional Education* **4(52)** 83-89

[7] Sherstyugina M A 2014 Increasing the productivity of poultry when using premixes *Agricultural J.* **7** 234-237

[8] Cerrate S, Corzo A 2019 Lysine and energy trends in feeding modern commercial broilers *Int. J. of Poultry Science* **18(1)** 28-38

Improving the environmental safety of milk and the productivity of dairy cows

L V Khoroshevskaya[1], A P Khoroshevsky[2], A A Mosolov[3], Yu V Starodubova[3], M V Frolova[3], N A Karabalina[3] and M I Slozhenkina[3]

[1] Department of Veterinary and Production Control of JSC firm "Agrokompleks" named after N I Tkachev, 1 Stepnaya Street, Vyselki, Krasnodar territory, 353100, Russia
[2] Poultry Department of LLC "Russian Grain Ufa", 1 Elevator Street, Ufa, 450018, Russia
[3] Volga Region Research Institute of Manufacture and Processing of Meat-and-Milk Production, 6 Rokossovsky Street, Volgograd, 400131, Russia

E-mail: a-leksha@mail.ru

Abstract. In recent years, close attention of scientists around the world has been focused on research on avoiding the widespread use of antibiotics of various types in the agricultural industry and on the transition to the use of environmentally friendly, natural additives and enzymes in the diets of animals and poultry, aimed at increasing the immune defense of the body capable of resisting aggressive the influence of the pathogenic microbial and viral background of the environment, capable of producing environmentally friendly products for human consumption. One of the simple and effective solutions for maintaining the health of agricultural animals and birds in the current conditions is the systemic use of prebiotics, probiotics, symbiotics, phytobiotics and natural enzymes as an additive to the diet, which are active agents for stimulating and enhancing immunity due to microbial pool of the gastrointestinal tract, which leads to an improvement in the safety of livestock and an increase in the production indicators of the industry. The experiments have shown the positive effect of the new probiotic feed additive on the milk productivity and quality of milk of cows by improving the breakdown of fiber and proteins, activating the cellulolytic microflora of the digestive tract of animals.

1. Introduction

The large-scale growth in the production of livestock and poultry products is directly related to the growth in the number of livestock and poultry around the world. To prevent the emergence and spread of various bacterial diseases in large volumes of livestock of large industrial enterprises, the veterinary service has to resort to the widespread use of feed and therapeutic forms of antibiotics.

At the same time, the often unsystematic use of antibiotics in veterinary medicine has now led to an aggravating crisis in the field of antibiotic therapy, both in animal husbandry [1, 2, 3] and in human treatment [2, 4], since the residual low concentrations of antibiotics in animal husbandry products ensured the rapid adaptation of pathogens living in the human body to them. The fact that the crisis is of a planetary nature and threatens all of humanity was especially seriously discussed when COVID-19 emerged. In a pandemic, Escherichia coli, enterococci, staphylococci, Friedlander's bacillus, acinetobacters, pseudomonas and enterobacteria, then there are conditionally pathogenic bacteria that live in our bodies and the environment, as a rule, without causing any problems, in conditions of infection in patients with weakened immunity against the background of COVID-19 in the case of resistance to antibiotics, they cause dangerous nosocomial infections, up to death [1, 5, 6].

Residual concentrations of antibiotics are dangerous not only for human health, but also pose a serious problem for the dairy industry, as they can disrupt the technological process, sharply reducing the number or suppressing the viability of the starter microflora, changing the useful and taste qualities of the finished product [7, 8]. As a result, serious economic and legal problems arise for the manufacturer of the products.

In order to reduce mastitis in highly productive dairy cows and stabilize somatic cells in milk of dairy cattle to increase the yield of premium milk on the farms of Agrocomplex Novokubansky, Krasnodar territory, an experiment was conducted on the use of a new ecologically clean probiotic feed additive (a mixture of dried yeast cultures and "Laktuvet"), which promotes the activation of the cellulolytic microflora of the digestive tract in ruminants, which in turn makes feed fiber more accessible, improves the breakdown of neutral detergent fiber, leads to an increase in proteins of microbial origin and reduces the breakdown of feed proteins in the proventriculus and metabolic acidosis in ruminants, eliminating the need for feed antibiotic.

The end result of a synergistic combination of probiotic yeast cultures and the drug "Laktuvet" in the composition of the studied feed additive should lead to an increase in milk yield and an increase in the quality of milk from the studied animals.

2. Materials and methods

Place of the experiment: MTF "Pobeda". For the experiment, two experimental groups of animals of the second phase of the dry period were formed, 47 heads in each group (registration by individual numbers).

The first, experimental group was fed a diet with the introduction of a new feed additive in the composition of the compound feed in an amount of 0.4% of the diet for 30 days during the second half of the dead wood (KK-60-3-2-681) and another 30 days from the beginning of lactation in the amount of 1.25% of the diet (KK-60-4-2-682).

The second experimental group was fed a diet, with the introduction of the studied new feed additive into the compound feed for 30 days only during the onset of lactation in the amount of 1.25% of the diet (KK-60-4-2-682).

The control group of animals was fed a standard diet at the same time of their physiological state. Holstein cows were used for the experiment. The experience period is September-November 2020.

The evaluation criterion from the use of a new feed additive was carried out according to the following parameters:

- consumption of feed by dry matter;
- general condition of animals in the prenatal and postnatal period;
- biochemical blood test of experimental animals (n = 15 animals);
- study of the level of ketone bodies with test strips on the 3rd and 7th day of calving;
- the number of manifestations of mastitis when animals are milked;
- the content of somatic cells in milk during the experiment on the 40th day after the end of the experiment to determine the effect of the test drug on the subsequent lactation period after its introduction into the diet;
- dynamics of the disease, treatment, retirement of animals during the test period;
- productivity of the tested animals during the period of the experiment;
- dynamics of reaching the peak of lactation;
- quality indicators of milk (milk fat, protein, level of somatic cells);
- economic justification of the results from the use of the tested probiotic feed additive.

3. Results and discussion

Biochemical blood tests of some cows from each group, by individual selection before the start of the experiment, were carried out on a PE-5300VI spectrophotometer manufactured by Ekohim LLC using Vital diagnostic kits. The obtained results of a biochemical blood test before the start of the experiment revealed in both experimental groups and the control group a discrepancy between the normative ratio of calcium and phosphorus in the blood, an excess of the content of urea, glucose in the blood serum, the total reserve alkalinity in the blood was below the level of 46%, which indicates an overfeeding of women. animals and whether they have acidosis.

Biochemical studies of the blood of cows in both experimental groups at the end of the experiment showed the presence of a standard level of calcium, phosphorus, bilirubin, glucose, urea, carotene, iron, cholesterol, magnesium, reserve blood alkalinity, ALT, AST, protein, which indicates a good physiological state of the experimental animals. , without pathological conditions by the end of the experiment.

Biochemical studies of the blood of cows in the control group revealed some of the animals with signs of acidosis and ketosis in animals. The eatability of feed in all three groups was good, no residues of feed dispensed according to the standard were observed.

Calving in both test groups passed without complications, the average live weight of newborn calves in the first group was 27.6 kg in the second group - 28.2 kg, in the control group - 30.2 kg with a weight spread from 27.0 to 32.0 kg, as a result, it was necessary to involve specialists in obstetrics for two heifers and the treatment of mastitis in one of the cows.

The number of somatic cells, milk fat, and the level of protein in milk were monitored every ten days for the entire period of the experiment and 40 days after the completion of the experiment. Commissioned sampling of milk to determine the content of somatic cells was taken from each section of the milk tanker. The study was carried out using a viscometric milk analyzer "Somatos-V" designed to control the quality of milk and determine the number of somatic cells in milk by the conditional viscosity in a licensed laboratory of the agricultural holding.

Throughout the experiment, the content of somatic cells in the milk of cows in the control group was significantly higher than in the experimental groups (Table 1). Cows in both test groups reached peak performance faster than the control group with higher performance at peak performance. This can be explained by the fact that the tested probiotic supplement in the body of experimental cows provided an increase in rumen biomass, where bacterial protein is an additional source of lysine and methionine for the body and replaces part of the dietary protein, which led not only to improved rumen health, increased milk fat, and increased productivity. but also a decrease in the consumption of concentrates, an increase in the consumption of roughage, a decrease in the cost of feed and dairy products in general.

Table 1. Results of the experiment.

Name	1 decade of experience			2 decade of experience			3 decade of experience			40 days after the end of the experiment		
	Group			Group			Group			Group		
	I	II	C	I	II	C	I	II	C	I	II	C
Milk fat level	4.5	4.4	4.4	4.1	3.9	3.8	3.9	3.8	3.7	3.8	3.8	3.7
Milk protein level	3.0	3.0	3.0	3.1	3.1	3.0	3.1	3.1	3.1	3.1	3.1	3.2
The presence of somatic cells	141	146	235	139	171	181	162	175	244	145	149	247
Acidity of milk, °T	16	16	16	16	16	16	16	16	16	16	16	16
Milk yield per cow, kg	26.7	25.6	22.4	27.8	27.4	25.7	33.4	32.6	29.7	33.0	32.3	29.4

Name	1 decade of experience			2 decade of experience			3 decade of experience			40 days after the end of the experiment		
	Group			Group			Group			Group		
	I	II	C	I	II	C	I	II	C	I	II	C
The amount of milk yield by group for the period, total, ton	12.6	12.0	10.5	13.1	12.9	12.1	15.7	15.3	13.9	1320	1292	1176
Including:												
premium milk, ton	12.6	12.0	8.28	13.1	12.9	12.1	15.7	15.0	11.5	1320	1292	869
premium milk, %	100	100	78.8						82.7			73.9
first grade, ton	-	-	2.0	-	-	-	-	-	2.4	-	-	307
first grade, %	-	-	19.0	-	-	-	-	-	17.3	-	-	26.1
second grade, ton	-	-		-	-	-	-	-	-	-	-	-
second grade, %	-	-		-	-	-	-	-	-	-	-	-
non-graded, ton	-	-	0.22		-	-	-	-	-	-	-	-
non-graded, %	-	-	2.2	-	-	-	-	-	-	-	-	-

In addition, the strong immune system of the tested animals in both groups made it possible to qualitatively milk the animals after calving, not a single animal had an udder disease in the form of mastitis, which made it possible in these groups to do without the use of medicinal antibiotics.

Stabilization of somatic cells in milk throughout the entire experimental period provided an increase in the yield of premium milk in the experimental groups not only for the period of the experiment, but also in the next 40 days of control after the end of the experiment. Treatment of one head of a cow in the control group after calving due to mastitis with the use of an antibiotic and the period of its elimination from the animal's body became the reason for obtaining non-grade milk in the first decade of the experiment in the control group, which is reflected in the economic indicators for the group based on the results of the experiment (Table 2).

Table 2. Economic feasibility of experience.

Name	I group		II group		Control	
	Milk yield, ton	Revenue, RUB	Milk yield, ton	Revenue RUB	Milk yield, ton	Revenue RUB
Total for the group for the test period, t including:	1361.2	47642	1332.2	4662.7	1200.4	4042.7
Premium milk, ton	1361.2	47642	1332.2	4662.7	887	3104.5
First grade, ton	-	-	-	-	310.9	932.7
Non-graded, ton	-	-	-	-	2.5	5.5

Selling price of premium milk - 35.0 rubles a liter, first grade - 30 rubles, second - 27.5 rubles, non-graded - 22.0 rubles. The average cost of 1 liter of milk on the farm at the time of the experiment was 19.7 rubles. Application of the test program gave rise in the price of the diet for the first test group by 107,600 rubles, for the second test group - by 82,260 rubles for the entire period of the experiment.

Net profit based on the results of the experiment, minus the cost of milk and additional costs for the tested additive, was obtained:

- 1 (experimental) group – 2072.8 thousand rubles;
- 2 (experimental) group – 1956.0 thousand rubles;
- 3 (experimental) group – 1677.1 thousand rubles.

4. Conclusion

The experiments have shown the positive effect of the new probiotic feed additive on the milk productivity and quality of milk of cows by improving the breakdown of fiber and proteins, activating the cellulolytic microflora of the digestive tract of animals. At the same time, the possibility was revealed to refuse the use of feed and medicinal antibiotics in milk production.

5. Acknowledgements

This work was carried out under the grant of the Russian Science Foundation 21-16 00025, SSI NIIMMP. Grant sponsors were not directly involved in the development, analysis, or writing of this article.

References

[1] Mahase E 2020 Antimicrobial resistant infections keep rising despite fall in GP prescriptions *British Medical J.* **371** 4502
[2] *Global Action Plan to Combat Antimicrobial Resistance* 2015 (Geneva: World Health Organization) p 29
[3] Khoroshevskaya L V 2016 New approaches to increasing the productivity of animals and poultry based on the use of non-traditional biologically active substances *Veterinary Medicine and Feeding* **3** 20-21
[4] Ganina V I 2015 Microbiological safety of milk raw *Dairy Industry* **11** 22-23
[5] Namazova-Baranova L S 2017 Antibiotic resistance in the modern *Pediatric Pharmacology* **10** 341-354
[6] Frisch M 2021 Basic facts you need to know about reducing the intake of antibiotics *TSENOVIK. Agricultural Review* **3** 36-41
[7] Slyusa A 2020 Milk business without antibiotics, seriously and for a long time *Compound Feed* **6** 66-67
[8] Shcherbinin S 2021 The basis of biological safety on the farm *TSENOVIK. Agricultural Review* **3** 48-49

IOP Conf. Series: Earth and Environmental Science **965** (2022) 012021 doi:10.1088/1755-1315/965/1/012021

Eco-friendly ways to feed pigs without antibiotics

V I Komlatsky[1], G V Komlatsky[1], A A Mosolov[2], A N Struk[2] and M E Spivak[3]

[1] Kuban State Agrarian University named after I T Trubilin, 13 Kalinin Street, Krasnodar, 350044, Russia
[2] Volga Region Research Institute of Manufacture and Processing of Meat-And-Milk Production, 6 Rokossovsky Street, Volgograd, 400131, Russia
[3] Volgograd State Agrarian University, 26 University Avenue, Volgograd, 400002, Russia

E-mail: kubanagro@list.ru

Abstract. The reduction in the use of feed antibiotics in animal husbandry to prevent a number of diseases and increase the productivity of pigs predetermined the need to search for alternative means in feeding that could be similar to antibiotics in terms of their effect on the pigs' organism. In this regard, the widespread use of acidifiers based on organic acids. Due to the biological characteristics of pigs, piglets aged 28-30 days after weaning from sows produce an insufficient amount of hydrochloric acid in the body. The introduction of acidifiers helps to increase the digestibility of feed, and has an antimicrobial effect. A positive effect on the increase in live weight gain and the safety of piglets was shown by the results of the use of feed concentrate "Furor-T" for growing and fattening. This feed additive, obtained using a special technology from natural organic substances, is a mixture of fulvic and humic acids. The introduction of concentrate into feed in an amount of 5 to 20 g per 1 kg of feed in different periods of growth increases the average daily gain by 10-13% and ensures 100% safety of the livestock.

1. Introduction

The efficiency of pork production with the achievement of the delivery weight of pigs of 110-115 kg at the age of 165-175 days is possible under a number of conditions, among which the main role, along with genetics and technology, belongs to feeding [1]. For the formation of bone and muscle tissues, animals, especially during the period of rearing and fattening, require a balanced diet in a certain ratio of nutrients and the capabilities of the digestive system of animals, as well as the level of metabolic processes, an age-appropriate diet [2, 3].

The use of intensive production technologies can negatively affect the immune status of newborn piglets, which makes them vulnerable to disease. The change in feeding especially affects when weaning piglets from sows. Therefore, it is especially important to know the potential of the organism at an early age and in time to increase its protective properties [4]. The problem of increasing the safety of piglets as a very important indicator of the economy of the entire industry is relevant here. The need to obtain 27-32 business piglets from a sow per year is becoming a necessary indicator of modern pig breeding. It is especially important to ensure the preservation of piglets at early weaning, when they are faced immediately with a number of stress factors associated with weaning from a sow, changing feeding and environmental conditions that affect the metabolism and subsequent productive qualities of pigs [5, 6].

IOP Conf. Series: Earth and Environmental Science **965** (2022) 012021 doi:10.1088/1755-1315/965/1/012021

The safety of piglets largely depends on the body's resistance to the effects of opportunistic microflora and the supply of mother's milk, which carries immune protection. Sows' milk is a powerful factor in energy nutrition and the establishment of acid-base balance in the stomach of piglets. Until recently, the most common technique in the fight against pathogens was the use of feed antibiotics [7]. However, they inhibit not only pathogenic, but also beneficial microflora, and the effectiveness of their application decreases due to the formation of antibiotic-resistant pathogen strains.

In addition, antibiotics cause “dysbiosis" due to a failure in the development of microflora of the gastrointestinal tract, which is one of the causes of diarrhea in “monogastric” animals. Violation of the microflora balance during “dysbiosis” entails a temporary decrease in the absorption of nutrients, which reduces the growth rate of the livestock [8]. We must not forget that the use of antibiotics in animal husbandry has been banned in European countries since 2006, which will subsequently lead to restrictions on the export of products for countries that ignore this ban.

Monitoring the results of scientists' research in the field of pig feeding indicates the need to find new ways to optimize diets and feeding technologies using additives of natural origin.

2. Materials and methods

The purpose of the research was: 1) to evaluate the effectiveness of use as an acidifier of formic acid on 3-breed young growth and fattening (Landrace × Yorkshire × Duroc) at early weaning from sows in the conditions of industrial technology; 2) justification of the feasibility and development of the scheme of application dry fodder organic concentrate "Furor-T" in feeding pigs. The objects of research were piglets in growing and fattening. The research was carried out at the educational and production complex "Piglet" of the Kuban Agrarian University.

In the course of the research, general methods of scientific knowledge were used, as well as zootechnical methods of conducting experiments.

3. Results and discussion

Acidifiers based on organic acids containing one or more carboxylic groups are among the most widespread in pig breeding [9]. These include formic, propionic, lactic, butyric, other acids, as well as mixtures thereof. In this case, it should be noted their multifunctional impact. The presence of acidifiers in the diet stabilizes acidity in the stomach, disinfects feed and water, promotes protein hydrolysis, stimulates the development of beneficial microflora and destroys pathogenic microflora [10]. It should be noted also the gustatory, genetically determined addictions of pigs, which eat acidified feed with great appetite.

Different acids have different properties, so their use must be carefully studied. Combining the properties of acids and the remoteness of their action, it is possible to achieve a synergistic effect and an effective effect on bacteria, mold fungi, yeast, etc. Thus, the selection of acidifiers is not only an important, but also a necessary factor in animal feeding. the activity of enzymes involved in the synthesis of proteins during the growth of pigs.

One of the main vectors for the use of acids is the normalization of acidity in the stomach due to its increase, as a result of which microorganisms either lose their ability to reproduce or die. At the same time, the ability to change the acidity in the gastrointestinal tract is different for different acids and depends on the buffering capacity of the acids used.

Especially important is the use of acidifiers during the feeding period of suckling pigs and young animals, when the synthesis of gastric juice is still at an insufficient level, and there is a risk of disturbances in the functions of the digestive system, due to the lack of sow milk, which not only provides the piglets' need for nutrients a is a natural immune modulator. The creation of an acidic environment in the gastrointestinal tract solves several problems. Thus, the use of acidifiers in feeding suckling pigs is aimed at adapting them to conditions after weaning, when their digestive tract is not yet fully developed.

During the suckling period, piglets experience intensive growth and development of digestive organs and tissues. Due to insufficient secretion of hydrochloric acid in the stomach, the digestion of proteins does not occur to the full extent, which can create a favorable environment for the growth of pathogenic microorganisms in the intestine. In addition, a feature of piglets is the tendency to overeat, which also leads to indigestible feed and the creation of conditions for the development of pathogenic intestinal microflora. Piglets acquire the ability to produce the required amount of hydrochloric acid only by 12 weeks.

After weaning of piglets, the diet changes and their saturation with the plant part. Due to the feeding stress experienced by the piglets, they have a poor appetite and, as a result, do not gain weight well. Low feed intake leads to villous atrophy and reduced nutrient absorption and available feed energy. It is common knowledge that feed composition and feeding patterns have a large impact on animal growth rates and pork quality. Given the fact that in industrial pig breeding, breeds with high growth rates are used, animals already at a physiologically very early age must consume and digest a huge amount of feed. Therefore, the introduction of appropriate acids into the feed is required to normalize the pH in the stomach and stabilize digestion.

Increased acidity in the stomach contributes to the suppression of unwanted bacteria and molds. It should be especially noted that bacteria do not develop resistance against organic acids, which allows them to be used successfully for a long time.

After weaning piglets from sows, the pH of the stomach contents rises, a large number of E. coli bacteria accumulate in it, which causes severe and prolonged diarrhea, as a result of which the loss of piglets can reach 12-15%. In addition, another pathogenic microflora begins to develop intensively, and the substances produced by it damage the intestinal epithelium, reducing its absorption capacity. When organic acids are used, an acidic environment is created in the stomach at pH = 4.5-5.0, which suppresses pathogens and reduces the amount of toxic metabolites they produce.

To combat them, it is advisable to use formic acid, which has strong antibacterial properties. This is especially true when using a liquid type of feeding to suppress yeast growth. Due to the anatomical structure of the gastrointestinal tract and the physiological characteristics of digestion, the greatest efficiency is observed when using acids in feeding piglets - weaners.

Another important aspect of using acidifiers is the activation of digestive enzymes that improve the digestibility of feed. With the introduction of acids, the process of protein breakdown is accelerated, which increases the digestibility of feed and the growth rate of young pigs. In an acidic environment, the activity of enzymes increases 2-3 times, which improves the absorption of nutrients, and also creates a protective barrier against infections. At the same time, due to the stabilization of the acid barrier of the stomach, the entry of microorganisms into the small intestine is reduced. It must be remembered that too high a concentration of acids can adversely affect the state of beneficial microflora, for example, lactic acid bacteria, therefore, it is necessary to closely control the acidifier dose.

An important property of organic acids is that when the development of pathogenic microorganisms is suppressed, the beneficial microflora of the gastrointestinal tract is not affected. Organic acids, as a natural alternative to antibiotics, improve the absorption of nutrients, which helps to increase animal growth and reduce their incidence.

Despite the available information on the use of acidifiers, there is insufficient information on the use of formic acid in the feeding of three-breed piglets, which prompted the production experiment. It should be noted that this acid is approved for use in organic animal husbandry, which also arouses additional interest in its study.

For the experiment, a 3-breed young growth was taken (Landrace × Yorkshire × Duroc). Two groups of weaning pigs were formed. Piglets of the experimental group were given formic acid in an amount of 0.5% to the feed weight in addition to the basic diet (BD). The results of the experiment are shown in Table 1.

It was found that the introduction of formic acid into the diet of weaned pigs in an amount of 0.5% to the feed weight increases the average daily gain by 14.0%, the average weight at the end of the experiment by 10.1%, safety by 3.0%. Compared with the control, the number of cases of gastrointestinal diseases decreased by 2.0%.

The trends in the development of industrial pig breeding require the search for new solutions to optimize the feeding and rearing of rearing and fattening pigs. In the context of the ban on the use of antibiotics, feed innovations are directed towards the use of various groups of additives of natural origin, including those based on organic acids.

Promising is, in our opinion, the use of organic fodder concentrate "Furor-T" of natural origin, consisting of a mixture of fulvic and humic acids and a number of substances of organic origin.

Table 1. Zootechnical indicators of piglets.

Indicators	1 group (control)	2 group (experiment) (+ 0.5% formic acid to BD)
Piglet age at the beginning of the experiment, days	30	30
Piglet age at the end of the experiment, days	80	80
Number of piglets in each group:		
at the beginning of the experiment, heads	30	30
at the end of the experiment, heads	28	29
Average live weight at the beginning of the experiment, kg	8.35 ±1.19	8.3 ± 1.21
Average live weight at the end of the experiment, kg	30.8 ± 1.12	33.9 ± 1.14
The amount of feed consumed per 1 head per day, kg	0.67 ± 0.10	0.71 ± 0.12
Average daily gain, g	449.0 ± 2.20	512.0 ± 2.24**
Safety, %	94.0	97.0
Number of piglets suffering from digestive tract disorders, %	3	1

It should be noted that the artificially obtained sodium, potassium, and ammonium humates are widely used in crop production as growth stimulants. At the same time, due to the chemical extraction technology, these substances have an alkaline pH level of more than 8 units. In crop production, this is due to the need to prevent soil acidification.

In contrast to them, feed concentrate "Furor-T" is produced using a unique technology without the use of chemicals and salts and does not contain heavy metals. Its production technology completely excludes heat treatment and the use of alkali and hydrochloric acid. Its high efficiency is achieved due to the content of biologically active substances that restore the electrical potential of cells, increase the permeability of cell membranes. In addition, fulvic acids are powerful chelates capable of absorbing heavy metals and removing them from the body.

To assess the effectiveness of the use of "Furor-T" an experiment was carried out to determine the intensity of growth and safety of piglets. In conditions of high intensity of growth and development of piglets with weaning weight of 8.7-9.2 kg, we have formed groups of analogous piglets. Piglets growing from the 30th to the 40th day were given a concentrate in the amount of 5 g/head once a day together with the feed, from the 41st to the 62nd day the dose was increased to 20 g/head. The fattening animals were given the concentrate in the amount of 15 g/head from the 63rd to the 120th day, and in the amount of 20 g/head from the 121st to the 153th day. During the experiment, the growth rate and safety of the piglets were monitored. The results of the experiment are shown in Table 2.

Table 2. Growth intensity piglets.

Indicator	Control group	Experienced group
Number of heads	50	100
Live weight of rearing pigs, kg:		
at the beginning of the experiment	8.78	8.62
at the end of the experiment	31.5	34.3
Average daily gain, g	366	414
Live weight of fattening pigs, kg:		
at the beginning of feeding	31.5	34.3
at the end of feeding	112	122
Average daily gain in fattening, g	874	964
Safety, %	98.0	100

It was found that the average daily gain in growing up in the experimental group was 13.1% higher than in the control group. This indicator was more than 10% higher in the experimental group compared to the control in fattening pigs. In addition, 100% safety of the livestock was recorded in the experimental group.

4. Conclusion

The results of the experiment showed that the introduction of organic acidifiers into the main ration of feeding pigs does not cause a negative effect and contributes to an increase in live weight gains in livestock with high safety of animals. It is likely that due to the suppression of a number of pathogenic and opportunistic microorganisms and the enhancement of the action of enzymes involved in the breakdown of the plant part of the feed, the animals used the nutrients of the diet more efficiently. The introduction of a concentrated feed concentrate of organic origin "Furor-T" into the diet provided an increase in average livestock gains in comparison with the control by 13% in rearing pigs and by 10% in fattening. This was facilitated by the fulvic and humic acids of the preparation, which activate the assimilation of feed nutrients in the gastrointestinal tract of piglets.

5. Acknowledgments

This work was carried out under the grant of the Russian Science Foundation 21-16 00025 SSI NIIMMP. Grant sponsors were not directly involved in the development, analysis, or writing of this article.

References

[1] Gorlov I F 2014 Innovative technologies for managing living systems in the production of high-quality livestock products *Proc. of the Nizhnevolzhsky Agrouniversity Complex: Science and Higher Professional Education* **3(35)** 104-118
[2] Submire L I 2013 The role of acidifiers in increasing animal productivity *Compound* **10** 73-76
[3] Savchenko O V 2005 *Influence of the Cidifier on Productive Qualities of Young Pigs on Fattening* (Troitsk) p 18

[4] Kryukov V S and Tarasenko V N 2011 Biological and practical aspects of the use of organic acids in the feeding of pigs *RatsVetInform* **1** 29-36
[5] Yasinskaya T P 2017 The efficiency of use of acidifiers in the rations of piglets-notes *Agriculture - Problems and Prospects* (Grodno) pp 321-325
[6] Bissing P, Tkconi A 2020 Organic acids and phytogenic additives in diets relations *Pigs* **6** 23-25
[7] Berryos R and Mendoza S 2019 Alternative Antibiotics for piglets in the first period of reducing *Compound* **2** 67-70
[8] Bulgakov A M, Kuznetsov D V and Zhukov V M 2017 Improving the efficiency of use of feed for pigs with introducing into their composition of various forms of acidifiers *Bulletin of the Altai State Agrarian University* **9(155)** 141-149
[9] Polozoyuk O N and Lapina T I 2019 The effect of biologically active substances on the interior indicators of piglets in the early period *Agricultural Scientific J.* **7** 68-72
[10] Krivosheev R A and Niyazov N S 2015 The reproductive ability of sows and the safety of piglets when using the feeder of the feed (sodium diformate) *Problems of the Biology of Productive Animals* **2** 90-96

IOP Conf. Series: Earth and Environmental Science **965** (2022) 012022 doi:10.1088/1755-1315/965/1/012022

Environmentally friendly protein supplements in poultry feeding

E A Lipova, O Yu Bryukhno, S Yu Agapov, S I Nikolaev, V N Agapova and M A Ryabova

Volgograd State Agrarian University, 26 University Avenue, Volgograd, 400002, Russia

E-mail: lipova.elenka@mail.ru

Abstract. Research has been carried out to study the use of a new feed product prepared on the basis of ginger oil cake and feed concentrate from vegetable raw materials "Sarepta" in feeding poultry meat cross Ross-308. These feed products allow to increase the live weight, the digestibility of nutrients in the diet and improve meat productivity. The results of weighing of broiler chickens showed superiority in live weight of the II experimental group - 2612.76 g, where protein-vitamin-mineral concentrate BVMC (S) was used as an additive in compound feed in the amount of 7.5, 10 and 12%, respectively, during the start, growth and finish. The live weight of the I experimental group was 2512.54 g, where protein-vitamin-mineral concentrate BVMC (R) was used as an additive, the dosages were similar to the II experimental group. The weight of the control group chickens was 2370.80 g, respectively. The conducted experience showed that the chickens of the experimental groups exceeded the control analogues in the digestibility of dry matter by 0.27-0.57%, crude protein - by 0.45-1.48%, crude fiber - by 0.7-2.53%, crude fat - by 0.94-2.29%. The results of anatomical cutting of the carcass showed the superiority of the experimental groups.

1. Introduction

The meat of broiler chickens occupies a special place in the structure of the Russian meat market. Industrial poultry farming is the industry of our country that is able to ensure food security [1, 2]. For the production of a large amount of high-quality chicken meat, it is important to comply with all the norms and recommendations for a particular cross [3, 4].

Broiler chickens have undergone significant changes over the past 20 years, advances in genetics and breeding have allowed to increase the growth rate of farm animals, they have become more effective in muscle growth with less fat. The new century is characterized by fundamental research in the field of biochemistry of nutrition [5, 6, 7].

As a result of these changes, there was a need for an accurate balance of feed nutrients, as well as a certain need for high-quality protein [8, 9].

Intensification of broiler chicken meat production can be achieved through the use of protein supplements; their use will also reduce feed costs per unit of production.

The aim of the study was to study the effectiveness of the use of new feed additives in the feeding of broiler chickens.

2. Materials and methods

To study ginger oil cake and feed concentrate "Sarepta" as a filler for a protein supplement, we created experimental batches of two feed products protein-vitamin-mineral concentrates BVMK (R) and BVMK (S). These feed products were created for their use in feeding broiler chickens.

In order to study the prepared experimental batches of protein supplements, the scientific and economic experience on the basis of the Volgograd State Agrarian University, on broiler chickens of the Ross-308 cross was set up. A day-old chickens similar in live weight and development was selected to complete the three experimental groups. There were 50 chickens in each group. The experiment lasted 42 days. Broiler chickens of the control and experimental groups were fed the basic diet. The difference was that 7.5, 10 and 12% of the standard protein-vitamin-mineral concentrate filler of which was sunflower oil cake were added to the diet of the control group during the start, growth and finish, respectively. Broiler chickens of the experimental groups were also given 7.5, 10 and 12% of protein-vitamin-mineral concentrate, respectively, during the start, growth and finish, in the I experimental group - protein-vitamin-mineral concentrate BVMC (R), the filler of which was ginger oil cake, in the II experimental group - protein-vitamin-mineral concentrate BVMC (S), the filler of which was feed concentrate "Sarepta".

3. Results and discussion

The weight of a broiler chicken is the most important indicator when growing. Broilers are characterized by the most intense growth dynamics [10, 11]. During the experiment, a control weighing of experimental chickens was carried out every seven days, starting from the first day. The data obtained during weighing are shown in Figure 1.

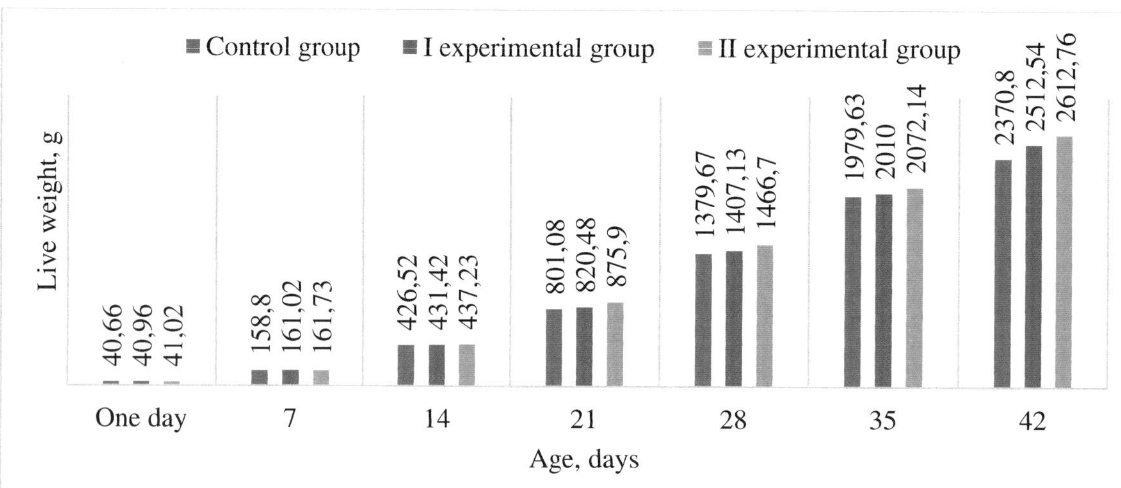

Figure 1. Dynamics of live weight of experimental broiler chickens, g.

The results of weighing the experimental broiler chickens showed superiority in live weight of the II experimental group - 2612.76 g. The live weight of the I experimental group was 2512.54 g, and the control group was 2370.80, respectively.

According to the live weight data, we calculated the total increase in the control group, it was the smallest and amounted to 2330.14 g, in the I and II experimental groups it was 2312.54 g and 2312.76 g, respectively (Figure 2). The safety in all experimental groups was 100%.

One of the important indicators determining the nutritional value, productive effect of feed and additives is the digestibility of nutrients in the diet [12, 13, 14]. The results on the amount of consumed and digested nutrients and the digestibility coefficients of these substances were obtained during a balance experiment on broiler chickens. The results of the balance experiment are presented in Table 1.

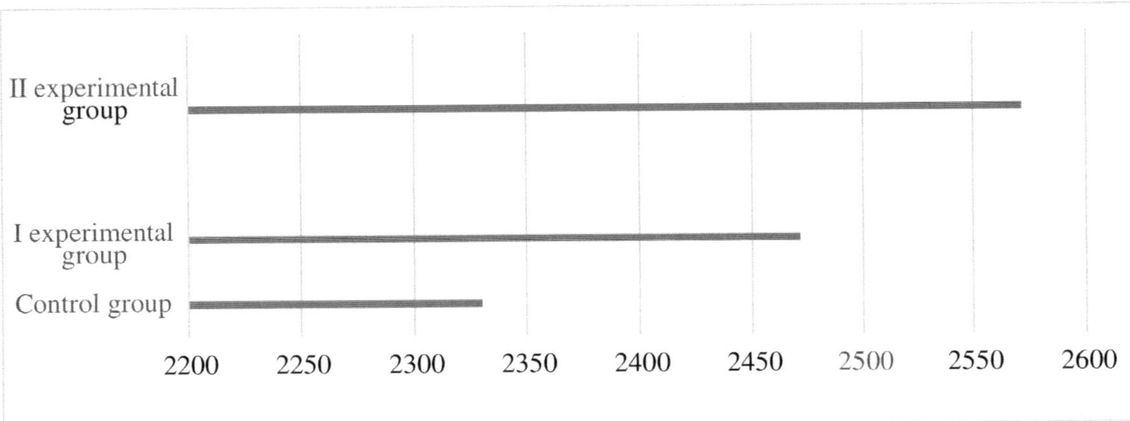

Figure 2. Total growth of experimental broiler chickens, g.

Table 1. Coefficients of digestibility of nutrients in diets, % (M ± m).

Indicator	Group		
	Control	I experimental	II experimental
Dry matter	77.55 ± 2.12	77.77 ± 2.17	78.12 ± 2.23
Organic matter	80.07 ± 3.03	80.52 ± 2.11	81.55 ± 3.22
Crude protein	78.13 ± 2.39	78.68 ± 1.12	80.50 ± 1.20
Crude fiber	18.92 ± 1.57	19.62 ± 1.42	21.45 ± 1.48
Crude fat	77.76 ± 2.50	78.70 ± 2.64*	80.05 ± 2.33
Nitrogen-free extractive substances	91.72 ± 3.81	92.14 ± 3.80	93.41 ± 3.65

* $P > 0.95$.

The conducted experience in determining the digestibility of feed nutrients showed that the chickens of the experimental groups exceeded the control analogues in the digestibility of dry matter by 0.27-0.57%, crude protein - by 0.45-1.48%, crude fiber - by 0.7-2.53%, crude fat - by 0.94-2.29%.

The analysis of the results of the digestibility experiment revealed the advantage in the digestibility of nutrients of the II experimental group, where BVMC (S) made on the basis of the feed concentrate "Sarepta" was used as a protein supplement.

The use of new supplements in feeding poultry with an intensive growth rate determines the study of blood [2, 6]. It is possible to determine the usefulness of feeding, the state of health according to the hematological profile. To study the hematological profile of the poultry we took blood samples. The hematological status data are presented in Table 2.

Table 2. Hematological status of experimental broiler chickens, (M ± m).

Indicator	Group		
	Control	I experimental	II experimental
Erythrocyte, 10^{12}/l	3.00 ± 0.16	3.30 ± 0.14	3.26 ± 0.12
Leukocyte, 10^{9}/l	32.30 ± 0.66	31.27 ± 0.67	32.69 ± 0.51
Total protein, g/l	53.23 ± 0.73	53.77 ± 0.58	54.16 ± 0.53
Albumin, g/l	24.07 ± 0.24	24.11 ± 0.19	24.41 ± 0.31
Glucose, mmol/l	11.30 ± 0.15	11.83 ± 0.16*	12.40 ± 0.32***
Calcium, mmol/l	3.95 ± 0.07	4.10 ± 0.04	4.18 ± 0.07*
Phosphorus, mmol/l	1.68 ± 0.01	1.75 ± 0.03*	1.81 ± 0.09

*$P > 0.95$, ***$P > 0.999$.

IOP Conf. Series: Earth and Environmental Science **965** (2022) 012022 doi:10.1088/1755-1315/965/1/012022

The number of erythrocytes in the blood of experimental broiler chickens in all three groups, including the control group, is within the physiological norm. However, in the control group and the I experimental group there was an equal number of erythrocytes - $3.30 \cdot 10^{12}$/l. The content of erythrocytes in the II experimental group was at the level of $3.26 \cdot 10^{12}$/l.

The protein mass is necessary for the fast-growing organism of broiler chickens [1]. Determination of total protein in the blood of broiler chickens revealed that the largest amount of protein compared to the control group was in the II experimental group and amounted to 54.16 g/l. The indicator of total blood protein was 53.23 g/l in the control group and 53.77 g/l in the I experimental group.

Glucose can be attributed to the energy element of the blood. It is an important nutrient that is used to form amino acids. The amount of glucose was 11.3 mmol/l in the control group, 11.83 mmol/l in the I experimental group and 12.40 mmol/l in the II experimental group.

Calcium and phosphorus in animal husbandry are important substances because with their deficiency or excess, diseases such as osteomalacia and bone deformation may occur, which significantly affects the productivity of poultry. A high index of calcium and phosphorus was observed in the II experimental group and amounted to 4.18 mmol/l and 1.81 mmol/l, respectively. The amount of calcium and phosphorus was 3.95 mmol/l and 1.68 mmol/l in the control group, 4.10 mmol/l and 1.75 mmol/l in the I experimental group, respectively.

The introduction of protein-vitamin-mineral concentrates BVMK (R) and BVMK (S) into the feed for broiler chickens did not have a negative effect on the hematological profile.

Most of the developed feed additives are aimed at obtaining more products. A full assessment of the meat productivity of the experimental poultry can be carried out only after its slaughter and anatomical cutting of carcasses.

Anatomical cutting was carried out according to the Federal Scientific Center "All-Russian Scientific Research and Technological Institute of Poultry Farming" of the Russian Academy of Sciences method. To do this, 3 heads were selected from each group, corresponding to the average live weight. The following indicators were taken into account: live weight, gutted carcass weight, slaughter yield, the ratio of edible parts of the carcass to inedible. The data of anatomical cutting of carcasses are presented in Table 3.

Table 3. Results of anatomical cutting of carcasses of experimental broiler chickens (M ± m).

Indicator	Group		
	Control	I experimental	II experimental
Pre-slaughter live weight, g	2355.87 ± 4.05	2502.05 ± 2.25***	2598.12 ± 3.12***
Gutted carcass weight, g	1680.30 ± 1.12	1799.52 ± 1.22***	1880.05 ± 2.78***
Slaughter yield, %	71.32 ± 0.11	71.92 ± 0.05***	72.36 ± 0.03***
Total muscle mass, g	1018.58 ± 1.06	1092.54 ± 1.05***	1120.04 ± 1.73***
including breast, g	535.55 ± 1.44	568.31 ± 0.95***	592.26 ± 1.63***
Edible parts of the carcass, g	1134.35 ± 1.27	1246.73 ± 1.59***	1318.74 ± 1.63***
Inedible parts of the carcass, g	545.95 ± 0.03	552.79 ± 0.27***	561.31 ± 0.56***
including bones	476.76 ± 0.03	531.20 ± 0.23***	562.08 ± 0.67***
Edible parts to live weight, %	48.15 ± 0.06	49.83 ± 0.05***	50.76 ± 0.02***
Inedible parts to live weight, %	23.17 ± 0.05	22.09 ± 0.05***	21.60 ± 0.04***
The ratio of edible parts of the carcass to inedible	2.08 ± 0.002	2.26 ± 0.007***	2.35 ± 0.006***

** $P > 0.99$, ***$P > 0.999$.

The results of anatomical cutting of carcasses showed that the lowest slaughter yield was in broiler chickens of the control group and amounted to 71.32%. The slaughter yield in the I experimental group was 71.92%, which is higher than in the control group by 0.6%. The slaughter yield in the II experimental was 72.36%, which is higher than in the control group by 1.04%.

One of the significant indicators characterizing meat productivity is the ratio of edible parts of the carcass to inedible. This indicator was the lowest in the control group - 2.08. In the I experimental group this indicator was 2.26, which is higher than in the control group by 8.7%, in the II experimental group - 2.35, which is higher than in the control group by 13.0%.

4. Conclusion

Thus, the introduction in the diet of broiler chickens of 7.5, 10 and 12% of protein-vitamin-mineral concentrates BVMC (R) and BVMC (S) for the growing periods allowed to increase their growth rate to 10.2%.

The results of weighing of broiler chickens showed superiority in live weight of the II experimental group - 2612.76 g, where protein-vitamin-mineral concentrate BVMC (S) the filler of which was feed concentrate "Sarepta" was used as an additive in compound feed in the amount of 7.5, 10 and 12%, respectively, during the start, growth and finish. The live weight of the I experimental group was 2512.54 g, where protein-vitamin-mineral concentrate BVMC (R) the filler of which was ginger oil cake was used as an additive, the dosages were similar to the II experimental group. The weight of the control group chickens was 2370.80 g, respectively. The conducted experience showed that the chickens of the experimental groups exceeded the control analogues in the digestibility of dry matter by 0.27-0.57%, crude protein - by 0.45-1.48%, crude fiber - by 0.7-2.53%, crude fat - by 0.94-2.29%.

The results of anatomical cutting of the carcass showed the superiority of the experimental groups. The slaughter yield in the II experimental was 72.36%, which is higher than in the control group by 1.04%. The slaughter yield in the I experimental group was 71.92%, which is higher than in the control group by 0.6%.

References

[1] Shkalenko V V, Struk A N, Tyubina A G and Dyuzheva N A 2018 Influence of biologically active additives on hematological indicators of laying hens of the haysex brown cross *AgroEkoInfo* **3(33)** 34

[2] Nikolaev S I, Karapetyan A K and Korneeva O V 2019 Influence of premixes and protein-vitamin-mineral concentrates on hematological parameters of agricultural poultry *Proc. of the Nizhnevolzhsky Agrouniversity Complex: Science and Higher Professional Education* **2(54)** 229-238

[3] Lipova E A, Karapetyan A K and Sherstyugina M A 2014 Application of protein-vitamin-mineral concentrates in poultry feeding *Proc. of the Nizhnevolzhsky Agrouniversity Complex: Science and Higher Professional Education* **1(33)** 173-176

[4] Budtueva O B, Struk M V, Pleshakova I G and Pleshakov D V 2018 Use of feed additive "Nutovit" in the diets of laying hens *Proc. of the Nizhnevolzhsky Agrouniversity Complex: Science and Higher Professional Education* **1(49)** 237-243

[5] Nikolaev S I, Cheprasova O V and Shkalenko V V 2015 *The Use of Additives in Feeding Cattle and Poultry* (Volgograd: Volgograd State Agrarian University) p 112

[6] Zlepkin D A, Zlepkin A F, Zlepkin V A and Gashuk R A 2016 Morpho-biochemical composition of blood of broiler chickens and their safety as a criterion for assessing the biological activity of tryptophan *Proc. of the Nizhnevolzhsky Agrouniversity Complex: Science and Higher Professional Education* **4(44)** 141-146

[7] Chehranova S V, Dyuzheva N A, Ponomarchenko I A and Kornilova V A 2018 Increasing the productivity of breeding birds of the Haysex brown cross when using premixes in feeding *Proc. of the Nizhnevolzhsky Agrouniversity Complex: Science and Higher Professional Education* **4(52)** 302-309

[8] Nikolaev S I, Lipova E A, Sherstyugina M A and Shkrygunov K I 2013 Application of protein-vitamin-mineral concentrates in feeding broiler chickens *Proc. of the Nizhnevolzhsky*

Agrouniversity Complex: Science and Higher Professional Education **4(32)** 120-125

[9] Karapetyan A K, Danilenko I Yu, Frizen D V and Rudnikov V N 2019 Premix based on seed products of oilseed processing in compound feeds for poultry *Bulletin of the Altai State Agrarian University* **2(172)** 115-121

[10] Karapetyan A K, Sherstyugina M A, Lipova E A and Shevchenko O S 2014 Development and use of biologically active additives in poultry feeding *Proc. of the Nizhnevolzhsky Agrouniversity Complex: Science and Higher Professional Education* **2(34)** 123-126

[11] Nikolaev S I, Cheprasova O V and Letyagina A A 2017 Physiological indicators and egg productivity of laying hens when using non-traditional feed products in diets *Proc. of the Nizhnevolzhsky Agrouniversity Complex: Science and Higher Professional Education* **4(48)** 190-196

[12] Nikolaev S I, Struk M V and Struk A N 2017 The effectiveness of the use of premixes based on oilseed processing products in feeding laying hens of the parent herd *Polythematic Network Electronic Scientific J. of the Kuban State Agrarian University* **131** 1668-1680

[13] Nikolaev S I, Lipova E A, Sherstyugina M A and Shkrygunov K I 2013 The effectiveness of the use of biologically active substances in the diets of broiler chickens *Proc. of the Nizhnevolzhsky Agrouniversity Complex: Science and Higher Professional Education* **4(32)** 115-120

[14] Nikolayev S I, Struk V N and Karapetyan A K 2013 Premixes in the feeding of broiler chickens *Vestnik OrelGAU* **5(44)** 46-49

IOP Conf. Series: Earth and Environmental Science **965** (2022) 012023 doi:10.1088/1755-1315/965/1/012023

Influence of sodium dimethylglycinate on environmental safety and meat productivity when growing broiler chickens

Z B Komarova, E N Tarasov, M I Slozhenkina, A V Rudkovskaya, M V Frolova, A A Mosolov and A N Struk

Volga Region Research Institute of Manufacture and Processing of Meat-and-Milk Production, 6 Rokossovsky Street, Volgograd, 400131, Russia

E-mail: niimmp@mail.ru

Abstract. The aim of the studies was to evaluate the effect of adding sodium dimethyl glycinate in broiler diets, as part of a feed additive to increase the digestibility of feed nutrients, improve health and increase the meat productivity of broilers. During the experiment, the prevalence of live weight was established by 77 g (3.81%) in the experimental group, with a reduction in feed costs per 1 kg of gain by 0.05 kg. Safety in the experimental group was 100%, and in the control - 97.5. The European efficiency index increased by 35.18 points and amounted to 389.35. An increase in the digestibility of nutrients obtained from feed by the body of broilers: dry matter - by 1.44, protein - by 1.25, fat - by 2.74, nitrogen-free extractive substances - by 1.67%. The mass of the pectoral muscles increased by 78 g, which confirms the high digestibility of nutrients in the feed. Sodium dimethylglycinate has also been shown to improve the immune system and reduce oxidative stress in broiler chickens. Activation of enzymes of the antioxidant status was observed under the influence of the studied additive: superoxide dismutase by 8.32%, ceruloplasmin by 9.47%, and the level of malondialdehyde decreased by 5.63%.

1. Introduction

Despite a slight decrease in the level of production of eggs and poultry meat in 2020 by 2.5 and 6.4%, respectively, due to the cessation of supplies of breeding eggs and day-old chickens from a number of European countries, due to the closure of borders due to the pandemic, many enterprises remain production volumes and quality of products at the proper level. To increase productivity, safety of birds, as well as safety and quality indicators of eggs and meat, poultry farmers are increasingly using feed biologically active additives in bird feeding that can stimulate poultry growth, activate metabolic processes, and increase the digestibility of feed nutrients.

One of these additives is sodium dimethylglycinate, a tertiary amino acid that plays an important role in various biological processes, including the cellular metabolism of choline and betaine, being as a source of glycine for the synthesis of glutathione [1, 2].

The body of chickens is practically unable to synthesize glycine. In this regard, at a young age, glycine is considered an essential amino acid in chickens due to the insufficient rate of biosynthesis [3, 4].

The need for glycine by the body of birds is explained by the participation of this acid in the synthesis of purine bases and the structure of heme [5]. Sodium dimethylglycinate can be metabolized in liver mitochondria, converting both of its methyl groups as a result of tetrahydrofolate transmethylation into free glycine [1, 6]. Sodium dimethylglycinate is completely absorbed in the form of a physiological compound, has antioxidant properties, thereby reducing the effect of oxidative stress and changes in the levels of blood plasma metabolites associated with oxidative stress. In addition, it can increase the ability of digestive enzymes and improve the ability of the intestinal brush border to come into contact with nutrients, which subsequently aids in the digestion and absorption of nutrients [7].

In this regard, in our studies, we studied the effect of a feed additive containing sodium dimethylglycinate as an active ingredient.

2. Materials and methods

The studies were carried out in the conditions of the State Scientific Institution RIPPMDP and the Research Center of LLC «MegaMix» on broiler chickens of the Ross 308 cross. The experiment involved two groups of chickens, 50 heads each. Broiler chickens from the control group received granulated feed, the experimental group received the studied feed additive as part of the feed in the amount of 1.2 kg/t of feed.

The studied feed additive contains sodium dimethylglycinate as the main component, due to which the feed conversion improves, the safety and productivity of broiler chickens increases.

The containment conditions were the same as recommended by Aviagen®. Live weight was recorded on a weekly basis, and safety was carried out daily. At the end of the experiment, during slaughter and anatomical cutting, the morphological and varietal composition of the carcasses was carried out, according to State Standard R 52702-206 «Chicken meat (carcasses of chickens, chickens, broiler chickens and their parts). Technical conditions».

The digestibility and use of nutrients in feed during the physiological experiment were determined using the methodological recommendations of the Federal Scientific Center "All-Russian Scientific Research and Technological Institute of Poultry Farming" of the Russian Academy of Sciences (2004). The nutritional value of feed and excrement was assessed according to State Standard R-51417-99.

Determination of immunoglobulins in the blood of experimental broiler chickens was carried out according to the Mancini method; superoxide dismutase - on an SF-46 spectrohatometer; ceruloplasmin - on the Cobas Mira analyzer; malondialdehyde - by the method M. Uchiyama and M. Mihara [8].

3. Results and discussion

Throughout the experiment, broilers of both groups developed in accordance with the standard indicators of the Ross 308 cross (Table 1), however, the chickens of the experimental group significantly prevailed, in terms of live weight, over their peers from the control group: at 21 days - by 2.60% ($P < 0.05$), 28 days - by 4.06% ($P < 0.01$), 35 days - by 77 g (3.81%; $P < 0.05$).

Feed consumption per 1 kg of gain in the experimental group decreased by 0.05 kg. It should also be noted a high level of safety in both groups: 100% in the experimental group, and 97.5 in the control group. When calculating the European efficiency index, a higher value of this indicator was established in the experimental group - 389.35 units, exceeding the control by 35.18.

The beneficial effect of sodium dimetryl glycinate on the digestibility of feed nutrients, according to Clapés and Infante [9], Cools et al [10], Kalmar et al [11], associated with the surfactant properties of dimethylglycine esters. In our studies, in the experimental group, a high assimilation by the body of broilers of nutrients obtained with feed, in particular fat, was established (Figure 1).

Table 1. Changes in live weight of broiler chickens during rearing (n = 50).

Age, days	Control	Experimental
0	41.6 ± 1.03	
7	179.6 ± 1.56	181.4 ± 1.25
14	443.8 ± 3.87	456.7 ± 4.98
21	885.4 ± 6.68	908.4 ± 7.83*
28	1402.3 ± 13.23	1459.2 ± 12.73**
35	2021.5 ± 17.46	2098.6 ± 15.93**
EIE - European Efficiency Index	354.17	389.35
Feed costs per 1 kg gain, kg	1.59	1.54
Safety, %	97.5	100

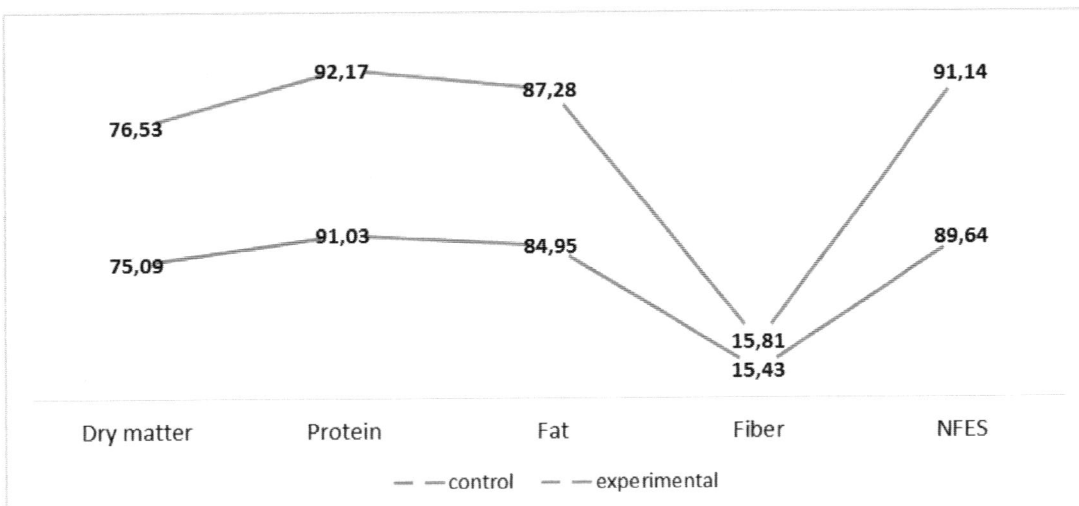

Figure 1. Digestibility of feed nutrients, % (n = 3).

The fat digestibility of the chickens from the experimental group was 2.74% higher than the control ($P < 0.01$). The digestibility of protein and nitrogen-free extractive substances was also higher than in the control group by 1.25 ($P < 0.05$) and 1.67% ($P < 0.05$), respectively, which is consistent with studies conducted by a number of authors [12, 13, 14, 15].

This positive effect supports the hypothesis that sodium dimethylglycinate acts as an emulsifying agent, as proved by Vanhauteghem et al [16].

Anatomical cutting of carcasses, at the age of 35 days, demonstrated the effectiveness of the studied feed additive. The lethal yield in the experimental group exceeded the control values by 0.8% and amounted to 73.59%. An increase in the mass of pectoral muscles by 78 g ($P < 0.05$) confirms the high digestibility of feed nutrients and, in particular, protein, which positively affected the yield of 1st grade carcasses in the experimental group, which amounted to 66.8%.

Studies [9, 17] found that sodium dimethylglycinate improves the body's immune system and reduces oxidative damage by removing free radicals, thereby improving the health and growth performance of animals and birds.

As a result of the experiment, we recorded a significant excess, in the blood of the chickens of the experimental group, the concentration of immunoglobulins by 18.45% ($P < 0.01$) in relation to the control. The use of the studied feed additive led to the activation of enzymes of the antioxidant status of broiler chickens: superoxide dismutase by 8.32% ($P < 0.05$), ceruloplasmin by 9.47% ($P < 0.05$).

The level of substances active to thiobarbituric acid and, in particular, malondialdehyde decreased by 5.63% ($P < 0.05$) compared with the control.

4. Conclusion

The present study has demonstrated that sodium dimethylglycinate serves as an additive that can activate metabolic processes, improve the absorption of nutrients in feed and increase the meat production of broiler chickens.

5. Acknowledgments

This study was funded by the grant of the Russian Science Foundation 21-16 00025 SSI NIIMMP.

References

[1] Prola L, Nery J, Lauwaerts A, Bianchi C, Sterpone L, De Marco M, Pozzo L and Schiavone A 2013 Effects of N,N-dimethylglycine sodium salt on apparent digestibility, vitamin E absorption, and serum proteins in broiler chickens fed a high- or low-fat diet *Poultry Science* **92** 1221-1226

[2] Friesen R W, Novak E M, Hasman D and Innis S M 2007 Relationship of dimethylglycine, choline, and betaine with oxoproline in plasma of pregnant women and their newborn infants *J. of Nutrition* **137(12)** 2641-2646

[3] Klasing K C 2000 *Comparative Avian Nutrition* (Wallingford, UK: CAB International) p 352

[4] Baykovskaya E Yu, Abashkina E M and Manukyan V A 2021 Synthetic glycine in compound feed for broiler chickens *Poultry* **3** 13-16

[5] Kolberg N A and Sadovnikov N V The role of the liver in the metabolism of birds. Morphological changes in the poultry liver when using antihomotoxic therapy *VI International Veterinary Congress on Poultry* http://webmvc.com/vet-articles/birds/aviculture/rol-pecheni-v-obmene-veshchestv.php

[6] Slow S, McGregor D O, Lever M, Lee B M, George P M and Chambers S T 2004 Dimethylglycine supplementation does not affect plasma homocysteine concentrations in pre-dialysis chronic renal failure patients *Clin. Biochem* **37** 974-976

[7] Pere C and Maria R I 2009 Amino acid-based surfactants: Enzymatic synthesis, properties and potential applications *Cheminform* **20** 215-233

[8] Uchiyama M and Mihara M 1978 Determination of malonalde-hyde precursor in tissues by thiobarbituric acid test *Anal. Biochem.* **86(1)** 271-278

[9] Clapés P and Infante M R 2002 Amino acid-based surfactants: Enzymatic synthesis, properties and potential applications *Biocatalysis Biotransform* **20** 215-233

[10] Cools A, Maes D, BuyseJ, Kalmar I D, Vandermeiren J A and Janssens G P J 2010 Effect of N,N-dimethylglycine supplementation in parturition feed for sows on metabolism, nutrient digestibility and reproductive performance *Animal* **4** 2004-2011

[11] Kalmar I D, Cools A, Buyse J, Roose P and Janssens G P J 2010 Dietary N,N-dimethylglycine supplementation improves nutrient digestibility and attenuates pulmonary hypertension syndrome in broilers *J. Anim. Physiol. Anim. Nutr.* **94** 339-347

[12] Gorlov I F, Slozhenkina M I, Komarova Z B, Krotova O E, Golovin V V, Ivanov S M, Frizen D V, Rudkovskaya A V and Voronina T V 2019 Mineral additive in compound feed for broiler chickens of the cross "ROSS 308" *Poultry and Poultry Products* **6** 30-33

[13] Gorlov I F, Komarova Z B, Krotova O E, Kurmasheva S S, Frizen D V, Rudkovskaya A V, Nozhnik D N and Voronina T V 2019 The effectiveness of using a mineral feed additive in growing broiler chickens of the cross ROSS 308 *Promising Agricultural and Food Innovations. Materials of the Int. Scientific and Practical Conf.* ed I F Gorlov pp 171-175

[14] Friesen V G, Ivanov S M, Gorlov I F, Slozhenkina M I, Komarova Z B and Voronina T V 2020 Effect of feed additive INNOVIT E 60 on indicators of antioxidant status and resistance of broiler chickens *Agricultural and Food Innovations* **1(9)** 39-46

[15] Komarova Z B, Pilipenko D N and Ivanov S M 2011 Features of the physiological state of laying hens using modern feed additives *Proc. of the Nizhnevolzhsky Agrouniversity Complex: Science and Higher Professional Education* **3(23)** 107-111
[16] Vanhauteghem D, Janssens G P J, Lauwaerts A, Sys S, Boyen F, Kalmar I D and Meyer E 2012 Glycine and its Nmethylated analogues cause pH-dependent membrane damage to enterotoxigenic Escherichia coli *Amino Acids* **43** 245-253
[17] Bai K, Jiang L, Zhu S, Feng C, Zhao Y, Zhang L and Wang T 2019 Dimethylglycine sodium salt protects against oxidative damage and mitochondrial dysfunction in the small intestines of mice *Int. J. of Molecular Medicine* **43(5)** 2199-2211

IOP Conf. Series: Earth and Environmental Science **965** (2022) 012024 doi:10.1088/1755-1315/965/1/012024

The determination of the hormonal program efficiency in estrus synchronization of high-yield cows for sustainable agricultural development

V A Galchenko, S P Pereryadkina, V D Kocharyan, K A Bakanova, N S Faizulina and O V Budtuev

Volgograd State Agrarian University, 26 University Avenue, Volgograd, 400002, Russia

E-mail: meshcheryakova-vika@mail.ru, pereryadkina.svetlana@mail.ru

Abstract. At present, the improvement of methodological approaches and the search for more rational methods of hormonal processing for estrus synchronization in cows remains one of the urgent tasks of agricultural science and veterinary practice. As a result of the studies carried out, the influence of hormonal preparations on the dynamics of the corpus luteum lysis and the growth of follicles in the cows ovaries was determined using ultrasound diagnostics. As a result of ultrasound examination, it was found that the double use of Prosolvin promotes lysis of the corpus luteum. This is also evidenced by the low concentration of progesterone in the serum of cows. After the application of Gonadin ultrasound images show an active formation of follicles, the size of which gradually increases and reaches the maximum size before artificial insemination.

1. Introduction

At present, the development of dairy cattle breeding is a profitable direction, both from a livestock point of view and from an economic point of view, since dairy products are one of the sources of human nutrition [1, 2]. The success and development in this industry largely depends on the work of the animal insemination operator: identifying cows in heat, fruitful insemination and obtaining healthy offspring [3, 4]. But it is not always possible to predict the correct time of insemination or to identify estrous cows, and in some cases there may be a lack of workers on the farm [5]. Thus, the method of synchronizing estrous cows and heifers is gaining popularity. This method allows you to effectively solve reproduction tasks in a short time, namely, to manage the sexual cycle, clearly plan the time of calving, organize the fertilization of cows with possible minor pathologies, as well as facilitate the work of a livestock breeder and artificial insemination operator [6]. Estrus synchronization in cows and heifers allows solving the following tasks: to inseminate a large number of animals in a short time; to postpone the period of mass calving in dairy farming for economic purposes; organize the fertilization of animals in cases where the identification of estrus is difficult or impossible due to a number of production reasons; shorten the service period [7]. Thus, further improvement of methodological approaches and the search for more rational methods of hormonal treatment for estrus synchronization and therapy of cows remains one of the urgent tasks of agricultural science and veterinary practice [8].

IOP Conf. Series: Earth and Environmental Science **965** (2022) 012024 doi:10.1088/1755-1315/965/1/012024

The purpose of our work is to determine the dynamics of the corpus luteum induction and the growth of follicles in the ovaries of the cows under the estrus synchronization program, using ultrasound diagnostics.

2. Materials and methods

The production experiments were carried out on Holstein cows on the farm of LLC Agricultural enterprise "Donskoye" in the Kalachevsky district of the Volgograd region in the period 2020-2021. The animal welfare conditions in these farms meet zoohygienic, veterinary and sanitary requirements.

To determine the state of the ovaries in cows under the hormone program, an experimental group of 16 cows was formed. The selection of animals for the experimental group was carried out according to the principle of analogs, taking into account their age, physiological state, and clinical examination data.

The scheme for the estrus synchronization, developed by us, was used in the experimental group (Figure 1). The hormone program was started on Monday, the cows were injected intramuscularly with 2.5 ml of Gonadin (gonadotropin-releasing hormone) at a random stage of the estrus cycle, then a week later they were re-injected with Gonadin intramuscularly at a dose of 2.5 ml. The purpose of this injection is to induce ovulation of the dominant follicle (and thus start a new follicular wave) or to be injected when a new follicular wave has already formed spontaneously (therefore, the injection does not essentially affect the stage of the estrous cycle).

Seven days later, Prosolvin (prostaglandin) was injected intramuscularly at a dose of 1 ml, and after 24 hours the injection of Prosolvin was repeated to regress any corpus luteum present. The seven day interval was used to actually ovulate the follicle with the initial injection of GnRH, then the corpus luteum would grow large enough to respond to PGF2α.

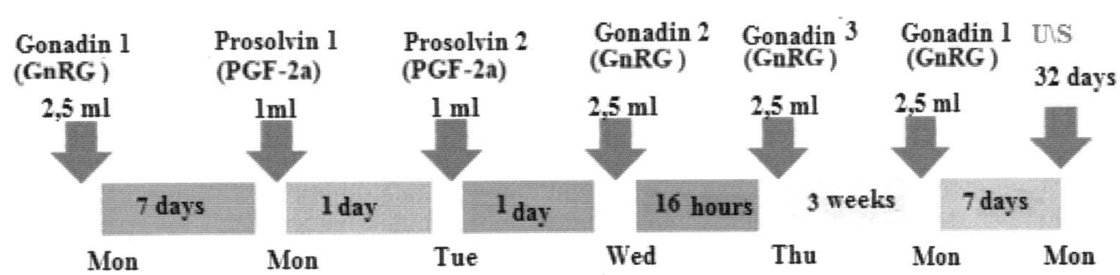

Figure 1. Program for estrus synchronization.

Then, a day later, the cows received another 2.5 ml of intramuscular injection of Gonadin to ovulate the new dominant follicle. Nine days between two GnRH injections is enough for a new follicle to appear and grow to the appropriate size for a response to GnRH. Then, after 16 hours, artificial insemination was carried out and 2.5 ml of Gonadin were injected to increase the fertility of cows and prevent early embryonic mortality.

Ultrasound diagnostics of the corpus luteum and follicles in the ovaries of cows was carried out after lunchtime milking for 9 days, from the beginning of the hormone program to artificial insemination. The evaluation of estrus synchronization programs efficiency was carried out for the number of pregnant animals after 36 days, using rectal ultrasound.

For ultrasound examination of the ovaries, an Easi-Scan Curve ultrasound scanner from BCF Technology was used. Linear matrix converters with a frequency range of 5.0 to 7.5 MHz. Transrectal ultrasound investigation was performed by inserting an ultrasound transducer into the rectum. This made it possible to visualize and evaluate the architecture of the ovaries. At the beginning of each ultrasound examination, cows were fixed in headlocks. Excess feces were removed from the rectum to facilitate examination of the reproductive tract. The lubricated transducer was then placed in the palm and gently inserted into the rectum. The probe was pressed tightly against the ventral rectum and the examination was started. After taking pictures of the ovaries, the transducer was carefully removed

IOP Conf. Series: Earth and Environmental Science **965** (2022) 012024 doi:10.1088/1755-1315/965/1/012024

from the rectum. The ultrasound results were interpreted in conjunction with farm records, herd data and visual observations.

To determine the effectiveness of Prosolvin hormonal preparation, we performed an analysis of serum for the concentration of progesterone in 6 selected animals that were under the estrus synchronization program. Blood samples were collected in clean micro-centrifuge tube from subcaudal vena after artificial insemination of the cows. The serum was separated by centrifugation at 4 °C with a speed of 3000 rpm for 10 min and was stored at a temperature of -20 °C for a hormonal assay. Progesterone was determined by radioimmunological analysis.

3. Results and discussion

As a result of 7 day ultrasound examination of synchronized cows, a large corpus luteum with a lacuna in the middle was observed on ultrasound images, which gradually decreased in size. Luteal tissue manifested itself as distinctly echogenic areas in the ovarian stroma. The granularity of the luteal tissue is visible (cells are enveloped in septa with blood vessels that feed and carry out the role of a carrier of hormones formed here). The lacuna was most often located in the center of the corpus luteum and filled with luteal fluid. Sometimes the lacuna was located on the edge of the corpus luteum, but the connection with the corpus luteum is always visible and the wall is lined with luteal cells. Unlike the vesicular follicle, which is not associated with the corpus luteum, the lacuna is combined with it into one formation. The lacuna is usually less than 1/3 of the corpus luteum size.

After the intramuscular injection of Prosolvin in a dose of 1 ml done twice with an interval of 24 hours, a rapid regression of the corpus luteum was observed, while the lacuna disappeared and follicles began to form (Figure 2).

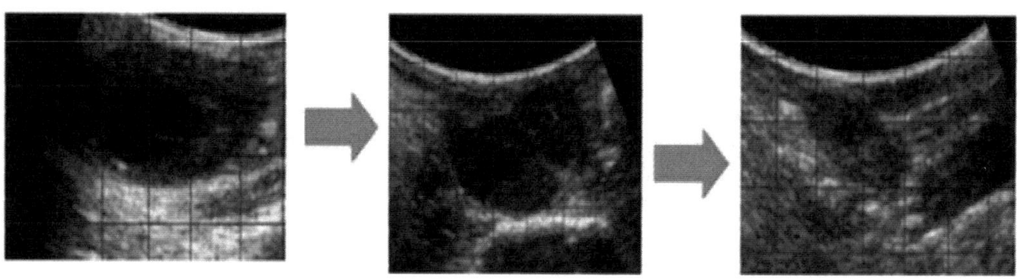

Figure 2. Ultrasound images. Dynamics of lysis of the corpus luteum in the ovaries.

Several small follicles and a small corpus luteum are visible. Such cows do not show estrus. But after the application of Prosolvin, prostaglandins F2a are released by the uterus, which act on the corpus luteum, destroying it. The production of progesterone stops and, as a consequence, the progesterone block is removed, leading to the formation of a new follicular wave and the formation of a dominant follicle.

After the administration of Gonadin at a dose of 2.5 ml, active follicular growth was observed in the ovaries. The growing follicle releases hormones that bring the cow in estrus. Visually, the follicle is visible as a cavity filled with follicular fluid with a thin wall (the closer the ovulation, the thinner), more often protruding above the surface of the ovary. When ovulation approaches, the inclusion of white (hyperechoic zones) crystals can be observed in the follicle wall (Figure 3).

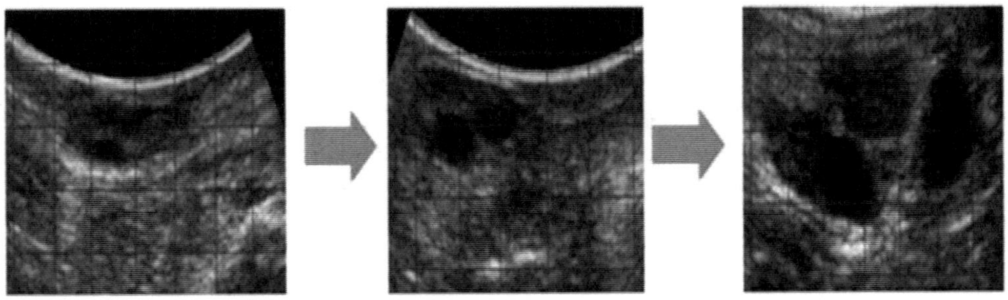

Figure 3. Ultrasound images. The dynamics of the growth of follicles in the ovaries.

As a result of the research of cows blood serum samples, we received the following data (Table 1).

Table 1. The content of progesterone in the blood serum of cows during the estrus synchronization.

No.	No. of animal	Milk yield, liters per day	Weight, kg	The day after calving	Progesterone ng/ml
1	4771	25.6	639	54	0.41
2	1801272	48.7	642	53	0.25
3	1703276	51.4	641	55	0.38
4	88850	53.6	642	56	0.35
5	93484	41.0	639	58	0.21
6	1801356	38.6	640	56	1.0

Thus, it can be seen from the data in the Table 1 that 72 hours after the injection of Prosolvin, the level of progesterone in the blood of cows reaches low levels, which is due to the massive lysis of the corpus luteum in the ovaries under the action of prostaglandin.

As a result, the following data were obtained for determining the efficiency of the estrus synchronization program, which are shown in Table 2.

Table 2. Evaluation of the efficiency of estrus synchronization programs.

Indicators		Synchronization scheme
Number of selected animals	cows	16
Number of cows with clinically expressed estrus	cows	14
Percentage of animals that came in estrus	cows	87.5%
Number of pregnant cows	cows	13
Percentage of the first insemination	cows	81.25 %

According to the information given in the table, it was found that the efficiency of the synchronization program is 81.25%. This is due to the fact that cows were injected with 2.5 ml of Gonadin before artificial insemination, to increase fertility and prevent early embryonic mortality.

4. Conclusion

After analyzing the ultrasound scan, it can be concluded that the double use of Prosolvin promotes lysis of the corpus luteum. This is also evidenced by the low concentration of progesterone in the serum of cows. Before the use of prostaglandin, the corpus luteum had a size of 2.4 cm, but after the intramuscular injection of Prosolvin, the corpus luteum regressed, its size began to decrease to 1.2 cm, then completely disappeared. After the application of Gonadin, an increase in the level of luteinizing hormone in the blood of animals begins, which contributes to folliculogenesis, ovulation of follicles and the manifestation of estrus.

References

[1] Shiriev V M, Aminova A L, Gubajdullin G Kh, Zejnalov O A and Tinku A A 2010 The effectiveness of the use of new generation bioregulators in the reproduction of cattle *Achievement of Science and Technology of the Agro-Industrial Complex* **1**
[2] Khamitova L F, Merzlyakova E A and Metlyakova A A 2015 Problems of herd reproduction *Scientific Notes of the Kazan State Academy of Veterinary Medicine named after N.E. Bauman* **222(2)** 234-236
[3] Pereryadkina S P, Avdeenko V S, Galchenko V A and Almtaev E A 2021 Morphological characteristics of ovaries in meet cattle at dysfunctional state of the gonad *Proc. of the Nizhnevolzhsky Agrouniversity Complex: Science and Higher Professional Education* **3(63)** 244-254
[4] Frolova S P and Zakharova M A 2008 Determination of the hormonal status of amniotic fluid *Proc. of the Nizhnevolzhsky Agrouniversity Complex: Science and Higher Professional Education* **4(12)** 144-145
[5] Derkho M A, Sled A N and Derkho A O 2021 Platelet hemostasis and its the relationship with cortisol and progesterone in cows at pregnancy *Scientific Notes of the Kazan State Academy of Veterinary Medicine named after N.E. Bauman* **246(2)** 60-66
[6] Fricke P M, Carvalho P D, Giordanoa J O, Valenzab A, Lopes G and Amundson M C 2014 Expression and detection of estrus in dairy cows: the role of new technologies *Animal* **8** 134-143
[7] Borchardt S, Schuler L, Wolf L, Wessenauer C and Heuwieser W 2018 Comparison of pregnancy outcomes using the Ovsynch or Cosynch protocol for the first temporary AI with liquid or frozen sperm in lactating dairy cows *Theriogenology* **107** 21-26
[8] Klis P, Piwczynski D, Sawa A and Sitkowska B 2021 Prediction of lactational milk yield of cows based on data recorded by AMS during the periparturient period *Animals* **11** 383

IOP Conf. Series: Earth and Environmental Science **965** (2022) 012025 doi:10.1088/1755-1315/965/1/012025

Improving the environmental safety of dairy products

V Yu Kontareva[1], S N Belik[2], E V Morgul[2], V V Kryuchkova[3], I F Gorlov[3], A A Slozhenkina[3] and N I Mosolova[3]

[1] Don State Agrarian University, 24 Krivoshlykov Street, Persianovsky, Rostov Region, 346493, Russia
[2] Rostov State Medical University, 29 Nakhichevan Lane, Rostov-on-Don, 344022, Russia
[3] Volga Region Research Institute of Manufacture and Processing of Meat-and-Milk Production, 6 Rokossovsky Street, Volgograd, 400131, Russia

E-mail: superbelik@mail.ru

Abstract. The authors studied the influence of prebiotic components on the biological value of enriched yogurt. Yogurt was produced by fermenting pasteurized normalized milk with yogurt starter containing freeze-dehydrated strains of Streptococcus thermophilus, Lactobacillus delbrueckii ssp. bulgaricus, Lactobacillus acidophilus, Bifidobacterium lactis and lactose with the addition of a prebiotic complex containing an inactivated yeast culture of Saccharomyces cerevisiae, dietary fibers, vitamins, amino acids, macro- and microelements. The prebiotic complex was introduced at the stage of fermentation of the mixture, at a concentration of 1%, previously dissolved in milk at 38...42 °C. The amino acid content was determined; the aminoacid score, the utility of the aminoacid composition and the utility coefficient were calculated. It was established that the biological value of proteins of yogurt is characterized by their high content, balance of four essential aminoacids and presence of three limiting aminoacids. Enrichment of yogurt with a prebiotic complex slightly increased its biological value and the ability to absorb essential aminoacids. Therefore, it is impractical to use this component to increase the biological value of the protein of traditional products. The main function of the prebiotic component of the studied yogurt is to stabilize the intestinal microflora, which determines its functional significance.

1. Introduction

Commensal microorganisms (microbiomes) play a key role in maintaining human health. Violation of the homeostasis of the microbiota - "dysbacteriosis" - is a key link in the development of a number of pathological conditions including obesity, diabetes, cardiovascular diseases and even cancer. Active use of antibiotics since the middle of the last century has led to a 22-times reduction in the overall mortality rate from infectious diseases (USA), but at the end of the 20th century, the level of septicemia almost doubled, and the mortality rate from infectious diseases increased by 50%, [1]. It was established that intestinal dysbiosis is accompanied by an increase in the production of inflammatory cytokines, dysfunction of the intestinal barrier and activation of cellular apoptosis, which may lead to the development of multiple organ failure. Therefore, the demand for the use of probiotics to restore and maintain normal intestinal microflora has increased significantly [1, 2, 3]. The results of numerous studies have shown such positive effects of probiotics as induction of antimicrobial peptides, production of antimicrobial factors, stimulation of mucus and IgA production, prevention of cytokine-induced apoptosis of intestinal epithelial cells, regulation of the development and functioning of immune and epithelial cells, antioxidant and antimutagenic activity, etc. [4, 5, 6].

It should be noted that the use of fibers or prebiotics is extremely necessary for dietary modulation of the human intestinal microbiota. Dietary fiber consumption contributes to extensive metabolic interactions between bacterial species present in the gastrointestinal microbial community. In the bacterial environment, cross-feeding occurs, while the products obtained as a result of fermentation of polysaccharides by one bacterial species provide substrates for the growth of other bacteria present in the community [7, 8]. Thus, prebiotics stimulate the growth of beneficial bacteria in the colon selectively. They are one of the promising groups of functional food ingredients that should have the following properties: they should not be degraded and absorbed in the upper gastrointestinal tract, they should be degraded by enzymes of microorganisms in the colon and selectively stimulate the growth of bifidobacteria and/or lactobacteria, they should have functional influence on the body and should be technologically stable in the production of food [9].

Nowadays, medicinal forms of probiotics, prebiotics and fecal transplants have taken a significant place in the pharmaceutical industry and occupy a fairly high price segment. At the same time, there is an available alternative natural way to restore and preserve normal microflora – this is the regular use of fermented-milk functional products [10]. It was found that the use of fermented-milk functional products enriched with probiotics and prebiotics in the diet not only contributes to the restoration of normal microflora after the use of broad-spectrum antibiotics, but also increases the nutritional and biological value of the base product [11, 12]. As functional ingredients, components of plant and animal origin are also widely used. They are the sources of such essential components of food as vitamins, vitamers, catechins, leucoanthocyanidins, flavanones, anthocyanins and anthocyanidins, etc. [13, 14, 15].

One of the promising complex prebiotics is a preparation containing inactivated yeast culture Saccharomyces cerevisiae (vini), dietary fibers (wheat bran), vitamins, aminoacids, macro - and microelements. This preparation was used to enrich yogurt [12], and stevia extract was used as a plant component. Stevia leaves contain glycosides that provide their sweet taste - stevioside and rebaudioside. These compounds are used by the food industry as sweeteners [16]. Stevioside and rebaudioside A make up respectively 5-10% and 2-4% of a/m (dry base) of the leaves. Other glycosides are represented in minor amounts (rebaudioside B, rebaudioside C, rebaudioside D, rebaudiosides E, rebaudiosides F, dulcoside F, rubusoside, and steviolbioside) [17, 18]. Rebaudioside (the sweetest compound of stevia) is a sugar substitute and is widely used in the food industry, stevioside is used in the complex treatment of diabetes, obesity, hypertension and caries prevention [19]. In addition, stevia leaves contain more than 100 phytochemical compounds with antioxidant and medicinal properties, including diterpenes, triterpenes, sterols, tannins, alkaloids, phytosteroids, saponins, flavonoids, ascorbic acid, beta-carotene, luteolin, quercetin, etc. [18]. Recently, the work on the creation of fermented milk products with high biological value has been actively carried out [20]. In this study, we set the task to assess the influence of the prebiotic complex and stevia on the biological value of yogurt.

2. Materials and methods

The biological value of the elaborated yogurt enriched with a prebiotic complex was estimated by the content of aminoacids, primarily essential ones, at the end of the intended expire date (after 10 days).

The aminoacid composition was determined by capillary electrophoresis using the capillary electrophoresis system "Kapel-105". Free aminoacids were converted to phenylthiocarbamyl derivatives (FTC-derivatives) using phenylisothiocyanate, and their ionic forms were separated in a quartz capillary under the action of an electric field. The FTC-derivatives were registered at a wave length of 254 nm in the corresponding buffer solution.

To determine the aminoacid score, the aminoacid content in the product was consistently compared with the standard FAO/WHO scale.

The aminoacid score was calculated according to the formula:

$$AC = \frac{AC_{iproduct}}{AC_{FAO/WHO}}, \quad (1)$$

where AC is the aminoacid score of the i-essential aminoacid, in unit fractions or in percents; $AC_{i\,product}$ – the content of the i-essential aminoacid in the product, g/100g of protein; $AC_{FAO/WHO}$ - the content of the i-essential amino acid in the FAO/WHO protein, g/100g of protein.

In order to assess the balance of the aminoacid composition of the protein, the utility of the aminoacid composition, which means the possibility of absorbing each aminoacid, was calculated:

$$\alpha_i = \frac{C_{\min}}{C_i}, \quad (2)$$

where α_i is the utility of amino acids, in standard units; C_{min} is the minimum aminoacid score, in unit fractions or in percents; C_i is the amino acid score of the i-essential aminoacid, in unit fractions or in percents.

The next stage of the calculations was the determination of the utility coefficient of the amino acid composition of proteins:

$$\mu = \frac{\sum_{k=0}^{n}(A_i \cdot \alpha_i)}{\sum_{k=0}^{n} A_i}, \quad (3)$$

where n is the number of amino acids; A_i – the amount of each aminoacid, in g/100g.

Data collection, processing and output was carried out using a personal computer with the operating system Windows ® 2000/XP. To assess the differences between the mean values in the comparison groups, the Student's t-test was used with an error probability of $P \leq 0.05$.

3. Results and discussion

Yogurt was produced by fermenting pasteurized normalized milk with yogurt starter containing freeze-dehydrated strains of Streptococcus thermophilus, Lactobacillus delbrueckii ssp. bulgaricus, Lactobacillus acidophilus, Bifidobacterium lactis and lactose (produced by "VIVO Industria" LLC) with the addition of the prebiotic complex "Eubikor" (produced by "SIC "NIR" LLC, St. Petersburg) at the stage of fermentation of the mixture, in a concentration of 1%, previously dissolved in milk at 38...42 °C. The technology is implemented in the following way. The preparation of raw materials is carried out (skimmed milk is pasteurized at a temperature of 82-86 °C with an exposure of 2-3 seconds; skimmed milk powder is added; the mixture is getting normalized). Then the mixture is homogenized at a temperature of 55-65 °C and a pressure of 15.0 ± 2.5 MPa; the milk mixture is pasteurized at a temperature of 90-94 °C with an exposure of 2-3 seconds and cooled to a temperature of 38-42 °C. Next, a starter consisting of strains of Streptococcus thermophilus, Lactobacillus delbrueckii ssp. bulgaricus, Lactobacillus acidophilus, Bifidobacterium lactis and lactose in an amount of 5% of the mixture weight is added to the milk mixture; prebiotic complex "Eubicor" in the amount of 1% of the mixture weight and dry stevia extract in the amount of 0.3% of the mixture weight are added afterwards. Obtained mixture is poured into consumer package (for example, a glass jar or a plastic container) and packed. It is fermented at a temperature of 38-42 °C to a clot acidity of 75 °T, the finished product is cooled and stored at a temperature of 2-6 °C. The obtained product samples were subjected to a quality assessment. During the experimental part of the work, generally accepted methods and techniques for determining the organoleptic, physicochemical and microbiological parameters of the finished product were used.

The results of the study of the aminoacid composition of yogurt enriched with a prebiotic complex are shown in Table 1.

IOP Conf. Series: Earth and Environmental Science **965** (2022) 012025 doi:10.1088/1755-1315/965/1/012025

Based on many years of FAO/WHO biomedical research. a criterion for determining the quality of protein was proposed – a standard that is balanced in essential amino acids and best meets the needs of the body. The amino acid limiting biological value is the one with the lowest score.

Biological value is an indicator of the quality of food protein reflecting the degree of correspondence of its amino acid composition to the body need for amino acids for protein synthesis. The biological value of food proteins is determined by various methods. One of the available methods is the calculation of the amino acid score.

Table 1. Aminoacid composition of yogurt enriched with prebiotic complex.

Amino acid	The actual value of indicators based on the results of research (g/100g)	
	Yogurt without prebiotic complex (control)	Yogurt enriched with prebiotic complex (experiment)
Arginine	0.59 ± 0.023	0.62 ± 0.021
Lysine	0.18 ± 0.006	0.19 ± 0.007
Tyrosine	0.19 ± 0.005	0.20 ± 0.006
Phenylalanine	0.24 ± 0.007	0.26 ± 0.008
Histidine	0.17 ± 0.005	0.18 ± 0.006
Leucine and Isoleucine	0.23 ± 0.006	0.24 ± 0.007
Methionine	0.10 ± 0.005	0.11 ± 0.005
Valine	0.23 ± 0.009	0.24 ± 0.009
Proline	0.31 ± 0.009	0.33 ± 0.009
Threonine	0.20 ± 0.005	0.21 ± 0.005
Serin	0.19 ± 0.005	0.21 ± 0.005
Alanine	0.34 ± 0.008	0.36 ± 0.009
Glycine	0.20 ± 0.007	0.21 ± 0.007
Aspartic acid	0.18 ± 0.004	0.19 ± 0.004
Glutamic Acid	0.16 ± 0.004	0.17 ± 0.005
Cysteine	0.038 ± 0.0004	0.04 ± 0.0004
The sum of all aminoacids	3.548	3.76
The sum of essential aminoacids	1.18	1.25
The sum of relatively essential aminoacids	0.76	0.8
The sum of non-essential aminoacids	1.608	1.71

A reference (ideal) protein is a hypothetical product whose composition perfectly satisfies the body's physiological need for essential aminoacids. The protein of a chicken egg and female breast milk are the closest to the ideal chemical composition.

The aminoacid composition of the reference protein proposed by the FAO/WHO Committee shows the content of each of the essential aminoacids in 100 g of protein (Table 2).

In a reference (standard) protein. the amino acid score of each essential acid is taken to be 100%. The amino acid limiting biological value is the one with the lowest score. The biological value of food proteins is determined by the first limiting amino acid.

Table 2. Amino acid scale.

Amino acid	Reference protein, g/100g of protein	Daily nutrition for adults, mg/kg of body weight
Isoleucine	4	10
Leucine	7	14
Lysine	5.5	14
Methionine + Cysteine	3.5	13
Phenylalanine + Tyrosine	6	14
Threonine	4	7
Valine	5	10

The results of the calculation of the biological value of the proteins of the elaborated yogurt with a prebiotic complex and yogurt without a prebiotic complex are presented in Table 3 and Table 4.

Table 3. Results of calculation of the aminoacid score of yogurt enriched with prebiotic complex.

Amino acid	Reference protein, g/100g of protein	The essential aminoacid content in the product, g/100g	The essential amino acid content, g/100 g of protein	Amino acid score, %
Isoleucine	4	0.24	5.5	50
Leucine	7			
Lysine	5.5	0.19	4.3	78.2
Methionine + Cysteine	3.5	0.11	2.5	71.4
Phenylalanine + Tyrosine	6	0.46	10.5	175
Threonine	4	0.21	4.77	117.5
Valine	5	0.24	5.5	110

Table 4. Results of calculation of the aminoacid score of yogurt without prebiotic complex.

Amino acid	Reference protein, g/100g of protein	The essential amino acid content in the product, g/100g	The essential amino acid content, g/ 100 g of protein	Amino acid score, %
Isoleucine	4	0.23	5.4	49
Leucine	7			
Lysine	5.5	0.18	4.28	77.8
Methionine + Cysteine	3.5	0.10	2.3	65.7
Phenylalanine + Tyrosine	6	0.43	10.2	170
Threonine	4	0.20	4.78	119.5
Valine	5	0.23	5.4	108

The limiting essential aminoacid is considered to be the aminoacid with the lowest Amino acid scope; in this case, the first limiting amino acid is isoleucine and leucine; the second limiting aminoacid is methionine and cysteine; the third one is lysine. In this case, we observe a slight increase in the biological value of the enriched product. Utility (α_i) should be understood as the ability to absorb each aminoacid. This indicator is the ratio of the minimum score (C_{min}) to the score of each aminoacid (S) for the studied products. The calculation results are presented in Table 5.

Table 5. Comparative analysis of the utility of amino acids.

Amino acid	Utility of amino acids, standard units	
	Yogurt enriched with prebiotic complex	Yogurt without prebiotic complex
Isoleucine + Leucine	1	1
Lysine	0.64	0.62
Methionine + Cysteine	0.7	0.75
Phenylalanine + Tyrosine	0.29	0.29
Threonine	0.43	0.41
Valine	0.46	0.45

Based on the table data. it can be concluded that the balance of the essential aminoacids of the studied products in relation to the protein reference is higher in the enriched product.

The utility coefficient of the aminoacid composition is of practical importance. since the possibility of utilization of aminoacids by the body is predetermined by the minimum score of one of them (Table 6).

Table 6. Utility coefficient of the amino acid composition of protein.

Amino acid	$A_i \cdot \alpha_i$	
	Yogurt enriched with prebiotic complex	Yogurt without prebiotic complex
Isoleucine + Leucine	0.24	0.23
Lysine	0.12	0.11
Methionine + Cysteine	0.08	0.08
Phenylalanine + Tyrosine	0.14	0.12
Threonine	0.09	0.08
Valine	0.11	0.1
The sum $A_i \cdot \alpha_i$	0.78	0.72
μ	0.54	0.52

4. Conclusion

As can be seen from the above. the biological value of the proteins of both traditional and enriched yogurt is characterized by their high content. balance of the three essential aminoacids and presence of three limiting aminoacids. Enriching yogurt with a prebiotic complex slightly increased its biological value and the ability to absorb essential aminoacids. Therefore. it can be concluded that it is impractical to use this prebiotic component to increase the biological value of the protein of traditional products. The main function of the prebiotic component of the studied yogurt is to stabilize the intestinal microflora. which determines its functional significance.

5. Acknowledgments

The authors are grateful to the Russian Science Foundation for the financial support in the implementation of this research according to the scientific project # 22-26-00138, SSI NIIMMP.

References

[1] Klingensmith N J and Coopersmith C M 2016 The gut as the motor of multiple organ dysfunction in critical illness *Crit. Care Clin.* **32** 203-212

[2] Marini J, Gattinoni L, Ince C, Kozek-Langenecker S, Mehta R L, Pichard C, Westphal M, Wischmeyer P and ean-ouis Vincentet J-L 2015 A few of our favorite unconfirmed ideas *Crit. Care* **19** S1

[3] Andrade M E, Araújo R S, de Barros P A, Soares A D, Abrantes F A, Generoso S de V,

Fernandes S O and Cardoso V N 2015 The role of immunomodulators on intestinal barrier homeostasis in experimental models *Clin. Nut.* **34(6)** 1080-1087

[4] Khailova L, Petrie B, Baird C H, Dominguez Rieg J A and Wischmeyer P E 2014 Lactobacillus rhamnosus G G and Bifidobacterium longum attenuate lung injury and inflammatory response in experimental sepsis *PLoS One* **9(5)** 97861

[5] Shimizu K, Ogura H, Asahara T, Nomoto K, Morotomi M, Tasaki O, Matsushima A, Kuwagata Y, Shimazu T and Sugimoto H 2013 Probiotic/synbiotic therapy for treating critically ill patients from a gut microbiota perspective *Dig. Dis. Sci.* **58(1)** 23-32

[6] Morgul E V, Kolmakova T S, Belik S N, Gorlov I F and Slozhenkina M I 2020 Genoprotective effects of probiotics *IOP Conf. Series: Earth and Environmental Science* **548(8)** 82065

[7] Tsai Y L, Lin T L, Chang C J, Wu T R, Lai W F, Lu C C and Lai H C 2019 Probiotics, prebiotics and amelioration of diseases *J. Biomed. Sci.* **26(1)** 3

[8] Holscher H D 2017 Dietary fiber and prebiotics and the gastrointestinal microbiota *Gut. Microbes.* **8(2)** 172-184

[9] Khramtsova G, Ryabtseva S A, Butkevich R E, Akhmedova V R, Rodnaya A B and Marugina E V 2018 Prebiotics as functional food ingredients: terminology, selection criteria and comparative evaluation, classification *Nutrition Issues* **1** 5-15

[10] Khan I T, Nadeem M, Imran M, Ullah R, Ajmal M and Jaspal M H 2019 Antioxidant properties of milk and dairy products: a comprehensive review of the current knowledge *Lipids Health Dis.* **18(1)** 41

[11] Kontareva V Yu, Belik S N, Morgul E V, Gorlov I F and Slozhenkina M I 2020 Yogurt enriched to correct intestinal microflora in dysbiosis *IOP Conf. Series: Earth and Environmental Science* **548(8)** 82051

[12] Kontareva V Yu, Belik S N, Morgul E V, Gorlov I F and Slozenkina M I 2020 The effect of prebiotic components on the quality of yogurt *IOP Conf. Series: Earth and Environmental Science* **548(8)** 82054

[13] Kryuchkova V V, Gorlov I F, Korneichuk K M, Belik S N and Mosolova N I 2020 Brine-ripened cheese enriched with vegetable ingredients: technology and quality *IOP Conf. Series: Earth and Environmental Science* **548(8)** 82063

[14] Serafini M and Peluso I 2016 Functional Foods for health: the interrelated antioxidant and anti-inflammatory role of fruits, vegetables, herbs, spices and cocoa in humans *Curr. Pharm. Des.* **22(44)** 6701-6715

[15] Kryuchkova V V, Gorlov I F, Belik S N and Kamlatsky A S 2020 Vegetable ingredients in functional fermented milk products *IOP Conf. Series: Earth and Environmental Science* **548(8)** 82092

[16] González-Moralejo S A 2011 Approach to understanding a natural sweetener alternative, Stevia rebaudiana bertoni: production, consumption and demand potential *Agroalimentaria* **17** 57-69

[17] Yadav A K, Singh D, Dhyani D and Ahuj P S 2011 A review on the improvement of S. rebaudiana [S. rebaudiana (Bertoni)] *Can. J. Plant* Sci. **91** 1–27

[18] Carrera-Lanestosa A, Moguel-Ordóñez Y and Segura-Campos M 2017 Stevia rebaudiana Bertoni: a natural alternative for treating diseases associated with metabolic syndrome *J. Med. Food* **20(10)** 933-943

[19] Rojas E *et al* 2018 Stevia rebaudiana Bertoni and its effects in human disease: emphasizing its role in inflammation. Atherosclerosis and metabolic syndrome *Curr. Nutr. Rep.* **11**

[20] Kryuchkova V V, Gorlov I F, Slozhenkina M I, Lomonova N V and Belik S N 2021 Functional curd product of increased biological value *IOP Conf. Series: Earth and Environmental Science* **677** 32073

IOP Conf. Series: Earth and Environmental Science **965** (2022) 012026 doi:10.1088/1755-1315/965/1/012026

The effect of mineral complexes on the growth intensity of young bulls for sustainable agriculture

S I Nikolaev, A V Randelin, A K Karapetyan, S N Kupriyanov, S V Chekhranova and E A Morozova

Volgograd State Agrarian University, 26 University Avenue, Volgograd, 400002, Russia

E-mail: kafedra-kormlenie@volgau.com

Abstract. The article presents the material on the study of the influence of mineral granular complexes developed from mineral raw materials at the enterprise of JSC "United Chemical Company "URALCHEM" according to different recipes on the growth intensity, clinical-physiological and hematological parameters of Ayshir bulls. It was found that the young animals that received a mineral granular complex, developed according to four recipes, surpassed the analogues from the control in live weight by 7.17, 5.44, 6.01 and 8.56%, average daily growth by 17.61, 13.19, 14.74 and 19.82%. In the blood of young animals of the experimental groups, in comparison with analogues from the control, an increase in the content of erythrocytes by 0.53-5.74% and hemoglobin by 2.8, 0.4, 1.6 and 3.3 g/l were noted. The total protein in their blood serum contained more by 4.79, 1.67, 2.47 and 8.97% and albumins, respectively, by 10.15, 6.25, 3.82 and 17.02%. The use of mineral granular complexes in the feeding of bulls is economically expedient. Additional products for 32844-49368 rubles were obtained for groups of bulls who consumed mineral granular complex. The complexes developed according to recipes No. 1 and 4 showed the highest efficiency.

1. Introduction

One of the important factors contributing to increasing the productivity of farm animals and preserving their health can be considered a full feeding that ensures their nutritional needs [1, 2, 3]. According to scientists [4, 5, 6, 7], in the diets of animals, especially highly productive animals, as a rule, there is a deficiency of individual or a whole complex of nutrients, which negatively affects the intensity of growth, productivity, increases feed costs and production costs.

Such substances include trace elements. Balancing rations for missing minerals can significantly increase the efficiency of livestock production [8, 9, 10, 11, 12].

In the works of a number of scientists, the high efficiency of the use of complex mineral feedings in animal husbandry, including mineral granular complexes, is noted [13, 14, 15, 16].

Since their use allows not only to balance the diet in terms of minerals, but also simplifies the technology of their use.

The purpose of the research was to study the effect of mineral granular complexes developed at the JSC "United Chemical Company "URALCHEM" on the growth intensity, clinical-physiological parameters, morphological and biochemical composition of the blood of bulls raised for meat.

2. Materials and methods

Experimental work on the study of the effectiveness of feeding mineral granular complexes to bulls based on domestic raw materials, including urea, monoammonium phosphate, chemical chalk, brucite, sodium chloride, was carried out on the farm of JSC "Agrofirm "Vostok" of the Nikolaevsky district of the Volgograd region.

In preparation for the experiment, a zootechnical analysis of feed was carried out. The study of the chemical composition of feed and mineral granular complex was carried out according to generally accepted and original methods.

The formation of experimental groups of bulls was carried out by the method of pairs of analogues. The experimental bulls were kept loosely separately into groups, the study was carried out according to the scheme presented in Table 1.

Table 1. Scheme of experience.

Group	n	Feeding features	Introduced mineral granular complex, g/head
Control	12	Basic diet (BD)	
Experimental:			
I	12	BD + mineral granular complex (developed according to recipe 1)	50.0
II	12	BD + mineral granular complex (developed according to recipe 2)	75.0
III	12	BD + mineral granular complex (developed according to recipe 3)	62.5
IV	12	BD + mineral granular complex (developed according to recipe 4)	75.0

The bulls of the experimental groups consumed a basic diet (BD) consisting of: alfalfa hay 1.3-1.5 kg, grain-legume haylage 2.5-5.0 kg, corn silage 5-9 kg, grain mixtures 1.5-2.5 kg and mixed feed 1.5-2.5 kg.

The nutritional value of the BD was 4.4-5.4 energy feed units (EFU), metabolic energy 44-55 MJ, dry matter 5.3-6.4 kg, crude protein 860-900 g, digestible protein 561-586, crude fiber 1134-1326 g, sugars 505-522 g, crude fat 210-267 g, calcium 40-49 g, phosphorus 25-28 g.

The bulls of the I, II, III and IV experimental groups were additionally introduced mineral granular complexes developed for young cattle by the JSC "United Chemical Company "URALCHEM" according to separate recipes (Table 2).

When studying the effect of feed additives on the body of young animals, special attention is paid to its growth and development. The live weight of the bulls was determined by their weekly weighing. The absolute and average daily increase was determined by the calculation method.

The duration of the scientific and economic experience was 125 days. For the research, 5 groups of experimental Ayshir bulls were formed at the age of 6 months with 12 heads each.

Table 2. Chemical composition and mineral raw materials used in the manufacture of mineral granular complexes.

Composition of components	Number of recipes			
	No. 1	No. 2	No. 3	No. 4
Urea, %	23.30	20.30	27.80	29.40
Monoammonium phosphate, %	37.20	34.80	27.80	29.40
Chalk, %	18.60	23.20	33.30	29.40
Brucite, %	7.00	7.20	11.10	11.80

Composition of components	Number of recipes			
	No. 1	No. 2	No. 3	No. 4
Sodium chloride, %	14.00	14.50	0	0
	1 kg of the complex contains:			
Nitrogen, g	152.29	135.55	161.80	171.11
Calcium, g	54.20	56.52	128.71	114.04
Phosphorus, g	99.47	93.06	74.34	78.62
Magnesium, g	32.67	33.59	51.76	54.97
Potassium, g	1.66	1.94	0.34	0.36
Sodium, g	71.53	89.22	0	0
Iron, mg	143.93	136.65	134.66	135.43
Copper, mg	0.41	0.41	0.52	0.51
Zinc, mg	4.09	4.21	5.86	5.81
Manganese, mg	13.88	13.43	15.61	15.80
Lead, mg	0.24	0.25	0.38	0.40
Chromium, mg	0.13	0.13	0.17	0.17
Nickel, mg	0.94	0.96	1.88	1.69

3. Results and discussion

During the research, it was found that mineral granular complexes introduced into the diet of animals had a growth-stimulating effect (Table 3).

Table 3. Indicators of growth intensity of experimental bulls (M ± m).

Group	Live weight indicators, kg		Growth	
	At the beginning of the experience	At the end of the experience	Gross, kg	Average daily, g
Control	179.2 ± 2.1	301.3 ± 3.1	122.1 ± 2.3	976.8 ± 32.9
Experimental:				
I	179.9 ± 1.9	323.5 ± 2.4	143.6 ± 1.8	1148.8 ± 35.1
II	179.5 ± 1.4	317.7 ± 2.8	138.2 ± 2.6	1105.6 ± 38.3
III	179.3 ± 1.7	319.4 ± 3.0	140.1 ± 2.0	1120.8 ± 31.7
IV	180.8 ± 1.5	327.1 ± 2.1	146.3 ± 1.6	1170.4 ± 30.4

So, when setting up an experiment, the live weight indicators of experimental bulls varied within the sampling error.

At the end of the experiment, the young animals of the experimental groups who received a mineral granular complex with a diet exceeded the analogues from the control in live weight with a significant difference. Thus, the bulls of the I experimental group, when removed from the experiment, had a live weight of 323.5 kg and exceeded the analogues of the control group by 22.2 kg or 7.17% ($P > 0.999$)., The live weight was greater than that of the control analogues by 16.4 kg or 5.44% ($P > 0.99$) in the bulls of the II experimental group, by 18.1 kg or 6.01% ($P > 0.99$) in the III experimental group and by 25.8 kg, or 8.56% ($P > 0.999$) in the IV experimental group.

The main indicators characterizing the intensity of growth of young animals are the gross and average daily growth in their live weight during a certain period of development. At the same time, the gross growth in live weight in young animals of the I experimental group was higher, compared with analogues from the control by 22.2 kg or 17.61% ($P > 0.999$), in the II experimental group - by 16.4 kg or 13.18% ($P > 0.99$), in the III experimental group - by 18.1 kg or 14.74% ($P > 0.99$) and in the IV experimental group - by 25.8 kg or 19.82% ($P > 0.999$).

IOP Conf. Series: Earth and Environmental Science **965** (2022) 012026 doi:10.1088/1755-1315/965/1/012026

The most informative indicator of the intensity of growth of young animals is the average daily growth in live weight. In our studies, the average daily growth of bulls was higher than that of analogues from the control group by 172.0 g or 17.61% ($P > 0.95$) in the I group, by 128.8 g or 13.19% ($P > 0.95$) in the II group, by 144.0 g or 14.74% ($P > 0.99$) in the III group and by 193.6 g or 19.82% ($P > 0.99$) in the IV group.

Consequently, additional feeding with the diet of the studied mineral granular complexes to the bulls allowed to increase their growth rates, in comparison with analogues from the control. At the same time, the highest effect of increasing their growth intensity was observed in young animals of the IV experimental group who received a mineral granular complex in addition to the diet, developed according to recipe No. 4 and the II experimental group who received a mineral granular complex, developed according to recipe No. 2.

It is known that the effect of unfavorable factors on the body of animals, including the consumption of certain feed products, can change their physiological state (Table 4).

Table 4. Clinical and physiological parameters of experimental animals (M ± m).

Indicator	Group				
	Control	Experimental			
		I	II	III	IV
Body temperature, °C	38.4 ± 0.9	38.6 ± 0.85	38.3 ± 1.1	38.6 ± 0.79	38.5 ± 0.95
Heart rate per minute	55.5 ± 0.82	54.3 ± 1.12	55.8 ± 0.85	54.9 ± 0.96	56.1 ± 0.8
Respiratory rate per minute	22.5 ± 0.88	22.0 ± 0.73	23.0 ± 0.92	22.0 ± 0.86	23.5 ± 0.78
Rumination in 2 minutes	2.8 ± 0.38	2.5 ± 0.42	2.9 ± 0.35	2.8 ± 0.31	3.0 ± 0.51

We studied the effect of the mineral complexes used on the clinical-physiological state and hematological composition of experimental young animals. It was found that the indicators of body temperature, heart rate, respiratory rate and rumination in experimental bulls corresponded to physiological norms.

On average, by groups, the body temperature of bulls varied from 38.4-38.6 °C, heart rate - 54.3-56.1 beats, respiratory rate - 22.0-23.5 beats per minute.

At the same time, there was a slight tendency to increase body temperature, respiratory rate and heart in animals that consumed a mineral granular complex with a diet.

The study of the morphological composition of the blood of experimental bulls showed that it is within the physiological norm (Table 5).

Table 5. Morphological composition of blood of bulls (M ± m).

Group	Indicator		
	Erythrocyte, 10^{12}/l	Leukocyte, 10^{9}/l	Hemoglobin, g/l
Control	7.48 ± 1.22	6.69 ± 1.20	99.8 ± 3.90
Experimental:			
I	7.80 ± 0.98	6.45 ± 1.22	102.6 ± 2.85
II	7.52 ± 1.64	6.58 ± 1.50	100.2 ± 3.76
III	7.68 ± 1.52	6.52 ± 1.36	101.4 ± 3.12
IV	7.91 ± 1.36	6.42 ± 1.61	103.1 ± 2.61

Thus, the content of erythrocytes in the blood of young animals of the experimental groups was higher than that of analogues from the control by 0.53-5.74%. Hemoglobin in the blood of bulls of the I, II, III and IV experimental groups contained more than in the control, respectively, by 2.8, 0.4, 1.6 and 3.3 g/l.

When studying the biochemical parameters of the blood of experimental young animals, certain differences were also established (Table 6).

Table 6. Biochemical blood parameters of experimental bulls (M ± m).

Indicator	Group				
	Control	Experimental			
		I	II	III	IV
Total protein, g/l	72.01 ± 1.64	75.46 ± 1.48	73.21 ± 1.83	73.79 ± 1.42	78.47 ± 1.36
Albumins, g/l	32.91 ± 1.72	36.25 ± 1.27	34.97 ± 2.32	34.17 ± 1.31	38.51 ± 1.83
Globulins, g/l	39.1 ± 2.15	39.21 ± 2.64	38.24 ± 2.41	39.62 ± 2.28	39.96 ± 2.32
Glucose, mmol/l	2.75 ± 0.16	2.82 ±0.14	2.76 ± 0.14	2.80 ± 0.16	2.94 ± 0.18
Calcium, mmol/l	3.21 ± 0.15	3.39 ± 0.14	3.25 ± 0.11	3.36 ± 0.13	3.41 ± 0.12
Phosphorus, mmol/l	1.71 ± 0.06	1.81 ± 0.06	1.72 ± 0.06	1.78 ± 0.06	1.86 ± 0.06

In terms of the amount of total protein in the blood serum, young animals of the I, II, III and IV experimental groups were superior to control analogues by 3.45 g/l or 4.79%, 1.20 g/l or 1.67%, 1.78 g/l or 2.47% and 6.46 g/l or 8.97% ($P > 0.95$), respectively. In terms of albumin content, the superiority in animals of the experimental groups was, respectively, 3.34 g/l or 10.15%, 2.06 g/l or 6.25%, 1.26 g/l or 3.82% and 5.60 g/l or 17.02% ($P > 0.95$). The content of globulins, calcium, phosphorus in the blood varied less significantly among the groups of young animals.

Thus, the introduction of mineral granular complexes into the diet of bulls had a positive effect on the morphological, biochemical and mineral composition of blood.

The most important indicator of the expediency of using feed additives, premixes of growth-stimulating agents, in feeding young animals raised for meat remains economic efficiency. We will calculate the economic efficiency of the use of mineral granular complexes in the diets of bulls (Table 7).

Table 7. Economic efficiency of the use of mineral granular complexes in the diets of bulls.

Indicator	Group				
	Control	Опытные			
		I	II	III	IV
Number of heads	12	12	12	12	12
Safety of livestock, %	100	100	100	100	100
Duration of the experience, days	125	125	125	125	125
Live weight of one head, kg:					
at the beginning of the experience	179.2	179.9	179.5	179.3	180.8
at the end of the experience	301.3	323.5	317.7	319.4	327.1
Gross body weight growth of 1 head during the period of experience, kg	122.1	143.6	138.2	140.1	146.3
The selling price of 1 kg of live weight, RUB	170.0	170.0	170.0	170.0	170.0
Additionally, a live weight gain was obtained, kg:					
per one head	-	21.5	16.1	18.0	24.2
per group	-	258	193.2	216.0	290.4
The cost of additional weight growth, RUB:					
per one head	-	3655	2737	3060	4114
per group	-	43860	32844	36720	49368

The calculation showed that the introduction of mineral granular complexes into the diet of bulls provided additional products and income. The cost of additional products amounted in the I, II, III and IV experimental groups, respectively, 43860, 32844, 36720 and 49368 rubles (based per 12 heads).

4. Conclusion

Consequently, the additional introduction of new mineral granular complexes into the diet of bulls contributed to an increase in their live weight by 16.4-25.8 kg and an average daily increase by 13.19-19.82%.

The hematological and clinical parameters of the experimental young animals varied within the physiological norm. At the same time, an increase in the concentration of erythrocytes, hemoglobin, total protein, calcium and phosphorus in the blood to the level of the upper limit of norms was found, which indicates an intensification of metabolic processes in the body. The use of granular mineral complexes in the diets provided additional products on average for 1 head of bulls from 2737 rubles to 4114 rubles. Mineral granular complexes developed according to recipes No. 1 and 4 showed the highest economic efficiency.

References

[1] Shagaliev F, Shamsutdinov D and Ardashirov S 2019 Mineral nutrition of ruminants *Animal Husbandry of Russia* **S1** 41-42
[2] Duborezov V M 2020 Feeding dairy cows according to detailed norms *Dairy and Meat Cattle Breeding* **4** 52-54
[3] Buryakov N and Hardik I 2021 About the balance of rations for dairy cattle *Compound Feed* **3** 42-46
[4] Abilov A I, Dunin M I and Bogolyubova N V 2021 Mineral exchange in bulls-producers of dairy breeds after a long winter period of exploitation *Animal Husbandry* **2** 20-25
[5] Faber V, Akmaliev T and Guseva O 2020 Mineral nutrition of ruminants *Animal Husbandry of Russia* **5** 30-33
[6] Merphy R 2019 Trace elements, feed and lack of elements in the diet: how the form of a trace element affects the quality of feed and animal health *Animal Husbandry of Russia* **4** 41-44
[7] Ardashirov S, Shagaliev F and Shamsutdinov D 2018 Macronutrients and livestock productivity *Animal Husbandry of Russia* **7** 51-53
[8] Sabitov M T, Malikova M G and Farkhutdinova A R 2019 Influence of the complex mineral and vitamin feed additive "Nadezhda" on the growth of calves *Dairy and Meat Cattle Breeding* **4** 31-34
[9] Gorlov I F, Slozhenkina M I, Randelin A V, Randelina V V, Sutorma O A, Garyaeva Kh B and Mosolova D A 2019 Effect of a new feed additive on meat productivity and quality of meat of bulls *Bulletin of Russian Agricultural Science* **4** 57-60
[10] Nikolaev S I, Karapetyan A K, Chekhranova S V, Danilenko Y I, Rabadanov S R and Struk M V 2019 Premixes and protein vitamin-mineral concentrates in livestock and poultry breeding: technological properties *Int. J. of Engineering and Advanced Technology* **8(6)** 5307-5312
[11] Gorlov I F, Slozhenkina M I, Mosolov A A, Garyaeva Kh B, Randelina V V and Randelin A V 2018 Influence of silicon-containing fodder additive on meat productivity of Kalmyk bulls on fattening *News of Gorsky State Agrarian University* **55(4)** 77-83
[12] Gorlov I F, Slozhenkina M I, Mosolova N I, Zlobina E Y, Randelin A V, Bondarkova E Y, Sherstyuk B A and Belyaev A I 2020 New synbiotic-mineral complex in lactating cows' diets to improve their productivity and milk composition *Iranian J. of Applied Animal Science* **10(1)** 31-43
[13] Sabitov M T 2021 Efficiency of using a complex mineral and vitamin feed additive in the diets of heifers *Animal Husbandry* **1** 18-25
[14] Farkhutdinova A R and Sabitov M T 2021 Influence of a complex mineral and vitamin feed additive for calves on the digestibility of nutrients *Dairy and Meat Cattle Breeding* **2** 40-45

[15] Gorshkov V 2019 Natural components - sources of energy, vitamins and minerals *Animal Husbandry of Russia* **S1** 23-24
[16] Nikolaev S I, Shkalenko V V, Randelin A V, Batrakova Yu M and Syuliev L A 2020 Influence of the mineral granular complex on meat productivity, quality indicators of meat of young pigs for fattening *Proc. of the Nizhnevolzhsky Agrouniversity Complex: Science and Higher Professional Education* **3(59)** 288-297

IOP Conf. Series: Earth and Environmental Science **965** (2022) 012027 doi:10.1088/1755-1315/965/1/012027

Nutritional value and environmental safety of bulls meat of the dairy direction

E V Karpenko[1, 2, 3], A A Kaidulina[1], I F Gorlov[1, 2, 4], E Yu Bondarkova[1], J S Grebennikova[1] and D A Mosolova[4]

[1] Volga Region Research Institute of Manufacture and Processing of Meat-and-Milk Production, 6 Marshal Rokossovsky Street, Volgograd, 400131, Russia
[2] Volgograd State Technical University, 28 Lenin Avenue, Volgograd, 400005, Russia
[3] Volgograd State University, 100 University Avenue, Volgograd, 400062, Russia
[4] Gustave Eiffel University, France

E-mail: niimmp@mail.ru

Abstract. The authors present the results of research, according to the comparative assessment, food and biological value of beef from the purebred bulls of the Holstein breed belonging to different lines. The genetic potential of animals had a certain impact on high-quality indicators of the resulting meat raw materials. It has been established that the maximum amount of protein in the long muscle of the back was in the bulls of the reflective line, respectively, 0.53 and 0.8% compared with the same indicator of other animal lines. Culinary and technological indicators of meat studied groups are defined. The calculation of the amino acid ski was above 100% in all experimental groups, which indicates the usefulness of the obtained meat raw materials. Intergroup differences are noted on the most optimal content of essential amino acids, which slightly affected the biological value of meat.

1. Introduction

One of the main food products for the population of our country is beef, due to the content in it of the required amount of protein of animal origin. The value of meat is determined by its chemical composition, namely the content of proteins, fats, vitamins, macro and microelements. The biological value of proteins of animal origin is undeniable due to the balanced amino acid composition and, accordingly, better assimilation by the human body [1, 2, 3, 4, 5].

At the same time, it should be noted that beef production lags behind the level of its consumption. If according to the norms, according to the development of the Institute of Nutrition of the Academy of Medical Sciences, it is necessary for each person to consume 40 kg of beef per year, then at present this norm is half satisfied.

Therefore, at present, the question of finding additional sources for increasing beef production in our country remains open. One of these sources can be dairy and combined cattle, and primarily the Holstein breed, which has a large distribution area in various regions of our country [6].

2. Materials and methods

The aim of our research was a comparative assessment of the nutritional and biological value of meat obtained from bulls of the Holstein breed of German selection of different linearity.

The experimental part of the work was carried out at LLC "Agricultural Enterprise "Donskoe", Volgograd Region. Three groups of Holstein gobies belonging to different lines were selected for the experiment: Reflection Sovering 198998 (group I), Vis Back Aydial 933122 (group II) and Montvik Chieftain 933122 (group III), 10 heads each. After removing the bulls from fattening at the age of 13 months, a control slaughter of 3 heads from each group was carried out.

The control slaughter of animals was performed to study the beef production. The chemical composition of the eye muscle of loin was examined in accordance with the guidelines. The content of complete and imperfect proteins as well as of essential amino acids was established using the capillary electrophoresis Kapel®-105M ("Lumex", Russia). The data obtained was compared to the reference protein standard of World Health Organization/Food and Agriculture Organization (WHO/FAO, 2007).

The data on different variables, obtained from the experiment, were statistically analyzed by Statistica 10 package (StatSoft Inc.). The significance of differences between the indices was determined using the criteria of nonparametric statistics for the linked populations (differences with $P < 0.05$ were considered significant: ***$P < 0.001$; **$P < 0.01$; *$P < 0.05$; ns = not significant at $P > 0.05$). Student's t-test and Wilcoxon test were applied for the statistical analysis. Regression and Correlation analyses were also computed to establish relationships among various parameters.

The water retention capacity (WRC) was determined by the classical method of Grau and Hamm.

The thermal weight loss (TWL) was expressed as a percentage of the difference between initial weight and after cooking of samples.

The culinary-technological properties (CTP) meat was expressed by the ratio of the storage capacity of the water (WRC) to heat weight loss (TWL).

The amino-acid score (AAS) was calculated as follows:

$$AAS,\% = \frac{\text{Evaluated protein amino acid content (g/100g protein)}}{\text{FAO scoring model amino acid content (g/100g protein)}} \cdot 100 \quad (1)$$

The coefficient of amino-acid score difference (CAAS) was estimated accordance with Nikitina et al. by the following formula:

$$CAAS,\% = \frac{\sum (C_i - C_{\min})}{\text{n}} \quad (2)$$

where C_i is the excess of amino-acid score, %; C_{min} is the minimum score of essential amino acids of evaluated protein relative to the physiological norm (FAO scoping model amino acid content), %; n is the number of essential amino acids.

The biological value (BV) was calculated by the following formula (3):

$$BV,\% = 100 - CAAS \quad (3)$$

MS Office 2010 package was employed for graphical presentation of the data [7].

Animal care and experimental studies were performed in accordance with the instructions and recommendations of Russian Regulations, 1987 (Order No. 755 on 08/12/1977 the USSR Ministry of Health) and "The Guide for Care and Use of Laboratory Animals (National Academy Press Washington, DC 1996). Efforts have been made in the studies to minimize animal suffering and reduce the number of samples used.

3. Results and discussion

Analysis of the results of the morphological composition of the carcasses of the experimental animals showed that the carcass weight of the bulls of group I was 3.22% higher than that of bulls of group II and 8% higher than that of animals of group III. With bulls of groups I and II, this difference was 10.9 and 14.9 kg, respectively. The pulp yield by groups was 82.3, respectively; 81.6; 81.5; the rest came from bones and tendons.

As a result of studies of the chemical composition of meat (Table 1), namely the longest muscle of the back, it was found that the dry matter was higher in animals of group I in comparison with animals from other groups by 0.29 and 0.8%, and there was also more protein by 0.53 and 0.8%, respectively. However, fat, on the contrary, was higher in animals of the II and III experimental groups. In terms of water-holding capacity and digestibility of meat, the animals belonging to the Reflection Sovering line had advantages. Thus, the belonging of bulls to one or another line affects the chemical composition of *the longissimus dorsi*, which directly affects the culinary and technological properties of meat.

Table 1. Chemical composition and culinary and technological properties of the long muscle of the backs of experimental bulls.

Indices	I group			II group			III group		
	Mean	Standard error of mean	P	Mean	Standard error of mean	P	Mean	Standard error of mean	P
Dry matter, %	24.65	0.52	ns	24.39	0.44	ns	23.85	0.40	ns
Protein, %	18.54	0.34	ns	18.01	0.29	ns	17.74	0.18	ns
Fat, %	5.04	0.06	ns	5.34	0.02	**	5.12	0.03	**
WRC, %	67.09	1.21	ns	66.44	1.71	ns	65.67	0.87	ns
TWL, %	32.64	0.15	*	34.98	0.24	**	35.12	0.33	**
CTP		2.05			1.89			1.86	

Analysis of the results of studies of the amino acid composition of the longest muscle of the back of experimental bulls, namely the essential amino acids, showed high indicators and compared with the recommended values of FAO/WHO, one can argue about the balance of muscle tissue of all experimental animals and their complete digestibility by the body (Table 2).

Table 2. The content of essential amino acids in the long muscle of the bulls of the Holstein breed of different linear affiliation.

Amino acid	Reference standard of FAO/WHO	I group			II group			III group		
		Mean	Standard error of mean	P	Mean	Standard error of mean	P	Mean	Standard error of mean	P
Valine	50	51.18	0.09	ns	51.02	0.05	*	52.66	0.11	*
Lysine	55	55.66	0.13	*	55.18	0.08	ns	55.11	0.09	ns
Leucine	70	71.22	0.12	*	70.58	0.09	**	70.87	0.16	*
Isoleucine	40	40.45	0.05	*	40.21	0.07	ns	40.16	0.04	**
Phenylalanine + tyrosine	60	60.25	0.13	ns	60.29	0.11	ns	63.94	0.14	*
Methionine + cystine	35	36.14	0.08	**	37.16	0.11	ns	35.77	0.09	**
Tryptophane	20	31.26	0.09	ns	31.14	0.10	*	30.69	0.11	*
Threonine	40	40.21	0.14	*	41.15	0.12	ns	40.01	0.12	*

However, it should be noted about the intergroup differences in the content of essential amino acids. Thus, the amino acid valine, which is necessary for tissue metabolism and maintenance of optimal nitrogen metabolism in the body, together with leucine and isoleucine, serves as a source of energy in muscle cells, and also prevents a decrease in serotonin levels. In our study, a comparative analysis showed that there was more valine in group III compared with peers in groups I and II by 2.81

and 3.11%, phenylalanine + tyrosine by 5.77 and 5.71%, and leucine and isoleucine, on the contrary. contained less by 0.49 and 0.41%, 0.72 and 0.12%, respectively.

The amino acid lysine is necessary for the development of the body, increases the regenerative properties of bone tissues and promotes growth, its higher content was found in group I, the difference in comparison with groups II and III was 0.9 and 0.49%, as well as for the essential amino acid tryptophan 0.38 and 1.86%, respectively, the latter plays an important role in metabolism, and directly affects the toughness of meat.

As for methionine + cystine and threonine, which are responsible for the activation of lipids, proteins, carbohydrates involved in the accelerated assimilation of other amino acids, the advantage was in group II.

The biological value of meat depends mainly on the content and ratio of its constituent essential amino acids [8]. Analysis of the amino acid score showed that no limiting amino acids were found in the meat proteins of the studied groups (Figure 1).

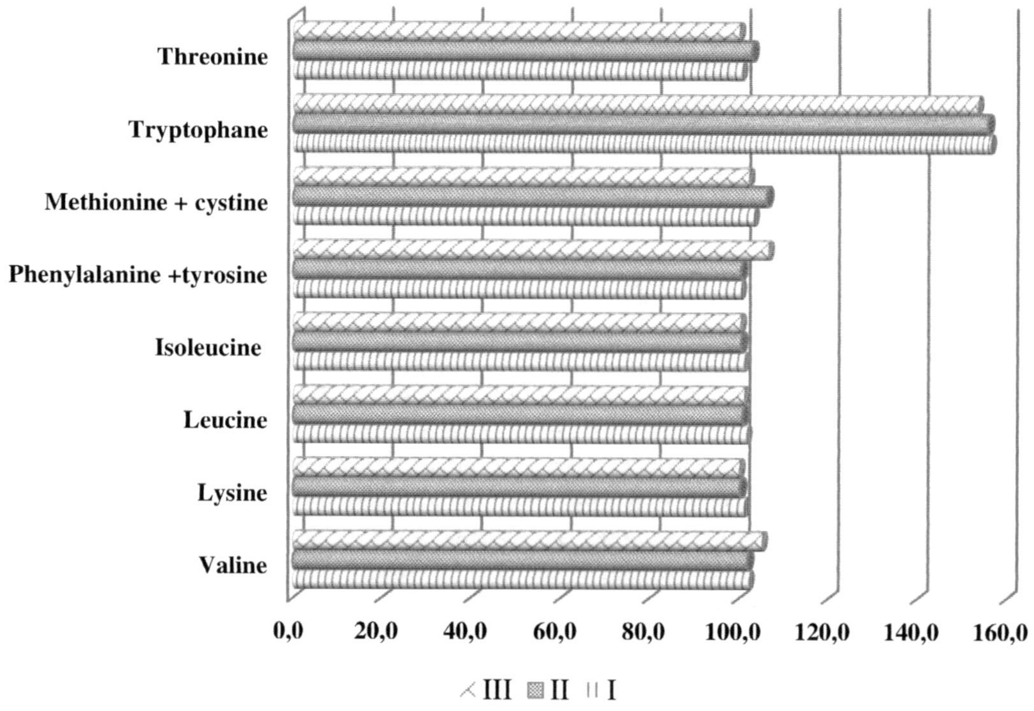

Figure 1. Amino-acid score, %.

To determine the biological value of meat, we calculated the coefficient of difference of amino acid score (CAAS), which shows the average amount of excess amino acid score of essential amino acids compared to the lowest score of any essential amino (Table 3).

It was found that a high average value of the amino acid score of the protein of the longest muscle of the back was in animals of group III, respectively, and a higher increase in the coefficient of difference in amino acid score, in comparison with groups I and II was 3.60 and 0.58%, 3.57 and 0.69%.

However, calculations of the biological value showed that bulls from group I had a more balanced ratio of essential amino acids. It is clearly seen that a significant increase in one of the amino acids led to a decrease in the biological value, in group II due to methionine + cystine and threonine, in group III - valine and phenalalanine + tyrosine.

Table 3. Biological value of proteins of the longest muscle of the back of Holstein gobies of different linearity.

Amino acid	I group			II group			III group		
	Mean	Standard error of mean	P	Mean	Standard error of mean	P	Mean	Standard error of mean	P
Valine	2.4	0.16	ns	2.0	0.13	*	5.3	0.19	ns
Lysine	1.2	0.09	***	0.3	0.01	**	0.2	0.04	**
Leucine	1.7	0.09	**	0.8	0.07	***	1.2	0.21	*
Isoleucine	1.1	0.06	ns	0.5	0.01	ns	0.4	0.01	ns
Phenylalanine + tyrosine	0.4	0.07	ns	0.5	0.08	**	6.6	0.14	**
Methionine + cystine	3.3	0.10	***	6.2	0.24	*	2.2	0.18	**
Tryptophane	56.3	1.01	ns	55.7	1.14	ns	53.5	0.88	ns
Threonine	0.5	0.06	***	2.9	0.11	*	0.0	0.10	***
Difference of amino acid score	66.9	1.12	ns	69.0	1.16	ns	69.4	1.15	ns
CAAS	8.37	0.13	ns	8.62	0.29	ns	8.68	0.19	ns
Biological value	91.63	1.45	ns	91.38	1.59	ns	91.32	1.65	ns

4. Conclusion

Thus, the results of the studies carried out showed high indicators of the food and biological value of meat of the studied groups, it can be described as a full-fledged chemical and balanced according to the composition of essential amino acids. In our study, differences in the quality characteristics of meat are established depending on the linear affiliation of the bulls. It has been proven that for an additional source of receiving beef can be used on the fattening of Holstein breed bulls.

5. Acknowledgments

This work was carried out as part of a state assignment of the Ministry of Science and Higher Education, No 1021032426917-2.

References

[1] Bartoň L, Řehák D, Teslík V, Bureš D and Zahrádková R 2006 Effect of breed on growth performance and carcass composition of Aberdeen Angus, Charolais, Hereford and Simmental bulls *Czech J. of Animal Science* **51(2)** 47-53

[2] Vostrikova N L, Lisitsyn A B, Chernukha I M. and Ivankin A N 2013 Investigations of proteins' usefulness in the different types of beef's muscle *Vsyo o myase* **2** 34-38

[3] Kryuchkova V V, Gorlov I F, Korneichuk K M, Belik S N and Mosolova N I 2020 Brine-ripened cheese enriched with vegetable ingredients: technology and quality *IOP Conf. Series: Earth and Environmental Science* **548(8)** 82063

[4] Kryuchkova V V, Gorlov I F, Belik S N and Kamlatsky 2020 As vegetable ingredients in functional fermented milk products *IOP Conf. Series: Earth and Environmental Science* **548(8)** 82092

[5] Gorlov I F, Slozhenkina M I, Bozhkova S E, Belova D S, Vorontsova E S 2020 Analysis of the cheese suitability of raw milk and the quality of fortified cheese products *Proc. of the Nizhnevolzhsky Agrouniversity Complex: Science and Higher Professional Education* **3(59)** 258-267

[6] Kibkalo L I, Kuravtsova T E, Kolesnikov A V, Gudymenko V I, Bugaev S P, Lyashuk R N and

Sein O B 2019 The influence of line accessory of bulls on meat quality *Int. J. of Advanced Biotechnology and Research* **10(1)** 26-32

[7] Gorlov I, Karpenko E, Zlobina E and Azhmuldinov E 2017 Comparative assessment of nutritional and biological value of beef from calves of various breeds *Engineering for Rural Development Proc.* 254-262

[8] Makaev Sh A 2020 Amino acid composition of the longest back muscle of bull calves of different factory lines of the kazakh white-headed breed *Proc. of the Orenburg State Agrarian University* **4(84)** 261-266

IOP Conf. Series: Earth and Environmental Science **965** (2022) 012028 doi:10.1088/1755-1315/965/1/012028

Environment and sheep wool quality indicators

P S Ostapchuk[1], V S Pashtetsky[1], E N Usmanova[1], T A Kuevda[1], E Yu Zyablitskaya[2], T P Makalish[2] and J S Saenko[2]

[1] Federal State Budget Scientific Institution "Research Institute of Agriculture of Crimea", 150, Kievskaya Street, Simferopol, 295493, Russia
[2] V I Vernadsky Crimean Federal University, Medical Academy Named after S I Georgievsky, 5/7 Lenin Boulevard, Simferopol 295051, Russia

E-mail: ostapchuk_p@niishk.ru

Abstract. The features of the skin histology and the determination of the quality indicators of wool of young female Tsigai breed sheep are presented in the article. A positive correlation was found between the fineness of the wool and the live weight ($p \leq 0.05$); between the shearing of natural wool and the fineness of the wool ($p \leq 0.05$); between the shearing of natural wool and the live weight ($p \leq 0.05$); between the washed wool and the fineness of the wool ($p \leq 0.05$) and the shearing of natural wool ($p \leq 0.05$). There is no reliable relationship between the fineness and the percentage of the washed wool shearing. However, the percentage of the shearing of washed wool is negatively related to the live weight ($p \leq 0.05$), the shearing of natural wool. The greatest thickness of the skin was on the back and leg are 968.07 and 968.24 microns respectively. It is assumed that the skin regeneration occurs due to the cells of the basal layer. This is due to the fact that the thickness of the basal layer decreases in the following order: side-back-leg-neck, which correlates with the degree of exposure to damaging factors on these areas.

1. Introduction

The development of sheep breeding for Crimea in scientific work should be solved on the basis of modern effective research methods that have been successfully tested by scientists in the southern regions of the Russian Federation [1,2, 3, 4, 5].

Sheep skin is an important object of study [6, 7]. Histological studies make it possible to judge the constitutional and breed affiliation, to carry out a biological assessment of various technologies for keeping animals, to assess the level of productivity, to assess the physical and technological qualities of sheepskins, the age of animals, etc. [8, 9, 10].

A significant influence on the formation of histological features of the skin is also due to genetic factors. Minor skin morphogenesis in the postnatal period is associated with the influence of various paratypical factors [11, 12]. Sheepskins are an important technological raw material as well [13].

Different breeds of sheep have different coats and are characterized by different morphogenesis at the histological level of the structural organization of the skin, according to many scientists. The study of breed differences in the structure of the skin and wool cover of sheep in different productivity directions is of theoretical and practical importance [14].

Since wool in sheep farming is considered the most energy-intensive component [15], it is important to widely implement a system of complex selection and technological techniques to improve the quality of fine wool produced [16].

Sheep of the Tsigai Breed is given an important place in the conditions of the Crimean steppe historically. Tsigai is also widely spread in areas of South Ukraine, in some regions of Russia, Kazakhstan, and Eastern Europe [17].

Intensive breeding of various breeds of farm animals often leads to the loss of unique, historically formed, genotypes. Not well-thought-out selection to increase the productive characteristics often leads to the loss of the hereditary characteristics of local breeds [18]. Thus, selection and breeding work should be carried out taking into account the preservation of existing animal breeds of the domestic gene pool and their valuable traits [19]. One of these features in the breed is the wool productivity of Tsigai Breed sheep, local to the Crimean Peninsula [20].

Based on the analysis of literary sources, the Tsigai Breed still occurs on the Eurasian continent. The animals are characterized by satisfactory indicators of live weight and average shearing of wool, achieved in the course of breeding work with herds in the last few decades. Breeders in the Tsigai Breed fixed the indicator of the fineness of wool fiber in the range from 23 to 40 microns. Due to the decrease in the relative economic value of wool in the total cost structure to 0.3%, breeders and practitioners in the area of distribution of the Tsigai Breed, are actively working to improve the wool quality [21].

The study of the histological characteristics of the skin and the determination of the characteristics of wool productivity is important in the selection work with the Tsigai Breed at the present stage; also it was the main goal of our research.

2. Materials and methods

The conditions for keeping sheep are pasture in the period from March to the first half of December. The rest of the period is stable maintenance. Therefore, the environment has an impact on the formation of the wool productivity of young female's sheep in the conditions of the Steppe Crimea, certainly.

The work was carried out on samples of wool of the Tsigai Breed young female (n = 30), taken for analysis in period 2019-2020, obtained in LLC "Southern Crimean Sheep Breeding". Skin samples were taken (n = 5) by biopsy. Accounting indicators: qualities of wool; histological indicators of the skin of the young female.

Samples of wool fibers were taken in the following places: 20-30 g of the sample from the sides of the trunk, 10-15 g of the sample from the back of the trunk, and 5-10 g of the sample from the leg and neck of the sheep. The studied parameters were also: shearing of natural and washed wool, the percentage of washed wool yield, the length of wool fibers, the thickness of wool fibers [22].

The percentage of the yield of clean (washed) wool was determined by the following formula (1):

$$B = \frac{M_2 \times 117}{M_1} \tag{1}$$

where B is the percentage of pure wool yield, %; M_2 is residual mass of the sample after washing, g; M_1 is the initial mass of the unwashed sample, g [22].

The material for histological examination in the Logos microwave *Milestone* histological processor was studied. The finished preparations were photographed on a *Leica Microsystems* DM 2000 microscope with N Plan 10×/0.25 and N Plan 40×/0.65 lenses. Skin layers were measured in the *ImageJ* program using the Stage Micrometer scale (TS-M1 P/N 106011) [23].

Statistical data processing was carried out with the calculation of the average value of the attribute, the statistical error, and the reliability of the difference according to the Student's criterion.

3. Results and discussion

The features of the distribution of the fineness of the wool fibers on different surfaces of the body of the young female are shown in Table 1.

The coarsest wool fibers are marked on the back, which significantly exceeds all other topographical areas of the body of the Tsigai Breed young female: in comparison with the data of the wool fineness on the back, this difference was 10.27 ($p < 0.05$) μm, on the neck was 12.4 ($p < 0.01$) μm and on the side was 14.1 ($p < 0.001$) μm (Table 1, Figure 1).

Table 1. The features of the distribution of the fineness of the wool fibers on different surfaces of the body of the Tsigai Breed young female

Statistical indicator	The wool fibers fineness, μm			
	leg	back	neck	side
$X \pm m_X$	35.3 ± 1.4	25.0 ± 1.3*	22.9 ± 1.1**	21.2 ± 0.8***
σ	4.5	4.4	3.6	2.6
C_v, %	12.8	17.5	15.5	2.3

Note: confidence levels calculated by the Student's criterion: * - $p \leq 0.05$; ** - $p \leq 0.01$; *** - $p \leq 0.001$.

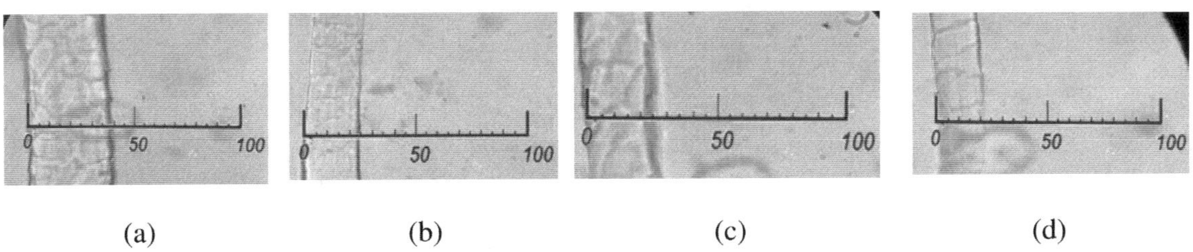

Figure 1. Micrographs of Tsigai Breed young female wool fibers from different topographic areas: (a) – from the leg (35–36 μm); (b) – from the back (24–25 μm); (c) – from the neck (21–22 μm); (d) – from the side (21 μm).

Indicators of wool productivity are shown in Table 2. The highest coefficient of variation is observed in the fineness of wool, therefore, the existence of the average level of variability causes to a certain extent, the efficiency of breeding work with Tsigai Breed herd the sheep towards the improvement of this important indicator.

Table 2. Indicators of wool productivity of the Tsigai Breed young female.

Statistical indicator	The average fineness of the wool, μm	Live weight, kg	Wool cut, kg		Percentage of cut of washed wool, %
			unwashed (natural)	washed	
$X \pm m_X$	26.1 ± 0.7	40.0 ± 0.5	4.0 ± 0.1	1.9 ± 0.03	57.2 ± 0.6
σ	2.4	1.7	0.3	0.1	1.8
C_v, %	9.1	4.2	6.9	4.2	3.2

The calculated indicators of the interactions of the studied characters of wool productivity and quality of the wool are shown in Table 3.

Positive and significant correlation between wool fineness and live weight ($p \leq 0.05$) was stated; positive and reliable production of natural wool correlated with wool fineness ($p \leq 0.05$) and live weight ($p \leq 0.05$), and washed wool cut with only with wool fineness ($p \leq 0.05$) and natural wool cut ($p \leq 0.05$). There is no reliable connection between the fineness and the percentage of cut of washed wool. However, the percentage of cut of washed wool is negatively related to such indicators as live weight ($p \leq 0.05$), natural wool cut ($p \leq 0.001$) and washed ($p \leq 0.05$) wool cut.

Table 3. Correlation of indicators of wool productivity indicators of quality of wool of Tsigai Breed young female.

	X2	X3	X4	X5
X1	0.68 ± 0.24*	0.60 ± 0.27*	0.68 ± 0.25*	-0.34 ± 0.31
X2		0.59 ± 0.27*	0.49 ± 0.29	-0.58 ± 0.27*
X3			0.92 ± 0.13***	-0.87 ± 0.17***
X4				-0.61 ± 0.26*

Note: confidence levels calculated by the Student's criterion: * - $p \leq 0.05$; ** - $p \leq 0.01$; *** - $p \leq 0.001$: * - $p \leq 0.05$; ** - $p \leq 0.01$; *** - $p \leq 0.001$. Symbols of the indicators studied in the correlation: X1 - wool fineness; X2 - live weight; X3 - cut of unwashed (natural) wool; X4 - cut of washed wool; X5 - percentage of cut of washed wool.

The skin of the studied Tsigai Breed young female at the histological level has a typical edifice. Figure 2 shows the characteristics of the main layers of the skin.

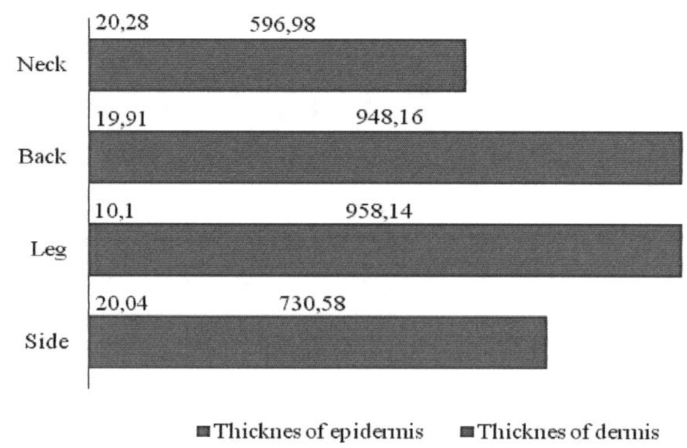

Figure 2. Thickness of the main skin layers.

The greatest thickness of the skin on the back and leg was 968.07 and 968.24 microns respectively. The skin was thinner on the side (751.62 ± 12.6 microns), and the thinnest is on the neck (617.18 ± 10.5 microns). The dermis was proportionally reduced at the same time.

The epidermis consists of keratinized, granular, spinous, and basal layers. Below the epidermis is the dermis, extending papillae into the epidermis. Among the connective tissue fibers of the dermis, there are hair follicles with sebaceous glands adjacent to them. There are also sweat glands. The thickness of the skin layers varies depending on the topography (Table 4).

Table 4. Characteristics of the skin layers of the Tsigai Breed young female (n = 5).

	Layer			
	keratinized (stratum corneum)	granular (stratum granulosum)	spinous (stratum spinosum)	basal (stratum basale)
Side	2.93±0.36▲■	4.15±0.44▲	11.52±0.55▲	1.79±0.17▲*
Leg	1.55±0.17●	1.64±0.14●*	6.78±0.14●■*	0.59±0.05●■
Back	1.71±0.17●	2.61±0.14	14.62±0.39▲	1.03±0.08▲*
Neck	2.04±0.16	3.77±0.52▲	14.47±0.67▲	0.58±0.04●■

Note: confidence levels calculated by the Student's criterion: $p \leq 0.05$. There is a significant difference in comparison with the: ● – side; ▲ – leg; ■ – back; * - neck.

The thickness of the epidermis was not directly related to the thickness of the skin as a whole. The thickness of the epidermis on the leg was the smallest, amounting to 10.10 ± 0.29 microns; however, in the other areas, the thickness of the epidermis was close to 20 microns. These differences are statistically significant with an error probability of less than 5%. The reticular layer of the skin of the leg and neck was characterized by a more friable arrangement of fibers.

The epidermis in the studied areas consisted of four layers. The greatest contribution to the thickness of the epidermis in all cases was the spinous layer. It was the smallest on the leg (6.78 ± 0.14) and significantly different from other areas. The granular layer has a second thickness; it is on the leg proved to be the most subtle. Thus, the regeneration of the skin occurs due to the cells of the basal layer. Its thickness decreased in various areas in the following order: side-back-leg-neck. This is adequate to the extent of the impact of damaging factors on these areas.

The thickness of the basal layer did not correlate with the thickness. However, the greatest thickness was also reached in the side area, amounting to almost 3 microns, which was significantly different from the back and leg area, where its thickness was the smallest.

In general, the most distinctive structure was the skin in the limb area. Most of all, it was different from the skin in the side area (Figure 3, Figure 4).

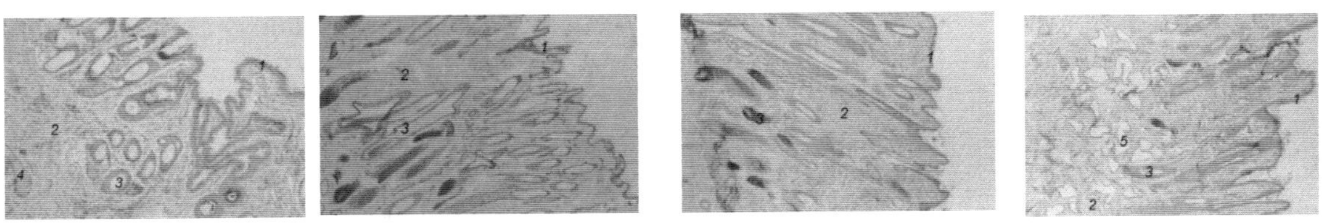

(a) (b) (c) (d)

Figure 3. Young female sheep skin in different areas of the body: (a) is side, (b) is leg, (c) is back, (d) is neck. Hematoxylin-Eosinstaining. Magnification is 100×. 1 - epidermis; 2 - dermis; 3 - hair follicle; 4 - sebaceous gland; 5 - sweat gland.

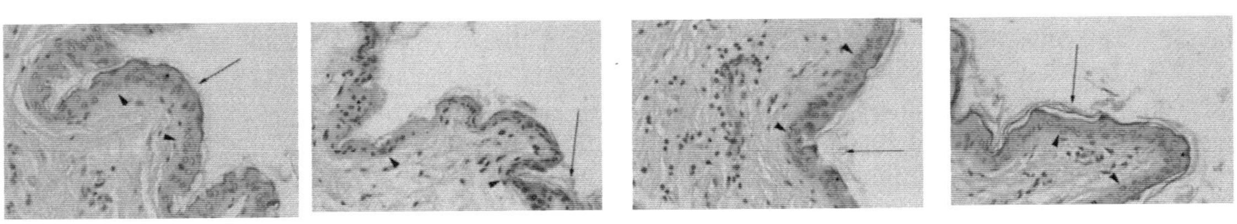

(a) (b) (c) (d)

Figure 4. Young female sheep skin epidermis in various areas: (a) is side, (b) is leg, (c) is back, (d) is neck. Hematoxylin-Eosin staining. Magnification is 400×. The epidermis has 4 layers: horny (long arrow), granular (*), spiny (°) and basal, or sprout (short arrow).

4. Conclusion

A positive correlation was established between the fineness of the wool and the live weight of young female sheep of Tsigai Breed ($p \leq 0.05$). A similar correlation pattern is proved between the haircut of natural wool and the fineness of the wool; between the haircut of natural wool and the living weight; between the washed wool and the fineness of the wool and the haircut of natural wool ($p \leq 0.05$). There is no reliable relationship between fineness and the percentage of washed wool per haircut. However, the percentage of washed wool shearing is negatively related to live weight and natural wool shearing ($p \leq 0.05$-$p \leq 0.001$). The greatest thickness of the skin was on the back and leg: 968.07 and 968.24 microns, respectively. It is assumed that the regeneration of the skin occurs due to the cells of the basal layer. This is due to the fact that the thickness of the basal layer decreases in the following order: side-back-leg-neck, which correlates with the degree of influence of damaging factors on these areas.

5. Acknowledgments

This work was supported by the State Program FSBSI "Research Institute of Agriculture of Crimea" (State Registration Number 0562-2019-0001).

References

[1] Skorykh L, Afanasyev M, Kisyuk V and Konik N 2017 *Engineering for Rural Development* **16**
[2] Degtyarev D Y, Emelyanov S A and Skorykh L N 2016 *RJPBCS* **7(5)**
[3] Degtyarev D Y, Skorykh L N, Kovalenko D V, Emelyanov S A and Konik N V 2016 *RJPBCS* **7** 4
[4] Afanasyev M A, Skorykh L N, Kovalenko D V and Omarov A A 2020 *IOP Conf. Series: Earth and Environmental Science* **488** 1
[5] Katkov K A, Skorykh L N, Pashtetsky V S, Ostapchuk P S and Kuevda T A 2019 *South of Russia: Ecology, Development* **14** 3
[6] Fourneau M, Canon C, Van Vlaender D, Collins M J, Fiddyment S, Poumay and Deparis O 2020 *Herit Sci* **8** 78
[7] Begembekov K N, Alzhaxina N E 2014 *Research Results* **3**
[8] Mobini B 2012 *J. Vet. Adv.* **2** 5
[9] Tuncer S S, Uslu S, Taş A and Şireli H D 2018 *Austral J. of Veterinary Sciences* **50** 1
[10] Gogaev O K, Thukvatulin G S, Kokoev H P, Vaneeva B B and Margieva F T 2018 *RJPBCS* **9** 5
[11] Pogodaev V A, Arilov A N, Mangutov S N and Khechev E V 2021 *IOP Conf. Series: Earth and Environmental Science* **624**
[12] Pogodaev V A, Sergeeva N V and Zavgorodnyaya G V 2019 *Bulletin of NSAU* **2**
[13] Dmitrik I I, Selionova M I and Gadzhiev Z K 2017 *Agricultural Bulletin of Stavropol Region* **1** 25
[14] Jackson N, Maddocks I, Watts J, Scobie D, Mason R, Gordon-Thomson C and Moore G 2020 *Genetics Research* **102** E4.
[15] Kolosov Yu A, Klimenko A I and Aboneev V V 2014 *Sheep, Goats, Woolen Production* **2**
[16] Kolosov Yu A, Aboneev V V, Abdulmuslimov A M and Kiselev A S 2020 *Agrarian Science* **2**
[17] Ostapchuk P S, Emelyanov S A 2018 *Vestnik J. Vest. SAU* **1**.
[18] Zabelina M V, Skrynnikov A P 2015 *Sheep, Goats, Woolen Production* **3**
[19] Ul'yanov A N, Kulikova A Ya 2015 *Sheep, Goats, Woolen Production* **3**
[20] Katkov K A *Bulletin of Agrarian Science* 2020 **2** 83
[21] Ostapchuk P, Yemelianov S, Skorykh L, Konik N and Kolotva N 2018 *RJPBCS* **9** 3
[22] Zavgorodnyaya G V, Dmitrik I I, Sidortsov V I, Pavlova M I and Mikryashova G N 2013 *Method of Complex Assessment of Runes of Pedigree Sheep of Fine-Wool Breeds* (Stavropol)
[23] Timofeev A V, Agapov M M, Vaganova A N and Simonov V S 2015 *Synopsis of a Laboratory Assistant-Histologist. Fundamentals of the Histological Process* (St. Petersburg: ELBI-SPb) p 64

IOP Conf. Series: Earth and Environmental Science **965** (2022) 012029 doi:10.1088/1755-1315/965/1/012029

Milk productivity and quality depending on the composition of the diet

A T Varakin[1], A A Ryadnov[1], M A Stepurina[1], E S Vorontsova[2, 3] and O O Datchenko[4]

[1] Volgograd State Agrarian University, 26 University Avenue, Volgograd, 400002, Russia
[2] Volga Region Research Institute of Manufacture and Processing of Meat-and-Milk Production, 6 Rokossovsky Street, Volgograd, 400131, Russia
[3] All-Russian Research Institute of Irrigated Agriculture, 9 Timiryazev Street, Volgograd, 400002, Russia
[4] Samara State Agrarian University, 2 Uchebnaya Street, Ust-Kinelsky, Samara region, 446442, Russia

E-mail: varakinat58@mail.ru

Abstract. The aim of the research was to study the effect of using wet feed wheat grain harvested with sodium acetate in the diets of lactating cows instead of dried grain on milk productivity and milk quality. The content of crude protein in wheat grain at the natural humidity of the experimental variant was significantly higher than in the grain storage of the control variant prepared by the traditional method of canning - fire drying. The data of the experiment on lactating cows indicated that the inclusion of wet flattened wheat grain with sodium acetate in the diet instead of dried flattened grain had a positive effect on the dairy productivity of cows and milk quality. In comparison with the control group, the advantage of cows of the experimental group in terms of the average daily milk yield was 7.8%, and in terms of fat content by 0.21% and protein by 0.04%. The consumption of energy feed units per 1 kg of natural milk in the control group of cows was 1.02 and in the experimental group - 0.95 and the tested grain fodder – 290 and 269 g. The economic efficiency of milk production in the experimental group increased by 14.5%.

1. Introduction

The expansion of the range of effective feed products makes it possible to more fully realize the genetically determined productivity potential of farm animals, reduce the cost and increase the profitability of production [1, 2, 3].

To a large extent, the implementation of the productive qualities of cattle is facilitated by the peculiarities of the modern technologies used for its maintenance [4, 5]. Dairy productivity and the quality of cow milk also depend on different breeding [6]. At the same time, the productivity and physiological indicators of livestock are significantly influenced by the provision of the completeness of its feeding [7, 8, 9].

It should also be noted that in physiological studies, a great importance is attached to the study of blood composition indicators in animals [10, 11].

The productivity of farm animals is greatly influenced by the energy value of the diets used [12, 13].

Also, when organizing animal feeding, along with other nutrients, their protein nutrition is important [14, 15, 16, 17].

The main indicators of the quality of the feed used are the energy value, their protein nutrition value, the content of mineral elements and vitamins in them. Moreover, in balancing the diets of lactating cows in terms of energy and basic nutrients, great importance is attached to concentrated feeds, and in particular grain fodder. Preservation of grain fodder in a wet state with chemical preparations serves is a progressive and rational way of its preparation and storage. Wet canned grain is better eaten by animals and absorbed after its flattening; therefore, the productive effect of this feed increases.

Chemical canning technology has been developed to preserve grain with high humidity. For example, production tests for preserving of wet grain carried out using propionic acid and carbon ammonium salts have shown their high efficiency. In comparison with thermal drying of grain and active ventilation, energy is consumed during its chemical preservation by 1.5-2 times less [18].

It should be noted that the assessment of the economic component also plays an important role in the production of animal products [19, 20].

At the same time, it is relevant to carry out research on the use of wet grain fodder harvested with a preservative additive - sodium acetate, in dairy cattle breeding.

The purpose of the research was to study the effect of the use of wet feed wheat grain harvested with sodium acetate in diets instead of dried grain of this crop on milk productivity and milk quality of lactating cows.

The objectives of the research were to determine the quality, chemical composition and nutritional value of wet wheat grain prepared with sodium acetate and dried grain of this crop and their influence in a comparative aspect on the productivity and quality of milk of lactating cows, their physiological indicators, economic assessment of milk production.

2. Materials and methods

When performing a laboratory (reconnaissance) experiment, a glass container with a capacity of 1 dm^3 was used. In the first container fodder grain was placed for storage – dried wheat with a humidity of 14% (I control variant); in the second container – wheat with a humidity of 25% (II control variant); in the third container – wheat with a humidity of 25% treated with a new preservative – sodium acetate at a dose of 9-10 g per 1 kg of wet grain based on the dry substance of the preservative (III experimental variant). The container was sealed with lids and left for storage.

In the laboratory experiment, the quality indicators of harvested fodder grain were studied using generally accepted methods. The quality of harvested wheat grain was determined by the color, smell, structure of feed particles, the content of protein, calcium, phosphorus. Along with this, biochemical studies were carried out with the determination of acetic, lactic and butyric acids in grain preserved with sodium acetate.

For the scientific and economic experiment, feed grain was prepared according to the scheme presented in the Table 1.

Table 1. Feed grain harvesting scheme.

Variant	Fodder crop	Features of the variants
I	Winter wheat (grain)	Control – dried grain
II	-//-	Treated with sodium acetate in an amount of 10 kg per 1 ton of wet grain per dry preservative substance

The wet wheat grain delivered to the current from the combine was immediately treated with sodium acetate at the prescribed dose. The processed grain was sent to the storage (warehouse). The grain mound of processed wheat was covered with a gas- impermeable material (polyethylene wrap). The condition of the grain treated with the preservative was monitored. To prevent moisture condensation, grain aeration and storage ventilation were periodically carried out (II experimental variant).

Dried wheat grain was stored in the same warehouse (I control variant).

Two groups of lactating cows with 8 heads each were formed for the research. Cows were selected into groups according to the principle of pairs of analogues, taking into account the breed, age, live weight, lactation by count, productivity for previous lactation, calving and insemination time, average daily milk yield and fat content in it. Scientific and economic experience was conducted on full-aged animals (3-5 lactation) with an average live weight of 590 kg.

The scientific and economic experiment on lactating cows was carried out for 150 days. In the preliminary period of the experiment, lasting 20 days, the similarity of the composition of the selected animals in groups was checked. At the same time, the animals of both groups were fed a basic diet using dried fodder flattened grain. In the transition period of the experiment, lasting 10 days, cows of the I control group were given a basic diet using dried flattened grain fodder, and the II experimental group was given a basic diet, which included wet flattened grain with sodium acetate instead of dried grain (habituation). During the main period of the experiment, lasting 90 days, the cows of the I group were fed a basic diet using dried fodder flattened grain, and the II group - the basic diet, which included wet fodder flattened grain with sodium acetate instead of dried grain. In the final period of the experiment, lasting 30 days, cows of the II experimental group were transferred to the basic diet using fodder dried flattened grain.

During the research work, the animals of both groups were under the same conditions of maintenance and care. Milking of cows was carried out twice a day. During the experiment, the indicators of dairy productivity of cows were studied every week with an assessment of the quality of milk produced from each animal.

Against the background of scientific and economic experience, physiological studies were conducted to study the digestibility and use of nutrients in diets for 3 animals from each compared group. Control over the physiological state of cows was performed by studying hematological parameters in 5 animals from each group at the beginning and at the end of the experiment.

The economic efficiency in the comparative aspect when using dried feed grain and wet grain harvested with the addition of a preservative additive - sodium acetate in the diets was determined by the actual costs of preparing feed and producing livestock products.

3. Results and discussion

Sodium acetate used as a preservative is a by-product of chemical production, which is a powdery substance of white color with a yellowish tinge, practically odorless. This product is characterized by the following chemical composition, wt.%: dry matter of sodium acetate - 56-58, admixture of polytetramethylene ethylene glycol - 0.2-2 and water - the rest. It contains crystalline hydrate water, stable during storage for a year, safe to handle. Sodium salt of acetic acid - sodium acetate is environmentally friendly.

Acetic acid is a natural metabolite. In the rumen of the cow's stomach, it is formed up to 1-1.5 kg or more per day, which contributes to the formation of energy in the body [21].

After two months of storage of feed wheat grain, a glass container was opened and its organoleptic evaluation, biochemical and chemical analysis was performed. Organoleptic evaluation of the quality indicators of harvested feed grain of all variants was carried out on a commission basis. The studies found that in the II control variant, grain with a moisture content of 25%, harvested without preservative, became moldy after storage. It had more than 50% of spoiled grain with a brown color and a musty smell, unsuitable for feeding. According to the studied indicators, in the I control variant, the dried grain with a moisture content of 14% and in the III experimental variant, the grain with a moisture content of 25%, harvested with a preservative - sodium acetate, was evaluated as benign. In the two above-mentioned variants, the grain was characterized by a marketable appearance, structure, had no mold and a sharp moldy smell.

An objective assessment of the quality of the grain under study was performed according to laboratory studies (Table 2).

Table 2. Chemical composition of feed grain, %.

Indicator	Grain harvesting variants	
	I control (dried)	III experimental (wet grain with sodium acetate)
Crude protein	11.5	12.5
Phosphorus	0.30	0.27
Calcium	0.04	0.11
Organic acid content:		
acetic acid	-	0.14
dairy	-	1.88
oil	-	0.03

As a result of the research, it was revealed that in wet feed grain preserved with sodium acetate, there was an optimal accumulation of organic acids. The mechanism of the preservative action of sodium acetate (sodium salt of acetic acid) is that it suppresses and inhibits enzymes due to acidification of the feedstock. It should be noted that in the I control variant, with the traditional method of grain conservation – fire drying, the protein content decreases, compared with the III experimental variant, with chemical canning of wet grain, by 1.0% (absolute). At the same time, a cheap preservative, sodium acetate, was used for canning, which allows preserving the nutritional value of wet feed grain.

According to the data obtained in the laboratory experiment, wet wheat grain prepared with the use of sodium salt of acetic acid - sodium acetate at a dose of 10 g/kg of wet grain per dry substance of the preservative has a better quality than dried grain. Therefore, when conducting scientific and economic experience, this method of canning was adopted when harvesting wet feed grain.

In the scientific and economic experiment, after 2 months of storage of feed grain, average samples were selected to assess its quality according to organoleptic indicators and the results of chemical analysis.

Organoleptic assessment of harvested grain was carried out by commission. In general, according to organoleptic indicators, the grain of both harvesting variants was evaluated as high-quality. In the compared variants, the wheat was of normal color and had no mold and moldy smell.

An objective assessment of the quality of the prepared grain fodder was obtained on the basis of zootechnical analysis data (Table 3).

The results obtained indicate that the content of dry and organic matter in the fodder flattened wheat grain of the experimental variant at natural humidity was lower compared to the control. This is explained by the fact that the preparation of feed wheat of the control variant was produced by the traditional method of preserving grain - fire drying, and wheat was used for the preparation of feed grain of the experimental variant without preliminary drying (with higher humidity). Due to the same reason, compared with the experimental version, the content of fiber and nitrogen-free extractives in the wheat grain in the control was higher. This had a definite effect on the general (energy) nutritional value of the prepared grain fodder. In terms of energy nutrition value, the wheat of the control variant slightly exceeded the experimental variant.

The treatment of wet grain fodder with a preservative preparation (sodium acetate) had a positive effect on the quality of the harvested feed. The content of crude protein in wheat grain at the natural humidity of the experimental version was significantly higher than in the control group. In terms of fat content, the compared variants of grain harvesting had no differences.

As in the laboratory experiment, in the harvested grain of the experimental variant, a greater accumulation of lactic acid was noted, a lesser accumulation of acetic acid and a minimum accumulation of butyric acid. Wheat of the experimental variant had a pH of 4.0. Preservative - sodium acetate (sodium salt of acetic acid) has fungicidal and bactericidal properties. Due to its introduction, the grain is acidified to the optimal pH value and conditions are created that are disastrous for microorganisms that

IOP Conf. Series: Earth and Environmental Science **965** (2022) 012029 doi:10.1088/1755-1315/965/1/012029

contribute to mold formation and spoilage of unprocessed grain. After treatment with sodium acetate, it does not heat up, does not mold and does not germinate.

Table 3. Chemical composition and nutritional value of feed grain at natural humidity.

Indicator	Grain feed variants	
	I control	II experimental
Total moisture, %	13.90	24.80
Dry matter, %	86.10	75.20
Organic matter, %	84.20	73.50
Crude protein, %	11.72	12.45
Crude fat, %	1.98	1.98
Crude fiber, %	4.0	2.91
Crude ash, %	1.90	1.70
Nitrogen-free extractive substances, %	66.50	56.16
It is contained in 1 kg of grain fodder:		
energy feed units, EFU	1.14	1.0
exchange energy, MJ	11.41	10.04
digestible protein, g	98.45	104.58
calcium, g	1.74	1.60
phosphorus, g	3.31	2.89
sugar, g	19.5	16.5

*1 EFU = 10 MJ of exchange energy.

However, for a more full view of the quality of the harvested grain fodder, the results of the chemical composition of wheat at natural humidity were recalculated to an absolutely dry matter (Table 4).

Table 4. Chemical composition of flattened grain in terms of absolutely dry matter, %.

Показатель	Variants for harvesting grain fodder	
	I control (dried)	II experimental (with sodium acetate)
Dry matter	86.10	75.20
Dry matter content of feed:		
crude protein	13.61	16.55
crude fat	2.30	2.63
crude fiber	4.65	3.87
nitrogen-free extractive substances	77.23	74.69
crude ash	2.21	2.26
Sugar content, g/kg	22.70	22.0

In comparison with the control, the dry matter of the wheat grain of the experimental variant contained more crude protein by 2.94% and crude fat by 0.33%. The compared grain fodder variants differed slightly in the content of nitrogen-free extractives and sugars.

Thus, previously obtained data in laboratory experience are confirmed by the results of scientific and economic experience. The use of sodium acetate as a preservative makes it possible to harvest feed grain without preliminary drying, which was provided for by the purpose of the research.

The milk productivity of lactating cows was studied when using wet feed wheat grain preserved with sodium acetate in the diets in scientific and economic experience. The composition and quantity of succulent and coarse feeds in the diets of cows of both groups was the same. Animals of both groups were given the same amount of cereal concentrates. The difference in the rations was that the cereal concentrates were of different harvesting technologies. On average, during the experiment period, the daily ration of cows in the compared groups included leguminous hay - 4.4 kg, corn silage - 28.4 kg, legume haylage - 4.0 kg, feed molasses - 1.0 kg. At the same time, the animals of the I control group were fed 5.2 kg of dried flattened wheat grain prepared by the traditional method of preserving it - fire drying, and the II experimental group - 5.2 kg of wet flattened wheat grain harvested with a preservative - sodium acetate (a by-product of the chemical industry) at a dose of 10 kg per 1 ton of wet grain based on the dry matter of the preservative. To meet the needs of cows in mineral elements and vitamins in the diets, the necessary feed additives were used. In the diet of cows of the I control group, the content of energy feed units was 19.1, exchange energy - 191.2 MJ, dry matter - 18.47 kg, and the II experimental group - respectively 18.4, 184.2 MJ, 17.91 kg. Crude protein in the diet of animals of the I group contained 2366.3 g and the II group – 2404.2 g, digestible protein – 1617.7 and 1649.6 g, respectively, crude fiber – 3633.0 and 3576.3 g, starch – 3032.4 and 2954.4 g, sugars – 1163.5 and 1148.0 g, crude fat – 473.5 and 473.5 g, calcium – 179.7 and 179.0 g, phosphorus – 77.3 and 75.1 g, carotene – 923 and 925 mg.

During the period of the experiment, cows of the II experimental group differed in higher consumption of feed (with the exception of grain cereal concentrates and feed molasses) and nutritional substances of the diets. So, in comparison with the I control group, per 1 animal in the II experimental group, more dry matter was consumed by 18.0 kg (1.14%), crude protein – by 12.35 kg (6.10%), digestible protein – 8.56 kg (6.17%), sugars – 1.25 kg (1.23%), crude fat – 1.92 kg (4.76%), calcium – 0.86 kg (5.71%), carotene – by 4.95 g (6.40%). At the same time, the difference in the consumption of energy feed units, exchange energy, starch and phosphorus between the compared groups was insignificant.

The obtained results of the performed studies indicate that the inclusion in the diet of wet flattened wheat grain harvested with sodium acetate, instead of dried flattened grain of the same crop, positively affected the dairy productivity of cows and milk quality (Table 5).

Table 5. Dairy productivity of cows in the main period of the experiment, kg.

Indicator	Groups of animals	
	I control	II experimental
Average daily milk yield of natural milk	17.90 ± 0.43	19.30 ± 0.50*
% to control	100	107.8
Average content in milk, %:		
fat	3.77 ± 0.04	3.98 ± 0.03***
protein	3.30 ± 0.03	3.34 ± 0.05

*P < 0.05, ***P < 0.001.

So, during the main period of scientific and economic experience in terms of average daily milk yield of natural milk, the advantage of cows of the II experimental group, in comparison with analogues of the I control group, was 1.40 kg (7.8%; P > 0.95).

Fat and protein content are important indicators in characterizing the quality of milk. These indicators in animals of the II experimental group were higher than in the I control group, respectively, by 0.21 (P > 0.999) and 0.04 absolute percent.

During the main period of the experiment (90 days), the average yield of natural milk from each cow of the I control group was 1611.0 kg and the II experimental group - 1737.0 kg. At the same time, more milk by 126.0 kg (7.8%) received from each cow of the II group in comparison with analogues of the I group.

At the same time, the average daily milk yield in terms of basic fat content (3.4%) was 19.85 in cows of the I control group, and in the II experimental group – 22.59 kg. Hence, the animals of the experimental group had an average daily milk yield of basic fat content higher by 2.74 kg (13.8%) than the control analogues.

According to the results obtained in the experiment, cows of the compared groups had differences in indicators of feed costs for the production of a unit of production. Thus, the consumption of energy feed units per 1 kg of natural milk in the I control group of cows was 1.02 and in the II experimental group - 0.95, and the tested grain fooder - 290 and 269 g, respectively.

In physiological studies on the digestibility and use of nutrients of diets by lactating cows of the compared groups, higher studied indicators were revealed in the experimental group, compared with analogues from the control. Indicators of morphological and biochemical composition of blood in animals in the compared groups were within the physiological norm.

Upon completion of the research, an assessment of the economic efficiency of milk production was carried out using wet flattened wheat grain (with sodium acetate) as part of the main diet instead of dried flattened grain of this crop. During the 90 days of the main period of scientific and economic experience, an average of 1786.5 kg of milk was milked per cow in the I control group and 2033.1 kg of milk with a basic fat content of 3.4% in the II experimental group. According to this indicator, the II group of cows had an advantage with a difference of 246.6 kg (13.8%). The performed calculations showed that the cost of milk produced, minus additional costs, on average per animal was 53250.0 rubles in the I control group and 60961.0 rubles in the II experimental group.

The economic efficiency (net income) from the introduction of wet wheat grain harvested with sodium acetate into the diets of lactating cows per 1 animal in the II experimental group was 7711.0 rubles, which is 14.5% more compared to the I control group.

Thus, in order to increase the economic efficiency of milk production, it is advisable to prepare wet wheat grains with sodium acetate and include lactating cows in the diets.

4. Conclusion

As in the laboratory experiment, in the scientific and economic experience in the harvested grain of the experimental variant with sodium acetate, a greater accumulation of lactic acid was revealed, a lesser accumulation of acetic acid and a minimum accumulation of butyric acid. Due to the introduction of sodium salt of acetic acid into the wheat of the experimental vaiant, the grain is acidified to an optimal pH value of 4.0 and conditions are created that are disastrous for microorganisms that contribute to mold and spoilage of unprocessed grain. After treatment with sodium acetate, it does not heat up, does not mold and does not germinate.

In the scientific and economic experience, after 2 months of storage of feed grain, according to organoleptic indicators, the wheat of both harvesting variants was assessed as high-quality, had a normal color and did not have mold and mildew smell. Studies of the chemical composition showed that the treatment of the experimental variant of wet grain fodder with a preservative preparation (sodium acetate) had a positive effect on the quality of the prepared feed. The content of crude protein in wheat grain at natural humidity of the II experimental variant was significantly higher than in the grain storage of the I control variant, prepared by the traditional method of preserving grain - fire drying. In terms of fat content, the compared grain variants had no differences.

Recalculation of the results of the chemical composition of wheat at natural humidity to an absolutely dry matter gave a more full view of the quality of the harvested grain fodder. In comparison with the control, the dry matter of the wheat grain of the experimental variant contained more crude protein by 2.94% and crude fat by 0.33%. The compared grain fodder variants differed slightly in the content of nitrogen-free extractives and sugars. The use of sodium acetate as a preservative makes it possible to harvest feed grain without prior drying.

The inclusion in the diet of wet flattened wheat grain harvested with sodium acetate, instead of dried flattened grain of this crop, positively affected the dairy productivity of cows and the quality of milk. According to the average daily milk yield of natural milk, the advantage of cows of the II experimental group, in comparison with analogues of the I control group, was 7.8%. The results of the quality of the milk produced indicated that the fat and protein content of the animals of the II experimental group was higher than in the I control group, respectively, by 0.21 and 0.04 absolute percent.

IOP Conf. Series: Earth and Environmental Science **965** (2022) 012029 doi:10.1088/1755-1315/965/1/012029

The best indicators of milk production efficiency with the use of wet flattened wheat grain (with sodium acetate) as part of the main diet instead of dried flattened grain of this crop have been established. The consumption of energy feed units per 1 kg of natural milk was 1.02 in the I control group of cows and 0.95 in the II experimental group, and the tested grain fodder – 290 and 269 g, respectively. The economic efficiency of milk production in the experimental group was 14.5% higher compared to the control group.

References

[1] Varakin A T, Kulik D K, Salomatin V V and Kharlamova E A 2019 The use of feed from irrigated lands for boars-producers with the introduction of mineral additives into the rations *Irrigated Agriculture* 1 63-66

[2] Miroshnikov S A, Yausheva E V, Sizova E A, Miroshnikova E P and Levakhin V I 2015 Comparative assessment of effect of copper nano- and microparticles in chicken *Oriental J. of Chemistry* **31(4)** 2327-2336

[3] Zoteev V S, Pisarev E I, Nikolaev S I, Salomatin V V and Varakin A T 2018 False flax cakein mixed feed for the fattening of lactating cows and dairy stores *Research J. of Pharmaceutical, Biological and Chemical Sciences* **9(5)** 1422-1428

[4] Baimishev Kh B 2018 Influence of technology of growing heifers on the morphology of their ovary *Bulletin of the Samara State Agricultural Academy* **3** 34-39

[5] Valitov Kh Z and Karamaev S V 2012 *Productive Longevity of Cows in Conditions of Intensive Milk Production Technology* (Samara: Samara State Agricultural Academy) p 322

[6] Lyashenko V V and Sitnikova I V 2013 Milk productivity and milk quality of Holstein first-calf cows of different breeding *Animal Husbandry* **9** 18-19

[7] Kulikov V M 1986 Feeding dairy cattle in the Lower Volga region *Scientific Bases of Full-Fledged Feeding of Farm Animals* (Moscow: Agropromizdat) pp 73-81

[8] Simonov G A, Kuznetsov V M, Zoteev V S and Simonov A G 2017 Organization of full-fledged feeding of dairy cows of the Sakhalin region *Scientific and Practical Ways to Improve Environmental Sustainability and Socio-Economic Support of Agricultural Production* (Solenoe Zaymishche: Caspian Research Institute of Arid Agriculture) pp 1369-1371

[9] Gorlov I F, Levakhin V I, Radchikov V F, Tsai V F and Bozhkova S E 2015 Effect of feeding with organic microelement complex on blood composition and beef production of young cattle *Modern Applied Science* **9(10)** 8-16

[10] Lents C A, Rempel L A, Klindt J, Wise T, Nonneman D and Freking B A 2013 The relationship of plasma urea nitrogen with growth traits and age at first estrus in gilts *J. of Animal Science* **91(7)** 3137-3142

[11] Nascimento C F, Branco R H, Bonilha S F, Cyrillo J N, Negrão J A and Mercadante M E 2015 Residual feed intake and blood variables in young Nellore cattle *J. of Animal Science* **93(3)** 1318-1326

[12] Dvalishvili V G 1989 Norms of energy and protein nutrition for meat and wool sheep producers Optimization *of Feeding of Farm Animals* (Dubrovitsy, Moscow region: All-Russian Research Institute of Animal Husbandry named after L K Ernst) vol 53 pp 80-88

[13] Kalashnikov A P, Fisinin V I, Shcheglov V V, Pervov N G and Kleymenov N I 2003 *Norms and Rations for Feeding Farm Animals: A Reference Guide* ed A P Kalashnikov, V I Fisinin, V V Shcheglov and N I Kleymenov (Moscow) p 456

[14] Varakin A T, Kulik D K and Salomatin V V 2018 The use of camelina oil cake for the production of mutton *Bulletin of the Samara State Agricultural Academy* **3** 39-43

[15] Varakin A T, Salomatin V V, Kharlamova E A and Varlamova T A 2018 Efficiency of milk production using flax and rapeseed cakes *Bulletin of the Samara State Agricultural Academy* **3** 30-34

[16] Nikolaev S I, Struk V N, Chekhranova S V and Nikishenko A V 2017 Influence of mustard protein-containing feed concentrate "Gorlinka" on milk productivity of cows *Proc. of the*

Nizhnevolzhsky Agrouniversity Complex: Science and Higher Professional Education **4(48)** 205-212

[17] Fenwick G R 1982 The assessment of a new protein source – Rapeseed *Proc. Nutr. Soc.* **41** 277-288

[18] Vladimirov V L 1997 Reducing the loss of nutrients during storage of feed with the help of chemical preservatives *Theory and Practice of Feeding Farm Animals and Feed Technology* ed N I Strekozov, A P Kalashnikov and V V Shcheglov (Dubrovitsy, Moscow region: Russian Academy of Agricultural Sciences; All-Russian Research Institute of Animal Husbandry named after L K Ernst) pp 20-25

[19] Varakin A T, Kulik D K, Salomatin V V and Ryadnov A A 2018 Productivity of fattened rams in a natural pasture *Proc. of the Nizhnevolzhsky Agrouniversity Complex: Science and Higher Professional Education* **3(51)** 236-241

[20] Uhtverov A M, Baymishev Kh B, Khakimov I N, Kokhanov M A, Grigoryev V S and Varakin A T 2018 Duration of economic use of breeding sows with the different levels of fat depth and early maturity *Research J. of Pharmaceutical, Biological and Chemical Sciences* **9(5)** 935-942

[21] *Chemicalization in the Branches of the Agro-Industrial Complex* 1990 ed A V Postnikov (Moscow: Rosagropromizdat) p 223

IOP Conf. Series: Earth and Environmental Science **965** (2022) 012030 doi:10.1088/1755-1315/965/1/012030

Organic pig farming as part of green economy

G V Komlatsky[1], M I Slozhenkina[2], R V E`lizbarov[1], A A Mosolov[2], M V Frolova[2] and A S Miroshnik[2]

[1] Kuban State Agrarian University named after I T Trubilin, 13 Kalinin Street, Krasnodar, 350039, Russia
[2] Volga Region Research Institute of Manufacture and Processing of Meat-and-Milk Production, 6 Rokossovsky Street, Volgograd, 400131, Russia

E-mail: niimmp@mail.ru

Abstract. Organic agriculture is an important area of high-quality food production and one of the stages of transition of agricultural production to green economy rails. At the moment, assortment on the Russian market of organic products is limited and there are no meat products in general. The aim of this study is to analyze the state of domestic pig breeding and the possibilities of conversion a certain segment of the industry to organic production. The methodological basis was the results of researches by domestic, foreign scientists, as well as our own studies. In the course of the work, general methods of scientific knowledge were used, which ensured reliability of the obtained results. Taking into account global trend towards healthy nutrition, Russian pig breeding has all prerequisites for conversion to organic production. Due to the high marginality of organic pig breeding, the presence in Russia of large labor and land resources suitable for organic production, there are all the prerequisites for organic pork to enter the world agrifood market. The implementation of this direction will require small business development, creation of a certification system for meat and meat products, training of qualified organic zootechnicians, as well as development and improvement of technologies.

1. Introduction

Organic agriculture is an important area of high-quality food production and one of the stages of conversion of agricultural production to green economy rails. In Russia, policy towards development of production and provision of the population with safe food products was proclaimed by the Food Security Doctrine of 2020, which develops provisions of the National Security Strategy [1].

The global organic market is currently valued at 80 billion dollars. In 84 countries, the production of such products is regulated by relevant laws. It should be noted that in Russia a similar law Federal Law No. 280 "On Organic Products" came into force on January 1, 2020, and even earlier similar laws were adopted in a number of regions of the country, including Krasnodar territory [2, 3].

The concept of organic agriculture, which is declared by the International Federation of Organic Agriculture Movement (IFOAM), is based on four basic principles, which can be formulated as follows:

- the principle of health, based on the position that human health is inseparable from well-being of the ecosystem, therefore, any activity should be focused on maintaining and improving the health of both humans and animals;

- the principle of ecology mean production of food products within scope of a natural ecological system and natural life cycles using organic substances and natural components;
- the principle of justice is focused on social part of economy and assumes food security of all countries and access to quality products for all people, while animals should also have decent living conditions, taking into account their physiological needs and instincts;
- the principle of care based on position that increasing efficiency and productivity of agricultural production should not occur at expense of health and well-being of both people and nature.

The purpose of the work is analysis of current state of domestic pig breeding and possibilities of conversion a certain segment of the industry into organic production.

2. Materials and methods

The methodological basis was results of researches by domestic, foreign scientists in the field of organic animal husbandry, as well as our own studies.

The analysis of status of the industry and prospects for ecologization was carried out using official data from state authorities and analytical materials from industry unions.

In the research work, general methods of statistical, scientific knowledge and methods of conducting zootechnical experiments were used. The cumulative use of generally accepted methods ensured reliability of the obtained results.

3. Results and discussion

Organic animal husbandry is a production that minimizes the use of synthetic feed, additives and growth stimulators. At the same time, the main requirement is to obtain products with biologically valuable and functional qualities, taking into account physiological and behavioral needs of animals (provision of high-quality organic feed, compliance with optimal number of animals per unit area, etc.) [4, 5].

The need to maintain a healthy lifestyle contributes to development of organic animal husbandry, because animal food constantly affects human health [6]. Therefore, high requirements are impose on the production of organic meat products at all stages of the technological process. Disorders at any stage of production process can negatively affect quality and safety of products. So, stress in pig breeding contributes to the accumulation of hormones unnecessary for humans in pork. It should be noted that narrowly targeted selection for high meat content led to the appearance of meat defects - pale, soft, exudative meat with reduced technological qualities and increased processing losses. According to the conclusion of Danish scientists, this defect is a consequence of "poor adaptation syndrome of pigs", which is also called stress syndrome. It should be noted that Denmark is the European leader in pig breeding. Danish pig producers in 1998 voluntarily adopted a ban on the use of antibiotics-growth stimulants on fattening farms. They introduced the mandatory use of toys in meat processing plants for pigs before slaughter, and much more. This is probably why the main buyer of pork in this country is Japan, whose residents are very responsible for their health. This branch of animal breeding significantly contributes to welfare of the country, where on average about two pigs per resident, and pig production account for 10% of export. In Germany, a program called the Animal Welfare Initiative has been running since 2015, aimed at obtaining the highest quality meat by creating the most comfortable living conditions for animals.

In Russia, organic meat is still consumed mainly by prosperous residents of big cities: Moscow, St. Petersburg, Sochi. High prices and competition from uncertified fresh products are holding back organic production. Such products in our country are 3-7 times more expensive than ordinary ones, while in the USA and EU countries this value is only 30-50%. More than 70% of the Russian organic meat market is imported, with actually no pork. Meanwhile, according to forecasts, by 2030 organic meat products market may reach 5 billion euros [7].

Analysis of prospects for development of organic pig breeding showed that Russians traditionally prefer pork meat (more than 40%). This is due to both gastronomic preferences of compatriots, high taste and nutritional value of meat. Pig meat has a high calorie content (in 1 kg of pork 3160 kcal, respectively - 13230 kJ), and in terms of vitamin B1 content it surpasses even black and gray bread (0.2-0.3 mg%). Also, this meat contains other vitamins and a large amount of minerals. Consumer properties of meat are determined content of biologically complete proteins, which are a source of essential amino acids. The digestibility of pork nutrients is 90-95%, the fat is more easily digestible than any other animal fat, and is necessary in production of sausages.

2018 was the turning point for the industry, when from an importer of pork in 2012, Russian pig breeding in 6 years reached full self-sufficiency in pork meat (Table 1).

Table 1. Pork production in the Russian Federation, thousand tons.

	Year					
	2005	2010	2015	2018	2019	2020
Pork production, thousand tons	1520	2337	3083	3744	3937	4319
Self-sufficiency, %	60	66	91	100	100	100

It should be noted that according to preliminary results of 2020, pork is practically the only driver of growth in production of all types of meat. Other sources of animal protein show little or no gain. Saturation of the domestic meat market with pork is one of prerequisites for finding ways to improve quality. The conversion of certain sector into organic production is obvious and understandable. It should be noted that pork is a relatively inexpensive product; it consistently ranks third in retail and wholesale prices after mutton and beef. With an indicator of meat consumption of 77 kg per person per year, part of pork is more than 28 kg and this value continues to grow. According to statistics, over the past 5 years, annual consumption of pork has increased by more than 0.5 million tons due to decrease in its price, while the upward trend is predicted to continue.

As the experience of foreign countries shows, organic products, as a rule, are obtained on family-type farms. In Russia, at the same time, a large share of pork is produced by large holdings and more than 80% of pigs are kept in large agricultural enterprises (Table 2).

Table 2. Pork producers in the Russian Federation, thousand tons.

Form of organization	Year	
	2019	2020
Agricultural organizations	3420.8	3823.6
Farm	40.4	38.0
Households	475.6	455.5
Total	3936.8	4319.1

Since organic pig breeding is a fairly new direction in animal husbandry, this niche has not yet been filled in the world meat market, and there has been no competition among producers.

Taking into account achieved complete self-sufficiency in pork, the issue of expanding export of this type of animal protein is relevant, which requires improving quality indicators to level of importing countries. Taking into account the global trends of growth in demand for organic food, production of organic pork may become one of the export items. Natural-climatic and production potentials allows to significantly increase volume of production and export of organic pig products.

Another important consumer of organic pork is baby food enterprises, where are severe regulatory requirements for quality of raw materials, especially for the presence of a number of growth-stimulating drugs. Therefore, production of pork for baby food is a separate segment of organic, eco-friendly, specializing in production of high-quality, and inexpensive products. This can be explained, in our opinion, both by complexity of the organization of the production process itself, and by the requirements for keeping, feeding and treating livestock. At least 50% of the feed must be of its own production or produced by other farms conducting organic production.

Organic pig breeding, taking into account the environmental factor, is a paradigm of sustainable development with low feed consumption and low emissions with high efficiency, which corresponds to the basic tenets of green economy [8; 9]. It should be noted that, both in the world and in Russia, organic animal husbandry is less developed than organic crop production. At least 50% of animal feeds must be of its own production or produced by other farms conducting organic production.

By cultivating quality feed, an optimal livestock density rate and a pig breeding system that ensures animal welfare and minimizes stressful situations, an optimal production model can be developed. Formation of harmonious relationships between soil, vegetation and animals, compliance of physiological and behavioral animal needs is the basis of organic pig breeding, because manure is an excellent fertilizer for obtaining high yields of cereals and soybeans. To maintain ecological balance in the European Union, the following standards for keeping pigs using organic technology have been adopted (Table 3) [10].

Table 3. Maximum number of animals per 1 ha in the European Union, grown by organic technology.

Animal species	Heads per hectare/year*
Piglets	74.0
Sows	6.5
fattening pigs	14.0
other pigs	14.0

*Calculated from the equivalent amount of generated manure and dung not exceeding 170 kg nitrogen/ha.

The total density of animals should ensure compliance with nitrogen intake limit of no more than 170 kg per 1 hectare of agricultural area per year. No more than 250 sows should be kept on territory of one household; at a final feeding point of herd - at the same time there should be no more than 1000 heads. A prototype of such a production works at the Kuban Agrarian University: training and production plant «Pyatachok» for 220 sows, where the production technology agree with the organic production system.

The requirements of the standard for organic agriculture regulate the use of certified natural feed, and only allow the use of veterinary drugs in emergency cases. Such requirements are quite feasible already at the present time in connection with the introduction of chips for animals, scanners that read not only the origin, but also the quality of rations when raising pigs.

The production of organic pork without the use of stimulants, the company imposes increased requirements for compliance with technological regulations. It is possible to obtain high performance in raising livestock at lower costs only with properly organized feeding and keeping of animals [11, 12, 13, 14, 15, 16]. In pig breeding, the sucking period is very important, when there is an intensive growth and development of digestive organs and tissues, as the basis for further assimilation of feed [17]. At this stage, it is necessary to provide young animals with optimal conditions, prepare piglets for future life, and adapt them to new living conditions. The problem of the period of weaning piglets from sows lies in the change of diets and their saturation with the plant part. This is especially true at early weaning (21-30 days), as well as for low-weight piglets in multiple litters (more than 12 heads) [18].

During this period, the condition and development of the intestines are critical to future health and productivity. As a rule, after weaning, piglets have a poor appetite, as a result of which they do not gain weight and do not reveal their productive potential. Low feed intake immediately after weaning results in villous atrophy and reduced nutrient absorption and available feed energy.

By the time of weaning, piglets, as a rule, have a poorly developed gastrointestinal tract, as a result of which an insufficient amount of hydrochloric acid is released for the digestion of feed. Due to physiological characteristics, the ability to produce hydrochloric acid in the required volume appears in piglets only by 12 weeks. It should also be noted that after weaning, a favorable environment is created for the development of pathogenic microflora, and the substances produced by it damage the intestinal epithelium, reducing its absorption capacity. At the same time, the pH of the stomach contents rises, a large number of E. coli bacteria accumulate, which is the cause of severe and prolonged diarrhea, as a result of which the loss of piglets can reach 12-15%.

Until recently, one of the methods used in animal husbandry to combat disease-causing microflora was the use of antibiotics. However, now all over the world their use is sharply limited. With regard to organic pig breeding, there is a strict ban. Therefore, to limit the growth of pathogenic and opportunistic microorganisms, it is advisable to use organic acids and their salts as an additive to feed [19; 20]. Due to the anatomical structure of the gastrointestinal tract and the physiological characteristics of digestion, the greatest efficiency is observed when using acids in feeding weaned pigs. With the introduction of organic acids, an acidic environment is created (pH = 5-5.5), in which pathogenic microorganisms die, and beneficial microflora and lactic acid bacteria develop normally, suppressing pathogens. At the same time, the acidic environment increases the activity of digestive enzymes by 2-2.5 times, which improves the digestibility of feed [21, 22]. The introduction of acids accelerates the process of protein breakdown, increasing the digestive activity and growth rate of young pigs. Thus, organic acids from the feed are an alternative to antibiotics. Preparations based on organic acids are safe to use, mix well with feed and practically do not interact with them.

In pig breeding, monocarboxylic acids are often used, containing one carboxyl group: formic and propionic acids in an amount of 0.5% by weight of feed. Compared to organic ones, they are in a partially dissociated state, and their dissociation is allowed to use formic acid - a fairly harmless substance registered in the register of food additives (E236). Compared to other organic acids, formic acid has a high acidic value - 17.3 meq/kg (for citric and lactic acid, this indicator is 9.5-9.4 meq/kg). When formic acid enters the stomach, the pH of the medium decreases to 4.0-4.5, and a barrier is created for the reproduction of pathogenic microflora. At the same time, reaching the optimum pH value increases the digestibility of the feed.

To confirm the feasibility of using formic acid for acidifying feed for weaned pigs, a research and production experiment was carried out. Piglets from 28 days of age, after weaning from sows and transferring to clean and warm pens of the nursery section, where they were kept for 7 weeks, were given acidified feed. Feeding was carried out from group feeders of a round type, one for two pens. For the experiment, two groups of piglets, 30 heads each, were formed according to the principle of analogues in terms of age, sex, origin and state of health. The experimental groups were fed with feed according to the diet (SK-5 OR), the pigs were housed in two adjacent pens. Piglets of the experimental group, in addition to the basic diet (OR), together with the feed (SK-5), were given formic acid in an amount of 0.5% by weight of the feed. The results of such studies are presented in Table 4, from which it can be seen that, with the same amount of food eaten, the experimental animals by the age of 84 days had a live weight of 32.8, versus 29.6 kg in the control, which is 10.1% higher.

Table 4. Efficiency of pig feeding.

Indicator	The control group	Experimental group (+ 0.5% formic acid to OR)
Number of pigs in group: at the beginning of the experiment, heads	30	30
at the end of the experiment, heads	28	29
Live weight at the beginning of the experiment, kg	8.42 ± 0.21	8.39 ± 0.18
Live weight at the end of the experiment, kg	29.60 ± 0.32	32.80 ± 0.24
Amount of feed consumed per 1 head per day, kg	0.69	0.70
Average daily gain, g	380.2 ± 25.4	436.4 ± 21.3
Survival, %	94	97
Number of piglets suffering from digestive tract disorders, %	3	1

In the course of a scientific experiment, it was found that the introduction of an acidifier - formic acid in the diet of piglets - weaners in an amount of 0.5% by weight of feed increases the growth rate. There was an increase in average daily gains by 14.7%; the safety in the experimental group was 3% higher compared to the control group of piglets. The number of gastrointestinal diseases among the animals of the experimental group decreased by 2%. It should also be noted that in the experimental group there were fewer fights and clashes between animals, which indicates a more comfortable ethological environment. Thus, the use of organic acid contributes to the improvement of health status, increases the intensity of growth and, as a result, the financial condition of the industry.

Due to the fact that the use of synthesized vitamins and stimulants is not allowed in organic animal husbandry, it is promising to include an organic feed concentrate "Furor" on the basis of natural humic acids in the diet [23]. On the basis of the UPK "Pyatachok" of the Kuban Agrarian University, a research and production experiment was carried out to study the effect of the feed concentrate "Furor". Piglets of the experimental group on rearing during the period of growth from the 30th to the 40th day, once a day, "Furor" was added to the feed in the amount of 5 g/head, and from the 41st to the 60th day of growth - 10 g/head. Piglets on the feed from the 63rd to the 120th day, the concentrate was introduced into the feed in an amount of 15 g/head, and from 121 to 153 days, 20 g/head. The additive was introduced in the following scheme: 4 days of application, 4 days off. The results of the experiment indicate that the use of an environmentally friendly feed additive for piglets for growing and fattening in doses selected depending on age has a positive effect on the growth rate and safety of animals (Table 5).

Table 5. Intensity of growth and safety of piglets when using organic feed concentrate "Furor".

Group	Average daily growth rate for growing, g	Average daily gain in fattening, g	Survival, %
Control	366	874	98
Experimental	414	964	100

It was found that the average daily gain for growing up in the experimental group was higher than in the control group, and 13.1%, and for fattening - by 10.3%. Safety in the experimental group was 100%, and in the control - 98%. The data obtained make it possible to recommend the use of fodder organic concentrate of natural origin "Furor" in feeding pigs and rearing and fattening.

Fulfillment of the requirement for the use of own-made feed in feeding cannot completely prevent the ingress of mycotoxins into the feed - the products of the vital activity of fungi developing on grain at high humidity and temperature. The consumption of feed with mycotoxins negatively affects the development of piglets, and in some cases can even lead to their death. For the purpose of contamination of feed, natural adsorbents based on bentonite clays can be used. It was found that the introduction of bentonite into the diet in an amount of 1% completely adsorbs mycotoxins. In addition, with bentonite, animals receive a whole range of trace elements necessary for the full growth and development of pigs [24].

The organic livestock management system includes measures to prevent environmental pollution and minimize the risk of harmful effects. Meanwhile, the intensification of production is inextricably linked with the growth of waste generation. At the same time, it should be noted that pig breeding is a potentially hazardous industry for the environment due to the large amount of organic matter and the emission of greenhouse gases. The main pollutants are nitrogen (N) and phosphorus (P) compounds. Long-term disposal of industrial pig waste leads to the accumulation of mobile phosphates in the soil to abnormally high values. Therefore, an important problem in pig breeding is the processing of pig waste. One of the ways of utilization is the processing of manure in biogas plants [25]. However, at the current level of technology development, biogas production is a rather expensive enterprise, and in the conditions of our country, with cheap electricity, it is inexpedient.

One of the ways to reduce the pressure on the agroecosystem is the production of pork in family farms for 5-6 thousand pigs. With such production, an ecological balance is achieved, when the manure transported to the fields, being included in the cycle, contributes to an increase in the amount of humus in the soil and an increase in yield by 8-11%. It should not be forgotten that, when running a family business, interest in the results of one's work increases, the socio-economic situation is stabilized due to the employment of the population. At present, in the domestic scientific literature, there are practically no works on scientific and methodological substantiation and increasing the efficiency of the functioning of pig-breeding enterprises of small forms of management. Meanwhile, in the leading countries of the world, it is this form of production that is dominant in pig breeding.

Improvement of the ecological situation can be ensured both by organizational and economic measures, and by monitoring compliance with the technology of raising animals. The main requirement of environmental legislation is the constant reduction of environmental impact through the introduction of the best environmentally friendly available technologies [26]. Recommendations for their use are set out in the information and technical reference book 41-2017 "Intensive breeding of pigs" [27, 28].

4. Conclusion

Growth in consumption of organic products is a global trend, while in Russia this segment of the meat market has significant growth potential. The country has labor resources, as well as a sufficient amount of land suitable for organic farming without long conversation period. The transformation of pig farming into organic production reduces the burden on the environment, while increasing the profitability of the business and the quality of the products produced. The state of the industry with 100% self-sufficiency of the population in pork creates the preconditions for the entry of organic meat products into the world agri-food market and profit from its export. The natural, climatic and production potential makes it possible to significantly increase the volume of production and export of organic pig products with government support from producers and the promotion of Russian products on the international meat market. The implementation of this direction will require the development of family pig breeding, the creation of a product certification system, the training of qualified specialists and the improvement of technology that meets the requirements of organic production. One of the ways to improve the safety of weaned pigs is to use acidifiers to replace antibiotics. To prevent mycotoxicosis in pigs, it is advisable to use natural bentonite clays as adsorbents in feeding.

5. Acknowledgement

This work was carried out under the grant of the Russian Science Foundation 21-16 00025 SSI NIIMMP. Grant sponsors were not directly involved in the development, analysis, or writing of this article. The authors would like to thank their colleagues for their contribution and support to the research. They are also thankful to all the reviewers who gave their valuable inputs to the manuscript and helped complete the paper.

References

[1] Doctrine of Food Security of the Russian Federation 2020 approved by the Decree of the President of the Russian Federation of January 21, 2020 No. 20

[2] Federal Law No. 280-FZ " On Organic Products and on Amendments to Certain Legislative Acts of the Russian Federation" 2018

[3] Law of the Krasnodar Territory "About production of organic agricultural products" of November 1" 2013 No. 2826-KZ

[4] Komlatsky V I and Vorokov V Kh 2017 Organic technologies in animal husbandry *Innovations in Increasing Agricultural Productivity Animals* pp 35-37

[5] Komlatsky G V 2012 Production of environmentally friendly pork *Pig Breeding* **4** 74-76

[6] Gorlov I F 2004 *Food Production in Accordance with the Requirements of the Concept of a Healthy Diet* (Volgograd: Polytechnik) p 381

[7] Gorlov I F, Slozhenkina M I, Korotkova A A, Mosolova N I, Zlobna E Yu and Khramova V N 2015 *System Technologies in Ensuring Quality of Food Products* (Volgograd: Volgograd State Technical Univercity) p 192
[8] Komlatsky G V 2012 Environmental aspects of pork production *Proc. of the Kuban State Agrarian University* **3(36)** 251-255
[9] Shalavina E V and Vasiliev E V 2017 Current ecological problems of pig production in Russia *Agricultural Engineering* **92** 166-173
[10] Regulation on the Production and Labeling of Organic Products and the Repeal of Regulation (EEC) No. 2092/91 Council Regulation (EC) 834/2007 of 28.06.2007
[11] Mikhailova O A 2017 Modern aspects of European pig breeding: trends, problems and prospects *Biology in Agriculture* **4(17)** 13-25
[12] Makarov D Y, Ruzheinikov F V, Shkalenko V and Nikolaev D V 2012 Lactulose-containing feed additives and their effect on the productivity of experimental gilts *Pig Breeding* **5** 61-62
[13] Filatov A S, Nikolaev D V, Kremenscova E A and Shkalenco V V 2014 Reproductive qualities of the canadian breeding sows in pig company khk "Krasnodonskoye" *Pig Breeding* **2** 14-15
[14] Nikolaev D V, Syulyev L A and Baranikov V A 2013 Effect of new biologically active feed supplements on hematologic indices of young pigs *Bulletin of Altai State Agrarian University* **6(104)** 69-71
[15] Nikolaev D V 2016 Development of methods for intensification of pork production and improving its quality due to optimizing genetic and paratypical factors *Proc. of the Nizhnevolzhsky Agrouniversity Complex: Science and Higher Professional Education* **1(41)** 154-161
[16] Slozhenkina M I, Gorlov I F, Miroshnik A S, Nikolaev D V, Mosolova N I and Radchikov V F 2021 Technology rejection from antibiotics in pig breeding through the use of probiotics *IOP Conf. Ser.: Earth Environ. Sci.* **677** 32015
[17] Komlatsky V I 2013 State and prospects for the development of pig breeding in the leading countries of the world *Proc. of the Kuban State Agrarian University* **5(44)** 175-177
[18] Komlatsky V I, Velichko V A and Velichko L F 2021 Technological methods for increasing the safety of suckling pigs from multiple sows *Polythematic Network Electronic Scientific J. of the Kuban State Agrarian University* **165(1)**
[19] Podobed L I 2013 The role of acidifiers in increasing the productivity of animals *Compound Feed* **10** 14-16
[20] Kryukov V S and Tarasenko V I 2011 Biological and practical aspects of the use of organic acids in pig feeding *RatVetInform* **1** 29-36
[21] Komlatsky V I, Velichko V A and Velichko L F 2019 Influence of feed concentrate Furor on the growth rate of pigs *Proc. of the Kuban State Agrarian University* **80** 209-214
[22] Miroshnik A S, Slozhenkina M I, Gorlov I F, Frolova M V, Nikolaev D V and Mosolova N I 2021 The use of biologically active supplements as an alternative for application of antibiotics in production of pork *IOP Conf. Ser.: Earth Environ. Sci.* **677** 032009
[23] Nikolaev D V, Slozhenkina M I, Gorlov I F, Miroshnik A S, Mosolova N I and Evdokimov I A 2021 Innovative technology of pork production with the use of phytobiotics *IOP Conf. Ser.: Earth Environ. Sci.* **677** 032006
[24] Kononenko S I and Dzagurov B A 2016 Productivity, digestive metabolism in young pigs with the addition of bentonite *Polythematic Network Electronic Scientific J. of the Kuban State Agrarian University* **118(4)**
[25] Kononenko S I, Ledin N P and Muradova E L 2013 Production of biogas and fertilizers on a livestock farm *Bulletin of Agrarian Science of the Don* **1(2)** 45-56
[26] Tikhomirov A I, Fomin A A 2019 Development of organic livestock: problems and opportunities *Int. Agricultural J.* **3(369)** 77-80
[27] Komlatsky V I 2017 The best available technologies are a compromise between business and nature *Pig Breeding* **7** 15-21

[28] Komlatsky V I 2017 The best available technologies as an element of ecological modernization of pig breeding *Polythematic Network Electronic Scientific J. of the Kuban State Agrarian University* **126(2)** 1261702041

IOP Conf. Series: Earth and Environmental Science **965** (2022) 012031 doi:10.1088/1755-1315/965/1/012031

Using protein concentrate in livestock and poultry diets for sustainable agriculture

S I Nikolaev, A D Imangaliev, Yu M Batrakova, V V Shkalenko, M A Ryabova and E A Morozova

Volgograd State Agrarian University, 26 University Avenue, Volgograd, 400002, Russia

E-mail: a.k.karapetyan@bk.ru

Abstract. The paper presents the main results of the experiment on the use of high-protein concentrate "Gorlinka" in the diets of dairy cows and laying hens on their productive qualities. It was found that in terms of the main indicators of nutritional value (protein, fat, nitrogen-free extractives, ash elements), as well as in terms of amino acid composition and mineral substances, the studied concentrate surpasses the traditionally used in feeding farm animals and birds, oil cake and meal from sunflower seeds. The results of studies have shown that the groups of animals that received protein concentrate with the diet were in the lead. The use of protein concentrate "Gorlinka" in the feeding of dairy cows and laying hens increased the average daily milk yields of cows by 5.47-7.12%. There were also positive changes in the mass fraction of protein in milk by 0.03-0.04%, in dry matter by 0.04-0.09%, in the number of eggs laid by laying hens by 216-822 pieces, in the average egg weight by 0.77-1.66 g. Thus, the use of protein concentrate "Gorlinka" in feeding dairy cows and laying hens had a positive effect on the quality of the products (milk and edible eggs), while economic indicators improved.

1. Introduction

The fundamental tasks of the development of agriculture and food security in Russia are the development of livestock and poultry industries, as well as an increase in the volume of products produced with its further export. The solution of this issue will ensure the growth of the competitiveness of the agro-industrial complex of the Russian Federation, will contribute to the revival of rural areas, will allow to regulate the food security of the state [1, 2]. The decisive role in this direction primarily belongs to the balanced feeding of animals and poultry, which is the most expensive item of expenditure [3, 4, 5].

In this regard, it is necessary to constantly search for more affordable and cheaper feed ingredients (food waste processing products, drought- and heat-resistant grain varieties), which are not inferior in nutritional value to traditional feeds [6, 7, 8, 9].

The prospect of our research consists in the search for cheaper alternative sources of feed protein of local origin and the enrichment of diets with various biologically active additives.

Increasing the productivity of dairy cows and egg-producing hens due to the introduction of the protein concentrate “Gorlinka” into the diet has become the goal of our research.

2. Materials and methods

Before the establishment of scientific and economic experiments, identification was carried out by the elementary composition of feed components - meal and oil cake from sunflower seeds, as well as protein concentrate "Gorlinka" (developed by scientists of the Department of Feeding and Breeding of Farm Animals of the Volgograd State Agrarian University together with LLC "Rodos").

After that, a series of experiments was carried out, in which the effect of the use of the protein concentrate in the feeding diets of dairy cows and egg-producing chickens on their productive indicators and the quality of the products obtained was studied. The research was carried out at the enterprises of the Volgograd region (Volgograd State Agrarian University, laboratory "Analysis of feed and livestock products", CJSC poultry farm "Volzhskaya", LLC "Donagrogaz") and Moscow city (Research Center "Cherkizovo").

In the course of scientific and economic experiments, the dairy productivity of cows and egg laying hens were taken into account. The scheme of the studies performed is shown in Figure 1.

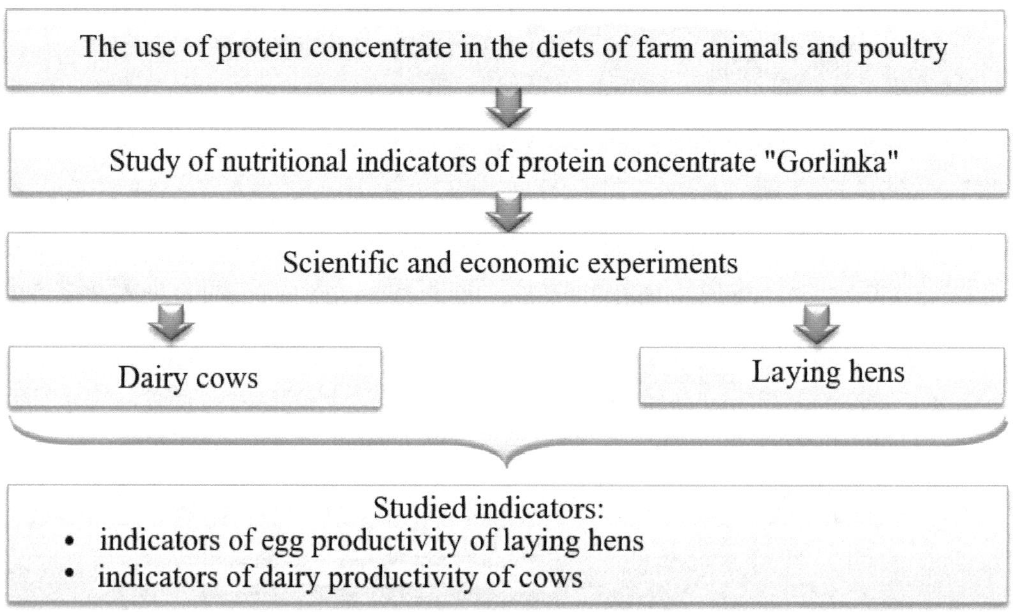

Figure 1. General scheme of research.

Based on the data obtained on the nutritional value of protein concentrate, oil cake and meal from sunflower seed, the following can be concluded. The studied feed additive showed its superiority over sunflower oil cake and meal in such indicators as: crude protein by 2.1% and 0.6%, respectively, crude fat by 0.4% and 5.6%, nitrogen-free extractives by 13.2% and 0.8%, crude ash by 1.4% and 0.2%, total amount of studied amino acids by 2.63% and 2.9%. At the same time, it was revealed that the studied concentrate had the leading position in terms of the main mineral elements, such as calcium, phosphorus, magnesium, iron, as well as vitamins of group B and vitamin D.

To conduct a scientific and economic experiment, animals were selected into groups by the method of pairs of analogues. When forming experimental similar groups, the age of animals, the main indicators of productivity, live weight, physiological condition, etc. are taken into account. 10 cows of the same age were selected for each of the groups.

The experiment was carried out according to the scheme presented in Table 1. Lactating cows of the I control group were given a diet, the concentrate part of which included sunflower oil cake in the amount of 1 kg per head per day. Animals of the II and III experimental groups were fed diets in which oil cake from sunflower seed was replaced by 50% and 100% of the studied protein concentrate.

Table 1. Scheme of the experiment on black-and-white cows.

Indicator	Groups of cows		
	I control	II experimental	III experimental
Number of cows in the group	10	10	10
Cow feeding conditions	The basic diet (BD) using oil cake from sunflower seed	BD + protein concentrate "Gorlinka" instead of 50% of sunflower oil cake	BD + protein concentrate "Gorlinka" instead of 100% of sunflower oil cake

Then, studies were conducted to study the effect of protein concentrate on the egg productivity of poultry. The experimental hens were kept in identical conditions in accordance with the guidelines for working with the Haysex Brown cross poultry and the recommendations of Federal Scientific Center "All-Russian Scientific Research and Technological Institute of Poultry Farming" of the Russian Academy of Sciences. The selection of young and adult laying hens into groups was carried out according to the generally accepted method of analogues.

Young poultry from the first day of life were grouped into 4 groups, of which there was one control (I) and three experimental (II, III, IV). The number of young animals in the group was 100 heads, adult birds was 60 heads. In terms of time, the experiment on young livestock lasted 120 days, and on an adult bird lasted 52 weeks. The scheme of the experiment on the egg poultry of the Haysex Brown cross is presented in Table 2.

The diet of the I control group consisted of compound feed, in which there was a greater amount of corn and wheat. During the experiment, protein concentrate was injected into poultry of the II, III and IV experimental groups as part of compound feed instead of 50%, 75% and 100% of sunflower seed meal, respectively.

Table 2. Scheme of the experiment on the egg poultry of the Haysex Brown cross.

Groups of hens			
I control	II experimental	III experimental	IV experimental
Feeding features			
Basic diet (BD)	BD + replacement of 50% of sunflower seed meal with protein concentrate "Gorlinka"	BD + replacement of 75% of sunflower seed meal with protein concentrate "Gorlinka"	BD + replacement of 100% of sunflower seed meal with protein concentrate "Gorlinka"

3. Results and discussion

Throughout lactation, the effectiveness of the productive use of animals can be assessed by milk productivity, which in particular can be judged by the average daily milk yield (Table 3) [10].

The average daily milk yields from the cows of the experimental groups differed somewhat. The use of the studied high-protein concentrate in the concentrate part of the diet had a certain effect on these differences. The best results for this indicator were observed in cows whose diets included the studied concentrate, 27.56 kg and 27.98 kg versus 26.12 kg in the I control group.

As you know, the percentage of fat in milk is a genetically stable and difficult-to-change indicator. In our experience, there were no differences in this indicator. However, in terms of protein content in milk, there were slight differences in favor of cows fed protein concentrate "Gorlinka", in the II and III experimental groups this indicator was 3.26% and 3.27%, respectively, against 3.23% in the I control group. A similar trend was observed in the percentage of dry matter in cow milk. The difference in favor of the experimental groups versus the I control group for this indicator was 0.04% and 0.09%.

Table 3. Indicators of dairy productivity of cows.

Indicator	Groups of cows		
	I control	II experimental	III experimental
Average milk yield per day, kg	26.12 ± 1.69	27.56 ± 1.78	27.98 ± 1.73
Fat in milk, %	3.61 ± 0.13	3.61 ± 0.13	3.61 ± 0.13
Protein in milk, %	3.23 ± 0.03	3.26 ± 0.03	3.27 ± 0.04
Dry matter in milk, %	12.99 ± 0.22	13.03 ± 0.25	13.08 ± 0.25
Dry skimmed milk residue, %	9.38 ± 0.08	9.42 ± 0.06	9.47 ± 0.06
Lactose in milk, %	5.38 ± 0.10	5.39 ± 0.10	5.42 ± 0.09
Ash in milk, %	0.77 ± 0.01	0.77 ± 0.01	0.78 ± 0.01

Partial or complete replacement of sunflower oil cake used in the basic diet with the studied concentrate had a positive effect not only on the quantity and quality of the products obtained, but also on economic indicators. In the II and III experimental groups, the profit from the sale of the received milk was higher compared to the I control group by 6028.00 rubles and 7810.00 rubles, respectively. An increase in the profitability of milk production amounted 46.67% in II experimental group and 49.15% in III experimental group.

At the end of the experiment to study the effect of protein concentrate on the egg productivity of poultry, the following results were obtained. The II experimental group of young poultry was the leader in live weight by 14.47 g compared to the I control group, in the III and IV experimental groups the live weight also exceeded the control by 44.85 g and 22.26 g. On average, eggs were obtained per laying hen in the II experimental group by 1.12% more than in the I control group, in the III experimental group by 4.25%, and in the IV experimental group by 3.07%. In terms of average egg weight, the II, III and IV experimental groups of hens were in the lead - from 1.2 to 2.6%, and there was also an increase in the content of protein, minerals and vitamins in eggs (the difference between the indicators is not reliable). The zootechnical indicators of the productive period of the egg poultry of the Haysex Brown cross are presented in Table 4.

Table 4. Zootechnical indicators of the productive period of the egg poultry of the Haysex Brown cross (M ± m).

Indicator	Groups of hens			
	I control	II experimental	III experimental	IV experimental
Live weight of young hens at the age of 4 months	1364.17 ± 33.87	1378.64 ± 33.67	1409.02 ± 32.83	1386.43 ± 34.19
Total eggs received	19350	19566	20172	19944
Eggs received per laying hen	322.5	326.1	336.2	332.4
Average egg weight, g	63.47 ± 1.90	64.24 ± 2.15	65.13 ± 1.96	64.91 ± 2.04
Feed costs per 1 kg of growth of young hens, kg	4.67	4.62	4.52	4.59
Feed costs per 1 kg of egg mass, kg	2.08	2.03	1.94	1.97

According to the results of the experiment, the feed costs per 1 kg of live weight growth of young hens in the groups that received protein concentrate as part of the feed were higher by 0.05 kg, 0.15 kg and 0.08 kg than in the I control group. The amount of compound feed consumed per 1 kg of egg mass in the hens of the experimental groups was less, in relation to the I control by 2.4-6.7%.

The presence of lead in the eggs of laying hens of all experimental groups was in the range from 0.01 to 0.03 mg/g, which is lower than the maximum permissible concentration. The concentration of cadmium, mercury, arsenic and cobalt was below the detection limit.

The difference in the cost of consumed compound feeds due to the use of protein concentrate "Gorlinka" for young hens in the II experimental group was 234.84 rubles, in the III experimental group was 389.34 rubles, and in the IV experimental group was 519.12 rubles. Additional profit was obtained by introducing the protein concentrate "Gorlinka" into the diet of laying hens instead of the traditionally used sunflower seed meal and it amounted to 2419.25-6162.82 rubles.

4. Conclusion

Based on the results obtained in the course of research, it can be argued that the greatest effect of production is achieved in animals and poultry that received protein concentrate with a diet: this allows increasing the level and quality of purchased products, as well as increasing the production of milk and edible eggs. The use of protein concentrate "Gorlinka" in diets increased the average daily milk yields of cows by 5.51-7.12% and egg productivity of laying hens by 1.12-4.25%. Additional profit when introducing protein concentrate instead of sunflower seed meal into the diet of hens was 2419.25-6162.82 rubles. Thus, the use of the provided protein concentrate is economically feasible.

References

[1] Shatskikh E V, Nufer A I and Drozdova L I 2020 Digestibility of dietary nutrients and morphohistological changes in the pancreas and duodenum of broiler chickens against the background of replacement of feed antibiotics with a complex phytobiotic drug *Agrarian Bulletin of the Urals* **12(203)** 61-67

[2] Nikolaev S I, Chekhranova S V, Karapetyan A K, Budtuev O A, Krikunov N A, Struk M V and Kornilova E V 2019 Changes in the physiological status of agricultural animals and poultry under the influence of biologically active additives *Advances in Animal and Veterinary Sciences* **7(1)** 100-105

[3] Baikovskaya E Yu, Abashkina E M and Manukyan V A 2021 Synthetic glycine in compound feed for broiler chickens *Poultry* **3** 13-16

[4] Egorov I, Egorova T, Shevyakov A and Kleinerman Yu 2021 High-quality fish meal in broiler chicken rations *Compound Feed* **1** 58-60

[5] Skvortsova L N 2020 Application of environmentally friendly additives with functional properties when growing broiler chickens *Polythematic Network Electronic Scientific J. of the Kuban State Agrarian University* **164** 55-69

[6] Saleeva I, Fisinin V and Lukashenko V 2019 Waste - in income *Animal Husbandry of Russia* **5** 8-9

[7] Lenkov T N, Egorova T A, Kashporov L M and Sysoeva I G 2020 Soy plus peas instead of soybean meal *Poultry and Poultry Products* **6** 31-34

[8] Buryakov N P, Buryakova M A and Zaikina A S 2021 Efficiency of application of protein concentrate in the rations of highly productive cows *Feeding of Farm Animals and Fodder Production* **2(187)** 15-25

[9] Nikolaev S I, Chekhranova S V, Karapetyan A K, Kantemirova M P, Shevchenko P A and Struk N V 2019 Improving the protein nutritional value of the diets for farm animals and poultry by introducing oil industry waste *Advances in Animal and Veterinary Sciences* **7(1)** 78-84

[10] Bakharev A A, Sheveleva O M, Chasovshchikova M A, Aleksandrova S S, Sukhanova S F and Koshchaev A G 2020 Milk yield and milk productivity of meat cow breeds of the Northern Trans-Urals *IOP Conf. Series: Earth and Environmental Science* **624** 012097

IOP Conf. Series: Earth and Environmental Science **965** (2022) 012032 doi:10.1088/1755-1315/965/1/012032

The influence of biologically active substances on safety of pigs

O N Polozyuk[1], M V Zabelina[2], E S Polozyuk[1], M E Rubanova[2], O V Kulagina[2] and L V Stupina[2]

[1] Don State Agrarian University, 24 Krivoshlykova Street, Persianovsky settlement, Oktyabrsky District, Rostov-on-Don, 346493, Russia
[2] Saratov State Agrarian University named after N I Vavilov, 1 Teatralnaya Square, Saratov, 410012, Russia

E-mail: polozyuk7@mail.ru, mvzabelina@mail.ru

Abstract. The results of the influence of the biologically active substance "Amylocin" on nonspecific factors of protection of the organism of sows in the second half of gestation and in the first decade after farrowing are presented. It was found that the cellular indications of the natural resistance of the organism throughout the experiment were higher in the experimental group of the lactating sows, that received the biologically active substance, in comparison with the gestating and lactating sows of the control group. Humoral indications of natural resistance of the organism: lysozyme activity of blood serum was higher in the experimental group of the lactating sows in comparison with gestating and lactating sows of the control group by 12.5 and 5.2%; On the contrary, bactericidal activity of blood serum was higher in group of sows during the second half of gestation by 18.9 and 24.9% than in the experimental and control groups of farrowing sows. There were no postpartum complications in the experimental group of sows, the safety of the obtained young growth was higher by 8.6% in comparison with the control group.

1. Introduction

The conditions for keeping, feeding and breeding pigs in individual farms are very different from the conditions of existence of pigs in their natural habitat, which sets a difficult task for scientists and specialists of agricultural enterprises in creating stable and unpretentious breeds of pigs that can increase the economic efficiency of this industry [1]. In this regard, research work is continuously being carried out to study the effect of biologically active substances in order to prevent diseases, due to which the growth and development of young growth is accelerated, the protective properties of the organism of sows are increased, the time for breeding of the obtained young growth is reduced, feed cost per unit of production is reduced and the profitability of this industry increases [2, 3]. For this, recently, not only feed antibiotics have been used, but also correctly selected natural or synthetic substances identical to natural ones, which strengthen health and prevent the development of diseases (parapharmaceuticals), which are biologically active substances (BAS), or substances that ensure the functional activity of the intestinal microflora (eubiotics) [4].

With the development of the microbiological industry, the number and varieties of dietary supplements are constantly increasing. Pigs as a biological object have their own characteristic features in the digestive system, which imposes certain rules in their normalized feeding [5]. Therefore, the purpose of our research was to study the effect of the probiotic "Amylocin" based on the bacteria of the type Bacillus of the bacillus B. subtillus and B. Amyloliquefaciens strains on the cellular and humoral blood indications of gestating and lactating sows, the presence of postpartum complications and the safety of the obtained offspring.

2. Materials and methods

The studies were carried out at JSC "Batayskoye" of the Azov district of the Rostov region on hybrid of large white and Landrace sows. For this purpose, an experimental and control group of main gestating sows, 15 heads each, were formed. During the periods of gestation and lactating, sows of all groups were kept under the same conditions, and they were fed according to the standards of Russian Research Institute of Animal Husbandry [6]. Sows of the experimental group, during the second half of gestation from day 80 to farrowing, were given 20 g of "Amylocin" each with feed. The sows of the control group did not receive the medicament.

To study the humoral and cellular factors of the organism's natural resistance, blood from gestating and lactating sows of the experimental and control groups was taken in the morning, at the same time, before feeding from the jugular vein at the beginning of the experiment (the 80th day of gestation) and on the 7th day after farrowing. Cellular and humoral indications of natural resistance were determined in the blood:

- lysozyme activity of blood serum - by the method of Dorofeychuk modified by Chebotkevich and Lyutinskiy (1998) photocolorimetric method using a daily culture of Micrococcus Ijsodeicticus;
- bactericidal activity of blood serum - according to Smirnova and Kuzmina (1966). Escherihia coli strain 04 was used as a test microbe;
- phagocytic activity of neutrophilic granulocytes, phagocytic index and phagocytic blood capacity according to Fedyuk et al;
- reaction of bacterial agglutination according to Birger (1982) [7].

The data obtained during the experiment were processed biometrically according to Plokhinsky (1970) and Merkuryeva (1970).

3. Results and discussion

Knowing of the characteristics of cellular and humoral factors of natural resistance of sows in different physiological states allows to interpret correctly and, if necessary, adjust the level of nonspecific organism protective properties to obtain healthy and viable offspring [8].

During the study, we were interested to find out which animals will have higher indications of nonspecific organism protective properties and how they will affect postpartum complications and the safety of the obtained offspring [9].

Throughout the entire observation period, when conducting studies on the effect of the probiotic "Amylocin", both during gestation and during lactation of the experimental animals, a decrease in response to external stimuli, depression, decrease or lack of the appetite, decrease in motor activity were not noticed. Aggressiveness or increased excitability towards the attendants were not observed. The gestation period proceeded without deviations from the physiological norm, i.e. abortion, toxicosis, appetite perversion were not observed. However, after farrowing, one of the sows in the control group had agalactia, and 60% of the obtained offspring of the other sow of the same group, were born dead. No postpartum complications were observed in the experimental group of the sows.

It is known that the immune system is one of the most important homeostatic factors of the organism. It largely determines the health of animals and their adaptive capabilities; in addition, it is an indicator system of ecological trouble and is sensitive to changes in the environment [10]. Analysis of Table 1 shows that the phagocytic activity of neutrophilic granulocytes on the fourth day after farrowing of the sows receiving the probiotic "Amylocin", after 30 minutes of phagocytosis was 3.2 ($P > 0.99$) and 7.5% ($P > 0.999$) higher than gestating sows and sows of the control group, that did not receive the biologically active substance, had.

The phagocytic number of the sows of the experimental group after farrowing was 9.5% higher than in the control group, but lower than of the gestating sows by 1.9%. The phagocytic capacity of the lactating sows of the experimental group also exceeded the gestating and lactating sows of the control group by 0.70 (P > 0.999) and 0.54 × 10^9/l (P > 0.999).

Table 1. Indicators of natural resistance of sows.

Indicators of natural resistance	Pregnant sows before administration of the drug	On the 7th day after farrowing	
		Experimental group	Control group
Cellular indicators of natural resistance			
Phagocytosis 30 minutes			
Phagocytic activity, %	47.10 ± 1.02**	50.50 ± 1.12	43.00 ± 1.04***
Phagocytic number, units	1.65 ± 0.06	1.62 ± 0.07	1.48 ± 0.04
Phagocytic blood capacity, 10^9/l	1.20 ± 0.03***	1.90 ± 0.01	1.36 ± 0.02***
Phagocytic index, units	0.76 ± 0.02	0.80 ± 0.03	0.60 ± 0.02
Humoral indicators of natural resistance			
Phagocytosis 120 minutes			
Phagocytic activity, %	43.00 ± 1.08*	47.20 ± 1.14	40.00 ± 0.96**
Phagocytic number, units	1.55 ± 0.10	1.43 ± 0.06**	1.35 ± 0.04***
Phagocytic blood capacity, 10^9/l	1.10 ± 0.01***	1.80 ± 0.02	1.03 ± 0.02***
Phagocytic index, units	0.67 ± 0.02	0.65 ± 0.02	0.51 ± 0.01**
PL	1.4	1.25	1.15
Bactericidal activity of blood serum, %	52.60 ± 1.40	33.70 ± 1.05***	27.70 ± 1.10***
Lysozyme activity of blood serum, %	14.90 ± 1.63***	27.40 ± 1.40	22.20 ± 1.32*

P > 0.95*, P > 0.99**, P > 0.999***.

After 120 minutes of phagocytosis, both in gestating and lactating sows of the experimental and control groups, the results obtained differed from the indicators of phagocytosis after 30 minutes in the direction of their decrease. However, after farrowing of the sows receiving the probiotic, the cellular indications of natural resistance were higher than phagocytic activity, phagocytic blood capacity and phagocytic index in control group by 7.20% (P > 0.99), 0.77 × 10^9/l (P > 0.999) and 0.14 units.

As we know, the system of organism natural resistance was formed in the process of evolution and the formation of new breeds of domestic animals. It determines their adaptability to exogenous and endogenous stresses, which are caused by the action of pathogens, unfavorable natural and climatic conditions, which is especially important in relation to animals that are kept in livestock buildings all year round. An important part in the natural protection of the organism is taken by the enzyme lysozyme, which is able to dissolve and destroy both positive and negative pathogenic microorganisms, thereby being a barrier to the penetration of pathogens into the organism [11].

Analyzing the obtained indicators of lysozyme activity of blood serum of the lactating sows of the experimental group, its increase in comparison with gestating and control lactating sows by 12.50 (P > 0.999) and 5.20% (P > 0.99), respectively. In the control group of the lactating sows, this indicator was 7.2% higher than in the group of gestating sows in favor of the sows in the control group. We associate these changes with a significant change of physiological processes in the organism of sows during farrowing, since the effect of lysozyme on bacteria is to dissolve specific mucopolysaccharides of the membranes of microbes, which also inhibits the growth and stops the vital activity of many bacteria [12].

The second studied indicator of humoral body defense (bactericidal activity of blood serum) was high in group of sows during the second half of gestation and amounted to 52.60%, which is 18.9% (P > 0.999) and 24.9% (P > 0.999) higher than in the experimental and control groups of lactating sows on day 7 after farrowing. In the control group of lactating sows, the difference in the bactericidal activity of thc blood serum was 5.20% (P > 0.95) less than in the experimental group that received the probiotic "Amylocin".

4. Conclusion

An increase in the cellular and humoral indications of the blood serum of lactating sows in the experimental group made it possible to increase their natural resistance of the organism, which had a positive effect on the obtained offspring, the safety of which increased by 8.6% and the absence of postpartum complications.

References

[1] Zhitnik I A, Fedyuk V V, Polozyuk O N and Fedyuk E I 2011 *Measures to Increase the Resistance of Pigs in the Conditions of the Rostov Region* (Persianovsky) p 11

[2] Karagodina N V, Polozyuk O N and Voitenko Yu S 2009 Influence of biogenic stimulators on natural resistance of pigs *Actual Problems of Pork Production in the Russian Federation* (Persianovsky) pp 111-113

[3] Fedyuk V V, Shatalov S V and Koshlyak V V 2007 *Natural Resistance of Cattle and Pigs* (Persianovsky) p 175

[4] Artemyeva O A, Perestrokova D A and Fomichev Yu P 2015 Biologically active drug as an alternative to the use of antibiotics against pathogenic microorganisms *Agricultural Biology* **50(4)** 513-519

[5] Polozyuk O N and Lapina T I 2015 Influence of conditions of keeping on fattening and meat qualities of animals *Agrarian Scientific J.* **2** 26-29

[6] Polozyuk O N, Kolesnikov I A and Polotovsky K A 2015 The influence of biologically active substances on the biochemical parameters of the blood of piglets *The use and Effectiveness of Modern Selection and Genetic Methods in Animal Husbandry* (Persianovsky) pp 152-154

[7] Birger M O 1982 *Clinical Laboratory Research* (Moscow: Medicine) p 582

[8] Galochkin V A, Ostrenko K S, Galochkina V P and Fedorova L M 2018 The relationship of the nervous, immune, and endocrine systems and nutrition factors in the regulation of animal resistance and productivity *Agricultural Biology* **53(4)** 673-686

[9] Polozyuk O N, Kolesnikov I A and Polotovsky K A 2016 Influence of biologically active substances on the physiological and biochemical status of purebred and crossbred piglets *Agrarian Scientific J.* **8** 23-25

[10] Fedorov Yu N 2004 Immunological monitoring in veterinary medicine. development trends, opportunities and reality *Agricultural Biology* **2** 3-9

[11] Chysyma R B, Fedorov Yu N, Makarova E Yu and Kuular G D 2015 Activity of humoral and cellular factors of natural resistance in local animal breeds in extreme natural and climatic conditions of the Republic of Tyva *Agricultural Biology* **50(6)** 847-852

[12] Maksimov G V, Polozyuk O N, Fedyuk E I and Kryshtop E A 2010 Natural resistance of pigs in the conditions of industrial cultivation technology *Veterinary Medicine* **9** 43-47

IOP Conf. Series: Earth and Environmental Science **965** (2022) 012033 doi:10.1088/1755-1315/965/1/012033

Improving the quality of broiler chicken meat without the use of antibiotics

M I Slozhenkina[1], A M Fedotova[1], O A Knyazhechenko[1], G V Fedotova[1, 2], A A Mosolov[1] and M V Frolova[1]

[1] Volga Region Research Institute of Manufacture and Processing of Meat-And-Milk Production, 6 Rokossovsky Street, Volgograd, 400131, Russia
[2] Russian State Economic University named after G V Plekhanov, 36 Stremyanny lane, Moscow, 117997, Russia

E-mail: niimmp@mail.ru

Abstract. The article presents impact study results of probiotic feed additives "Laktumin-1" and "Laktuvet-1" on growth, productive characteristics of broiler chickens of the "Cobb-500" cross and quality their meat. Study of meat quality and productivity of the chickens raised by technology without the use of antibiotics on the basis of experimental vivarium of Volga Region Research Institute of Manufacture and Processing of Meat-And-Milk Production was carried out in October 2020. The chickens were divided into three groups of one hundred chicks each and were raised up to 37 days of age. As a result of the studies, it was found that from the first ten-day periodof the experiment, broiler chickens receiving "Lactumin-1" and "Laktuvet-1" surpassed the analogs of the control group in live weight. In the experimental meat samples taken from their pectoral muscles, protein content was higher than in the control group.

1. Introduction

White meat is traditionally uses in human nutrition, including children and dietetic food, helps to normalize metabolism, quickly recover from surgery and serious traumas. Poultry meat is easily digestible, an excellent source of protein, considered an essential building block of a human body. This meat is economically available, easy to digest, and an excellent source of protein. This causes a stable increasing demand for it in the Russian Federation and increase in assortment of meat products based on it [1].

At the same time, modern mass breeding of chickens is associated with a large overcrowding of livestock, risks of infection and, as a result, a decrease in productivity and deaths of animals in large agricultural enterprises, complexes, holdings, which, in turn, necessitates the use of antibiotics. However, some producers, along with the use of antibiotic for prevention of seasonal diseases, use them as drugs to increase productivity of animals and poultry. Such a large-scale use of antibiotics in livestock production technology leads not only to increase of antibiotics content inside animals, but also to their transfer to meat and eggs. Ultimately, due to biomagnification, such drugs accumulate in human body, that increases risk of spread of pathogenic microorganisms, bacteria and viruses resistant to many types of antibiotics. That is why today the question of the use of antibiotic therapy in the practice of poultry farming to prevent various infections remains controversial, and the search of alternatives to these substances is becoming actual topic for research [2, 3, 4, 5].

Currently, a large amount of experimental data indicates that probiotics can function as such an alternative since they are immunomodulators, capable to correct dysbacteriosis, regulation of microbiological processes in the digestive tract, treatment and prevention of diseases of gastrointestinal tract of alimentary and infectious etiology [3, 6-10].

Special attention among probiotics deserves lactulose – a synthetic disaccharide consisting of galactose residues and fructose molecules, which are structural isomers of lactose. In small intestine, there are no disaccharidases, which are necessary for hydrolysis of lactulose, so it is not adsorbed reaches large intestine unchanged. At the same time, being a prebiotic, it has a positive effect on growth of beneficial intestinal microflora, in particular bifidobacteria and lactobacilli.

On the basis of lactulose, domestic research centers have developed a number of successfully used drugs. "Lactumin-1" is a feed additive for feeding farm animals and poultry, which is a composition of natural biologically active substances obtained by combining honey extract from fresh Jerusalem artichoke tubers, dry milk molasses with lactulose and succinic acid. It is an outwardly homogeneous mass amber or brown color, without foreign smell and taste, except for those caused by the components of the composition. The presence of a crystalline lactose precipitate is allowed. The developer of the feed additive "Lactumin-1" is Volga Region Research Institute of Manufacture and Processing of Meat-And-Milk Production, Volgograd. The manufacturer of the additive is the Scientific-Implementation Center "New Biotechnologies" LLC, Volgograd.

"Laktuvet-1" - is a composition of organic acids with a prevailing content of lactic - 5.2% and citric 2.3%, as well as nitrogen-containing peptide substances. The additive contains up to 97% of dry matters, including lactulose - 14.6%, lactose - at least 25.2%, monose (galactose, glucose) - at least 12.5%, calcium - 3.4-4.4%, phosphorus - 1.4-1.7%, potassium - 0.7-1.7%, magnesium - 0.5-0.7% and other macro- and microelements. The feed additive does not contain genetically engineered modified products. Content of harmful impurities does not exceed maximum permissible standards in force in the Russian Federation. The developers of feed additive "Laktuvet-1" are Engineering Center of Institute of Living Systems of North-Caucasus Federal University and Dairy company "Stavropolskiy". The bifidogenic additive "Laktuvet-1" is used as a feed additive in the process of raising farm animals and poultry at a recommended dosage of 0.2-0.4 g per kg of live weight.

Feed additives "Lactumin-1" and "Laktuvet-1" enable to activate natural mechanisms and physiological processes in gastrointestinal tract of animals and poultry, better assimilate feed, increase natural bioconversion of feed in their body, which will directly affect the growth rate.

In practice, there are already examples of the use of lactulose-containing preparations in the diet of poultry. A good increase in live weight was observed in the broiler chickens - by 2.9-3.2% higher than in the control group, which also made it possible to reduce feed costs per unit of live weight gain by 2.9-3.1% in the experimental groups.

In this regard, scientific research aimed at production of chicken without the use of antibiotics and their partial replacement with lactulose-containing preparations is gaining high relevance.

The aim of this work is to study the effect of probiotic feed additives "Laktumin-1" and "Laktuvet-1" on growth, productive characteristics of broiler chickens of the "Cobb-500" cross and quality their meat.

2. Materials and methods

Study of broiler meat quality produced by technology without the use of antibiotics on the basis of experimental vivarium of Volga Region Research Institute of Manufacture and Processing of Meat-And-Milk Production was carried out in October 2020 on broiler chickens of the Cobb 500 cross. The chickens were divided into three groups of one hundred chicks each and were raised up to 37 days of age.

Poultry were fed manually with dry complete compound feed. Chickens of the control group were given standard dry compound feed, experimental group I - to standard compound feed were added

"Lactumin-1" at 0.5% by weight of compound feed, the experimental group II - "Laktuvet-1" at 0.5% by weight of compound feed. The first two days chickens was allowed to eat plenty and then daily feed intake was controlled. The slaughter was executed on 37th day of the experiment. To study physical-chemical properties (content of protein, fat, moisture, ash) and amino acid composition of meat the samples of pectoral and femoral muscles were taken.

Meat analysis was carried out in laboratory of Volga Region Research Institute of Manufacture and Processing of Meat-And-Milk Production in according with following methods: GOST 31470-2012 "Poultry meat, by-products and semi-finished products from poultry meat. Methods of organoleptic and physical-chemical research"; GOST 33319-2015 "Meat products. Moisture Determination Methods"; GOST 23042-2015 "Meat and meat products. Methods for determination of fat content"; GOST 25011-2017 "Meat and meat products. Methods for determination of protein content"; GOST 31727-2012 "Meat and meat products. Method for determination of total ash content". To determine amino acid composition, the method of measuring mass fraction of amino acids was used on system Kapel-105M.

3. Results and discussion

The feed additives used in the experiment had a positive effect on live weight gain of broiler chickens, which is shown in Table 1.

Table 1. Dynamics of live weight.

Age, days	Index, g	Group		
		Control	Experimental I	Experimental II
1	Live weight	43.0 ± 0.14	42.9 ± 0.17	42.89 ± 0.22
10		191.0 ± 1.9	201.4 ± 2.1***	201.8 ± 1.7***
20		824.2 ± 11.6	880.4 ± 10.9***	890.3 ± 9.8***
30		1442.1 ± 17.4	1522.0 ± 16.7***	1530.2 ± 15.84***
37		2246.7 ± 19.35	2365.7 ± 18.64**	2386.7 ± 19.21**
1-37	Average daily gain	59.56 ± 0.14	62.35 ± 0.21***	63.35 ± 0.18***
	Relative growth rate	1.925	1.929	1.929

Starting from the first decade of the experiment, broiler chickens of experimental groups I and II surpassed the analogs of the control group by 5.44 ($P \geq 0.999$) and 5.65% ($P \geq 0.999$); 20 days - by 6.82 ($P \geq 0.999$) and 8.02% ($P \geq 0.999$); 30 days - by 5.54 ($P \geq 0.999$) and 6.11% ($P \geq 0.999$); 37 days - by 5.29 ($P \geq 0.99$) and 6.23% ($P \geq 0.99$), respectively. Over the entire period of the experiment, broiler chickens of the I and II experimental groups in terms of average daily gain were higher compared to the analogs of the control group - by 4.68 ($P \geq 0.999$) and 6.36%. The calculation of relative growth rate of broiler chickens up to 37 days of age was higher in the experimental groups compared to the control. After corresponding period of keeping, physical-chemical studies of chicken meat were carried out. The results are shown in Table 2.

Experimental meat samples taken from breast muscles of broilers of experimental groups I and II contained more protein than the control group by 0.54 and 0.89% ($P \geq 0.95$), respectively. However, moisture in meat samples obtained from broiler chickens of experimental groups I and II was less by 0.60 and 1.00% compared to the control group.

In the study of femoral muscles, it was found that broiler chickens of experimental groups I and II exceeded the analogs of the control group in protein content by 0.54 and 0.66% ($P \geq 0.95$), fat - by 0.03 and 0.07% respectively. However, in terms of moisture content in the femoral muscles, poultry of experimental groups I and II were inferior to the analogs of the control group by 0.58 and 0.75%.

Table 2. Physical-chemical characteristics (n = 3).

Index	Group		
	Control	Experimental I	Experimental II
	Pectoral muscles		
Protein content, %	23.71 ± 0.14	24.25 ± 0.23	24.60 ± 0.22*
Fat content, %	1.33 ± 0.03	1.35 ± 0.04	1.42 ± 0.05
Moisture content, %	73.7 ± 0.34	73.10 ± 0.24	72.70 ± 0.27
Total ash content, %	1.26 ± 0.08	1.30 ± 0.06	1.28 ± 0.07
	Femoral muscles		
Protein content, %	18.7 ± 0.12	19.24 ± 0.22	19.36 ± 0.17*
Fat content, %	8.35 ± 0.24	8.38 ± 0.33	8.42 ± 0.31
Moisture content, %	71.91 ± 0.37	71.33 ± 0.36	71.16 ± 0.29
Total ash content, %	1.04 ± 0.07	1.05 ± 0.06	1.06 ± 0.05

One of the most important indicators assessed when establishing quality of raw meat is determination of content of nonessential and irreplaceable amino acids. Table 3 shows essential amino acids content in femoral muscles of broiler chickens of the studied groups.

Table 3. Protein amino acid rate of femoral muscles.

Amino acid composition, g/100 g	Group		
	Control	Experimental I	Experimental II
Threonine	81.23	83.26	85.26
Tyrosine	120.34	124.23	126.34
Cystine	180.26	182.21	184.35
Valine	73.26	75.46	77.41
Methionine	95.64	95.73	96.28
Phenylalanine	136.67	138.19	140.21
Isoleucine	91.67	92.34	93.54
Leucine	88.75	89.13	91.24
Lysine	157.76	159.25	160.24
Tryptophan	143.53	145.18	147.62

As can be seen from the data in table 3, broiler chickens of experimental groups I and II, grown with the use of probiotic feed additives, surpass the analogs of the control group in terms of content of almost all studied amino acids in meat. The same dynamics was observed in study of pectoral muscle of broiler chickens.

When calculating the protein-quality indicator of the femoral muscle of broiler chickens - ratio of nonessential amino acids to essential amino acids, it was found that chickens of experimental groups I and II exceeded the analogs of the control group by 0.17 and 0.34, and in the case of pectoral muscle by 0.26 and 0.41 respectively.

Serum analysis showed that broiler chickens of experimental groups I and II exceeded analogs of the control group in terms of content of albumin and glucose. Globulins and phosphorus contents in the control group more than the analogs experimental groups I and II.

The organoleptic assessment carried out on a 5-point scale showed that taste and aroma of bouillon and chicken meat of experimental groups I and II exceeded the control group in terms of the overall score when assessing the broth by 0.13 and 0.21 points, and meat - by 0.16 and 0.26 points, respectively.

4. Conclusion

An experiment on growing broiler chickens of the Cobb-500 cross using lactulose-containing feed additives "Lactumin-1" and "Laktuvet-1" was carried out. As a result of the studies, it was found that from the first ten-day of the experiment, broiler chickens receiving "Lactumin-1" and "Laktuvet-1" surpassed the analogs of the control group in live weight by 5.44 and 5.65%; 20 days - by 6.82 and 8.02%; 30 days - by 5.54 and 6.11%; 37 days - by 5.29 and 6.23%, respectively; in the experimental meat samples taken from their pectoral muscles, the protein contained more than in the control group by 0.54 and 0.89%, and in femoral muscles - by 0.54 and 0.66%, respectively.

High biological value of the broiler chickens meat obtained in the experimental groups receiving feed lactulose-containing additives "Lactumin-1" and "Laktuvet-1", in comparison with the control group, was confirmed as a result of analysis of amino acid composition, organoleptic evaluation and calculation results of protein-quality indicator.

All foresaid allows us to recommend the use of biologically active feed additives in feeding broiler chickens to obtain high-quality meat.

5. Acknowledgements

This work was carried out under the grant of the Russian Science Foundation 21-16 00025 SSI NIIMMP. Grant sponsors were not directly involved in the development, analysis, or writing of this article.

References

[1] Krivoshlykov V S 2019 Price elasticity of demand in the food market **3(94)** 107-120

[2] Fisinin V I *et al* 2017 Antibiotic-free poultry production based on innovative nutritional programs with the involvement of probiotics *Problems of Nutrition* **86(6)** 114-124

[3] Miroshnik A S, Slozhenkina M I, Gorlov I F, Frolova M V, Nikolaev D V and Mosolova N I 2021 The use of biologically active supplements as an alternative for application of antibiotics in production of pork *IOP Conf. Ser.: Earth Environ. Sci.* **677** 032009

[4] Polomoshnov N A and Malysheva L A 2012 Efficiency of using probiotics for prevention of salmonellosis *Veterinary Pathology* **1** 52-56

[5] Kochish I I, Myasnikova O V, Martynov V V and Smolensky V I 2020 Intestinal microflora and expression of immunity-related genes in hens as influenced by prebiotic and probiotic feed additives *Agricultural Biology* **55(2)** 315-327

[6] Grin M S 2018 The use of lactulose in composition of compound feed KR-1 *Actual Problems of Intensive Development of Animal Husbandry* **22(1)** 178-184

[7] Ovcharova A N, Safronova O V and Polyakova L L 2019 Influence of probiotic lactobacilli on nonspecific resistance and physiological parameters of laboratory animals *Pharmacokinetics and Pharmacodynamics* **2** 28-31

[8] Miroshnik A S, Slozhenkina M I, Gorlov I F, Frolova M V, Nikolaev D V and Mosolova N I 2021 The use of biologically active supplements as an alternative for application of antibiotics in production of pork *IOP Conf. Ser.: Earth Environ. Sci.* **677** 032009

[9] Nikolaev D V, Slozhenkina M I, Gorlov I F, Miroshnik A S, Mosolova N I and Evdokimov I A 2021 Innovative technology of pork production with the use of phytobiotics *IOP Conf. Ser.: Earth Environ. Sci.* **677** 032006

[10] Makarov D J, Rujeinnikov F V, Shkalenko V V and Nikolaev D V 2012 Influence lactulosecontaining fodder additives on efficiency experimental young pigs *Pig Breeding* **5** 61-72

IOP Conf. Series: Earth and Environmental Science **965** (2022) 012034 doi:10.1088/1755-1315/965/1/012034

Environmental influence on the features of the formation of placental structures in cow

Yu I Velikorodnaya[1], S A Prikhodko[2], M A Ushakov[2], Zh Sh Ushakova[2], D S Sharonov[2] and V A Antonov[1]

[1] Federal State Unitary Enterprise "Research Institute for Hygiene, Toxicology and Occupational Pathology" at Federal Medical and Biological Agency, 12 Zemlyachki Street, Volgograd, 400048, Russia
[2] Volgograd State Agrarian University, 26 University Avenue, Volgograd, 400002, Russia

E-mail: alta-u@mail.ru

Abstract. Normal organization of placental structures at different stages of pregnancy is critical for the growth and development of the fetus. The aim of this study is to characterize the localization and ratio of epithelial and mesenchymal cells in the placenta of cows with physiological and complicated preeclampsia pregnancy using the immunohistochemical method. To date, it has been established that adverse environmental factors adversely affect the reproductive potential of animals regardless of their gender. The results of the study demonstrated that in the placenta of cows with the development of preeclampsia in the last stages of pregnancy, there was a significant decrease in the content of keratin filaments both in the trophoblasts of the villous tree of the fetus and in the epithelial cells of the maternal part in comparison with physiological values. In this case, the vimentin filaments also underwent partial degradation. Thus, the ratio of keratin and vimentin intermediate filaments can be used as a histological criterion for changes in the placental structures of cows during the development of preeclampsia.

1. Introduction

Successful gestation depends on both effective implantation and placentation, and an efficient blood supply between mother and fetus is vital in the first trimester. Therefore, any abnormalities in the functioning of the placenta can have a negative effect on the developing fetus, leading to intrauterine growth retardation, asphyxia, multiple organ failure, premature birth and perinatal death.

To date, it has been established that adverse environmental factors adversely affect the reproductive potential of animals regardless of their gender [1, 2, 3].

One of the most common complications of pregnancy in mammals is preeclampsia, a multisystem disease resulting from abnormal placentation due to insufficient invasion of trophoblasts into the spiral arteries of the hemochorial type of placenta [4, 5]. And if the pathogenesis of preeclampsia in mammals with hemochorial placentas is being studied quite actively, then the features of the initiation and development of this pathological state in mammals with other types of placenta remain outside the sphere of interest of researchers.

According to the modern classification based on a complex of morphological characters, the placenta of cows belongs to the synepitheliochorial type. An earlier assumption that the uterine epithelium

IOP Conf. Series: Earth and Environmental Science **965** (2022) 012034 doi:10.1088/1755-1315/965/1/012034

degrades during placentation, as a result of which the trophectoderm is directly attached to the maternal connective tissue, turned out to be untenable. It has now been established that the placenta of a cow is formed by the migration and fusion of a certain type of trophectodermal cells with epithelial cells of the uterus, thereby forming a hybrid fetal-maternal syncytium, represented by trophoblastic binuclear / giant cells [6, 7].

The trophoblast cells are represented by the placental epithelium, which develops rapidly during pregnancy, creating an organ responsible for the transport of nutrients and gas exchange between the blood of the mother and the fetus. To maintain the integrity of the cell structure in the cytoplasm, there is an internal framework of intermediate filaments of types I and II - keratins, which are necessary for the normal development and functioning of the placenta. It has been established that defects in the keratin cytoskeleton of trophoblasts (degradation or, conversely, the formation of aggregates) can lead to embryonic lethality [8, 9, 10]. In turn, placental stromal cells are of extraembryonic mesenchymal origin and are important for the formation and maintenance of villi, while cytoskeletal mesenchymal cells consist of another type of intermediate filaments - vimentin.

The aim of this study is to characterize the localization and ratio of epithelial and mesenchymal cells in the placenta of the synepitheliochorial type in physiological and complicated preeclampsia pregnancy as a morphological criterion for changes in the placental structures of cows with the development of a pathological condition.

2. Materials and methods

For pathomorphological examination, samples of placentas from cows with normal pregnancy and diagnosed preeclampsia were fixed in 10% neutral formalin. Then the samples were dehydrated and clarified, and embedded in a paraffin medium according to a conventional technique. Sections, 5 μm thick, were mounted on glass slides and stained with hematoxylin-eosin.

Paraffin sections of placenta samples 5 μm thick, subject to immunohistochemical and immunofluorescence studies, were mounted on glasses treated with poly-L-lysine («Menzel»).

Immunohistochemical reactions were performed using a peroxidase-polymer imaging system according to the manufacturer's instructions (LabVision). To block endogenous peroxidase, the sections after dewaxing were incubated for 20 minutes in 3% hydrogen peroxide. Antibodies were unmasked by boiling the sections at 100 °C in citrate buffer with pH = 6.0 for 10 minutes. The slides were incubated overnight at 4 °C with antibodies to keratin-18 (mouse, monoclonal, clone C-04) and vimentin (rabbit, monoclonal, clone EPR3776). Peroxidase was developed with 3-3-diaminobenzidine from the protocol kit. At the final stage of the reaction, the sections were counterstained with Mayer's hematoxylin. Preparations without incubation with primary antibodies in full compliance with the remaining stages of the protocol served as a negative control.

In an immunofluorescence study, sections after dewaxing were incubated in a glycine solution to reduce autofluorescence. Antibodies were unmasked by boiling the sections at 100 ° C in citrate buffer with pH = 6.0 for 10 minutes. Then the slides were placed horizontally in a humid chamber, 2 types of primary antibodies were applied: mouse and rabbit, and left overnight at 4 °C. The next day, slides with sections were washed in two shifts of 0.05 M Tris buffer for 20 minutes and first incubated for 1 hour with secondary antibodies conjugated to Alexa 594 (goat anti-rabbit, Abcam), then for another 30 minutes with secondary antibodies conjugated to FITC (goat anti-mouse, Abcam). Washed with buffer for a total of 30 minutes, then a few drops of embedding medium containing DAPI (Sigma) were applied to the sections. Micro-preparations were studied and photographed using an AxioScope.A1 (ZEISS) microscope equipped with a high-resolution digital camera AxioCamMRc5. The resulting photos were processed using the ZET pro 2012 (ZEISS) program.

IOP Conf. Series: Earth and Environmental Science **965** (2022) 012034 doi:10.1088/1755-1315/965/1/012034

3. Results and discussion

3.1. Results of the study of the content of intermediate filaments of the epithelial and mesenchymal phenotype in the placenta of a cow during the physiological course of pregnancy

The results of the carried out immunohistochemical study demonstrated that the cytokeratin skeleton of the cells lining the lacunae of the uterine septa occupied the entire space of the cytoplasm. Chorionic villi giant trophoblast cells also tested positive for the keratin-18 intermediate filament, but the protein was recorded predominantly on the surface of the cytoplasmic membrane (Figure 1).

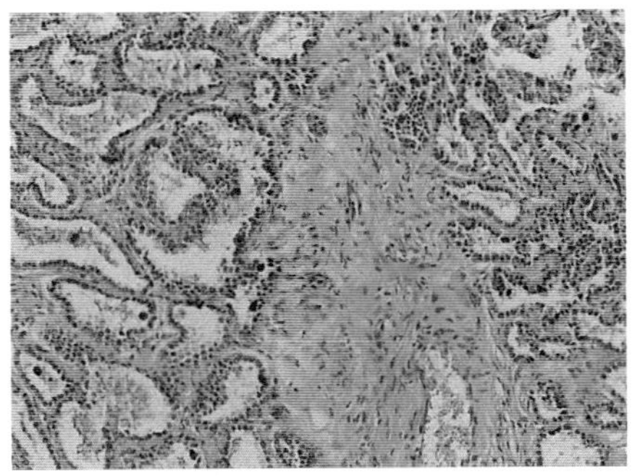

Figure 1. Immunolocalization of K-18 in the placenta of a cow during physiological pregnancy. The presence of keratin-18 in the cells of the lining of the uterine crypts, as well as in the trophoblasts of the chorionic villi. Antibodies to keratin-18. Method – «Applied Behavior Analysis». Staining with Mayer's hematoxylin. Total magnification 200.

The study of the immunolocalization of mesenchymal filaments revealed that vimentin + cells, mainly fibroblasts, constituted the connective tissue matrix of the uterine septa, and were also part of the stroma of the chorionic villi. In the central part of the villi, cells immunopositive to vimentin, forming the capillary walls - endotheliocytes, were also recorded (Figure 2).

A visual assessment of the number of capillaries showed that, according to morphological features, the bulk of the stem and terminal villi were mature and well vascularized, which ensured good intrauterine development of the fetus.

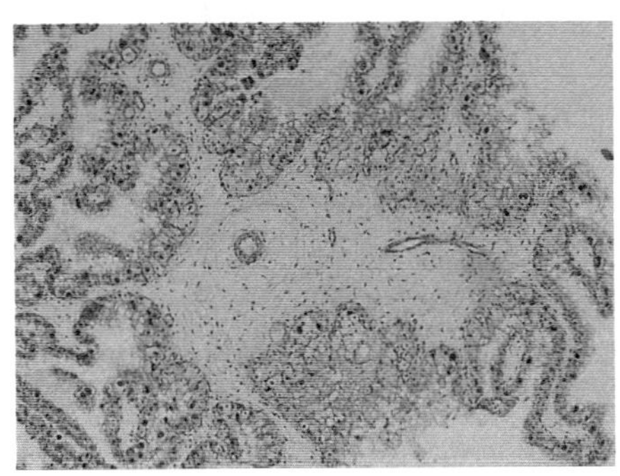

Figure 2. Immunolocalization of vimentin in the placenta of a cow during normal pregnancy. imentin-positive structures that are part of the uterine septa, as well as the stroma of the chorionic villi and capillary walls (indicated by arrows). Antibodies to vimentin. Method – «Applied Behavior Analysis». Staining with Mayer's hematoxylin.
Total magnification 200.

3.2. Results of the study of the content of intermediate filaments of the epithelial and mesenchymal phenotype in the placenta of cows during pregnancy complicated by preeclampsia

Histological examination of the maternal part of the placenta of cows with pregnancy complicated by eclampsia revealed hyperplasia of the uterine crypts. At the same time, the free space of the crypts formed by the septa turned out to be critically reduced, and the villi in such lacunae were almost completely reduced.

The study of keratin-18 revealed a sharp decrease in the amount of protein in the epithelial cells of the lining of the uterine crypts. A pronounced decrease in the content of epithelial-type filaments on the surface of trophoblasts was also noted (Figure 3).

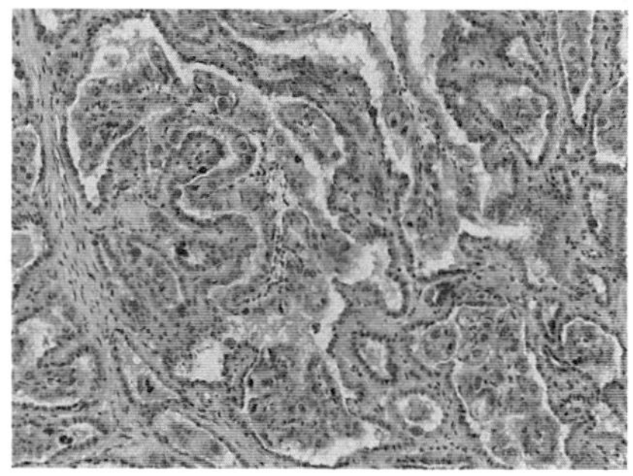

Figure 3. Immunolocalization of K-18 in the placenta of a cow in complicated pregnancy with eclampsia. Degradation of the keratin skeleton in cytotrophoblasts and lack of response to epithelial filaments in the cells of the lining of uterine crypts. Antibodies to keratin-18. Method – «Applied Behavior Analysis». Staining with Mayer's hematoxylin. Total magnification 200.

The results of the immunohistochemical reaction to vimentin confirmed the degradation of the connective tissue elements that made up the skeleton of the intermediate and terminal villi: only structureless masses and a small number of macrophage-like cells were recorded in the lumen of the lacunae. In the central part of the unchanged villi, a sharp decrease in the number of small capillaries was noted (Figure 4).

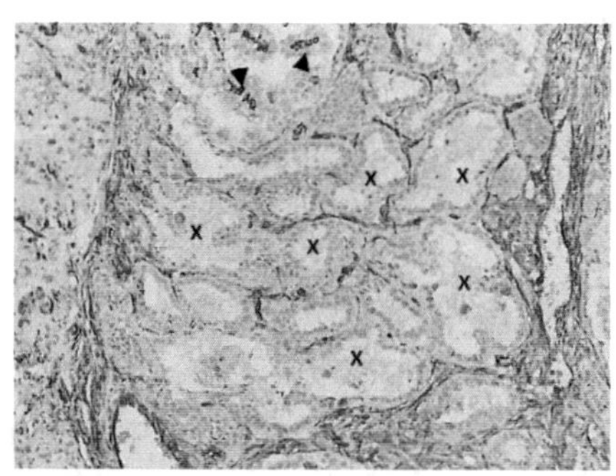

Figure 4. Immunolocalization of vimentin in the placenta of a cow in complicated pregnancy with eclampsia Reduction of terminal chorionic villi (indicated by crosses) and decreased vascularization of the remaining villi (indicated by arrows) Antibodies to vimentin. Method – «Applied Behavior Analysis». Staining with Mayer's hematoxylin. Total magnification 200.

3.3. Results of the study of the ratio of intermediate filaments of the epithelial and mesenchymal phenotype in the placenta of cows with physiological and complicated preeclampsia pregnancy

The results of studying the ratio of epithelial and mesenchymal elements using the immunofluorescent method of double staining of antibodies on one preparation confirmed that a significant decrease in keratin filaments both in trophoblasts of the villous tree of the fetus and in epithelial cells of the maternal part occurred in the placenta of cows during the development of preeclampsia in the last stages of pregnancy, compared to physiological values (Figure 5). In this case, vimentin filaments also underwent partial reduction, which indicated a premature degradation of the extracellular matrix.

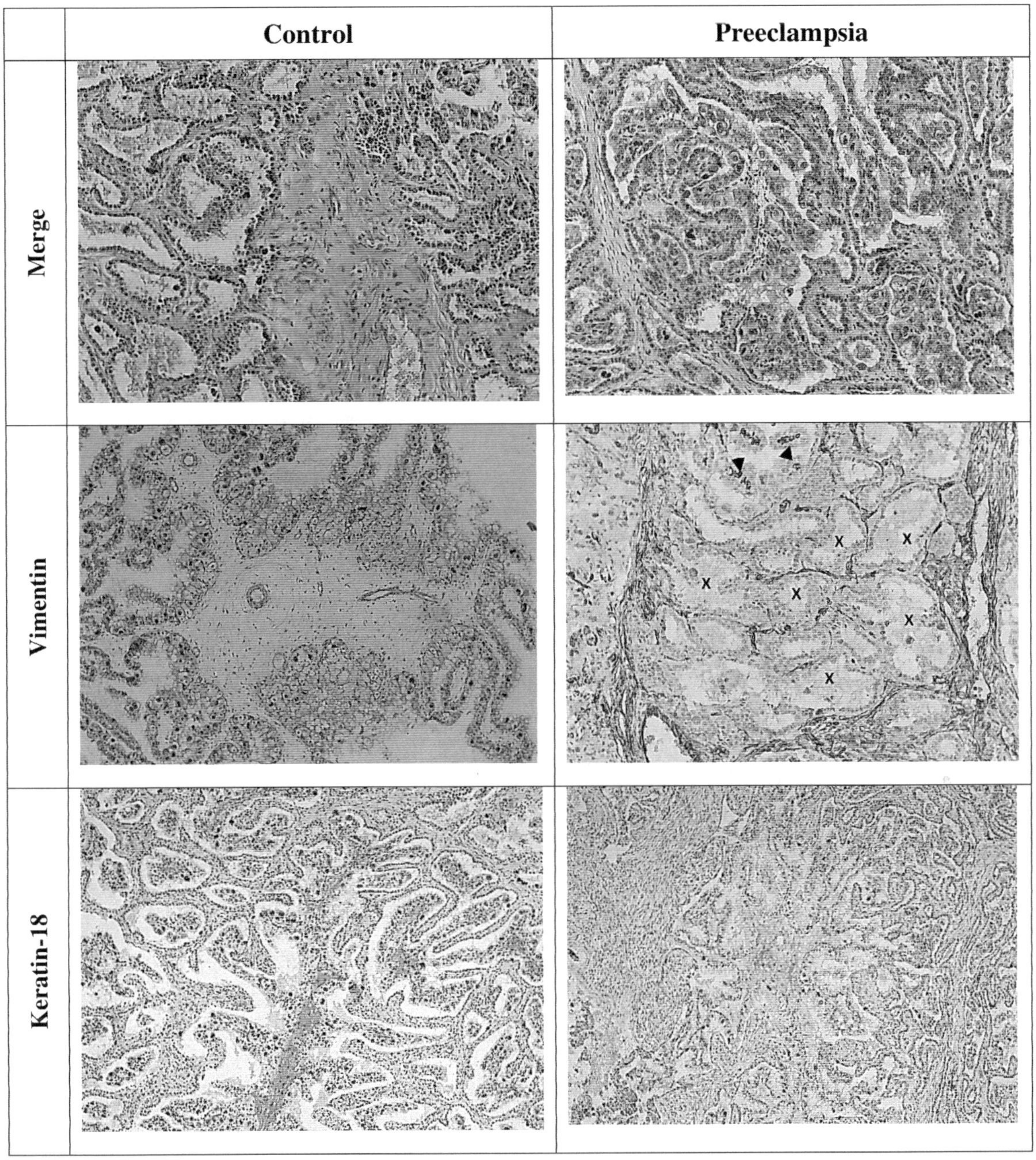

Figure 5. The ratio of epithelial and mesenchymal elements in placental structures during physiological pregnancy and complicated preeclampsia. Immunofluorescence method. Kernels are finished with DAPI. Total magnification × 200.

4. Conclusion

Understanding the patterns of development and involution of placental structures at different stages of gestation is of great importance for the early diagnosis of pathological conditions in farm animals, especially cattle. However, in the scientific literature there is a lack of experimental data on what molecular mechanisms underlie the pathogenesis of preeclampsia in cows, while the state of pathological retention of membranes has been studied in sufficient detail.

The analysis of the content of intermediate filaments of the epithelial and mesenchymal phenotype in the placental structures of a cow during the physiological course of pregnancy made it possible to identify the main morphological criteria in the structure of the organ:

1. The presence of a large number of small capillaries both in the chorionic villi and in the uterine septa.
2. Equilibrium ratio of mesenchymal and epithelial structures in the placenta during normal pregnancy.
3. Content of a large amount of keratin-18 intermediate filament in crypt lining cells, which play an important role in the formation of hybrid fetal-maternal syncytium.

On the other hand, the analysis of epithelial and mesenchymal filaments in the placenta of a cow during pregnancy complicated by preeclampsia demonstrated the following morphological signs of the pathological condition:

1. Reduction of vimentin filaments, accompanied by a decrease in vascularization of chorionic villi.
2. Degradation of keratin filaments involved in the formation of cell contacts between cytotrophoblasts of the villous tree of the fetus and epithelial cells of the maternal part of the placenta.

It should be noted that the processes of degradation of the extracellular matrix of the placenta also proceed normally, however, the time frame of these changes is shifted towards the end of pregnancy and, together with vascular disorders of the local type, they cause prenatal transformation of placental structures.

As our study has shown, for the physiological course of pregnancy, an appropriate balance of keratin and vimentin filaments in the structures of the placenta is required. Thus, it has been established that the keratin cytoskeleton in the trophoblasts of the mouse and human placenta is necessary to maintain the structural integrity of these cells and the development of the placenta, and keratin deficiency causes both defects in trophoblast differentiation and the integrity of the villi. In turn, vimentin filaments play a significant role in villous contractility and modulation of the intervillous space, thereby affecting the efficiency of blood circulation between the mother and fetus. Therefore, changes in the balance or simultaneous degradation of keratin and vimentin filaments in the structures of the placenta of cattle can be a criterion for the disruption of the placentation process with the subsequent development of preeclampsia.

Thus, our study showed that in the placenta of cows with preeclampsia, degradation of the cytokeratin skeleton, which is involved in the formation of cell contacts in the hybrid fetal-maternal syncytia, occurs, which ultimately disrupts the intrauterine development of the fetus. Therefore, the ratio of keratin and vimentin intermediate filaments can be used as a histological criterion for changes in the placental structures of cows during the development of preeclampsia.

References

[1] Guvvala PR, Ravindra JP and Selvaraju S 2020 Environ Sci Pollut Res Int. **27(4)** 3819-3836.
[2] Dobson H, Elsie J, Robert R and Smith F 2020 Anim Reprod **17(3)**
[3] Gathiram P and Moodley J 2016 *Cardiovasc. J. Afr.* **27** 71-78
[4] Haeger J D, Hambruch N and Pfarrer C 2016 *Theriogenology* **86** 306-312
[5] Jaquemar D, Kupriyanov S, Wankell M, Avis J, Benirschke K, Baribault H and Oshima R G 2003 *J. Cell. Biol.* **161** 749-756
[6] Attupuram N M, Kumaresan A, Narayanan K and Kumar H 2016 *Mol. Reprod. Dev.* **83** 287-297
[7] Boro P, Kumaresan A, Pathak R, Patbandha T K, Kumari S, Yadav A, Manimaran A, Baithalu R K, Attupuram N M and Mohanty T K 2015 *Anim. Reprod. Sci.* **157** 11-16

[8] Watson E D 2007 *Placenta* **28** 111-115
[9] Benedictus L, Thomas AJ, Jorritsma R, Davies CJ and Koets A 2012 *Am. J. Reprod. Immunol.* **67** 224-230
[10] Beham A, Denk H and Desoye G 1988 *Placenta* **9** 479-492

IOP Conf. Series: Earth and Environmental Science **965** (2022) 012035 doi:10.1088/1755-1315/965/1/012035

Vegetable oil various types together with enzyme preparation influence on broiler chickens' meat productivity and quality

V A Zlepkin[1], V V Salomatin[1], A A Ryadnov[1], N A Zlepkina[1], M N Mishurova[1], T A Ryadnova[1] and Yu A Kurskaya[2]

[1] Volgograd State Agrarian University, 26 University Avenue, Volgograd, 400002, Russia
[2] Smolensk State Agricultural Academy, 10/2 Bolshaya Sovetskaya Street, Smolensk, 214000, Russia

E-mail: zlepkin@mail.ru

Abstract. The article provides data on the vegetable oil various types together with the "CelloLux-F" enzyme preparation influence on the broiler chickens of the meat cross "Cobb-500" meat productivity and quality indicators. Chickens of the control group got compound feed with sunflower oil; the experimental group I received the same compound feed and additionally 100 g/t of the "CelloLux-F" preparation, the experimental group II received compound feed, which was added with mustard oil (instead of sunflower oil) and additionally 100 g/t of the "CelloLux-F" preparation, the experimental group III received compound feed, which was added with camelina oil (instead of sunflower oil) and additionally 100 g/t of the "CelloLux-F" preparation. It was found that young animals of the experimental groups, in comparison with the control group, the weight of the drawn bird increased by 4.2-9.2%, the edible carcass parts - by 4.3-9.6%, the pectoral muscles - by 5.42-10.96%. The most expedient, both from a zootechnical and from an economic point of view, is the use of camelina oil (instead of sunflower oil) together with the enzyme preparation "CelloLux-F", which will increase the poultry meat production profitability level by 18.02%.

1. Introduction

The problem of providing the population with high-quality competitive food products, including poultry meat, is considered the most important for the agricultural industrial complex. The solution to this problem is possible not only by increasing the poultry genetic potential, but also through scientifically based balanced feeding.

In this regard, in the market economy conditions, the issue of increasing the poultry breeding efficiency based on the rational use of local food means, research and production of new unused feed products with a high content of nutrients and energy, such as oilseed cakes (oilseed sunflower, camelina, rape, etc.), which combine well the best potential seed productivity with a high content of oil and protein [1, 2, 3, 4], is of great importance.

It was proved that high doses of protein not associated with the energy of the diet, lead to a negative effect, that is, to the decrease in the protein use, the most expensive component of the diet, since its part is used to replace the body's energy needs. To improve the protein use, it is necessary to increase the energy part of the diet by adding high-calorie components, such representatives can be fodder fats and vegetable oils [5].

Poultry feeding programs in Russia take into account the rather high use of wheat, barley, bran, peas, sunflower processing by-products, rapeseed, etc. in compound feed. At the same time, non-starchy polysaccharides contained in these ingredients have a negative effect on the compound feed energy use by poultry, as well as for the digestibility and assimilation of other nutrients. This, in turn, is reflected in the decrease in its productivity. [6, 7].

In recent years, the genetic potential of poultry productivity has significantly increased. However, the poultry productivity genetic potential realization is impossible without proper and high-quality feeding. An integral part of modern diets are enzymatic preparations that allow to increase the level of feed assimilation, optimally balance the feed and at the same time optimize the cost of rations through the use of more accessible, but difficult to digest components [8, 9, 10].

The purpose of our research was to study vegetable oil various types in combination with the "CelloLux-F" enzyme preparation influence on growth and development, digestibility and use of nutrients, morpho-biochemical parameters of blood, broiler chickens meat productivity and quality.

2. Materials and methods

To carry out a scientific, economic and physiological experiment by the method of analogs, 4 groups of broiler chickens of the meat cross "Cobb-500" (control and three experimental ones) at the age of one day were formed, 100 heads each. The period for breeding broiler chickens was 40 days. The experiment was carried out in the conditions of JSC "Poultry Farm Krasnodonskaya" in the Ilovlinsky district of the Volgograd region.

Throughout the experiment, the chickens were fed with 4 brands of compound feed (CF): CF-0 at the age of 1 to 4 days; CF-2 from 5 to 14 days; CF-5 from 15 to 28 days and CF-6 from 29 to 40 days.

In terms of the compound feed ingredients set in the control and experimental groups, they practically did not differ during all the breeding periods. The difference was that broiler chickens of the control group received a compound feed, which was supplemented with sunflower oil, during the breeding; broiler chickens of the experimental group I received the same compound feed as broiler chickens of the control group, but they additionally got 100 g/t of the "CelloLux-F" enzyme preparation; broiler chickens of the experimental group II received compound feed, which included mustard oil (instead of sunflower oil) and 100 g/t of the "CelloLux-F" enzyme preparation; broiler chickens of the experimental group III received compound feed, which included camelina oil (instead of sunflower oil) and 100 g/t of the "CelloLux-F" enzyme preparation.

In accordance with the adopted technology, broiler chickens were housed in a floor housing with deep-litter system. The experimental broiler chickens were kept in the building separately in groups in specially fenced off sections with a stocking density of 16.4 heads per 1 m^2 during 40 days with free access to water and feed. All microclimate parameters for all groups were the same and corresponded to the recommendations for breeding broiler chickens of the "Cobb-500" cross.

During the scientific and economic experience, the following indicators were taken into account and studied:

- livestock survival rate - with daily poultry inspection;
- broiler chickens live weight - by individual weighing on electronic scales;
- chemical composition of feed, metabolic products were determined according to the general method of zootechnical analysis;
- nutrients digestibility in the diet, nitrogen, calcium, phosphorus balance and use were determined in a digestible trial;
- morphological and biochemical blood tests were carried out on 6 broiler chickens from each group;
- meat productivity - by control slaughter and complete anatomical cutting of dressed chicken at the age of 40 days, 6 heads from the group (3 cockerels and 3 pullets) according to the All-Russian Scientific Research and Technological Institute of Poultry Breeding method;

- chemical composition and energy nutrition of the pectoral muscles (according to the All-Russian Scientific Research and Technological Institute of Poultry Breeding method);
- tasting assessment of meat and broth was determined in accordance with the All-Russian Scientific Research and Technological Institute of Poultry Breeding method.

3. Results and discussion

Live weight is considered the most important indicator of the broilers growth and development and one of their productivity results. The control over the movement of live weight makes it possible, even during the chickens' life, to judge the meat productivity and some processes associated with the development of the whole organism, allows to plan feed costs per unit of live weight gain and the economic efficiency of breeding the young birds. The growth rate of experimental broiler chickens can be judged by live weight at different age periods [11].

How the broiler chickens growth proceeded when using various types of vegetable oil in compound feeds in combination with the "CelloLux-F" enzyme preparation can be judged by the change in live weight during the entire breeding period (40 days), as well as by the average daily gain in poultry live weight.

As the research result, it was found that the broiler chickens live weight at the age of one day was 45.18-45.54 g. At the age of 7 days, the broiler chickens live weight between the groups differed slightly and varied within 157.2-159.6 g. At the age of 14 days, broiler chickens of the experimental group exceeded the control group by 9.8-20.3 g or 2.5-5.3%, at the age of 28 days - by 5.3-12.5 g, or 0.41-0.96%. At the end of the breeding period (40 days), the broiler chickens live weigh in the experimental groups was 2364.90-2484.40 g, exceeding the control group by 38.65-119.2 g or 1.7-5.1%, average daily increase - by 1.67-5.14%. For the entire breeding period (1-40 days) the survival rate in the control group was 94.8, in the experimental group, respectively, it was 95.8; 96.9 and 99.0%, which is more by 1.0; 2.1 and 4.2%, compared with the control group.

Compound feed containing the same amount of nutrients and energy has a different effect on the poultry productivity. This is largely due to the fact that the nutrients of the feed components have different digestibility and availability.

During the experiment, it was found that the feed dry matter digestibility coefficient in broilers of the experimental groups was higher than in the control group, respectively, by 1.32 ($P < 0.01$); 1.44 ($P < 0.01$) and 1.58% ($P < 0.01$), crude protein - by 1.69; 2.33 ($P < 0.01$) and 3.39% ($P < 0.001$), crude fat - by 0.16; 0.85 and 0.99% ($P < 0.05$), crude fiber - by 1.70 ($P < 0.01$); 3.81 ($P < 0.001$) and 4.72% ($P < 0.001$) and nitrogen-free extractive substances - by 1.69; 2.39 ($P < 0.001$) and 2.54% ($P < 0.001$).

In broiler chickens of the experimental groups, the nitrogen utilization coefficient was higher, in comparison with the control one, by 5.44, respectively; 6.40 and 7.44% ($P < 0.001$), calcium - by 1.03; 2.64 ($P < 0.05$) and 5.65% ($P < 0.001$), phosphorus - by 3.54 ($P < 0.001$); 6.22 ($P < 0.001$) and 6.98% ($P < 0.001$).

The determination of the blood constituent parts number quantitative and qualitative content is extremely important for the broiler chickens body health.

The hematological parameters in the experimental broiler chickens were within the physiological norm. At the same time, broiler chickens of the experimental groups exceeded the control group in terms of hemoglobin content, respectively, by 4.36; 5.88 and 6.87%, the number of erythrocytes - by 6.0; 7.33 and 8.33%, leukocytes - by 4.47; 5.59 and 4.0%. The total protein content in the experimental groups was 2.84 more; 3.79 and 5.68% than in the control one. The concentration of calcium was lower in the control group, in comparison with the experimental groups, by 7.45; 12.94 and 10.98%, phosphorus - by 8.03; 14.66 and 7.33%.

Meat productivity is the most important economically useful property of poultry. It is characterized by poultry live weight and meat quality at slaughter age.

It was found that the increase in the broiler chickens pre-slaughter weight in the experimental groups in relation to the control group was: in the experimental group I - 38.1 g or 1.6%, in e the experimental group II - 68.4 g or 2.9% and in the experimental group III - 118.5 g or 5.1%.

A similar trend was observed for drawn bird weight. Thus, the experimental groups in terms of drawn bird weight exceeded the control group by 67.8 g (4.2%), 106.1 g (6.5%) and 150.2 g (9.2%), respectively. The broiler chickens slaughter yield in the experimental groups was 1.74-2.85% higher than in the control group. The number of the 1-st grade carcasses in all experimental groups was high from 97.34 to 98.41%, which is 0.13-1.03% more than in the control group. The highest yield of meat of the 2-nd category was found in the experimental group I and amounted to 2.53%. Non-standard carcasses were noted only in the control group in the amount of 1 piece, which amounted to 0.44%.

A more detailed assessment of the broiler chickens meat productivity was carried out during the carcass anatomical dressing, followed by chemical analysis of the meat and assessment of its tasting qualities.

The results of carcasses anatomical dressing indicate that broiler chickens of the experimental groups exceeded the control one in terms of the edible parts yield by 56.8 (4.3%), 89.7 (6.8%) and 127.4 g (9.6%).), total muscle mass - by 44.95 (4.3%), 71.1 (6.8%) and 100.9 g (9.7%). In terms of the pectoral muscles mass, the superiority of the experimental groups was 5.42-10.96%, the thigh muscles - 4.35-10.12% and the drumstick muscles - 4.3-10.35%, compared with the control group.

The meat nutritional value is characterized by the content of nutrients in it - proteins and fats. It is determined by its chemical composition and calorie content.

During the broiler chicken carcasses anatomical dressing the compared groups, the pectoral muscles, crural, drumstick and truncus were taken to study their chemical composition and energy nutrition.

Characterizing the broiler chickens muscles chemical composition and energetic nutritional value of all groups, it was found that there was more protein, respectively by 0.80; 0.94; 0.81; 0.95 and 0.83; 1.00%, but less fat by 0.49; 0.18; 0.47; 0.19 and 0.50; 0.20% in the broiler chickens pectoral muscles (white meat) of the experimental groups, in comparison with the thigh and drumstick muscles. Therefore, nutritionally white meat has higher dietary properties than red meat.

Along with the physicochemical analysis, one of the important places belongs to the organoleptic assessment, which results are final and decisive in determining the meat quality.

The tasting assessment of the broth in the experimental groups ranged from 4.37 to 4.56 points and in the control group it was 3.90 points, and the assessment of cooked meat was 4.45-4.54 points, respectively, and 4.13 points in the control group.

Economic indicators of the vegetable oil various types using together with the "CelloLux-F" enzyme preparation effectiveness when growing broiler chickens indicate that the yield of meat products was: in the experimental group I by 5.76 kg or 3.79%, in the experimental group II - 11.90 kg or 7.83% and in the experimental group III - 17.97 kg or 11.83% more compared to the control group. Feed consumption per 1 kg of live weight gain was 1.88 kg in the control group, from 1.85 to 1.78 kg in the experimental groups, which is 1.6 - 5.3% less than in the control group. Profits from the meat products sale were received in the experimental groups by 864.5; 1089.8 and 1925.7 rubles more than in the control group. The level of broiler chickens meat production profitability in the experimental groups was 22.42; 24.39 and 31.12%, up 6.67; 8.64 and 15.37% higher than in the control group.

4. Conclusion

Summarizing the results of the studies, it can be noted that the use of mustard and camelina oils as the source of fats in broiler chicken fodder together with the "CelloLux-F" enzyme preparation (instead of sunflower oil) does not reduce feed intake, livestock survival rate, growth rate, allows to reduce the cost per unit of production and increase the meat production profitability level. The highest economic effect is achieved when camelina oil (instead of sunflower oil) is added to the broiler chicken feed together with the "CelloLux-F" enzyme preparation.

References

[1] Burlakova L V, Koshelev S N and Loshkomoinikov I A 2006 Oilcakes are important source of biologically active, energy-intensive, high-protein substances *Milk and Beef Cattle Breeding* **8** 21-24

[2] Terekhin G V 2006 Influence of camelina oil on the use of nutrients in the diet and productivity of broiler chickens of the "Competitor-3" cross *Scientific Conf. of Young Scientists of the Russian State Agrarian University - Moscow Agricultural Academy named after K A Timiryazev* pp 40-44

[3] Nikolenko L A, Boyko L V and Chernyshov N A 2006 Ripe cake and oil in poultry feeding *Feeding Farm Animals and Feed Production* **7** 64-65

[4] Lisitsyn A N 2011 By-products and waste of fat and oil production *Oils and Fats* **1(119)** 38-46

[5] Matyushkin V G 2004 Influence of fat and fatty acids ratio level on the chickens productivity *Poultry* **7** 15-16

[6] Lenkova T N, Egorova T A and Sysoeva I G 2016 New domestic enzyme *Poultry* **6** 17-20

[7] Egorov I A, Manukyan V A and Okolelova T M 2018 *Guidelines for Feeding Poultry* (Sergiev Posad) p 226

[8] Chegodaev V, Merzlyakova O and Zhdankova G 2004 Domestic production enzymes in poultry rations *Poultry Farming* **3** 28-29

[9] Lenkova T N, Egorova T A and Sysoeva I G 2016 New domestic enzyme *Poultry Farming* **6** 17-20

[10] Golovko A N 2012 Broiler meat productivity and quality *Poultry Farming* **9** 25-27

[11] Zlepkin A F, Zlepkin V A and Bayer T A 2014 The influence of the drug "Karcesel" separately and together with the "CelloLux-F" enzyme preparation on the rearing flocks reproductive organs growth and development intensity *Proc. of the Nizhnevolzhsky Agrouniversity Complex: Science and Higher Professional Education* **1(33)** 137-140

IOP Conf. Series: Earth and Environmental Science **965** (2022) 012036 doi:10.1088/1755-1315/965/1/012036

Protein and prebiotic feed additives: influence on the quality indicators of rabbit meat

A G Zolotareva[1], A N Struk[2], A A Mosolov[1], I F Gorlov[2], M I Slozhenkina[2] and O A Knyazhechenko[2]

[1] Volga Region Research Institute of Manufacture and Processing of Meat-And-Milk Production, 6 Rokossovsky Street, Volgograd, 400131, Russia
[2] Volgograd State Technical University, 28 Lenin Avenue, Volgograd, 400066, Russia

E-mail: genzol5@mail.ru

Abstract. Feeding is one of the key factors affecting the growth and development of rabbits, as well as the quality of the meat obtained. The article discusses the possibility of joint use of additives in the form of a suspension of chlorella, which contains a large amount of protein and vitamins, as well as "Lactuvet", which includes a powerful prebiotic lactulose. In the course of the study, the dynamics of the growth of animals was analyzed, an assessment of the hematological and biochemical parameters of blood, as well as the main physicochemical parameters of meat was carried out. As a result, it was concluded that the proposed additives have a positive effect on the weight gain of animals, as well as on the nutritional value of the resulting meat.

1. Introduction

Rabbit breeding is one of the most relevant areas of animal husbandry today. Interest in raising rabbits is largely due to the fact that rabbits are distinguished by their fertility and early maturity. This direction is actively developing in the south of the country. In addition, rabbit meat is distinguished by high organoleptic characteristics and nutritional properties. Their meat is white, juicy, tender, and contains a large amount of easily digestible protein [1].

The main question facing breeders is feeding. It is feeding that largely determines the yield of meat and its quality. Analyzing different diets of feeding in many, there is often a lack of any nutrients. In addition, antibiotics are often used by manufacturers to stimulate growth. This leads to the fact that antibiotics end up in the human body, which negatively affects health [2]. The issue of obtaining safe and high-quality meat is still relevant. Agricultural researchers are studying various feed additives that are as productive as antibiotics. And an important aspect here is also the production of meat with high nutritional and biological value [3, 4].

Chlorella is a type of green algae and is a supplement with high nutritional and biological value. Thus, the content of protein in the dry mass is about 40-50% of protein and includes all essential amino acids. In addition, it contains a large amount of carotenes, B vitamins, vitamins C, D, E [5, 6].

"Lactuvet" is a feed additive, the main function of which is to normalize the intestinal microflora and optimize digestion processes in farm animals. This effect is achieved due to the prebiotic lactulose contained in the supplement, which affects the pH-environment of the intestine, creating favorable conditions for the development of beneficial microflora [7, 8].

It is assumed that the combined use of these two feed additives will significantly increase the nutritional and biological value of meat by improving metabolic processes and better digestibility of nutrients contained in large quantities in the chlorella suspension.

2. Materials and methods

2.1. Setting up an experiment

A number of studies have been conducted to analyze the effects of additives. For this, the animals were selected and formed into 3 groups. For feeding, compound feed PK3-93 and additives were used as a basic ration, depending on the study group. The first group - control - did not receive any additives, the second - A - received an additive in the form of a suspension of chlorella in an amount of 0.5 l per 10 kg of feed, the third - B - together with a suspension of chlorella received an additive "Lactuvet" at the rate of 5 kg per 1 ton of feed. The experimental scheme is shown in Table 1.

Table 1. Experiment scheme.

Group	Number of animals	Diet
Control	10	PKZ-93
A	10	PC 3-93 + Chlorella
B	10	PC 3-93 + Chlorella + "Lactuvet"

2.2. Assessment of the growth and development of animals

The change in live weight was determined by weighing the animals at the beginning of the experiment, 45 days after the experiment was set up, and at the end of the experiment at the age of 120 days. Absolute and relative growth rates were calculated in accordance with generally accepted formulas.

2.3. Research of the main indicators of meat and blood of animals

At the age of 120 days, control slaughter of animals and assessment of physical and chemical parameters, as well as assessment of hematological and biochemical parameters of blood were carried out. The study of physical and chemical parameters, such as the content of protein and fat, was carried out on the basis of the Volga Research Institute. The determination of the mass fraction of total protein in the product was carried out according to GOST 25011-2017 using the complex for the determination of nitrogen and protein by the Kjeldahl method "Keltran". The study of the mass fraction of fat was carried out according to GOST 23042-2015. The mass fraction of moisture and dry matter was determined according to GOST 33319-2015, the mass fraction of ash - according to GOST 31727-2012. The study of hematological and biochemical parameters of blood was carried out using equipment BiochemSa instrument (High Technology, inc., USA).

2.4. Statistical data analysis

The analysis of the obtained data was carried out using the Microsoft Office software package. The article uses generally accepted standardized methods of analyzing the objects under study. Ethical approval. The authors confirm that the experiments on animals were in accordance with the principles of the European Convention for the Protection of Vertebrates Used for Experiments or Other Scientific Purposes.

3. Results and discussion

3.1. Analysis of animal growth and development

To determine the effectiveness of the proposed feed additives, the dynamics of the live weight gain of rabbits throughout the experiment was analyzed. The results of the studies are presented in Table 2.

Table 2. Dynamics of live weight of rabbits.

Indicator	Group		
	Control	A	B
Live weight, g: The			
beginning of the experiment (age 45 days)	962.4 ± 4.17	957.2 ± 1.48	969.5 ± 3.5
Mid-experiment (age 90 days)	2190.2 ± 7.51	2317.4 ± 9.34	2389.1 ± 7.9
End of the experiment (age 120 days)	3012.5 ± 9.8	3243.8 ± 12.73	3303.6 ± 9.6
Absolute increase	2050.1 ± 5.65	2286.6 ± 11.3	2334.1 ± 6.28

The results of the study show that the experimental group of animals had higher growth and development rates in comparison with the control group. Thus, the absolute increase was higher by 11.5% and 13.3% in groups A and B, respectively, in comparison with the control group. This difference can be explained by the high protein content in the chlorella suspension fed in both experimental groups. A higher indicator in group B compared to group A suggests that the combined use of chlorella and "Lactuvet" supplements has a positive effect on metabolic processes.

3.2. Analysis of the main indicators of meat and blood of animals

Nutritional value is the main indicator for meat, which characterizes its quality, in particular the protein content in it. The results of the study are presented in Table 3.

Table 3. Physical and chemical parameters of meat.

Indicator	Group		
	Control	A	B
Mass fraction of protein, %	26.51 ± 0.1	26.85 ± 0.07	27.2 ± 0.15
Mass fraction of fat, %	0.25 ± 0.05	0.75 ± 0.25	0.75 ± 0.05
Mass fraction of moisture, %	71.2 ± 1.1	72.5 ± 1.8	72.5 ± 0.6
Mass fraction of dry matter, %	28.8 ± 0.4	29.5 ± 0.2	29.5 ± 0.6
Mass fraction of total ash, %	1.99 ± 0.27	1.92 ± 0.26	2.25 ± 0.3

During the analysis, an improvement in the main quality indicators of the prototypes was noted in comparison with the control one. Thus, the mass fraction of protein increased in comparison with the control group by 1.2% and by 2.6% in groups A and B, respectively. The mass fraction of fat in the experimental samples also increased significantly in comparison with the control. The amount of moisture in the test samples increased by 1.4%. The growth of these indicators indicates an increase in the nutritional value of the meat of the experimental groups. Thus, we can talk about the effectiveness of the use of additives, especially chlorella, which contains a large amount of protein. In addition, higher indicators of group B in comparison with group A, in particular, an increase in the amount of protein, indicates that "Lactuvet" contributes to the improvement of metabolic processes and normalization of the intestinal microflora of animals. To assess the health status of the animals, a blood test was conducted before slaughter. It is the blood that characterizes the main metabolic mechanisms in the body, as well as its normal functioning. The results of the study are presented in Table 4.

The analysis of the main hematological parameters allowed us to judge that the level of red blood cells and hemoglobin increased in comparison with the experimental sample. This indicates an improved supply of oxygen to the animal's body, which in turn affects the speed of metabolic processes. The level of white blood cells in the control group significantly exceeds the norm (<10 109/l), which indicates possible inflammatory processes occurring in the body. The level of white blood cells in the experimental group is within the normal range, indicating the normal functioning of the immune system. The indicators of total protein in the blood serum of the experimental groups slightly exceed those of the control group and are also within the normal range. The level of calcium and phosphorus in the experimental groups is higher in comparison with the control group. Thus, calcium is higher by 16.6%

and 38.8%, respectively, for group A and B. The phosphorus level increased by 9.9% and 24.2%, respectively, for groups A and B compared to the control group.

Table 4. Results of animal blood tests.

Indicator	Group		
	Control	A	B
Red blood cells, 10^{12}/l	2.71	2.9	2.75
Hemoglobin, g/l	98.0	105.0	99.0
White blood cells, 10^{9}/l	15.0	4.9	5.9
Total protein, g/l	62.02	62.57	62.53
Calcium, mmol/l	0.9	1.05	1.15
Phosphorus, mmol/l	0.91	1.0	1.13

4. Conclusion

Analyzing the results of research, we can say that the use of a protein supplement in combination with a prebiotic in the diet of rabbits had a positive effect not only on the growth dynamics of animals, while not having a negative impact on their body, but also on the quality indicators of the resulting meat. Thus, a significant increase in weight was observed in the experimental groups in comparison with the control, in particular in the group where the additives were used together. There was also an increase in the nutritional value of the meat of the experimental groups of animals in comparison with the control. All this indicates that chlorella, as a source of protein, has a positive effect on the quality of meat, and "Lactuvet" affects the intestinal microflora and contributes to the ongoing metabolic processes.

5. Acknowledgments

This work was carried out under the grant of the Russian Science Foundation 21-16 00025 SSI NIIMMP. Grant sponsors were not directly involved in the development, analysis, or writing of this article.

References

[1] Tinaev N I, Zhvakina A R and Kharlamov K V 2016 Morphological composition of carcasses and cuts in purebred and hybrid young rabbits *Rabbit Breeding and Fur Farming* **2** 10-13

[2] Volchenkova E S and Frolov D I 2018 The possibility of using chlorella suspension as an alternative to antibiotics in animal husbandry *Innovative Technology and Technology* **2** 19-22

[3] Gorlov I F, Semenova I A, Knyazhechenko O A, Mosolov A A and Karpenko E V 2020 Assessment of the impact of new complex feed additives in the production of rabbit meat *IOP Conf. Series: Earth and Environmental Science* **548** 82053

[4] Semenova I A, Gorlov I F, Knyazhechenko O A, Mosolov A A, Slozhenkina M I and Zolotareva A G 2021 Improving rabbit meat productivity: the effect of antioxidant feed additives on meat quality *IOP Conf. Series: Earth and Environmental Science* **677** 32067

[5] Frolova M V, Moskovets M V, Ptitsyna L A and Toropov A Yu 2021 Chlorella suspension as a biostimulator in feeding of young cattle *Agrarian and Food Innovations* **2** 34-39

[6] Sidorenko Yu A and Maslyuk A N 2019 The practice of using algae in animal feeding and its effectiveness *Youth and Science* **12** 43

[7] Khramtsov A G *et al* 2021 Protein-carbohydrate feed additives with lactulose in the diets of lactating cows *Veterinary Medicine* **3** 59-64

[8] Slozhenkina M I, Gorlov I F, Khramtsov A G, Komarova Z B, Frolova M V, Kurmasheva S S and Rudkovskaya A V 2021 Growing broiler chickens using new feed additives based on lactulose *Poultry and Poultry Products* **1** 17-20

IOP Conf. Series: Earth and Environmental Science **965** (2022) 012037 doi:10.1088/1755-1315/965/1/012037

Opportunities for the development of domestic agricultural production in the context of global challenges and threats

M A Kholodova[1], O A Kholodov[2], M I Slozhenkina[3], D A Mosolova[3], M O Vasilieva[3] and O P Shakhbazova[4]

[1] Federal State Budget Scientific Institution "Federal Rostov Agricultural Research Centre", 1 Institutskaya Street, Rassvet village, Aksai district, Rostov region, 346735, Russia
[2] Rostov State University of Economics, 62 Ostrovsky Lane, Rostov-on-Don, 344000, Russia
[3] Volga Region Research Institute of Manufacture and Processing of Meat-and-Milk Production, 6 Rokossovsky Strret, Volgograd, 400131, Russia
[4] Don State Agrarian University, 24 Krivoshlykov Street, Persianovsky, Oktyabrsky district, Rostov region, 346493, Russia

E-mail: niimmp@mail.ru

Abstract. The article presents the opportunities for the development of domestic agricultural production in the context of global challenges and threats. The article studies qualitative and quantitative characteristics of economic growth in agriculture in Russia. Particular attention is paid to the trend of innovation and investment activity in agriculture. Tendencies of innovation processes development in separate branches of economy are regarded as the instrument of realization of global competitive advantages of domestic agriculture. The article proves that modern functioning of the branch occurs contrary to natural laws of development. Researches have shown that under conditions of global cataclysms the development of the branch occurs mainly due to its plant-growing orientation and large-scale exhaustion of natural, labor, financial, material and technical resources. It is proved that the external well-being of agricultural commodity producers is a result of the tightened belts. It has been noted that the paradox of functioning of domestic agricultural production is in the fact that, not providing the simple reproduction of consumed resources, the branch shows the profitability. It is proved that domestic agricultural production with high production potential can demonstrate more intensive development, based on the search for breakthrough innovative technologies.

1. Introduction

The objective processes of successful development of post-industrial society assume the predominance of the service sector over the real economy, its gradual displacement from the structure of GDP, including agricultural production, in conditions of improving the material well-being of the population. This criterion is an indicator of the degree of economic development of individual states. The specific weight of the cost of agricultural production in the structure of GDP of the world powers is from 3.0-5.0%. Particularly in the USA it is 1.1%, in Germany - 0.6% [1], in Russia - 3.4%, while in India its share reaches 17.0%, in the poorest countries of Africa - 50-70.0% [2]. However, the coronovirus pandemic has forced the entire world community to rethink the fact that under conditions of global crises and threats, the importance of agricultural production becomes paramount.

In these circumstances, it is impossible not to agree with the opinion of Rousseau: "The only way to keep a state independent from anyone is agriculture. If you have all the wealth in the world, if you have nothing to eat, you depend on others... Trade creates wealth, but agriculture ensures freedom".

2. Material and methods

The research methodology is based on the standard set of methods of economic analysis necessary to identify opportunities for the development of domestic agricultural production in the context of global challenges and threats. The information-empirical basis of the study was the data of Rosstat.

3. Results and discussion

The policy of food embargo and the implementation of export-oriented strategy stimulated the growth of investment in innovative agricultural production "from farm to counter". Thus, between 2014 and 2019, the level of investment increased 2.6 times from $2.9 billion to $7.6 billion [3]. Attracted financial resources were mainly aimed at processing and logistics technologies, the use of agro-biotechnology, bioenergy and bio-materials, "smart" farms, new farming systems (Figure 1).

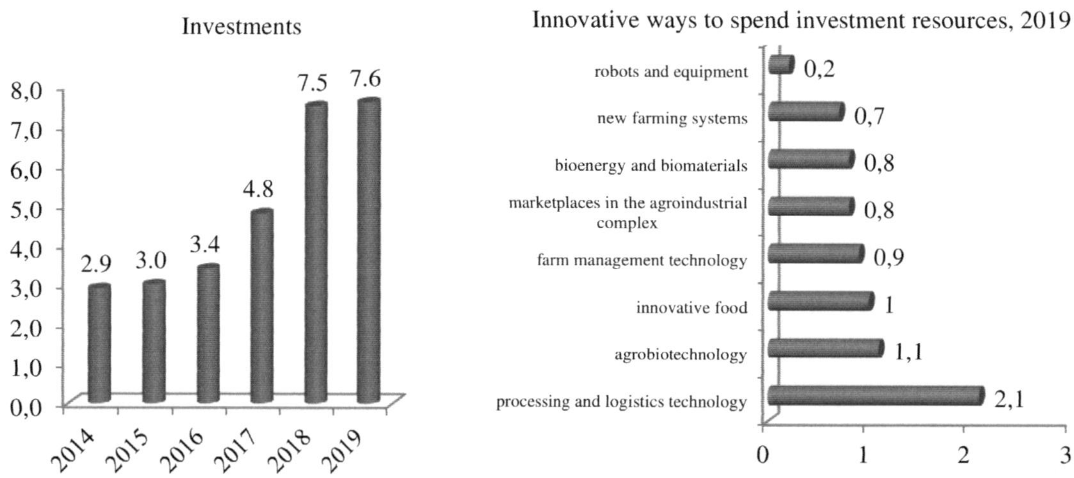

Figure 1. Dynamics of investment activity in the innovation sector of agriculture in Russia for 2014-2019, $ billion [1].

Despite the positive growth dynamics of investment activity in innovative technologies of agricultural production, the main source of their funding still remains the own funds of large enterprises-integra tors, who understand the commercial benefit and can bear such significant costs, as well as credit resources (Figure 2).

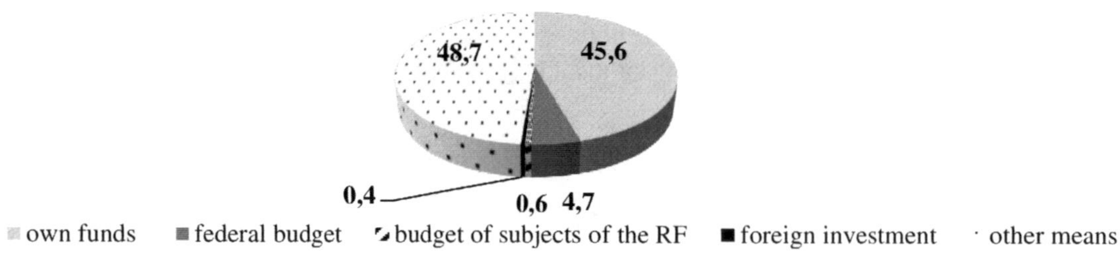

Figure 2. Sources of financing innovative activity in 2019, % [4, 5].

The research showed that in 2019 the share of innovative goods in the total volume of sales of the agricultural sector was 1.9% (for comparison, this indicator was 13.7% and 6.0%, respectively, in services and industry (Figure 3). The share of organizations that carry out technological innovations in the agricultural sector reached 6.4% (against 2.1% in the service sector and 18.5% in industry); the volume of innovation costs in their total structure - 1.2%.

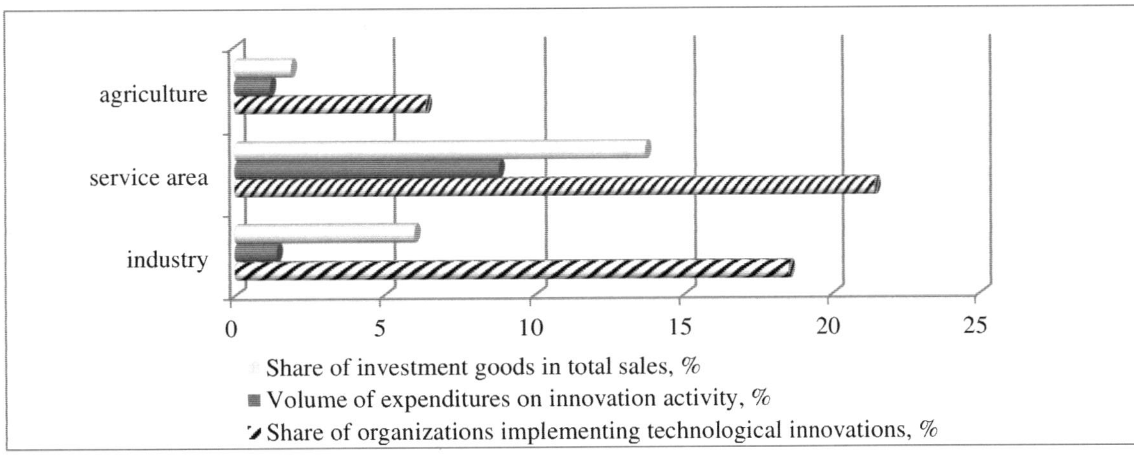

Figure 3. Trends in the development of innovation processes in individual sectors of the Russian economy in 2019, % [4, 5].

Studies have shown that the poor financial condition of a significant part of economic entities, together with the deficiencies of organizational-economic mechanism of economic management do not provide the formation of a new innovation-technological way of life in agriculture in Russia. Thus, there is a high level of depreciation of the main fleet of agricultural machinery in the domestic crop production sector (Figure 4).

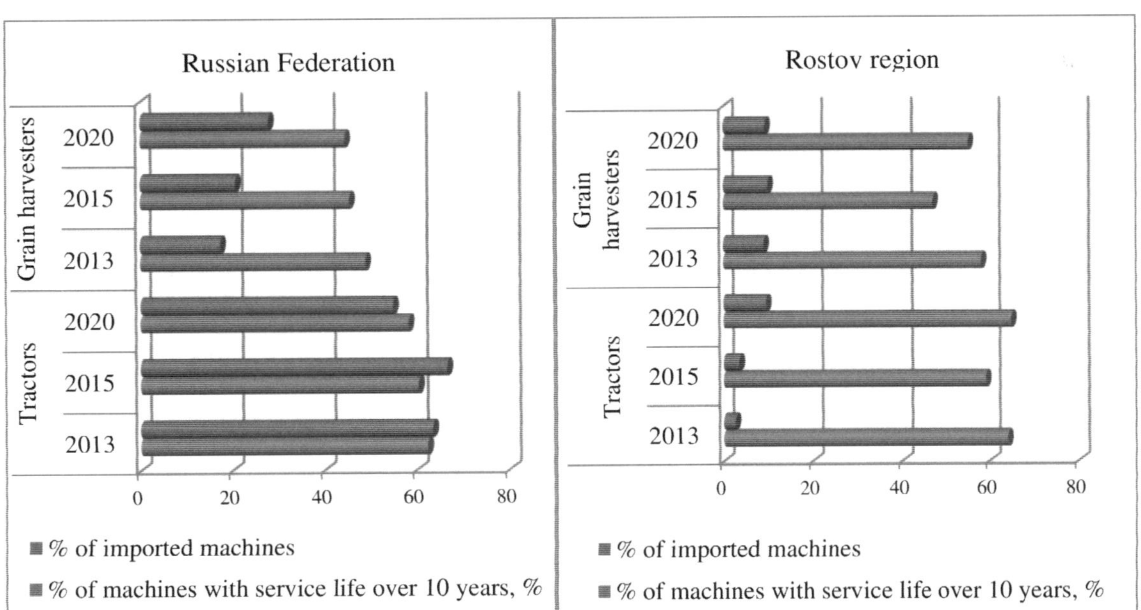

Figure 4. Characteristics of the machine and tractor fleet of the agrarian sector in Russia 2013-2020 [6].

In 2020 the number of tractors with service life over 10 years in domestic agricultural producers was 58.2%, grain harvesters - 45.1%, forage harvesters - 42.7%, which significantly reduces its productivity, increases crop losses, entails significant costs for the restoration and maintenance of machinery and tractor fleet [7].

Other reasons include significant dependence of domestic production on imported components and technologies, as evidenced by the high proportion of imported agricultural machinery. Particularly tractor park is equipped by 54.9% with foreign technique, share of imported grain and forage harvesters is 27.9% and 28.6% respectively, which creates a serious threat to national independence of the state, and not the one mentioned in Doctrine of food security, but shadow and very dangerous [7].

The study of the structure of machinery in the Rostov Oblast, one of the major agricultural regions of the country, the leader in the production of grain and oilseeds, shows that an important problem for the agricultural sector is still the insufficient level of provision with basic means of mechanization, their ultimate depreciation. Thus, in 2020, there were 23620 tractors and 10919 grain harvesters in Rostov region farms (Figure 5).

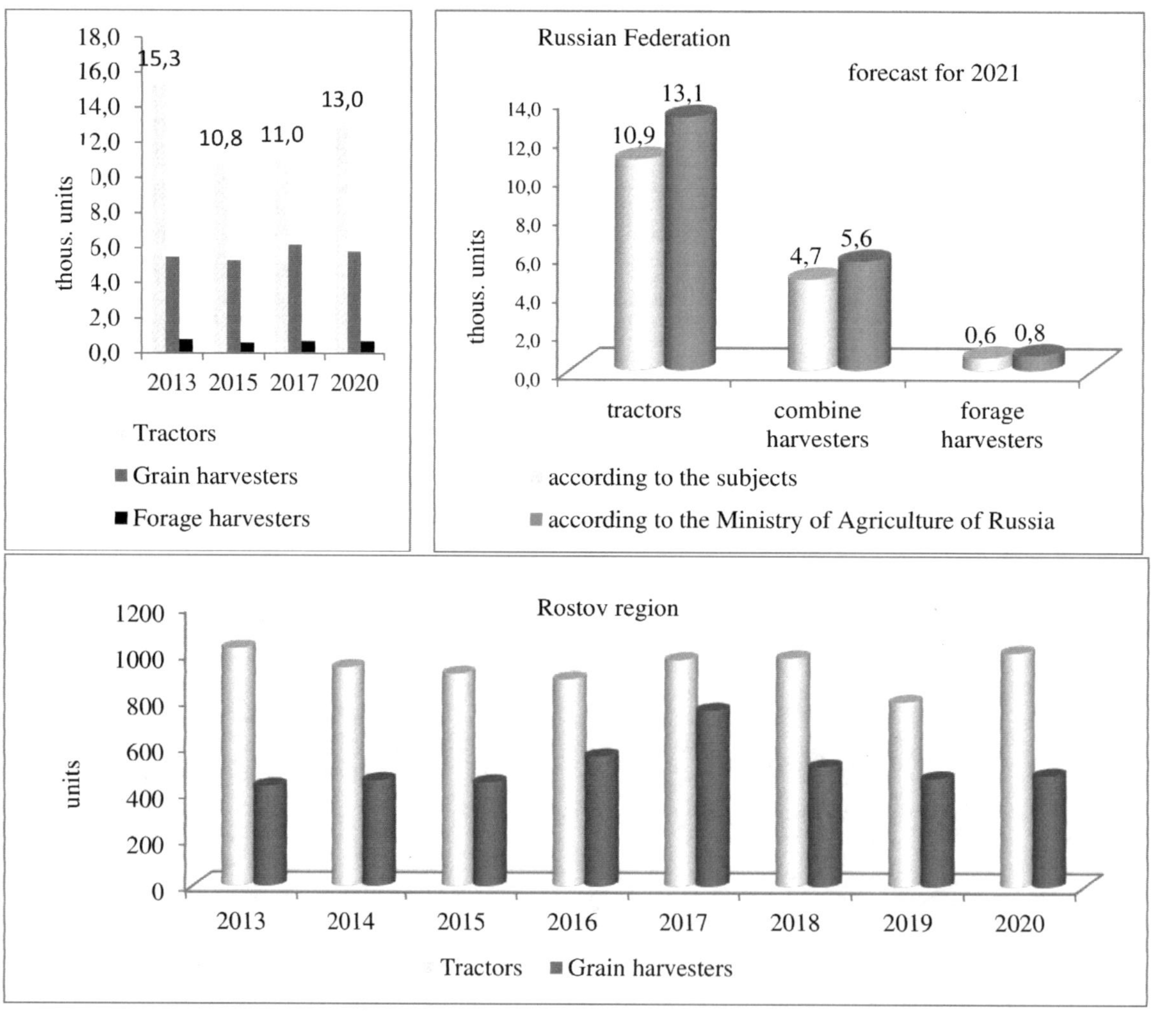

Figure 5. Dynamics of Agricultural Machinery Acquisition in Agricultural sector of Russia by all categories of farms in 2013-2020, forecast for 2021, thousand units.

Despite the positive trend in recent years, in 2020 the share of agricultural machinery beyond the depreciation period remains high: 64.8% for tractors, 55.1% for grain harvesters, 52.2% - forage harvesters. Provided that the region produces modern agricultural machinery that is in high demand among agricultural producers not only in the region, but also in the whole of Russia [7, 8, 9].

Thus, against the background of significant wear and tear of the material and technical park, small amounts of fertilizer application (Figure 6), reduction of reclaimed land area, record harvesting of crops is observed (Figure 7).

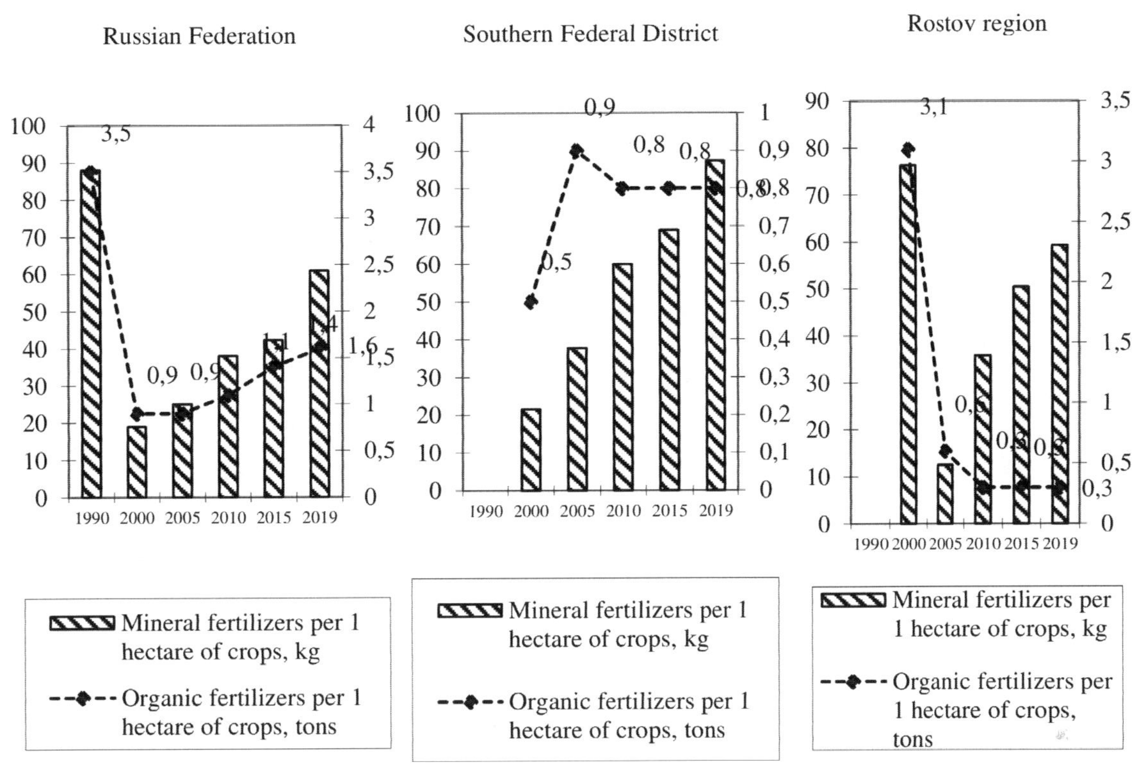

Figure 6. Volumes of mineral and organic fertilizers agricultural producers in Russia for 1990-2019 [6].

During the period of sanctions and epidemiological crisis, the high level of productivity of the southern fields in conjunction with the use of high-yielding seeds allowed to achieve a high gross harvest of crops (Figure 7).

Increasing the production of export-oriented crops, among which cereals and sunflowers, has become customary for agricultural producers in the southern zone, in particular, in Rostov region.

It should be noted that such record gross yields of grains and sunflowers was not even up to the 90-ies of the last century. Undoubtedly, the attractive prices on the world agricultural market for grain and oilseeds encourage agricultural producers to increase production volumes of export-oriented crops in the whole country and in the south of Russia. It should be noted that Rostov Region accounted for 30.0% of Russian food exports in 2020.

One of the most vulnerable branches of agricultural production in the conditions of the change of the development paradigm of agricultural production is the native cattle breeding. Realized programs of agricultural development at federal and regional levels promoted creation of definite organizational-economic conditions for development of sub-branch and stimulated stable tendencies of satisfactory economic growth in cattle breeding.

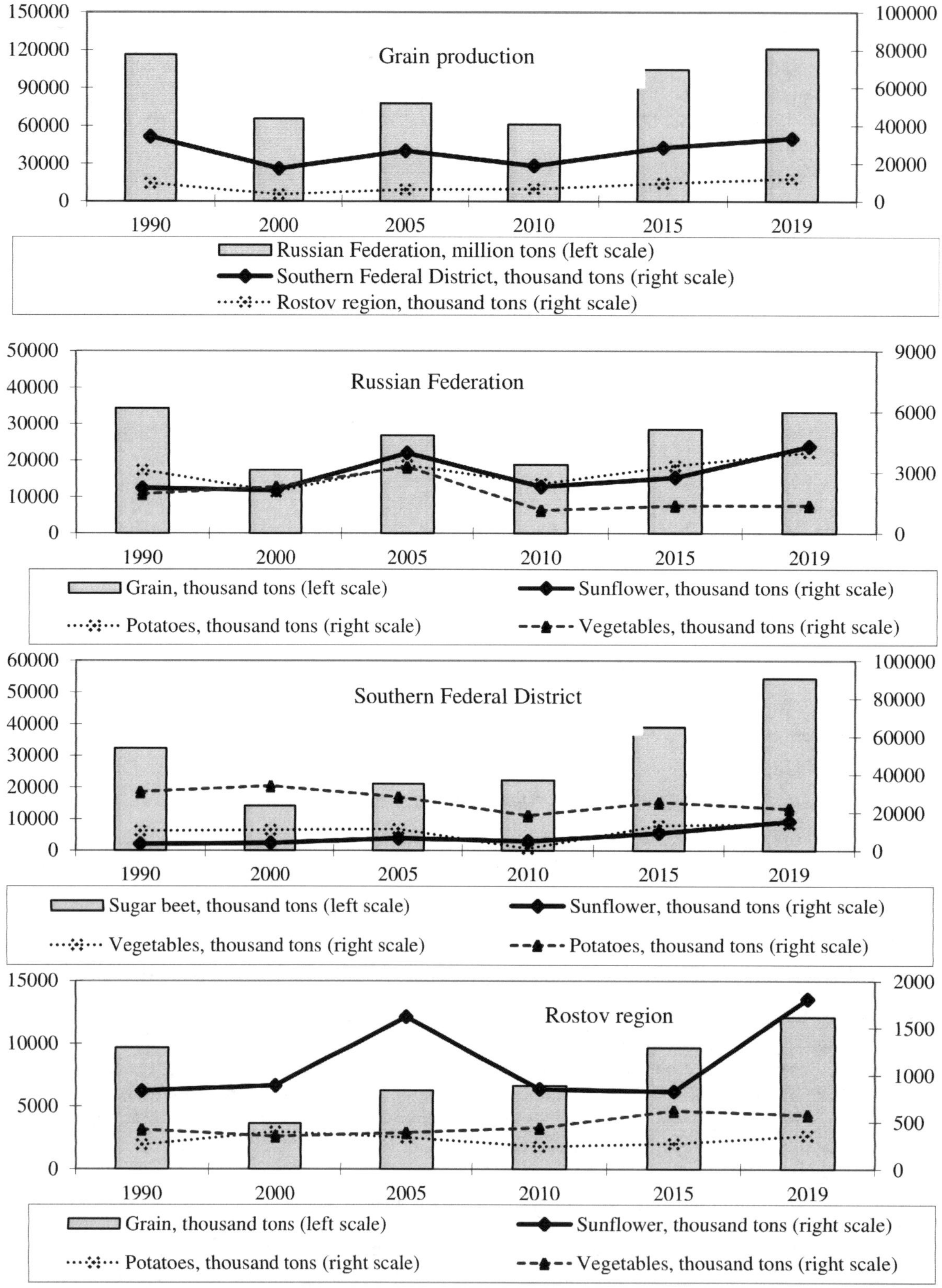

Figure 7. Dynamics of production of main types of crop production in Russia for 1990-2019 [6].

Thus, in recent years, the growth rate of livestock production was in 2014 - 2.0%, in 2015 - 3.1 %, in 2016 – 1.5%, in 2017 – 2.8%, 2018 - 1.3%, 2019 - 1.5% (Figure 8) [6]. However, the situation concerning the products of meat and dairy cattle breeding has not changed radically. Studies have shown that under the food embargo, the greatest lag in the provision of the country's population with basic types of food is in the production of beef and mutton. Thus, in the industry of meat cattle breeding the share of domestic production of beef was 30.0% of the recommended norm, of mutton – 50.0%, respectively, which is significantly below the indicators that determine the food security of the country. Milk production is 84.4% of the rationally recommended norm of consumption [5].

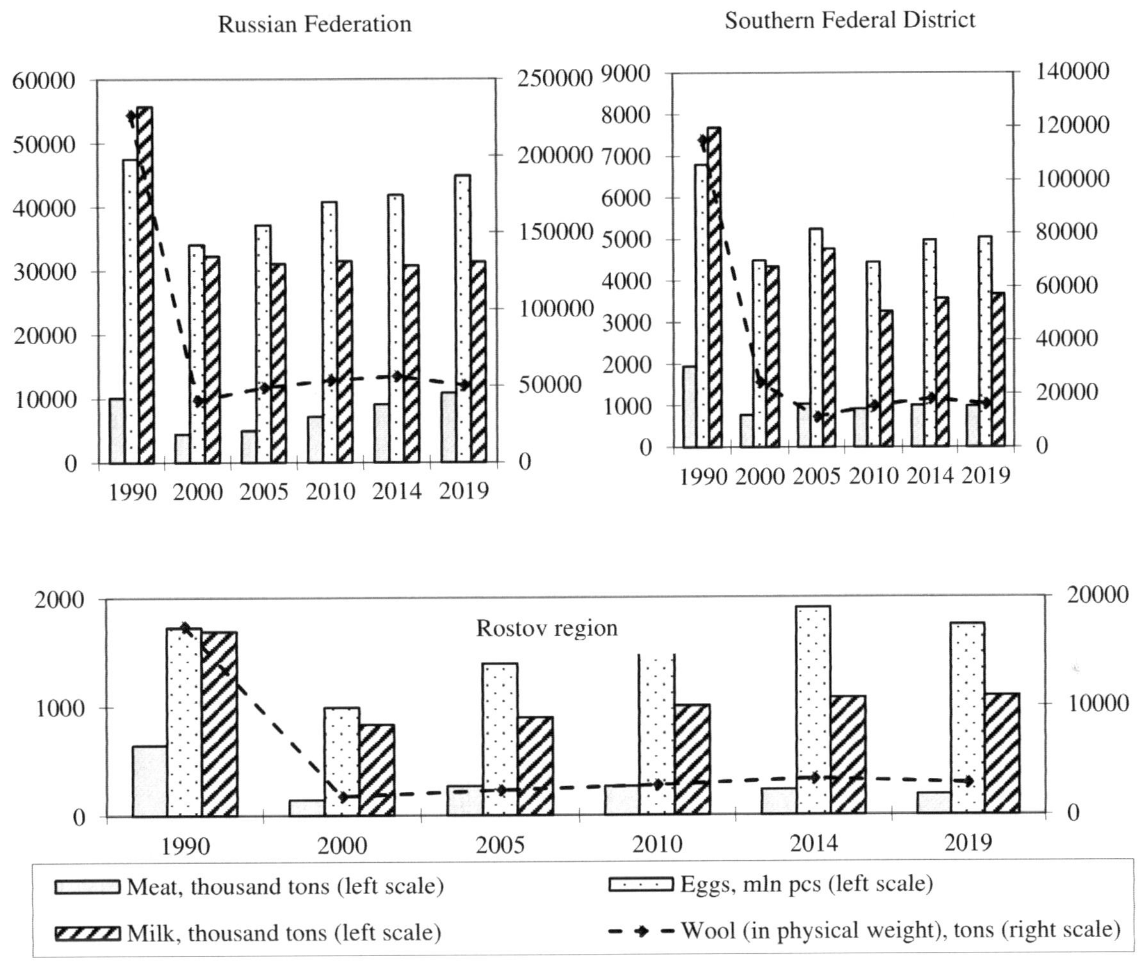

Figure 8. Dynamics of production of major livestock products cattle breeding in Russia [6].

Production of certain types of livestock products in the country as a whole and in the south of Russia did not reach the pre-reform level. In particular, milk production in the Russian Federation was 56.3% of the 1990 level, wool - 22.1%, eggs - 94.4%, respectively (Figure 8). Thanks to the formation and successful functioning of modern pig and poultry complexes operating on an innovative technological basis, the volume of meat production in 2019 was 107.6% of the 1990 level.

A somewhat different picture with the production of major livestock products is emerging in the Southern Federal District, where meat production was 51.0% of the 1990 level, wool - 14.1%, eggs - 74.2%, milk - 47.8%, respectively [6].

In the breadbasket of southern Russia - Rostov region, meat production in 2019 was 28.8% of the 1990 level, wool - 16.8%, milk - 64.7%, respectively. Egg production in the region exceeded the pre-reform level by 1.1%.

Studies have shown that the level of profitability of agricultural organizations, signaling sufficient financial sustainability, only taking into account state support barely reaches the level necessary to obtain borrowed financial resources (Figure 9). In general, the official financial well-being of agrarians is the result of tight belts to look successful in the face of fierce competition, both in the domestic and foreign agro-food market. The paradox of the functioning of domestic agricultural producers is that, not providing a simple reproduction of consumed resources, agrarians show profitability.

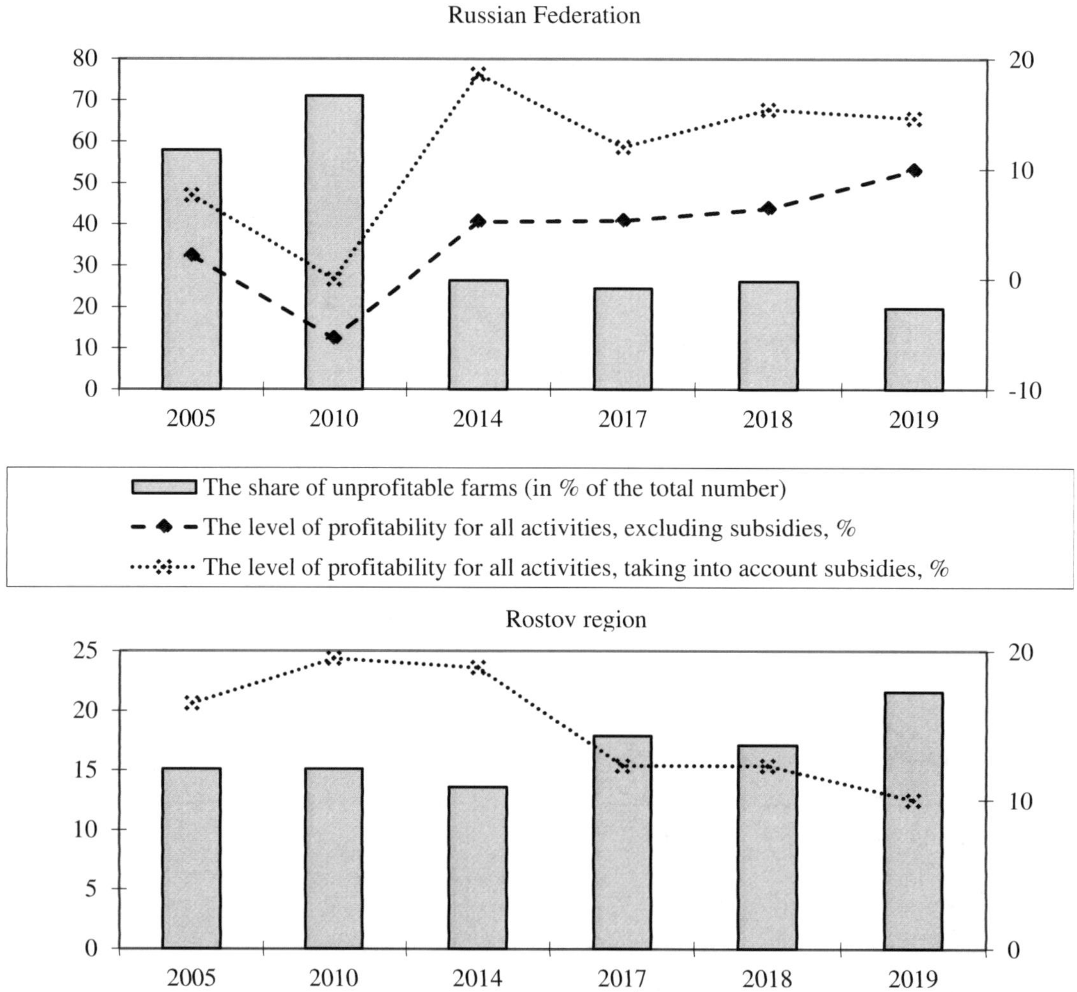

Figure 9. Results of the financial and economic situation of agricultural enterprises of the system of the Ministry of Agriculture of Russia [6].

So, the sector of crop production due to the high yield of agricultural crops demonstrates a fairly high balanced financial result (Figure 10) both in the country as a whole and in the south of Russia. It should be noted that the lack of large-scale cultural and technical activities in the management of agriculture, in the long term can lead to the opposite financial effect.

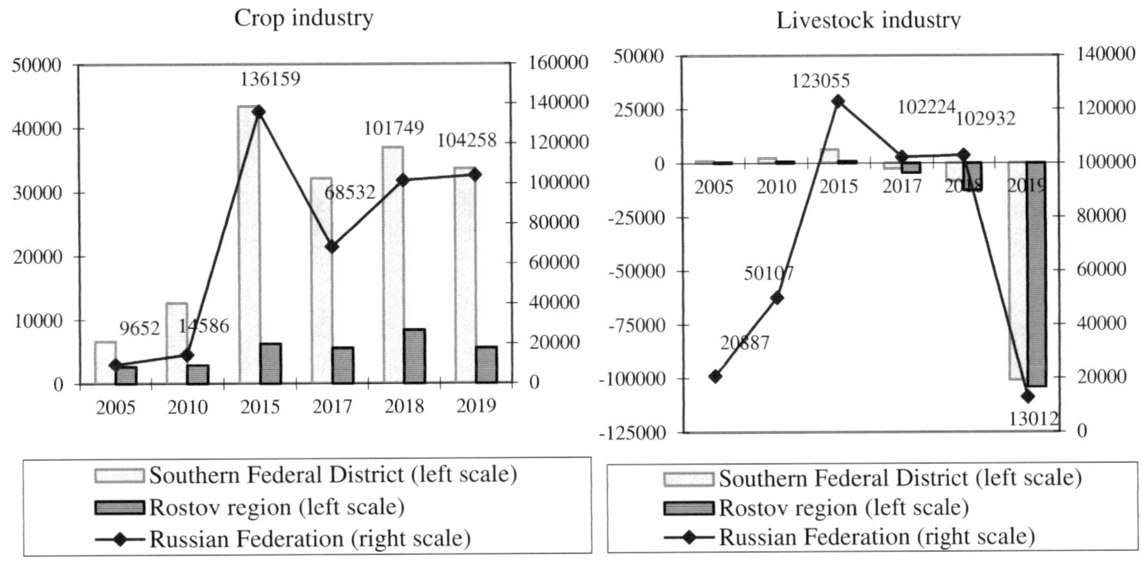

Figure 10. Sectoral cross-section of the balanced financial result of production and economic activity of agricultural organizations in Russia for 2005-2019, million rubles [6].

Negative balancing result in Russian livestock industry is caused by bankruptcy of large agrarian formations in poultry farming, in particular, in Rostov region, example of which is "EuroDon" group of companies LLC and "Belaya Ptitsa-Rostov" LLC, which not so long ago were the key producers of domestic poultry meat market.

It should be noted that under the conditions of global challenges and threats in agricultural production in Russia there is a decrease in the average annual number of workers, which is not a disaster for the industry, as this trend in the era of digital transformation is typical for both the national and global economies.

However, a serious unresolved socio-economic problem of domestic agricultural production is the low level of wages of domestic farmers. The setting of sectoral priorities, including the creation of favorable conditions for the expansion of the presence of domestic agricultural producers in the domestic market and the increase in export potential have not contributed to the improvement of rural workers' material well-being [10, 11, 12, 13]. Thus, in 2019 the average wage in Russian agricultural organizations did not reach 60.0% of the average in the economy (Figure 11), while in 1990 the wage of workers in state and collective farms was 103.5% of the average in the economy.

Differentiation of economic development of agricultural production by regions of Russia shows that in the south of the country the ratio of average wages in agriculture to the average economy exceeded 85.0%. At the same time the growth rate of wages in agriculture barely covered the inflationary increase of the general price level. That is why at present there is no point to speak of an increase in the living standards and wages of the rural population. This fact significantly aggravates the already difficult situation in the sector. In particular, lack of prestige of agricultural labor does not contribute to the formation of labor force reproduction. In the conditions of overdue reset of the industry there is an acute shortage of experienced zootechnicians, machine operators, agronomists, specialists with higher and secondary education, especially in such priority areas of activity as IT-technology and biotechnology, seed production and breeding [2, 10].

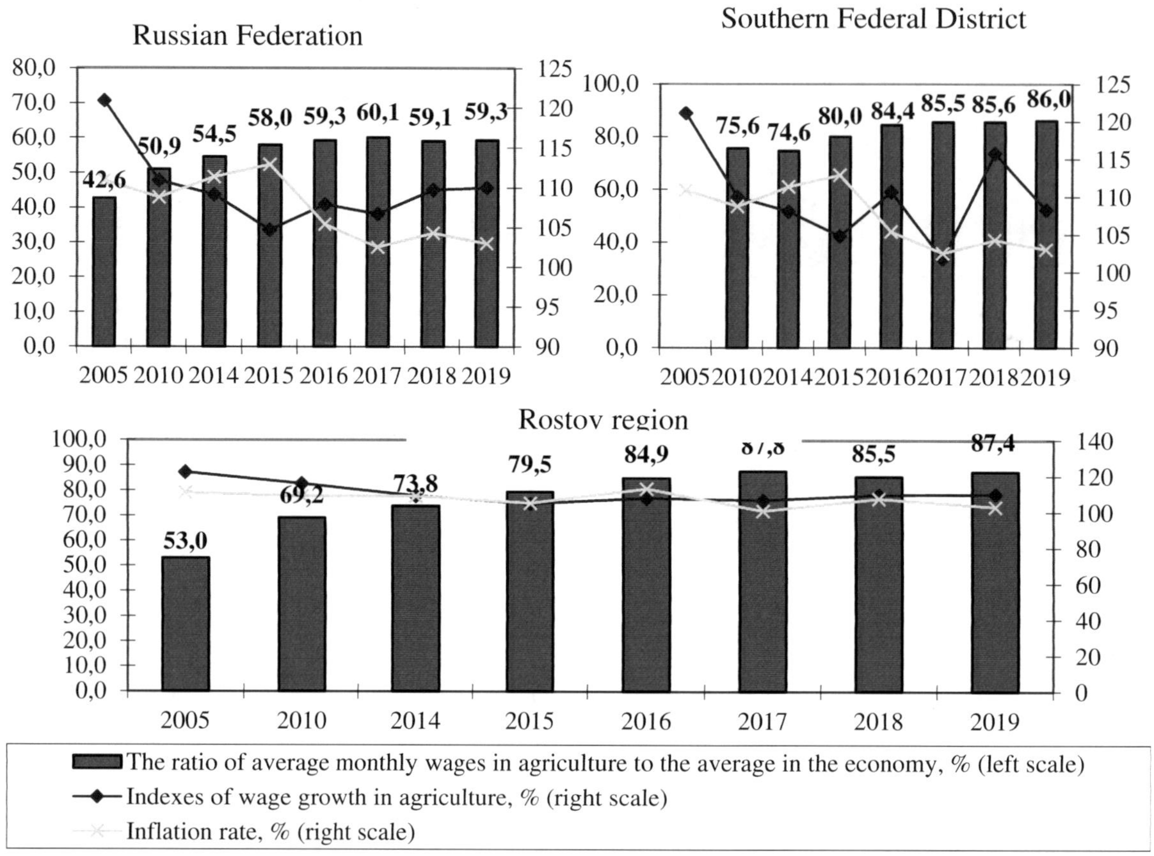

Figure 11. Qualitative indicators of labor remuneration assessment in agriculture Russia's agriculture for 2005-2019, rubles [6].

The current graduates of agrarian educational institutions do not want to return to the village, which lacks developed infrastructure. For example, in the Rostov region in 2020, the proportion of housing with central water supply was 64.1%, the level of gasification - 61.1%. Many settlements do not have their own medical and obstetric centers, there are no educational and sports institutions, there are villages that are not connected with the district center by roads with hard cover, in rural areas there is no stable Internet connection [6].

4. Conclusion

The conducted research showed that the domestic agriculture, which has a high production potential, learned to stay afloat in times of crisis, in the medium term can demonstrate a more intensive development, based on the search for breakthrough technologies, innovative and relevant solutions in the presence of sufficiently effective government regulation.

5. Acknowledgments

This research was carried out under a grant from the Russian Science Foundation No. 21-16-00025 NIIMMP.

References

[1] Semin A N 2019 Priority directions of agrarian and economic research of scientific and technological development of agroindustrial complex of Russia *Economics of Agricultural and Processing Enterprises* **1** 2-11

[2] Golubev A V 2020 Possibilities of development of plant breeding in Russia in conditions of global challenges *Agrarian Scientific J.* **11** 4-10
[3] Kholodova M A 2019 The role of the regional agro-industrial complex in the formation of food security of the country *Regional Agro-Systems: Economics and Sociology* **1** 3
[4] Gokhberg L M, Ditkovsky K A and Evnevich E I 2020 *Indicators of Innovation Activity* (Moscow: National Research University Higher School of Economics) p 336
[5] Orlova N V, Serova E V and Nikolaev D V 2020 Innovative development of the agroindustrial complex in Russia *XXI April Int. Scientific Conf. on the Problems of Economic and Social Development* (Moscow: National Research University Higher School of Economics) p 128
[6] *Federal State Statistics Service* 2021 http://www.gks.ru
[7] Tarasov A N and Kholodova M A 2018 Technical modernization of agricultural production: problems and solutions *Economics of Agricultural and Processing Enterprises* **8** 38-45
[8] Usenko L N, Usenko A M, Uryadova T N, Bashkatova T A and Belyaeva S V 2017 Monitoring methodology for socio-economic development of a region (through the example of the south of Russia regions) *Espacios* ` 24
[9] Golubev A V and Golubov I I 2019 Domestic innovations in the Russian quail breeding *J. of Adv. Research & Control Systems* **11** 2063-2067
[10] Tarasov A N, Isaeva O V, Kholodova M A, Pavlushkina O I and Chernaya A E 2018 *Agriculture in the South of Russia: Trends, Development Forecasts* (Rostov-on-Don: All-Russian Research Institute of Economics and Standards - branch of the Federal State Budgetary Scientific Institution "Federal Rostov Agrarian Scientific Center") p 80
[11] Glazyev S Y 2020 On the implementation of macroeconomic policy in accordance with the development goals of the country, set by the President of Russia *Proceedings of the Free Economic Society of Russia* **221(1)** 69-78
[12] Glazyev S Y 2019 Methodology of advanced economic development: how to solve the problem of a breakthrough in economic development set by the President of Russia *Proceedings of the Free Economic Society of Russia* **218(4)** 124-132
[13] Kleiner G B 2020 Systemic reset of Russian economy: key directions and prospects *Proceedings of the Free Economic Society of Russia* **223(3)** 111-122

IOP Conf. Series: Earth and Environmental Science **965** (2022) 012038 doi:10.1088/1755-1315/965/1/012038

A new generation of soy proteins for sustainable food production

A L Alekseev[1], T V Alekseeva[1], L F Obrushnikova[2], O A Knyazhechenko[2], Yu V Starodubova[2] and M I Slozhenkina[2]

[1] Don State Agrarian University, 24 Krivoshlykov Street, Persyanovsky, Oktyabrsky district, Rostov region, 346493, Russia
[2] Volga Region Research Institute of Manufacture and Processing of Meat-and-Milk Production, 6 Rokossovsky Street, Volgograd, 400131, Russia

E-mail: knyazhechenko71@gmail.com

Abstract. One of the promising directions in the creation of meat products is the development and production of products of combined composition, combining a balanced complex of nutrients necessary for the body. In recent years, soy processing products have been most intensively used as a vegetable component in the production of meat products. Protein additives obtained by processing soybeans are widely used in the meat processing industry due to their high functional and technological characteristics, a positive effect on the organoleptic characteristics and nutritional value of the product. The purpose of the work is to study the nutritional value and technological parameters of the soy protein concentrate "Arkon-S". The theoretical part of the research was carried out on the basis of the Department of Food Technologies of Don State Agrarian University, the development of pilot batches of products - in the production conditions of the sausage factory in Rostov-on-Don. The developed technology for the production of cooked sausages with soy concentrate "Arkon-S", the functional properties and nutritional value of which, in combination with economic feasibility, allow it to be used as a protein ingredient in the production of combined meat products.

1. Introduction

Soybeans are the most common among pulses and oilseeds. It serves as a raw material for a wide range of food products, and the high content of protein and valuable food components allows it to be used as an inexpensive and healthy substitute for meat [1, 2, 3].

In recent years, soy processing products have been most intensively used as a vegetable component in the production of meat products. They are most widely used in sausage production, to give products the appropriate texture, increased nutritional properties, high taste and functional qualities while reducing the cost of basic raw materials [4, 5].

One of the features of soybean processing products is the binding of fat and water in the minced meat systems and providing a stabilizing effect. Soy products better retain water, proteins, and better balance food in terms of nutritional value and physiological needs of the body [6, 7, 8].

Soy contains in its composition a significant amount of high-value protein, dietary fiber, minerals that determine its nutritional value and attract numerous researchers to find the most effective ways to use soy and its processed products in order to create new types of functional products and increase their nutritional value [9].

Soy proteins are traditionally produced in various forms: soy flour, soy protein concentrate, and soy protein isolate. Currently, a new generation of soy proteins has appeared on the market - functional concentrates Arkon-S, Danpro-HVX, Danpro-S-760, which are obtained as a result of acid washing and high-temperature treatment [10].

The purpose of the work is to study the nutritional value and technological parameters of the soy protein concentrate "Arkon-S".

2. Materials and methods

The theoretical part of the research was carried out on the basis of the Department of Food Technologies of Don State Agrarian University, the development of pilot batches of products - in the production conditions of the sausage factory "VEPOZ", Rostov-on-Don.

As objects of research, we used:

- raw meat - trimmed beef first grade, trimmed semi-fat pork, lateral fat;
- high-functional soy protein concentrate "Arkon-S" (Netherlands, SEZ No. 77.99.04.916.D.008850.12.02);
- other ingredients according to the recipe.

Experimental studies provided for the use of modern physicochemical, structural-mechanical, functional-technological, microbiological and organoleptic research methods, according to State standards.

In addition, the study was conducted using methods of statistical data analysis, comparison, analogy and systematization.

3. Results and discussion

At the beginning of the research, the assessment of the main physical and chemical parameters of the soy protein concentrate "Arkon-S" (Netherlands, SEZ No. 77.99.04.916.D.008850.12.02) was carried out.

Soy protein concentrate "Arkon-S" contains: 68-70% protein, 7.8-8.0% moisture, 0.9-1.0% fat, 4.9-5.0% ash, does not contain cholesterol, has a neutral taste, high solubility, water binding and fat emulsifying ability. The fatty acid composition of soy protein concentrate "Arkon-S" is presented in Table 1.

Table 1. Fatty acid composition of soy protein concentrate "Arkon-S".

Name	Content, mg/100 g
The amount of lipids	95-97
Triglycerides	80-83
Peaceful acids (amount)	15-18
- palmitic	10-12
- stearic	3,0-4,0
Monounsaturated	13-15
- oleic	13-15
Polyunsaturated	48-50
- linoleic	41-43
- linolenic	5-7

The research methodology provided for the study of the effect of soy protein concentrate "Arkon-S" on the technological properties of model minced meat systems. As a control sample, minced meat was used, consisting of trimmed beef 1 grade, trimmed semi-fat pork and side lard.

Soy concentrate "Arkon-S" was added to the minced meat in the form of a pre-prepared gel. The gel was prepared in a cutter, with the addition of 4 parts of cold water, until a glossy shine was formed. Soy concentrate was introduced into the model minced meat systems, replacing part of the trimmed beef of the 1st grade in an amount from 10% to 30% (Table 2).

Table 2. Recipes of experimental model minced meat.

Ingredients	Control	Replacement level		
		10%	20%	30%
Unsalted raw materials, kg (per 100 kg of raw materials)				
Beef trimmed 1 grade	60.0	50.0	40.0	30.0
Pork trimmed semi-fat	25.0	25.0	25.0	25.0
Lateral fat	15.0	15.0	15.0	15.0
Soy concentrate "Arkon-S"	-	2.0	4.0	6.0
Hydration water	-	8.0	16.0	24.0
Total	100.0	100.0	100.0	100.0

In experimental samples there was an increase in pH from 6.03 to 6.11, which is associated with the presence of soy protein concentrate "Arkon-S" (pH 7.1), the control sample had the lowest pH value of 5.8. The moisture binding capacity of the test samples exceeds the control data, which is caused by a decrease in the amount of myofibrillar proteins involved in the structure formation of the finished product and an increase in the moisture content due to the preliminary hydration of the soy concentrate (Table 3).

Table 3. Moisture binding capacity of minced meat systems with different levels of soy protein concentrate in the formulation.

Replacement level	Moisture binding capacity, % of total moisture
Control	79.16
10% protein concentrate	85.53
20% protein concentrate	87.25
30% protein concentrate	88.19

In the production conditions of the sausage factory "VEPOZ" (Rostov-on-Don), prototypes of combined sausages with different levels of replacement of meat raw materials with soy protein concentrate "Arkon-S" were developed. As a control, we used the "Separate" sausage recipe (State Standard 33673-2015 "Cooked sausage products: Technical conditions"), the protein additive was introduced into the recipe of the experimental samples (Table 4).

Table 4. Recipe of prototypes of combined sausages with different levels of substitution of meat raw materials.

Ingredients	Control	Replacement level		
		10%	20%	Control
Unsalted raw materials, kg (per 100 kg of raw materials)				
Beef trimmed 1 grade	60.0	50.0	40.0	30.0
Pork trimmed semi-fat	25.0	25.0	25.0	25.0
Lateral fat	15.0	15.0	15.0	15.0
Soy protein concentrate "Arkon-S"	-	2.0	4.0	6.0
Hydration water	-	8.0	16.0	24.0
Total	100.0	100.0	100.0	100.0
Food additives, spices, materials, g (per 100 kg of unsalted raw materials)				
Sugar	150.0	150.0	150.0	150.0
Table salt	2500.0	2500.0	2500.0	2500.0
Sodium nitrite	6.4	6.4	6.4	6.4
Black pepper	100.0	100.0	100.0	100.0

Ingredients	Control	Replacement level		
		10%	20%	Control
Allspice	100.0	100.0	100.0	100.0
Fresh garlic	120.0	120.0	120.0	120.0

The organoleptic assessment carried out indicates that the assessment indicators of the prototypes using the soy protein concentrate "Arkon-S" correspond to the characteristics of the Otdelnaya boiled sausage, and in some positions, in particular, the consistency and juiciness, slightly exceed the values of the control sample (Table 5).

Table 5. Organoleptic characteristics of cooked sausages.

Test samples	Appearance	Color	Flavor	Taste	Consistency	Juiciness	Average rating
Control	4.5	4.6	4.3	4.7	4.3	4.3	4.45
10%	4.5	4.6	4.4	4.7	4.5	4.3	4.50
20%	4.5	4.5	4.3	4.7	4.5	4.3	4.46
30%	4.3	4.4	4.2	4.5	4.2	4.3	4.31

With an increase in the amount of concentrate to 30%, these indicators decrease, the smell and taste are unusual for this type of product, they have a characteristic soy taste. Table 6 shows the physicochemical indicators of the quality of the prototypes of the combined sausages.

Table 6. Physicochemical indicators of the quality of prototypes.

Test samples	Content, %				
	Moisture	Protein	Fat	Ash	Carbohydrates
Control	65.16 ± 1.32	11.10 ± 0.20	20.21 ± 0.18	2.33± 0.01	1.2 ± 0.02
10%	68.71 ± 1.37	11.41 ± 0.22	15.22 ± 0.16	2.53± 0.02	2.04± 0.01
20%	68.74 ± 1.30	11.45 ± 0.23	15.17 ± 0.12	2.55 ± 0.03	2.18 ± 0.01

In comparison with the control, there was a decrease in the mass fraction of fat, which is a positive factor in the light of modern trends in the field of healthy nutrition. The results of the analysis of the chemical composition of cooked sausages indicate that the formulations containing from 10% to 20% soy concentrate have a fairly high biological value.

The value of the yield of finished products is one of the criteria that determine the economic feasibility of the production of cooked sausages, which, at the same time, is closely related to the functional properties of the proteins of the minced meat system, which determine the juiciness of the finished product. The results of determining the yield and weight loss of sausages are presented in Table 7.

Table 7. Yield and weight loss of cooked sausages.

Indicators	Control	10%	20%
Output, %	113.3 ± 1.7	117.6 ± 1.8	117.8 ± 1.8
Weight loss, %	13.1 ± 0.3	11.3 ± 0.3	11.2 ± 0.3

The results of determining the yield indicate that the maximum value of the investigated indicator for the sample containing 20% soy concentrate "Arkon-S".

IOP Conf. Series: Earth and Environmental Science **965** (2022) 012038 doi:10.1088/1755-1315/965/1/012038

4. Conclusion

On the basis of the studies carried out, the expediency of using the hydrated soy concentrate "Arkon-S" in the production technology of combined sausages has been confirmed. In the experimental samples there was an increase in pH from 6.03 to 6.11, which is associated with the presence of soy protein concentrate "Arkon-S" (pH 7.1), the control sample had the lowest pH value - 5.85 compared to the experimental. In comparison with the control, in the experimental samples there was a decrease in the mass fraction of fat, an increase in the total content of protein and carbohydrates, formulations containing from 10% to 20% of soy concentrate have a fairly high biological value. Organoleptic evaluation of model minced meat systems showed that the use of soy protein concentrate "Arkon-S" in an amount of no more than 20% does not cause deterioration in performance; with an increase in the amount of isolate up to 30%, their decrease occurs, the smell and taste are unusual for this type of product, they have a characteristic soy taste. The use of soy protein concentrate "Arkon-S" in the technology of combined sausages made it possible to increase the water-holding capacity, reduce losses during heat treatment and the consumption of raw meat.

5. Acknowledgments

This work was carried out as part of a state assignment SSI NIIMMP of the Ministry of Science and Higher Education, No 1021032427412-9.

References

[1] Aleshin V N, Lobanov V G, Naydenova A A, Kornen E P and Butina E A 2010 Directed modification of technologically functional properties of protein additives *Izvestiya Vuzov. Food Technology* **5-6** 33-35

[2] Vasyukova A T and Homeland E V 2012 Research of functional properties of minced meat *Food Industry. Agro-Iindustrial Complex* **6** 13-18

[3] Gavrilov A F and Crooks V O 2009 Functional properties of soybeans as the main consumer characteristics of their processing products *Technics and Technology of Food Prod*uction **2** 16-19

[4] Gorlov I F, Slozhenkina M I and Bushueva I S 2013 Improvement of consumer properties of meat products due to biologically active substances *Storage and Processing of Agricultural Raw Materials* **5** 32-33

[5] Merenkova S P and Savostina T V 2014 Practical aspects of the use of vegetable protein additives in the technology of meat products *Bulletin of the South Ural State University. Series: Food and Biotechnology* **1** 23-29

[6] Morozov A 2012 Meat products of emulsion type and based on protein-lipid composites for healthy nutrition *X Int. Scientific-Practical Conf. Technologies and Products of Healthy Nutrition Functional Food Products* (Moscow) pp 188-193

[7] Rogov I A, Oreshkin E N and V N Sergeev 2017 Medico-technological aspects of the development and production of functional food products *Food Industry* **1** 33-35

[8] Statsenko E S 2019 Development of food additive technology based on soybean raw materials of biotechnological modification *Technics and Technology of Food Production* **3** 367-374

[9] Kritikos S, Papanikolaou K, Draganidis D, Jamurtas A Z and Fatouros I G 2021 Effect of whey vs. soy protein supplementation on recovery kinetics following speed endurance training in competitive male soccer players: a randomized controlled trial *J. of the Int. Society of Sports Nutrition* **18(1)** 23

[10] Acosta-Domínguez L, Cocotle-Ronzón Y, Alamilla-Beltrán L and Hernandez-Martinez E 2021 Effect of a cryogenic treatment in the microstructure, functional and flow properties of soy protein isolate *Food Hydrocolloids* **119** 106871

Evaluation of the quality, environmental safety and biological value of functional curd product

V V Kryuchkova[1], I F Gorlov[1], M I Slozhenkina[1], N V Lomonova[2], S N Belik[3] and P V Skripin[2]

[1] Volga Region Research Institute of Manufacture and Processing of Meat-and-Milk Production, 6, Rokossovsky Street, Volgograd, 400131, Russia
[2] Don State Agrarian University, 24 Krivoshlykov Street, Persianovsky, Rostov region, 346493, Russia
[3] Rostov State Medical University, 29 Nakhichevansky Lane, Rostov-on-Don, 344022, Russia

E-mail: kverav@yandex.ru

Abstract. The authors conducted studies to assess the quality and calculation of biological value of the developed curd product. Estimation of organoleptic parameters was carried out with use of tasting scale of curd and curd products estimation, on the basis of which the profilogram was built and it was established that the developed curd product received the maximum quantity for taste and smell, appearance and consistency. Nutritional value of the experimental sample increased due to introduction of whey proteins and sesame seeds: protein (by 2.0%), appearance of carbohydrates (0.3%) and dietary fibers (0.05%). In the experimental sample the number of lactic acid microorganisms considerably exceeds both the normal values and the values of the control sample, which proves its probiotic properties and indicates a high physiological value. The calculated amino acid scoring of the eight essential amino acids is quite high and ranges from 105.9% valine to 191.0% tryptophan, which is well balanced and significantly exceeds the ideal protein. The biological value of the experimental sample with sesame seeds and whey proteins was 65.2%, the utility coefficient was 0.77, which indicates its high biological value and high balance of the amino acid composition.

1. Introduction

According to the estimates of global analytical agencies, the most promising area of development of the food industry at present is the production of functional foods. Their consumption level is approximately 150,000 tons per year in Europe and 1,500 tons per year in Russia. The relevance of healthy nutrition is confirmed by studies indicating a direct correlation between the immune status of the person and the food he consumes. Manufacturers expand the range of preventive products using various functional ingredients: dietary fiber, vitamins, antioxidants, polyunsaturated fatty acids, pro- and prebiotics [1, 2, 3].

The authors developed a functional cottage cheese product. Microflora of mesophilic lactic acid bacteria was used as a probiotic component. To improve the nutritional and biological value of the product a combination of vegetable (white sesame seeds) and dairy (whey concentrate) raw materials was used.

Sesame seeds (Sesamum indicum) contain 40-60% oil, represented by equal parts of oleic (from 35 to 54%) and linoleic (from 39 to 59%); besides, sesame seeds contain 10% palmitic acid and 5% stearic acid [4] contains a large amount of fats and proteins, vitamins A, B, E, C, PP, mineral substances (calcium, phosphorus, iron, potassium, magnesium, etc.). Sesame contains sesamine lignan, which serves as a powerful antioxidant, it is used in the prevention of cancer, reduces blood cholesterol levels [5]. Sesamol is a natural phenolic compound and the main lignan isolated from sesame seeds and sesame oil. There is evidence that sesamol acts as a metabolic regulator with antioxidant, antimutagenic, anti-hepatotoxic, anti-inflammatory, rejuvenating and chemopreventive properties and has anti-cancer effects. [6] The inhibitory effect of consumption of sesame seeds and their derivatives on oxidative stress in people with systemic arterial hypertension, dyslipidemia and type 2 diabetes was established [7]. Consumption of sesame seeds for 8 weeks promotes the absorption and accumulation of vitamin E isoforms - alpha - and gamma-tocotrienols, as well as gamma-tocopherol in adipose tissue and skin ($P < 0.05$) [8]. The use of sesame seeds in the diet improves spatial cognitive abilities, increases the concentration and activity of superoxide dismutase, catalase and glutathione peroxidase, and reduces the level of malonic dialdehyde in the hippocampus. It increases neuronal integrity, decreases the apoptotic index, decreases the expression of pro-apoptotic protein Bax and increases the expression of anti-apoptotic protein Bcl-2 in CA1 hippocampal neurons, which allows using this product in neurodegenerative diseases and encephalopathies of various genesis [9, 10, 11].

Reducing the amount of protein in food rations negatively affects human health, reduces its resistance to adverse external influences [12, 13], and leads to impaired absorption of riboflavin, nicotinic and ascorbic acids. In this regard, special attention was paid to non-traditional sources of animal protein, one of which is whey. The main component of whey protein that has an important nutritional value is whey protein, which is formed after casein precipitation from milk by acid at pH 4.6-4.7. On average, serum contains 0.6% proteins. The composition of whey protein is represented by the following types of proteins: β-Lactoglobulin, α-Lactalbumin, bovine serum albumin, lactoferrin, glycomacropeptides, immunoglobulins, lactoperoxidases and protease-peptones, and other valuable food components [14]. Whey protein has not only an important nutritional value, but is also used in dietary corrections of pathological processes. The use of whey protein in the diet of cancer patients significantly alleviates cancer cachexia syndrome. The usefulness of whey protein is confirmed by its high leucine content and ability to modulate IGF-1 (insulin-like growth factor) concentration, which prevents musculoskeletal hypotrophy [15]. Protein fraction of whey protein - α-lactalbumin suppresses hunger, promotes weight loss in obesity, and improves glucose tolerance in diabetic patients [16]. Fraction of whey protein - lactoferrin improves gut microbiota, reduces liver lipidosis and mesenteric fat, and increases glucose tolerance [17]. Lactoferrin supplementation for 8 weeks 27, reduced the degree of visceral obesity in patients without changing caloric intake. Reduced caloric intake with the whey protein-based diet is associated with increased circulating concentrations of lower intestinal satiety hormones, including glucagon-like peptide-1 (GLP-1) and pancreatic peptide [18, 19].

Thus, the use of sesame seeds and whey proteins in the production of functional curd product designed for all consumer groups is relevant and appropriate. In view of the above, the purpose of the work is to study the qualitative indicators of the developed curd product and evaluation of its biological value.

2. Materials and methods

The objects of research were samples of cottage cheese products developed with the addition of functional ingredients and the control sample - cottage cheese product developed in accordance with State Standard 31453-2013 “Cottage cheese. Technical conditions”.

When organizing and conducting research used a set of common, standard and modified methods of research: determination of the mass fraction of fat - State Standard R ISO 2446-2011; determination of protein - Kjeldahl method according to State Standard 34454-2018; thermogravimetric method to determine the mass fraction of dry substances in cottage cheese products according to State Standard

28561-90; vitamin C content - titrometric method according to State Standard 24556-89; organoleptic evaluation was carried out in accordance with State Standard R ISO 22935-2-2011 and State Standard 33630-2015; microbiological indicators were determined according to State Standard 32901-2014. The content of amino acids, vitamins and minerals was determined by capillary electrophoresis using a drop-105M system. Statistical processing of the obtained data and construction of mathematical models was carried out using the software package Statistica 6.0.

3. Results and discussion

The authors evaluated the quality of curd products and determined organoleptic, physico-chemical, microbiological and safety indicators, vitamin-mineral and amino acid composition and calculated biological value.

To assess the organoleptic characteristics, we used a tasting scale for evaluating curd and curd products, presented in Table 1.

Table 1. Tasting scale for evaluating cottage cheese and cottage cheese products.

Product performance, overall score	Discount, point	Score, point
Taste and smell		
Very good: clean, sour milk, with no extraneous flavors and odors, with a slight taste of pasteurization, for curds using dried milk - with a taste of dried milk, for curds and cottage cheese products using fillers with the flavor and smell of the filler introduced.	0	5
Good: With no more than one of the following signs: sour taste, whey flavor, nutty flavor (for curds without fillers), slightly foddery	1	4
Satisfactory: the simultaneous presence of not more than 2 of the following signs: characteristic for cottage cheese taste and smell expressed weakly, there is a yeasty smell and taste, excessively sour, slight bitterness.	2	3
Bad and very bad, depending on the severity of the defect: taste and smell - not characteristic of the product, there are pronounced foreign flavors and smells (at least one): feed, musty, oxidized, bitter, rancid, ammonia, metallic, etc.	3	2
Appearance and consistency		
Very good: homogeneous, soft, slightly smeared, crumbly, with or without perceptible particles of milk protein. In a product with filler the presence of particles of filler. For a non-fat product, slight whey secretion	0	3
Good: the presence of no more than 2 of the following signs: smeary, slight grittiness, slight flouriness. For a nonfat product, noticeable whey separation.	1	2
Satisfactory: the presence of not more than 2 of the following features: loose, excessively crumbly, slightly heterogeneous, perceptible grittiness, weakly rubbery. For nonfat product - considerable separation of whey	2	1
Bad and very bad: depending on the severity of malformations: pronounced rubbery, lumpy, rough, crumbly, excessively smeared, slimy, strong secretion of whey, etc.	3	0
Colour		
Satisfactory: white to light creamy, uniform in mass or due to the color of the added filler.	0	1
Unsatisfactory: uneven throughout the mass	1	0

Note: The maximum number of points is 10, of which: 5 - taste and smell, 3 - appearance and consistency, 1 - color, 1 - packaging and labeling.

Table 2 shows the organoleptic characteristics of the developed curd drinks.

Table 2. Organoleptic characteristics of developed curd products.

Indicator name	Cottage cheese products	
	Control sample	Experimental sample (with sesame seeds and whey proteins
Appearance and consistency	Soft, spreadable, slightly crumbly consistency	Soft, slightly crumbly consistency with the presence of single sesame seeds
Taste and smell	Pure, sour milk with a faintly pronounced taste of pasteurization	Pure, sour milk with a distinct flavor, harmoniously combined pasteurized milk and sesame seeds
Colour	Milky white uniform throughout	Light creamy, with isolated flecks of sesame seeds and whey proteins, evenly distributed throughout the mass

The table shows that the organoleptic characteristics are high enough, do not have a sharply expressed taste and smell, which will allow their use by different groups of people, both children and the elderly. On the basis of the results obtained, a profilogram of censoring organoleptic characteristics of curd products was built. It was established that the developed curd product got 10 points in comparison with the control sample, which got 8 points, the enriched product got higher marks for taste and smell and appearance and consistency.

The chemical composition, microbiological indicators, safety indicators and calculated energy value of curd products were determined (Table 3).

Table 3. Nutritional and energy value of cottage cheese products.

No.	Indicator name	Control sample	Experimental sample
1	Mass fraction of fat, not less %	20.0	20.0
2	Mass fraction of protein, not less %	8.0	10.0
3	Mass fraction of carbohydrates, g	-	0.3
	including mass fraction of dietary fiber, g	-	0.05
4	Mass fraction of moisture, not more %	72.0	69.65
5	Titratable acidity, °T	176	172
6	Phosphatase	not available	
7	Temperature when leaving the plant, °C	4 ± 2	
8	Energy value, kcal	212.0	221.2

Analysis of the table showed that the nutritional value of the experimental sample increased due to the increased protein by 2.0% due to the introduced whey proteins and the appearance in the experimental sample of carbohydrates (0.3%) and dietary fiber (0.05%) and sesame seeds, which give the product a functional orientation. The developed product is characterized by moderate acidity and increased caloric content by 9.2 kcal due to the appearance of carbohydrates and increased protein content.

Microbiological indicators and safety indicators are of great importance in the production of curd products, as they characterize the safety of curd products. The results are presented in Table 4 and Table 5.

Table 4. Microbiological parameters of curd products.

No.	Indicator name		Standard according to TR CU 033/2013	Control sample	Experimental sample
1	Lactic acid microorganisms, CFU/cm^3 (g), not less (at the end of storage life)		1×10^6	1.7×10^6	4.3×10^6
2	Weight of the product cm^3 (g) in which is not allowed	E. coli bacteria (coliforms)	0.1		
		S. aureus	1.0		
		Pathogens, including Salmonella	25	not detected	not detected
3	Yeast, CFU/cm^3 (g), not more		50	18	14
4	Molds, CFU/cm^3 (g), not more		50	16	15

Analysis of the Table 4 showed that the number of lactic acid microorganisms in the experimental sample significantly exceeds both the standardized values and the values of the control sample, which proves its probiotic properties and indicates a high physiological value. Disease-causing microorganisms such as E. coli bacteria, Staphylococcus aureus and Salmonella bacteria were not detected in the samples of sour-milk products. The number of yeast and mold cells does not exceed the values regulated by the Technical Regulation of the Customs Union TR CU 033/2013 "On safety of milk and dairy products" for this group of dairy products.

The study of safety indicators examined the presence of toxic elements, mycotoxins, pesticides, radionuclides, antibiotics, etc. (Table 5).

Table 5. Safety indicators of fermented dairy products.

Name Indicator	Permissible level	Index value, mg/kg (for radionuclides - Bq/kg)	
		Control sample	Experimental sample
Toxic elements			
Lead	0.1	0.04	0.03
Arsenic	0.01	0.01	0.01
Cadmium	0.01	0.01	0.01
Mercury	0.001	0.0001	0.0001
Pesticides			
Hexachlorocyclohexane (α-, β-γ-isomers)	0.05	0.03	0.03
DDT and its metabolites	0.05	0.01	0.01
Antibiotics			
Levomycetin	is not allowed	not detected	
Tetracycline group	is not allowed	not detected	
Streptomycin	is not allowed	not detected	
Penicillin	is not allowed	not detected	
Mycotoxins			
Aflatoxin M_1	0.0005	0.0001	0.0001
Radionuclides			
Cesium-137	100	76	74
Strontium-90	25	11	10

The results of the study of safety parameters of curd products, presented in Table 5, show that the content of toxic elements, nitrates, radionuclides and mycotoxins in the experimental and control samples meet the requirements of the Technical Regulations of the Customs Union "On safety of milk and dairy products" (TR CU 033/2013), their values do not exceed the allowable level. A very important factor for functional curd products is the presence of vitamins, minerals, amino acids, as well as prebiotic substances that give the product a functional orientation. The vitamin and mineral composition of curd products was determined for the final estimated shelf life. The results are presented in Table 6.

Table 6. Mineral composition of cottage cheese products, g/100 g of the product.

No.	Indicator name	Control (cottage cheese without additives)	Enriched cottage cheese product	Deviations, +/-
1	Calcium, g/100 g of product	0.13±0.02	0.16±0.01	+0.03
2	Potassium, g/100 g	0.47±0.01	0.52±0.01	+0.05
3	Phosphorus, g/100 g	0.29±0.01	0.35±0.01	+0.03
4	Magnesium, g/100 g	0.12±0.01	0.15±0.01	+0.03
5	Sulfur, g/100 g	0.22±0.01	0.24±0.01	+0.14
6	Selenium, g/100 g	-	00013±0.001	+0.0001
7	Zinc, g/100 g	0.036±0.02	0.037±0.01	+0.01
8	Copper, g/100 g	0.02±0.01	0.06±0.01	+0.004
9	Cobalt, g/100 g	0.002±0.01	0.003±0.01	+0.001
10	Iron, g/100 kg	0.036±0.02	0.037±001	+0.001
11	Iodine	-	0.001	+0.001

As can be seen from the table, the increase in the mineral composition occurs in the experimental curd product for almost all macro- and microelements due to the introduction of sesame seeds and whey proteins, so the increase in magnesium occurred by 25%, calcium - by 23%, phosphorus - by 21%, potassium - by 10%, sulfur - by 9%, appeared in enriched curd such important microelements as iodine and selenium, selenium is necessary for iodine assimilation and normal development of the body. The vitamin composition of curd products is presented in Table 7.

The vitamin composition of the cottage cheese product is quite diverse and contains both water-soluble and fat-soluble substances. Analysis of the table showed that the enriched product had an increase in vitamins of almost all types, especially for vitamins such as choline - by 40%, C - by 33.3%, B_1 - 20%. So, quite a lot of ascorbic acid in the product, which is the recognized leader among antioxidants, which helps to extend youth and preserve beauty. To cover the daily requirement of ascorbic acid is enough to eat 200 grams of this product every day. The composition of vitamins in cottage cheese also includes rutin, a substance needed to strengthen the walls of blood vessels, especially useful for the health of people with varicose veins and fragile capillaries.

Table 7. Vitamin composition of cottage cheese products, mg/100 g product.

No.	Control	Control	Enriched cottage cheese product	Deviations, +/-
1	Vitamin B_1 (thiamine)	0.04±0.02	0.048±0.01	+0.44
2	Vitamin B_2 (riboflavin)	0.25±0.01	0.253±0.01	+0.003
3	Vitamin B_3 (pantothenic acid)	0.20±0.01	0.21±0.01	+0.01
4	Vitamin B_6 (pyridoxine)	019±0.01	0.20±0.01	+0.01
5	Vitamin B_9 (folic acid)	0.04±0.01	0.05±0.01	+0.01
6	Vitamin PP (nicotinic acid)	4.00±0.01	4.05±0.001	+0.05
7	Vitamin C (ascorbic acid)	0.36±0.02	0.48±001	+0.12
8	Rutin	0.02±0.01	0.03±0.01	+0.01
9	Vitamin E (alpha-tocopherol)	0.02±0.01	0.023±0.01	+0.003
10	Vitamin D	0.036±0.02	0.037±0.01	+0.001
11	Choline	0.02±0.01	0.28±0.01	+0.08

Next, the amino acid composition of curd products was determined (Table 8).

Table 8. Amino acid composition of cottage cheese products.

Amino Acids	Control sample, g/100g protein	Experimental sample, g/100g protein	Deviations, +/-
	Essential amino acids		
Arginine	3.3±0.02	3.37±0.02	+ 0.07
Valin	5.4±0.02	5.51±0.02	+ 0.11
Histidine	2.4±0.01	2.52±0.01	+ 0.12
Isoleucine	5.0±0.02	5.13±002	+ 0.13
Leucine	9.1±0.02	9.37±0.02	+ 0.27
Lysine	8.5±0.02	8.71±0.02	+ 0.21
Methionine	2.6±0.01	2.65±0.02	+ 0.05
Threonine	4.5±0.02	4.59±0.02	+ 0.09
Phenylalanine	4.4±0.02	4.49±0.02	+ 0.09
Tryptophan	1.8±0.01	1.91±0.01	+ 0.01
Total essentials	47.0	48.58	+ 1.58
	Substitutable amino acids		
Alanine	3.4±0.02	3.51±0.02	+ 0.11
Aspartic acid	7.6±0.02	7.82±0.02	+ 0.22
Glycine	1.8±0.01	1.85±0.01	+ 0.05
Prolin	9.7±0.02	9.83±0.02	+ 0.13
Serin	5.3±0.02	5.39±0.02	+ 0.09
Tyrosine	4.7±0.02	4.77±0.02	+ 0.07
Cystine	0.9±0.01	0.97±0.01	+ 0.07
Total substitutable amino acids	33.4	34.65	+ 1.25
Total amino acids	80.4	83.23	+ 2.83

Analysis of the table showed that the experimental sample of cottage cheese product contains 3.7% more of all amino acids than the control sample, 3.4% more of essential amino acids and 3.74% more of substitutable amino acids. The results show that the product is rich in amino acids, both quantitatively and qualitatively.

As can be seen from the figure, each essential amino acid exceeds its values compared to the control sample, indicating that the enrichment of the curd product with sesame seeds and whey proteins leads to positive results, which contributes to a beneficial effect on the human body.

And in Table 8 we observe that there was a significant increase in the substitutable amino acids for all the items, which is also beneficial and beneficial for the human body when the enriched curd products are regularly consumed.

To assess the biological value of curd products, its amino acid composition is compared with the amino acid composition of the ideal protein by determining the amino acid score.

One accessible way to calculate the amino acid score is to calculate the ratio of the amount of each essential amino acid in the test protein to the amount of that amino acid in the ideal protein (formula 1):

$$\text{Amino acid grading} = \frac{\text{mg essential amino acids in 1 g of test protein}}{\text{mg essential amino acids in 1 g of ideal protein}} * 100 \qquad (1)$$

where *mg essential amino acids in 1 g of test protein* is the content of an essential amino acid in 1g of the protein under study, mg; *mg essential amino acids in 1 g of ideal protein* is content of the same amino acid in 1g of "ideal" (standard) protein, mg; 100 - conversion coefficient in percent.

The limiting amino acid of biological value is the one with the lowest scorption.

Amino Acid Score Difference Factor (ASDF), numerical characteristic of the imbalance of essential amino acids in relation to the physiologically necessary norm (standard), was calculated by formula 2:

$$ASDF = \frac{\sum_{i=1} \Delta ACD}{n}, \qquad (2)$$

where ΔACD – the difference of the amino acid scoring of an amino acid, which is determined by formula 3:

$$\Delta ACD = S_i - S_{\min}, \qquad (3)$$

where S_i – excess score i-th essential amino acid, %; S_{min} – is the minimum of the scores of the essential amino acid of the protein under study relative to the reference, %; n - amount of essential amino acids.

The value of biological value (BV) is determined by formula 4:

$$BV = 100 - ASDF. \qquad (4)$$

To assess the balance of essential amino acids in relation to the reference protein, a utility coefficient K_i is calculated according to formula 5.

$$K_i = \frac{AC_{\min}}{AC_i}, \qquad (5)$$

where AC_i is the amino acid score of the i-nonessential amino acid in the enriched product; AC_{min} is the amino acid score of the first limiting amino acid in the product.

The amino acid composition utility ratio (U) was calculated according to formula 6:

$$ASDF = \frac{\sum_{i=1}^{n} (A_i K_i)}{\sum_{i=1}^{n} A_{i,et}}, \qquad (6)$$

The results of calculations of the biological value of curd products are presented in Table 9.

The Table 9 shows that the amino acid score of the developed product is high enough for eight essential amino acids and ranges from 105.9% valine to 191.0% tryptophan. The limiting amino acid is phenylalanine at 93.5%.

Evaluating the quality of the proteins of the products produced by the amino acid scoring method, it can be noted that they are well-balanced. Thus, the optimal content of particularly scarce amino acids, such as leucine, lysine and tryptophan amino acid scoring is much higher than the ideal protein.

The calculation of biological value of the control sample of curd products was 67.4%, and the experimental sample with sesame seeds and whey proteins was 65.2%, which indicates a high biological value of the developed curd product.

Table 9. Results of calculations of indicators of biological value of curd products.

Name	AO reference, mg/1g protein	Curd product, (mg/1g)	Control sample						Functional product, (mg/1g)	Experimental sample					
			Amino acid grading, %	ΔACD	ASDF	BV	K_i	U		Amino acid grading, %	ΔACD	ASDF	BV	K_i	U
Valine	5.2	5.4	103.8	12.2			0.88		5.51	105.9	12.4			0.88	
Histidine	1.8	2.4	133.3	41.7			0.69		2.52	140. 0	46.5			0.67	
Isoleucine	4.6	5.0	108.7	17.0			0.84		5.13	111.5	18.0			0.84	
Leucine	6.9	9.1	131.9	40.2			0.70		9.37	135.8	42.3			069	
Lysine	5.5	8.5	154.5	62.9			0.59		8.71	158.4	64.9			0.59	
Methionine	2.8	2.6	92.9	1.2	32.6	67.4	0.99	0.77	2.65	94.6	1.1	34.8	65.2	0.99	0.77
Threonine	3.7	4.5	121.6	30			0.75		4.59	124.1	30.6			0.75	
Phenylalanine	4.8	4.4	91.7	0			1		4.49	935	0			1	
Tryptophan	1.0	1.8	180.0	88.3			0.51		1.91	191.0	97.5			0.49	
Total			1118.5	293.5			6.95		44.88	1154.9	313.0			6.90	

Calculation of the utilitarian coefficient of amino acid composition (U) shows that both the control and the experimental sample have the same value (0.77), due to the fact that all amino acids of the developed curd product, including the limiting amino acid (phenylalanine) increased by 1.87%. Utility coefficient characterizes the balance of the amino acid composition, the better-balanced amino acids in protein, the more rationally they can be used by the body.

Based on all of the above, it follows that cottage cheese product enriched with sesame seeds and whey proteins is a high-quality product, balanced by chemical, vitamin-mineral and, especially, amino acid composition, which together characterize its high biological value.

4. Conclusion

The authors conducted research to assess the quality of the developed curd product and calculated its biological value. When evaluating organoleptic indicators, it was found that the developed curd product got the maximum number of 10 points compared to the control sample, which received 8 points, higher scores were given to the enriched product for taste and smell, appearance and consistency. The chemical composition, microbiological indicators, safety indicators and calculated energy value of cottage cheese products were determined. It was found that the nutritional value of the experimental sample increased due to the increase of protein by 2.0% due to the introduced whey proteins and appearance of carbohydrates (0.3%) and dietary fiber (0.05%) in the experimental sample due to sesame seeds, which increased the energy value by 9.2 kcal. Evaluation of microbiological parameters made it possible to establish that the number of lactic acid microorganisms in the experimental sample significantly exceeds both normalized indicators and values of the control sample, which proves its probiotic properties and indicates a high physiological value. The study of vitamin and mineral and amino acid composition has shown an increase in all macro- and microelements in the developed product: magnesium - by 25%, calcium - by 23%, phosphorus - by 21%, potassium - by 10%, sulfur - by 9%; iodine and selenium are revealed. The vitamin composition of the cottage cheese product is quite diverse and the developed product showed an increase in all vitamins: for example, choline by 40%, C by 33.3%, B1 by 20%.

The amino acid composition of the enriched curd product increased by 3.7 %, 3.4 % more essential amino acids and 3.74 % more substitutable amino acids. When assessing the biological value, it was found that the amino acid content of eight essential amino acids is high enough and ranges from 105.9% valine to 191.0% tryptophan. The limiting amino acid is phenylalanine - 93.5%. The calculation of the biological value of the control sample of curd products was - 67.4%, and the experimental sample with sesame seeds and whey proteins was - 65.2%. Calculation of the utilitarian coefficient of amino acid composition shows that both the control and experimental samples have the same value (0.77). Utility coefficient characterizes the balance of the amino acid composition, the better the balance of amino acids in protein, the more rational they can be used by the body.

Based on all of the above, it follows that cottage cheese product enriched with sesame seeds and whey proteins is a high-quality product, balanced by chemical, vitamin-mineral and, especially, amino acid composition, which together characterize its high biological value.

Thus, the useful properties of the curd product can become the main factor for its promotion strategy in the market of functional foods, and the introduction of the developed technology at the dairy enterprises will expand the assortment of curd products aimed at solving the problem of satisfying the population with wholesome food products.

5. Acknowledgments

This work was carried out as part of a state assignment SSI NIIMMP of the Ministry of Science and Higher Education, No 1021032427412-9.

References

[1] Kontareva V Yu, Belik S N, Morgul E V, Gorlov I F and Slozhenkina M I 2020 Yogurt enriched to correct intestinal microflora in dysbiosis *IOP Conf. Series: Earth and Environmental Science* **548(8)** 82051

[2] Kryuchkova V V, Gorlov I F, Korneichuk K M, Belik S N and Mosolova N I 2020 Brine-ripened cheese enriched with vegetable ingredients: technology and quality *IOP Conf. Series: Earth and Environmental Science* **548(8)** 82063

[3] Kryuchkova V V, Gorlov I F, Belik S N and Kamlatsky 2020 As vegetable ingredients in functional fermented milk products *IOP Conf. Series: Earth and Environmental Science* **548(8)** 82092

[4] Wan Y, Li H, Fu G, Chen X, Chen F and Xie M 2015 The relationship of antioxidant components and antioxidant activity of sesame seed oil *J. Sci. Food Agric.* **95(13)** 2571-2578

[5] Dar A A and Arumugam N 2013 Lignans of sesame: purification methods, biological activities and biosynthesis-a review *Bioorg. Chem.* **50** 1-10

[6] Majdalawieh A F, Massri M and Nasrallah G K 2017 A comprehensive review on the anti-cancer properties and mechanisms of action of sesamin, a lignan in sesame seeds （Sesamum indicum） *Eur. J. Pharmacol.* **815** 512-521

[7] Gouveia Lde A, Cardoso C A, de Oliveira G M, Rosa G and Moreira A S 2016 Effects of the Intake of Sesame Seeds (Sesamum indicum L.) and Derivatives on Oxidative Stress: A Systematic Review *J. Med. Food* **19(4)** 337-345

[8] Ikeda S, Toyoshima K and Yamashita K 2001 Dietary sesame seeds elevate alpha- and gamma-

tocotrienol concentrations in skin and adipose tissue of rats fed the tocotrienol-rich fraction extracted from palm oil *J. Nutr.* **131(11)** 2892-2897

[9] Song Y, Ding W, Bei Y, Xiao Y, Tong H D Wang LB and Ai LY 2018 Insulin is a potential antioxidant for diabetes-associated cognitive decline via regulating Nrf2 dependent antioxidant enzymes *Biomed Pharmacother* **104** 474-484

[10] John J, Nampoothiri M, Kumar N, Mudgal J, Nampurath GK and Chamallamudi MR 2015 Sesamol, a lipid lowering agent, ameliorates aluminium chloride induced behavioral and biochemical alterations in rats *Pharmacogn. Mag.* **11(42)** 327-36

[11] Morgul E V, Belik S N, Morgul A R, Gorlov I F, Slozhenkina M I and Mosolov A A 2021 The possibility of using amino acid supplements for stress prevention *IOP Conf. Series: Earth and Environmental Science* **677** 32068

[12] Dudnikova O A, Khramtsov A G and Lodygin A D 2011 Study of nutritional, biological value and storability of lipid-protein concentrate *Food Engineering and Technology* **2(21)** 72-76

[13] Almeida C C, Alvares T S, Costa M P and Conte-Junior C A 2016 Protein and Amino Acid Profiles of Different Whey Protein Supplements *J. Diet Suppl.* **13(3)** 313-323

[14] Teixeira F J, Santos H O, Howell S L and Pimentel G D 2019 Whey protein in cancer therapy: A narrative review *Pharmacol. Res.* **144** 245-256

[15] Yamaguchi M and Takai S 2014 Chronic administration of bovine milk-derived alpha-lactalbumin improves glucose tolerance via enhancement of adiponectin in Goto-Kakizaki rats with type 2 diabetes *Biol. Pharm. Bull.* **37(3)** 404-408

[16] Zapata R C, Singh A, Pezeshki A, Nibber T and Chelikani P K 2017 Whey Protein Components - Lactalbumin and Lactoferrin - Improve Energy Balance and Metabolism *Sci. Rep.* **7(1)** 9917

[17] Chungchunlam S M, Henare S J, Ganesh S and Moughan P J 2015 Dietary whey protein influences plasma satiety-related hormones and plasma amino acids in normal-weight adult women *Eur. J. Clin. Nutr.* **69(2)** 179-186

[18] Pezeshki A, Fahim A and Chelikani P K 2015 Dietary whey and casein differentially affect energy balance, gut hormones, glucose metabolism, and taste preference in diet-induced obese rats *J. Nutr.* **145(10)** 2236-2244

[19] Kontareva V Yu, Belik S N, Morgul E V, Gorlov I F and Slozenkina M I 2020 The effect of prebiotic components on the quality of yogurt *IOP Conf. Series: Earth and Environmental Science* **548(8)** 82054

IOP Conf. Series: Earth and Environmental Science **965** (2022) 012040 doi:10.1088/1755-1315/965/1/012040

The kombucha infusion use prospects in as an environmentally friendly food product

V V Kleschevnikova, N A Zlepkina, E A Skorokhodov, A V Gorbunov, I A Avoyan and V A Chuchunov

Volgograd State Agrarian University, 26 University Avenue, Volgograd, 400002, Russia

E-mail: tpppgit36@mail.ru

Abstract. The Medusomyces gisevi (combucha) culture has long been widely used by the population for food purposes, as well as a natural prophylactic and medicinal agent. The detoxification, antioxidant, anti-inflammatory, immunostimulating, hypolipidemic properties etc. are among the proven properties of kombucha. Researches by a number of authors have proven a pronounced bacteriostatic and bactericidal effect of kombucha metabolites against a wide range of pathogenic microorganisms. In this work we conducted research on the fish pickling technology on the basis of vinegar obtained from the cultural liquid of the kombucha. We studied the organoleptic and physicochemical indicators of the experimental pickled fish batches, according to the results of which it was noted that in the course of autoleptic changes when pickling by the proposed method, the proteins undergo a stronger denaturation.

1. Introduction

The problem of nutrition has been and remains one of the most important for human society. There is a shortage of food products with a high content of complete protein rich in essential amino acids. It is known that a simple increase in food consumption cannot solve the problem of rational nutrition and effective assimilation of various products by human body. Therefore, the development of food and processing industries should be based not only on the knowledge of the chemical composition of the raw materials used, but also on the patterns of their formation, transformation and modification during technological processing and storage, as well as the degree of assimilation of finished products and their physiological value. Fish and fish products derived from it are one of the most important components of the protein nutrition of most people.

In this regard, particular attention is paid to the species composition of fish, their fishing, storage, delivery and processing methods. The priority direction of the fish processing industry in our country and abroad is the improvement of classical methods and technics of its processing, the development and use of modern equipment, the purposeful regulation of biological and physicochemical processes in fish occurring at different stages of its technological cycle of processing [1]. The main volume of commercial fish (over 60%) consists of three fish families - codfish, herring and anchovy, among which: pollack, hake, saury, mackerel, tuna, capelin, herring, anchovy and others. Among the pond commercial fish priority is given to carp, silver carp, trout, sturgeon, pikeperch, etc.

The chemical composition is characterized by the content of complete proteins - 14-22% on average, easily digestible biologically active fats - 0.2-33%, mineral substances - 1-2%, extractive substances - 1.5-3.9%, fat- and water-soluble vitamins, and other substances. Water accounts for 52–85% of the fish mass.

2. Materials and methods

The purpose of our research was to study the possibility of the fish pickling technology improving using vinegar obtained on the basis of a kombucha solution instead of acetic acid.

To study the vinegar application obtained from a kombucha solution in the technology of fish products pickling, such fishing mass objects were used as the white amur, the carp family (Ctenopharyngodon idella), which is famous for its taste and nutritional qualities, and the mackerel - Scomber family (Scomber scombrus), which range is widely represented in the commodity markets. The kombucha (Medusomyces gisevii) is a symbiotic culture in which various forms of acetic acid bacteria and yeast fungi co-exist [2]. This is a multilayer elastic lamellar structure, the nutrient medium for which, as a rule, is a sweetened tea solution. In the process of the microorganisms' vital activity that make up the kombucha in the presence of oxygen, enzymatic fermentation of the tea solution occurs with the formation of a specific product that can be consumed as a drink [3, 4]. The chemical composition of the kombucha drink has been studied several times, however, as noted in the research works of Danielyan [5], the published results often differ from each other, which may be due to the different microbiological composition of the Medusomyces gisevii samples, the concentration of the nutrient tea solution or other modifications of the nutrient medium, the methods and the time of enzymatic fermentation used [5, 6]. According to the data of Konovalov and Semenova [7], Hauser [8], Bauer-Petrovska and Petrushevska-Tozi [9], Danielyan [5], Velicanski [10], Vitas [11], the kombucha drink contains:

- phenolic compounds, polyphenols;
- flavonoids;
- organic acids (acetic, gluconic, glucuronic, citric, succinic, malic, vinic, malonic, oxalic, L-lactic, D-saccharic, pyroracemic and also usnic);
- sucrose, glucose and fructose;
- vitamins B_1, B_2, B_6, B_{12}, C;
- 14 amino acids, biogenic amines, purines, pigments, lipids, proteins, some hydrolytic enzymes;
- substances with antibacterial activity;
- carbon dioxide CO_2;
- ethyl alcohol;
- manganese, iron, nickel, copper, zinc, lead, cobalt, chromium, cadmium.

The kombucha cultivation at home and the use of the fermentation product of its cultural liquid as a drink is widespread among the population. Many people believe that this drink is a valuable prophylactic and therapeutic agent for various diseases and they regularly consume it [12, 13]. Considering the above, it should be recognized that the study of the Medusomyces gisevii fermented culture fluid effect and its other metabolites on the body internal environment, as well as the systematization of materials, are relevant.

Danielyan conducted a long-term study of the kombucha biological characteristics. According to the data obtained the Medusomyces gisevii cultural liquid, firstly, has a pronounced antimicrobial activity, which is associated with the presence in the studied liquid composition of antibacterial substances of a wide spectrum of action, which have both bacteriostatic and bactericidal properties. Secondly, under conditions of exposure to cultural fluid, there is an increase in the size and volume of bacterial cells, a change in their shape, vacuolization of the hyaloplasm, the appearance of granular inclusions in it. Thirdly, under the influence of the kombucha cultural fluid, there is a decrease in the intensity of redox processes in microbial cells, a decrease in their virulence and an increase in immunogenicity. Danielyan also showed that the kombucha cultural liquid at its oral intake does not have any toxic effect on the body. In addition, the author notes the presence of hypotensive activity in the studied fluid, the ability to regulate secretion and motility of the stomach and intestines [5].

IOP Conf. Series: Earth and Environmental Science **965** (2022) 012040 doi:10.1088/1755-1315/965/1/012040

Steinkraus et al linked the kombucha antimicrobial activity with the effect on acetic acid microorganisms, the main fermentation product of the cultural liquid [14]. The composition of acetic acid bacteria varies, so the substances they produce are heterogeneous. Some of them oxidise the ethyl alcohol produced by yeast into acetic acid, others convert sucrose into glucose and fructose and oxidise monosaccharides into gluconic acids. The formed acids are used by the yeast to synthesize vitamins required for the development of acetic acid bacteria. As a result of the combined vital activity of yeast fungi and acetic acid bacteria, the finished kombucha infusion has a sour taste and effervescence.

The body is positively affected by glucuronic acid, which has a detoxifying effect, while lactic acid destroys harmful intestinal microflora and normalizes its functions. Preparations made from the kombucha infusion have anti-bacterial activity, which they do not lose during long-term storage, as well as antibiotic properties. They can be successfully used as growth stimulators for animals, inhibitors of various microorganisms' vital functions and for many diseases treatment.

The preparation of vinegar based on kombucha is as follows: the kombucha is placed in a solution of not too strong tea in which sugar was previously diluted (sugar syrup) and left for 2-3 months, and not for 7 days, as in the case of obtaining a medicinal drink. After this period, the solution is poured into an enamel bowl and boiled for 30-40 minutes. Then the infusion is cooled and filtered through gauze folded in several layers. Filtered vinegar is poured into bottles. Bottles are sealed tightly and placed in a cool place. It is used when needed.

In the standard sense, pickling is a method of fish preserving using cooking salt, acetic acid and a set of spices. Products obtained by pickling are called marinades. There are two types of marinades: cold and hot. Hot marinades are prepared from previously cooked, fried or smoked fish; cold marinades are from fresh or salted fish. Cold marinades are the most widespread in the industry.

For the production of pickled goods are sent mainly salted semi-finished products. There are two methods of cold pickling: with pre-cooking of fish in a vinegar-salt solution and without pre-cooking.

The pickled fish becomes soft, the bones are easily separated, and the fish turns white and gets a sour taste.

To obtain control and experimental samples, white amur and mackerel were pre-salted with cooking salt.

The salting process is quite long. The salting rate in different periods is not the same. At first, when the difference in osmotic pressure is large, salting goes faster, then it slows down and stops altogether when the osmotic pressure drops to zero (the concentration of the salt solution in pickled brine and fish tissues is equalized.

The fish salting rate, i.e. the time required to obtain fish with a certain content (concentration) of salt, depends on the salt concentration in pickled brine (brine), the presence and nature of the skin cover, the chemical composition of fish tissue and salt, ambient temperature, the method of salting and the speed of mixing the solution and fish. Fish with tissues containing little moisture and a lot of fat is salted more slowly than nonfat fish, because salt and moisture are poorly soluble in fat [1].

The salt crystal size (grind) affects the dissolution rate and therefore the maintenance of a high salt concentration in the pickled brine. Usually, salt is used for salting so that its dissolution rate is faster than the salting rate of the fish. However, in some cases, salting is slower if the salt consists of only large crystals or, conversely, only small crystals. Salt of too fine grinding can delay fish salting due to strong dehydration and hardening of its surface, which can lead to fish spoilage. Salt grinding No. 2 and No. 3 and less often №1 are used for salting, the best No. 2 is 3-4 mm [15].

For the production of control and experimental samples we used cooking salt containing 39.34% sodium and 60.66% chlorine and grinding size No. 2.

Two experimental and two control fish samples were washed, dried, and then cut into 3-4 cm pieces. Then we salted the experimental and control batches and left them for 12 hours, after receiving a salty semi-finished product we washed it and poured cold water for 1 hour (to soak and remove unnecessary salinity of the samples), then we drained the water. After that, we proceeded to the pickling itself.

IOP Conf. Series: Earth and Environmental Science **965** (2022) 012040 doi:10.1088/1755-1315/965/1/012040

To give the marinades a piquant taste, spices were used, namely: allspice and whole black pepper, cloves, bay leaves, coriander. Each of the experimental and control batches were placed in a container in layers: spices, salted semi-finished product, which was laid loosely, without tamping. Subsequent layers were alternated in the same way. Further, according to the standard technology of pickled fish production, the control batches were poured into the spicy vinegar-salt solution with the acetic acid content of 3-4% at the ratio of the solution amount to the fish weight 2:1. In the experimental batches, we replaced acetic acid with vinegar obtained on the basis of kombucha solution.

Further, we placed the control and experimental fish batches in the refrigerating chamber with the temperature of +1…+4 °C for 8 hours. In industrial conditions with large batches the ripening of pickled fish should be carried out at a temperature of about 0 °C for 10-30 days depending on the salt and vinegar concentration.

The concentration of cooking salt was determined by argentometric method.

3. Results and discussion

After receiving the finished product, an organoleptic and physicochemical analysis of control and experimental batches was carried out (Table 1).

Table 1 - Organoleptic and physico-chemical indicators of control and experimental batches of pickled fish.

Indicators	Pickled fish batches			
	Control batch		Experimental batch	
	Mackerel	White amur	Mackerel	White amur
Appearance	The surface is clean with a color typical of pickled mackerel.	The surface is clean with a color typical of white amur. There is slight sediment of protein substances in the marinade.	The surface is clean with a color typical of pickled Mackerel.	The surface is clean with a color typical of white amur. There is slight sediment of protein sub-stances in the marinade.
Consistency	Tender, juicy	Dense, juicy	Tender, juicy	Dense, juicy
Taste and smell	Mackerel fish tastes moderately salty and sour. The smell is characteristic of spices and marinade.	The taste of white amur is salty and sour. The smell is characteristic of spices and marinade.	The taste of salted mackerel fish is peculiar, moderately salty and slightly sweet. The smell is characteristic of spices and marinade.	The taste of salted white amur fish is peculiar, moderately salty and slightly sweet. The smell is characteristic of spices and marinade.
Mass fraction of cooking salt, %	5.6	7.1	5.5	7.2

4. Conclusion

Analyzing the results of the conducted research we can say that the fish pickled in this way (using kombucha vinegar) turns out to be tenderer, juicy, acquires a peculiar aroma and taste. Pickled mackerel, in comparison with pickled white amur, has a delicate texture and a moderately salty taste. The cooking salt content in white amur is 1.5-1.7% higher than in mackerel, while the fish was salted with the same amount of salt per 1 kg of raw material.

References

[1] Ibragimova S A 2020 *Food Biotechnology. Section "Fish Production Biotechnology"* (Saransk: Publishing house of Mordovia University) p 70

[2] Jarrell J, Cal T and Bennett J W 2000 The kombucha consortia of yeasts and bacteria. *Mycologist.* **14** 166-170

[3] Dutta D and Gachhui R 2007 Nitrogen-fixing and cellulose-producing Gluconacetobacter kombuchae sp. nov., isolated from Kombucha tea *Int. J. Syst. Evol. Microbiol.* **57** 353-357

[4] Murugesan G S, Sathishkumar M, Jayabalan R, Binupriya A R, Swaminathan K and Yun S E 2009 Hepatoprotective and curative properties of Kombucha tea against carbon tetrachlorideinduced toxicity.*J. Microbiol. Biotechnol.* **19(4)** 397-402

[5] Danielyan L T 2005 *Kombucha and Its Biological Features* Moscow: Medicine p 176

[6] Jayabalan R, Marimuthu S and Swaminathan K 2007 Changes in content of organic acids and tea polyphenols during kombucha tea fermentation *Food Chem.* **102** 392-398

[7] Konovalov I N and Semenova M N 1955 To the physiology of the Tea mushroom *Bot. J.* **40** 567-570

[8] Hauser S P 1990 Dr. Sklenar's kombucha mushroom infusion-a biological cancer therapy. Documentation No. 18 *Schweiz Rundsch Med. Prax.* **79** 243-246

[9] Bauer-Petrovska B and Petrushevska-Tozi L 2000 Mineral and water-soluble vitamin contents in the kombucha drink *Int. J. Food Sci. Technol.* **35** 201-205

[10] Velicanski A S, Cvetkovic D D and Markov S L 2013 Characteristics of kombucha fermentation on medicinal herbs from Lamiaceae family *Roum. Biotechnol. Lett.* **18** 8034-8042

[11] Vitas J S, Malbasa R V, Grahovac J A and Loncar E S 2013 The antioxidant activity of kombucha fermented milk products with stinging nettle and winter savory *Chem. Ind. Chem. Eng. Q.* **19** 129-139

[12] Sreeramulu G, Zhu Y and Knol W 2000 Kombucha fermentation and its antimicrobial activity *J. Agric. Food. Chem.* **48** 2589-2594

[13] Jayabalan R, Subathradevi Marimuthu P, Sathishkumar M and Swaminathan K 2008 Changes in free-radical scavenging ability of kombucha tea during fermentation *Food Chem.* **109** 227-234

[14] Steinkraus K H, Shapiro K B, Hotchkiss J H and Mortlock R P 1996 Investigations into the antibiotic activity of tea fungus/kombucha beverage *Acta Biotechnol.* **16** 199-205

[15] Zlepkina N A and Gorbunov A V 2018 *Fish Processing Technology* (Volgograd: Volgograd State Agrarian University) p 28

IOP Conf. Series: Earth and Environmental Science **965** (2022) 012041 doi:10.1088/1755-1315/965/1/012041

Evaluation of quality indicators of fermented milk product based on sheep's milk enriched with vegetable ingredients

Y A Kolosov[1], N V Shirokova[1], N G Chamurliev[2], N N Kolosova[1], A S Shperov[2] and E S Vorontsova[2, 3]

[1] Don State Agrarian University, 24 Krivoshlykov Street, Persianovsky settlement, Oktyabrsky district, Rostov region, 346493, Russia
[2] Volgograd State Agrarian University, 26 University Avenue, Volgograd, 400002, Russia
[3] All-Russian Research Institute of Irrigated Agriculture, 9 Timiryazev Street, Volgograd, 400002, Russia

E-mail: zootexnia@volgau.com

Abstract. The purpose of this study is a comprehensive assessment of the quality of yogurt enriched with vegetable ingredients. The object of research is a fermented milk product, the main ingredient of the formulation is sheep's milk. The unique composition of sheep's milk makes it a promising raw material for obtaining food products with high biological and nutritional value. Milk thistle meal, ground walnut and chokeberry syrup were selected as vegetable fortifiers for the yogurt under study. Experimental studies were conducted in the Don State Agrarian University. Traditional yogurt based on cow's milk without fillers was selected as control samples. When analyzing the results of studies of the fatty acid composition of the samples of the fermented milk product, it was revealed that the prototype in terms of the content of some fatty acids exceeds the values of the control, due to the replacement of cow's milk and the introduction of vegetable ingredients. The study assessed the quality indicators of yogurt samples, such as organoleptic, physico-chemical, biochemical and microbiological. Based on the results obtained, it can be concluded that it is advisable to enrich the fermented milk product with vegetable ingredients.

1. Introduction

Recently, consumers have been forming an increased demand for quality food products. The quality of food is characterized as a set of properties that determine its suitability to meet certain needs in accordance with its purpose. Quality control of manufactured products is a process of obtaining and processing information about the object of research in order to determine and find its parameters within the required limits [1, 2]. Medical studies conducted by researchers in Russia in recent years have shown that, on the one hand, there is a decrease in the consumption of food sources of energy and protein in the diet of the population, on the other hand, against the prevailing nutritional background, many people are obese, which is a consequence of metabolic disorders. Consequently, the optimization of the population's nutrition structure is inextricably linked with an increase in the production of mass-consumption products with high nutritional and biological value, including those enriched with high-quality protein, vitamins and minerals [3, 4]. As a rule, the technological process of production of any fermented milk product, including yogurt, is controlled by organoleptic, physico-chemical, biochemical and microbiological indicators [5]. Organoleptic indicators are the main characteristics of consumer properties of food products, which are determined using the human senses. The main organoleptic indicators for fermented milk products are: appearance, taste and smell, consistency. The basis for the examination of the quality of fermented milk is the determination of physico-chemical indicators. These include temperature characteristics, the mass fraction of fat, protein and moisture. Biochemical methods are based on biochemical processes. As a rule, these methods are used to control the quality of raw materials used for production, as well as to assess the nutritional and biological value of products. Microbiological methods are based on the vital activity of microorganisms. These methods are used to control the quality of raw materials, to control technological processes, technological hygiene of equipment and finished products [6]. Guided by the above approaches, we have identified the area of our interests as the development of high-quality and healthy for human nutrition fermented milk products based on sheep's milk. Thus, the purpose of this study is a comprehensive assessment of the quality indicators of yogurt made from sheep's milk, enriched with vegetable ingredients, prepared according to the recipe developed by us.

2. Materials and methods

Most authors mainly use raw materials of animal and vegetable origin when developing the formulation of preventive products as a source of vitamins, protein and minerals, fats and carbohydrates. In some cases, a natural food component is used as a raw material, in others, the raw materials are enriched with special additives [7, 8].

The main ingredient in the formulation of the studied yogurt is sheep's milk. The unique composition of sheep's milk makes it a promising raw material for obtaining food products with high biological and nutritional value, with biologically active components that provide benefits to human health, since it contains 1.4 times more dry matter and 1.7 times more protein in its composition compared to cow's milk.

Milk thistle meal, ground walnut and chokeberry syrup were selected as vegetable fortifiers for the studied yogurt. Each of the listed vegetable ingredients contains a range of useful nutrients necessary for the normal functioning of the human body.

Experimental studies were conducted at the Department of Food Technology and Commodity Science of the Don State Agrarian University. Traditional yogurt based on cow's milk without fillers was selected as control sample, and yogurt based on sheep's milk enriched with vegetable ingredients was selected as experimental sample.

To implement the set tasks for the assessment of quality indicators, studies were conducted on the basis of standard and generally accepted methods:

- acceptance, sampling, preparation for the study – the State Standards GOST 3622-68, GOST 26809.1-2014;
- organoleptic indicators - the State Standard GOST ISO 5492-2014;
- mass fraction of fat - the State Standard GOST 5867-90;
- mass fraction of protein - the State Standard GOST 23327-98;
- mass fraction of dry skimmed milk residue - the State Standard GOST R 54668-2011;
- phosphatase or peroxidase - the State Standard GOST 3623-2015;
- titrated acidity - the State Standard GOST 3624-92;
- microbiological indicators - the State Standards GOST 10444.11-2013 (ISO 15214:1998), GOST 10444.12-2013 GOST 30347-2016, GOST 31659-2012 (ISO 6579:2002);
- mass fraction of fatty acids - the State Standard GOST 32915-2014, methodical instructions MU 4.1./4.2.2484-09. The triglycerides of fatty acids isolated from the samples were converted into methyl esters and the resulting mixture was analyzed on a chromatograph "Chromatek-Crystal 5000". The results were compared with the normative indicators presented in attachment A of the State Standard GOST R 52253-2004.

At the final stage of the experiment, an organoleptic analysis of the studied yogurt was carried out

IOP Conf. Series: Earth and Environmental Science **965** (2022) 012041 doi:10.1088/1755-1315/965/1/012041

using a quantitative descriptive method of analytical evaluation. Statistical processing was performed using the Microsoft Office 2013 software package (Microsoft, USA) and the Statistica7 application software package.

3. Results and discussions

The yogurt was produced by a thermostatic method, the sourdough was added individually to each container after bottling, the fermented clot was not mixed, which made it possible to achieve a denser structure of the product. Since the fermentation process is the most vulnerable stage of the technological process, the introduction of prepared plant components was carried out before pasteurization, in order to avoid bacterial contamination. At the initial stage, in accordance with the set goal, the organoleptic indicators of the control and experimental yogurt samples were evaluated according to the State Standard GOST 31981-2013 "Yogurts. General technical conditions", the results of which are presented in Table 1.

Table 1. Organoleptic quality indicators of control and experimental yogurt samples.

Indicator	Control sample	Experimental sample
Consistency and appearance	Homogeneous, with a broken clot moderately viscous	Homogeneous, with a broken clot moderately viscous. The presence of inclusions of insoluble particles characteristic of the introduced components is visible
Taste and smell	Pure, fermented milk, without foreign tastes and odors	Pure, fermented milk, with the appropriate taste and aroma of the added components
Colour	Milky white	Milk-cream

In terms of consistency, both samples were characterized by uniformity, and in appearance, the experimental sample was distinguished by the presence of inclusions of milk thistle meal and walnut. According to the taste range, the control sample can be characterized as having a characteristic pure, fermented milk flavor, the experimental sample is characterized by a fermented milk taste with a nutty flavor. The color of the control sample is marked as milky white, uniform throughout the mass, and the experimental sample is milky-cream.

Thus, the introduction of vegetable ingredients in the formula of the fermented milk product contributed to the appearance of a pleasant, harmonious taste and smell characteristic of the ingredients introduced.

The next stage was the evaluation of the physico-chemical quality indicators of yogurt samples, which are shown in Table 2.

Table 2. Physico-chemical parameters of samples.

Indicator	State Standard GOST31981-2013	Control sample	Experimental sample
Mass fraction of fat, %	0.5-10.0	2.5 ± 0.10	2.5 ± 0.13
Mass fraction of protein, %	2.8-3.2	3.1 ± 0.10	2.8 ± 0.10
Mass fraction of dry skimmed milk residue, %	8.5-9.5	8.7 ± 0.10	9.4 ± 0.10
Acidity, ° T	75-140	80 ± 0.10	85 ± 0.10
Phosphatase or peroxidase	lack of	-	-

The obtained values of physico-chemical indicators of the samples correspond the requirements of the State Standard GOST 31981-2013 "Yoghurts. General technical conditions". Thus, the experimental sample of the enriched fermented milk product has the most harmonious organoleptic characteristics, its titrated acidity is within the regulated limits.

At the final stage, the most important biochemical quality indicators, such as fatty acid and vitamin-mineral composition, were evaluated.

When analyzing the results of studies of the fatty acid composition of samples of fermented milk product (Table 3) it was found that the experimental sample in terms of the content of some fatty acids exceeds the values of the control sample. This, in our opinion, was the result of the use of a different dairy component instead of cow's milk and the introduction of vegetable ingredients.

Table 3. Fatty acid composition of samples, %.

Name of the fatty acid	Acceptable levels	Control sample	Experimental sample	Error rate (+/-)
		Saturated fatty acids		
Butyric C4:0	2.0-4.2	2.4	1.0	0.40
Caproic C6:0	1.5-30	1.8	0.7	0.40
Caprylic C8:0	1.0-2.0	1.2	0.5	0.40
Capric C10:0	2.0-3.5	2.8	1.0	0.40
Lauric C12:0	2.0-4.0	3.2	1.2	0.40
Myristic C14:0	8.0-13.0	9.9	3.8	0.40
Palmitic C16:0	22.0-33.0	29.9	15.4	2.20
Stearic C18:0	9.0-13.0	10.7	5.6	2.20
Begenic C22:0	до 0.1	0.1	0.1	2.2
Arachinic C20:0	до 0.3	0.2	0.3	0.40
		Monounsaturated fatty acids		
Decenic C10:1	0.2-0.4	0.3	0.1	0.40
Myristoleic C14:1	0.6-1.5	1.4	0.5	0.40
Palmitoleic C16:1	1.5-2.0	1.3	0.5	0.40
Oleic C18:1	22.0-32.0	27.5	24.1	2.20
		Polyunsaturated fatty acids		
Linoleic C18:2	3.0-5.5	3.0	36.8	0.40
Linolenic C18:3	до 1.5	0.2	0.7	0.40

Caprylic and capric acids together with butyric and caproic acids determine the organoleptic characteristics and quality of the finished dairy product under storage conditions.

The content of butyric acid in the control sample is at the level of 2.4%, in the experimental - 1.0%. No less important is caproic acid, it is contained in the control sample in an amount of 1.8%, and in the experimental - 0.7%.

The content of caprylic and capric acid in the control sample is 1.2% and 2.8%, respectively. In the experimental sample, the content of caprylic and capric acid is 0.5% and 1.0%, respectively.

One of the criteria for the naturalness of milk fat is the presence of decenic acid in its fatty acid composition. It is a minor component of milk fat, but its presence is mandatory. The content of decenic acid in the experimental sample ranges from 0.1%, in the control sample - 0.3%.

Methodical instructions MU 4.1./4.2.2484-09 sets the content of lauric acid in milk fat in the range from 2.0 to 4.4%. According to the data, the experimental sample of lauric acid contains 1.2%, and the control sample contains 3.2%.

The content of myristic acid, as well as lauric acid, characterizes the naturalness of milk fat. The range of its content in accordance with the requirements of MU 4.1./4.2.2484-09 should be 8.0-13.0% of the total of fatty acids. The data obtained indicate that the content of myristic acid in the experimental sample is relatively low – 3.8%. Its amount is reduced by the addition of vegetable additives, and hence other fats, since the proportion of any of the fatty acids in milk fat is related with to the content of other fatty acids that make up its composition. In the control sample, myristic acid is in the amount of 9.9%.

In accordance with the requirements of MU 4.1./4.2.2484-09, the content of myristoleic acid in milk fat should be within 0.6-1.6% of the total of fatty acids. In terms of the content of myristoleic acid, both samples meet the requirements of regulatory documentation. In the control sample it is 1.4%, and in the experimental sample it is 0.5%. According to the data obtained, the content of palmitic acid in the fat of the studied dairy products is equal to 15.4% in the experimental sample and 29.9% in the control. The content of palmitinoleic acid in the experimental sample varies within 0.5%, and in the control sample – 1.3%.

No less important as a fatty acid composition is the role of saturated milk fat acid – stearic acid. Stearic acid, together with other saturated acids, is responsible for the formation of the texture of finished dairy products made on the basis of milk fat. Its content in the experimental and control samples is 5.6% and 10.7%, respectively.

Oleic and linoleic acids are referred to the category of particularly important unsaturated milk fat acids. Their content generally determines the qualitative characteristics of finished dairy products. The content of oleic acid in the experimental and control samples is 24.1% and 27.5%, respectively. The content of linoleic acid, an essential polyunsaturated acid, in the experimental sample is 36.8%, and in the control sample - 3.0%. The composition of the product includes another essential polyunsaturated acid - linolenic acid. In the experimental sample, it is contained in an amount of 0.7%, in the control sample – 0.1%. Thus, the obtained values of fatty acids in the experimental sample compared the requirements, but differ significantly from their content level in the control sample, forming a higher-quality fatty acid composition.

The results of the study of the vitamin and mineral composition of the samples of the fermented milk product are presented in Table 4.

Table 4. Vitamin and mineral composition of samples, mg per 100 g.

Indicator	Control sample	Experimental sample
Vitamin A	0.046	0.37
Beta-carotene	0.007	0.39
Vitamin E	0.1	0.67
Vitamin K	0.0003	0.11
Vitamin B_1	-	0.03
Vitamin B_2	0.2	0.22
Vitamin B_3	0.2	0.2
Vitamin B_4	14.3	15.9
Vitamin B_5	0.4	0.43
Vitamin B_9	0.005	3.09
Vitamin B_{12}	0.0005	0.0005
Vitamin C	-	0.23
Magnesium	15.0	31.8
Zinc	0.4	1.2
Selenium	0.002	0.025
Copper	0.01	0.03

The analysis of the results showed that when vegetable ingredients were added, the amount of vitamins increased: A by 0.324 mg, E by 0.57 mg, K by 0.11mg, B_9 by 3.085mg, beta-carotene by 0.383 mg, and the experimental sample is characterized by the presence of vitamins such as B_1 and C.

The analysis also showed that in comparison with the control sample, the experimental sample contains more minerals, for example, the control sample contains 15 mg of magnesium, and the experimental sample contains 61.8 mg. Zinc is contained in the control sample in an amount of 0.4 mg, and in the experimental sample – 1.21 mg. The silene in the experimental sample is contained in an amount of 0.025 mg, which is 0.023 mg more than in the control, and copper in an amount of 0.03 mg, which is 0.02 mg more than the control.

The results of the studied microbiological parameters of the fermented milk product samples are shown in Table 5.

Table 5. Microbiological indicators of fermented milk products.

Indicator	The value of the indicator according to Technical Regulations of the Customs Union TR CU 033/2013	Actual value	
		Control sample	Experimental sample
Lactic acid microorganisms, CFU/cm^3(g), not less	1×10^7	1.6×10^7	1.6×10^7
Yeast, CFU/cm^3 (g), no more	50	20	35
Mold, CFU/cm^3 (g), no more	50	25	30
The amount of E. coli group bacteria (coliform), in 0.1g	Not detected	Not detected	Not detected

According to the requirements of the Technical Regulations of the Customs Union TR CU "On the safety of milk and dairy products", the value of lactic acid microorganisms in the product should be at least 1×10^7 CFU/cm^3 (g), the data of the study results showed that both samples meet the requirements.

The yeast and mold content in the control sample is 20 CFU/cm^3 (g) and 25 CFU/cm^3 (g), respectively, and the experimental sample is 35 CFU/cm^3 (g) and 30 CFU/cm^3 (g). Both samples in terms of yeast and mold content correspond the requirements of the regulations.

Pathogenic microorganisms, such as bacteria of the E. coli group, were not found in the samples of fermented milk products.

4. Conclusion

Thus, in the course of the study, the quality indicators of yogurt samples, such as organoleptic, physico-chemical, biochemical and microbiological, were evaluated. Based on the results obtained, it can be concluded that it is advisable to use sheep's milk as the main composite and enrich the fermented milk product with vegetable ingredients.

References

[1] Ayazbekova M A, Baygozhina G T and Suleimenova M Sh 2016 The quality of yoghurt from sheep's milk and pumpkin juice *Technical sciences - from Theory to Practice* **12(60)** 110-115

[2] Khisamova E D and Zainutdinova E E 2018 *Product Quality Assurance* (Kazan: Kazan University) p 170

[3] Anistratova O V and Soklakov V V 2020 Research of quality and safety indicators of yoghurt enriched with plant components *XXI Century: Results of the Past and Problems of the Present Plus* **9(2)** 62-66

[4] Demyanova T V, Shirokova N V, Skripin P V and Lodygin A D 2020 Justification of the use of plant components in the dairy industry *Actual Directions of Innovative Development of Animal Husbandry and Modern Production Technologies Food Products* (Persianovsky: Don State Agrarian University) pp 388-391

[5] Shlemen M M 2017 Research of technological features of yoghurt production from frozen sheep milk *Food Systems: Theory, Methodology, Practice* (Moscow: All-Russian Research Institute

of Refrigeration Industry – branch V M Gorbatov Federal Research Center for Food Systems of the Russian Academy of Sciences) p 384

[6] Demina E N, Simonenkova A P, Safronova O V and Sergeeva E Yu 2020 Comprehensive assessment of the quality of enriched yoghurt *Polzunovsky Bulletin* **1** 56-60

[7] Afanasyeva M M and Shirokova N V 2021 The relevance of the use of plant ingredients in the technology of fermented milk product *Innovative Technologies of Food and Feed* (Persianovsky: Don State Agrarian University) pp 6-9

[8] Shirokova N V and Afanasyeva M M 2020 Development of a formulation of an enriched fermented milk product with hepatoprotective properties *Actual Problems of Intensive Development of Animal Husbandry* (Bryansk: Bryansk State Agrarian University) pp 160-162

Consumer properties and environmental safety of a functional curd product

V V Kryuchkova[1], I F Gorlov[1], M I Slozhenkina[1], N I Mosolova[1], N V Lomonova[2], S N Belik[3] and P V Skripin[2]

[1] Volga Region Research Institute of Manufacture and Processing of Meat-and-Milk Production, 6 Rokossovsky Street, Volgograd, 400131, Russia
[2] Don State Agrarian University, 24 Krivoshlykov Street, Persianovsky, Rostov region, 346493, Russia
[3] Rostov State Medical University, 29 Nakhichevansky Lane, Rostov-on-Don, 344022, Russia

E-mail: kverav@yandex.ru

Abstract. The authors studied the composition and properties of sesame seeds and whey proteins. It was found that they have medicinal properties, high nutritional and biological value, contain a complex of essential minerals and vitamins necessary for the normal functioning of the body. The dose, method and technological stage of introducing sesame seeds and whey proteins in the production of curd products were studied, on the basis of which the technology of a functional curd product was developed and its consumer properties were determined. It was found that the introduction of sesame seeds and whey proteins increases the organoleptic characteristics, nutritional and physiological value of the curd product.

1. Introduction

Nutrition is practically the only means that prolongs the species life expectancy of a person by 25-40%. The diet of each person should include about 600 nutrients, 95% of which have therapeutic and prophylactic properties, but the imbalance and inadequacy of modern diets, as a rule, leads to disruption of metabolic processes in the body and, as a result, to the occurrence of diseases. An increase in the proportion of the population with various nutritional-related diseases, demographic and environmental problems, and a high epidemiological load have necessitated the development and implementation of a new nutraceutical direction - functional nutrition, which is based on functional foods (FP) [1, 2, 3]. In connection with the above, the development of new FP formulations from traditional raw materials is the main task facing the food industry and the scientific community.

In accordance with the requirements of State Standard R 52349 - functional food products are any modified food product intended for systematic use in the diet of all age groups of the healthy population, reducing the risk of developing diseases associated with nutrition, maintaining and improving health due to the presence in it composition of physiologically functional ingredients that can have beneficial effects on human health, in addition to the effects of the traditional nutrients they contain. Functional food ingredients include physiologically active, valuable and healthy ingredients with known physicochemical characteristics, for which beneficial properties have been identified and scientifically substantiated, and the daily physiological need has been established. As a rule, these are components of animal, plant, microbiological, mineral origin or identical to natural ones, which are part of a functional food product in an amount of at least 15% of the daily physiological requirement, per serving of the product [4].

The main principle of creating a new type of FP is to achieve the highest possible level of usefulness and guaranteed safety. This requires a deep study of the chemical composition and nutritional value of the components of the planned product and the identification of the biological effects of the main functional ingredients.

Currently, functional products are usually created on the basis of traditional technologies, using modifications that provide an increase in the content of useful ingredients to a level comparable to the physiological norms of their consumption (10-50% of the average daily requirement) [5]. The basis of the developed FP was cottage cheese, functional ingredients - white sesame seeds and whey protein concentrate.

One of the most widely demanded fermented milk products is cottage cheese. It is a national Russian protein product with a high biological and nutritional value. Cottage cheese proteins contain all essential amino acids (lysine, histidine, methionine, isoleucine, arginine, threonine, valine, leucine, phenylalanine, tryptophan), in the optimal amount and ratio, the bioavailability of these proteins is 95-97%. The curd contains a small amount of lactose (milk sugar). Compared to fats and proteins, milk lactose passes into cottage cheese to the least extent, since it is in the state of a true solution in milk and remains in the whey during protein coagulation [6, 7]. Among the minerals, a special role belongs to calcium and phosphorus, which are in the curd in a balanced ratio, which determines their high assimilation by the body. Ca and P are the main irreplaceable minerals that make up the bone tissue (98% Ca and 85% are found in bones). Calcium and phosphorus play an important role in the regulation of hemostasis (approximately 1% of calcium ensures the normal functioning of the blood coagulation system, the generation and transmission of nerve impulses, the contraction of muscle fibers, the activation of certain enzymatic systems and the release of certain hormones). Ca and P provide the regulation of cellular functions (if the amount and ratio of these ions in the cell is disturbed, muscle weakness, tetany, coma may occur). The optimal ratio of calcium to phosphorus intake is 2: 1. In addition to the high content of amino acids, the curd contains vitamins A, E, P; group B [7, 8]. Cottage cheese is necessary for the body for the normal functioning of all internal organs, especially bone tissue.

One of the herbal additives used in the enriched curd product technology under development is white sesame seeds. Sesame contains a large amount of oil, consisting of organic acids, saturated and polyunsaturated fatty acids, triglycerides and glycerol esters. The useful composition of sesame seeds includes carbohydrates, amino acids, proteins and vitamins A, B, E, C, PP. They are also rich in calcium, phosphorus, iron, potassium, magnesium and other mineral compounds [9]. The main biologically active components of sesame are sesamin and sesamol. These components have a pronounced antiproliferative and antioxidant effect. In this connection, they have an anticancer effect against many lines of cancer cells [10]. Sesame and its lignins reduce the risk of cardiovascular disease by lowering total serum cholesterol and low-density lipoprotein levels [11]. Sesame improves the condition of nails, human hair, has a positive effect on blood composition, normalizes metabolism and improves the functioning of the nervous system. The high content of calcium makes it indispensable for maintaining the health of bones and joints, as well as in the prevention of arthritis and osteoporosis [12, 13].

Another functional ingredient in the development of a functional curd product technology is whey proteins obtained from the processing of whey obtained in the production of curd. The main serum proteins are albumin and globulin. Albumin is a simple protein and is highly soluble in water. Under the action of rennet and acids, albumin does not coagulate, and when heated to 70 °C, it precipitates. Globulin is present in milk in a dissolved state; it is the carrier of immune bodies. Globulin also refers to simple proteins, coagulates when heated in a weakly acidic environment to a temperature of 72 °C. Albumin and globulin are blood plasma proteins [14]. The concentration of amino acids and peptides in the blood rises sharply within the first hour after taking a diet based on whey proteins. At the same time, the acid-forming function of the stomach does not change, which eliminates the disruption of its work and the formation of gases. The digestibility of whey proteins is extremely high - 96-98%. The use of whey proteins in nutrition leads to the normalization of blood coagulation and helps to reduce the formation of blood clots [15]. In addition, serum proteins are a source of biologically active peptides that can be used in the treatment of pathological processes such as cancer, HIV infection, hepatitis B, cardiovascular and neurodegenerative diseases, and osteoporosis [16, 17, 18].

Thus, the use of sesame seeds and whey proteins in the production of a curd product intended for all groups of consumers, as factors shaping consumer properties and quality, is relevant and appropriate.

Taking into account the above, the purpose of the work is to study the composition and properties of sesame seeds and whey proteins, to establish the dose, method and technological stage of their introduction, as factors that form new quality indicators of a curd product and affect its consumer properties.

2. Materials and methods

The objects of research are: raw milk in accordance with State Standard R 52054-2003; sesame seeds; whey proteins; samples of curd products, developed with the addition of functional ingredients, using various methods of introduction; control sample - curd product produced in accordance with State Standard 31453-2013 Curd. Technical conditions.

When organizing and conducting research, a complex of generally accepted, standard and modified research methods was used: determination of the mass fraction of fat - State Standard R ISO 2446-2011; protein determination - by the Kjeldahl method in accordance with State Standard 34454-2018; thermogravimetric method for determining the mass fraction of dry substances in curd products in accordance with State Standard 28561-90; vitamin C content by titrometric method according to State Standard 24556-89; organoleptic assessment was carried out in accordance with State Standard R ISO 22935-2-2011 and State Standard 33630-2015; microbiological indicators were determined according to State Standard 32901-2014. Statistical processing of the obtained data and construction of mathematical models was carried out using the Statistica 6.0 software package.

3. Results and discussion

As a result of the study, the organoleptic characteristics and nutritional value of the functional components were established. Organoleptic characteristics: sesame seeds used for food production are not heated, in a healthy state, have the color, taste and smell characteristic of normal sesame seeds - without musty, moldy and foreign taste and smell. Whey protein concentrate has a soft, spreadable, slightly crumbly consistency with the presence of perceptible particles of milk protein; pure fermented milk taste and smell with a pronounced taste of pasteurized milk; cream color, uniform throughout the mass.

It has been established that the nutritional value of functional ingredients is quite high, since fats (49.7%) and carbohydrates (23.6%) predominate in sesame seeds, incl. dietary fiber - 5.6%, and whey proteins - protein (25.8%) (Table 1).

Table 1. Nutritional information for sesame seeds and whey proteins.

№	Indicators	Indicator value	
		Sesame seeds	Whey proteins
1	Mass fraction of fat, g	49.7±0.19	1.8±0.02
2	Mass fraction of protein, g	17.7±0.13	25.2±0.18
3	Mass fraction of carbohydrates, g	23.6±0.18	-
	including mass fraction of dietary fiber, g	5.6±0.10	-
4	Mass fraction of moisture, g	9.0±0.11	73.0±0.14
	Mass fraction of moisture, g	100.0	100.0

The results of studying the amino acid composition of sesame seeds, casein and whey proteins are presented in Table 2.

Table 2. Amino acid content in sesame seeds, whey proteins and casein (g/100g protein).

Amino acids	The standard is chicken protein	Whey proteins	SV deviations from the standard, +/-	Casein	K deviation from the standard, +/-	Sesame seeds	SC deviation from the standard, +/-
			Essential amino acids				
Arginine	4.4±0.002	1.8±0.002	- 1.4	3.3±0.002	-1.1	10.7±0.002	+6.3
Valine	5.2±0.003	5.1±0.002	- 0.1	5.4±0.003	+0.2	5.0±0.003	-0.2
Histidine	1.8±0.002	2.0±0.002	+ 0.2	2.4±0.002	+0.6	2.7±0.002	+0.9
Isoleucine	4.6±0.003	6.2±0.002	+ 1.6	5.0±0.003	+0.4	4.4±0.002	-0.2
Leucine	6.9±0.003	12.8±0.002	+ 5.9	9.1±0.003	+2.2	7.6±0.003	+0.7
Lysine	5.5±0.003	10.3±0.002	+ 4.8	8.5±0.003	+3.0	3.1±0.003	-2.4
Methionine	2.8±0.002	2.0±0.002	- 0.8	2.6±0.002	-0.2	3.2±0.002	+0.4
Threonine	3.7±0.002	4.2±0.002	+ 0.5	4.5±0.003	+0.8	4.3±0.002	+0.6
Phenylalanine	4.8±0.003	3.8±0.002	- 1.0	4.4±0.003	-0.4	5.0±0.003	+0.2
Tryptophan	1.0±0.002	2.5±0.002	+ 1.5	1.8±0.002	+0.8	1.8±0.002	+0.8
∑ essential amino acids	40.7	50.7	+ 10.0	47.0	+6.3	47.8	+7.1
			Substitutable amino acids				
Alanin	4.7±0.002	5.0±0.002	+ 0.3	3.4±0.002	-0.7	4.4±0.002	-0.3
Aspartic acid	8.3±0.003	10.2±0.002	+ 1.9	7.6±0.003	-0.7	9.4±0.003	+1.1
Glycine	2.9±0.002	1.9±0.002	- 1.0	1.8±0.002	-1.1	7.8±0.002	+4.9
Proline	3.2±0.002	4.1±0.002	+ 0.9	9.7±0.003	+6.5	4.2±0.002	+1.0
Serine	5.6±0.003	3.9±0.002	- 1.7	5.3±0.003	-0.3	5.3±0.003	-0.3
Tyrosine	3.6±0.002	3.4±0.002	- 0.2	4.7±0.003	+1.1	4.0±0.002	+0.4
Cystine	2.1±0.002	3.7±0.002	+ 1.6	0.9±0.002	-1.2	1.8±0.002	-
∑ substitutable amino acids	30.4	32.2	+1.8	33.4	+3.0	36.9	+6.5
∑ all amino acids	71.1	82.9	+11.8	80.4	+9.3	84.7	+13.6

It was revealed that the content of all amino acids in whey proteins is 16.6% higher than in the standard, in casein - by 13.1%, in sesame seeds - by 19.1%, incl. there are more essential amino acids in whey proteins than in the standard - by 24.6%, in casein - by 15.5%, in sesame seeds - by 17.4%; Substitutable amino acids - by 6.0% more than in the standard, in casein - by 10.0% and sesame seeds - by 21.4%. The amino acid tryptophan is very important, in whey proteins it is 2.5 times more than in the standard, 1.8 times more than in sesame seeds and 39% more than in casein.

Sesame seeds in large quantities contain both essential and substitutable amino acids (in 100 g the daily allowance is 37.0% and 27.4%, respectively), while the content of all amino acids found in the seeds (in 100 g) exceeds 10% of the daily value. Sesame is especially rich in essential amino acids:

tryptophan (1.8 g), valine (1.1 g) and isoleucine (0.9 g); among the Substitutable amino acids, arginine (3.1 g), glutamic acid (4.7 g) and glycine (1.4 g) are dominant.

The amount of essential amino acids in whey proteins is 12.0% higher than in casein, including arginine and leucine by 3.7 g, lysine - by 1.8 g and isoleucine by 1.2 g, which is very important for curd products which are very important in nutrition for children, pregnant women, the elderly and athletes.

Sesame seeds contain 23.4 g of carbohydrates per 100 g of the product, of which dietary fiber accounts for 5.6 g. The content of dietary fiber in sesame covers 22.4% of the body's daily requirement, which is very important for a functional product (Table 3).

Table 3. The content of carbohydrates and fatty acids in sesame seeds.

Name	Carbohydrates		Fatty acid	
	Content, mg per 100 g of product	% of the daily value	Name	Content, g per 100 g of product
Mono- and disaccharides	2.0	4.0	Oleic 18: 1 (omega-9)	18.52
Glucose	0.10	1.0	Linoleic 18:2 (omega-6)	21.37
Fructose	0.07	0.2	Linolenic 18:3 (omega-3)	0.38
Sucrose	0.3	–	Palmitoleic 16:1 (omega-7)	0.15
Starch	10.20	–	Gadoleic 20:1 (omega-11)	0.070
Alimentary fiber	5.6	22.4	Total unsaturated	40.49
Pectin	0.4	8.0	Myristic 14:0	0.13
			Palmitic 16:0	4.44
Total content all carbohydrates	23.4	6.7	Stearic 18:0	2.09
			Total saturated	6.66
			Total	47.15

Sesame seeds are also characterized by a high content of unsaturated fatty acids (99%), especially linoleic acid (omega-6), which contains 213.7% of the daily value per 100 g. The ratio of omega-6 and omega-3 fatty acids is 5.7: 1, which is optimal for the human body. There are quite a lot of sesame seeds and oleic acid - a monounsaturated (omega-9) fatty acid (18.52 g), which protects blood vessels from the formation of cholesterol plaques and is a good prevention of atherosclerosis.

The vitamin and mineral composition of sesame seeds has been determined. The results are shown in Table 4.

The basis of the mineral composition of sesame seeds is made up of such minerals as: silicon - 663.3% of the daily value, copper - 410.0%, vanadium - 136.8%, nickel - 126.7%, manganese - 123.0%, calcium - 113.6%, iron - 96.7%, magnesium - 87.8%, phosphorus - 67.5%, zinc - 64.6%, boron - 52.9%, zirconium - 50.0%, selenium - 37.5%, molybdenum - 21.4%, cobalt - 20.0%, potassium - 19.2%, iodine - 8.9%, etc.

Analysis of the table showed that vitamins such as vitamin B1 predominate in sesame seeds - in 100 g 46.8% of the daily value; vitamin B6 - 39.5%, folic acid - 24.3%, vitamin PP - 22.6%, gamma-tocopherol - 16.3%, vitamin B2 (riboflavin) - 12.5%.

Thus, the conducted studies of the qualitative characteristics of sesame seeds and whey proteins allow us to conclude that they have high consumer properties, physicochemical indicators, the amount of essential amino acids in whey proteins exceeds the number of essential amino acids in all essential amino acids, except for valine and phenylalanine. It follows from this that the introduction of whey proteins

IOP Conf. Series: Earth and Environmental Science **965** (2022) 012042 doi:10.1088/1755-1315/965/1/012042

and sesame seeds into the formulation of a curd product will lead to an improvement in the consumer and functional properties of the fortified product.

Table 4. Vitamin and mineral composition of sesame seeds.

Name	Content, mg per 100 g of product	% of the daily value	Name	Content, mg per 100 g of product	% of the daily value
Macronutrients			Vitamins		
Potassium, mg	478.0 ± 0.003	19.2	Vitamin B1 (thiamine), mg	0.79 ± 0.002	46.8
Calcium, mg	1274.0 ± 0.003	113.6	Vitamin B2 (riboflavin), mg	0.25 ± 0.003	12.5
Silicon, mg	192.0 ± 0.002	663.3	Vitamin B3 (pantothenic acid)	0.05 ± 0.002	1.0
Magnesium, mg	348.0 ± 0.003	87.8	Vitamin B6 (pyridoxine), mg	0.79 ± 0.003	39.5
Sodium, mg	11.0 ± 0.002	0.8	Vitamin B9 (folic acid), mcg	97.0 ± 0.003	24.3
Sulfur, mg	16.0 ± 0.001	1.6	Vitamin E (alpha-tocopherol), mg	0.25 ± 0.002	1.7
Phosphorus, mg	483.0 ± 0.003	67.5	Vitamin PP (nicotine acid), mg	4.52 ± 0.003	22.6
Chlorine, mg	21.0 ± 0.002	0.9	Vitamin K (phylloquinone), mg	0.3 ± 0.002	0.3
Micro- and ultra-microelements			Biotin, mcg	1.9 ± 0.002	3.8
Boron, mcg	36.0 ± 0.002	52.9	Beta-carotene, mcg	5.0 ± 0.003	0.1
Vanadium, mcg	54.7 ± 0.003	136.8	Beta-cryptoxanthin, mcg	2.0 ± 0.002	0.04
Iron, mg	14.5 ± 0.002	96.7	Lycopene, mcg	16.0 ± 0.003	3.2
Iodine, mcg	13.0 ± 0.003	8.9	Choline, mg	25.6 ± 0.003	5.1
Cobalt, mcg	2.0 ± 0.002	20.0	Betaine trimethylglycine, mg	0.4 ± 0.002	0.04
Manganese, mcg	2380.0 ± 0.003	123.0			
Copper, mcg	4040.0 ± 0.003	404.0			

In addition to the composition and properties of functional ingredients in the production of fortified dairy products, the dose, method and technological stage of their introduction into the curd base after heat treatment is of great importance [19].

When using herbal ingredients in the production of dairy products, their thermal treatment is mandatory, which has several goals: to destroy foreign microorganisms, to almost completely inactivate enzymes and, if possible, to completely preserve the presence of vitamins and biologically active substances [20].

The mode of heat treatment of sesame seeds was established experimentally using different temperatures: 65 ± 2 °C, 75 ± 2 °C and 85 ± 2 °C with an exposure time of 25-30 minutes. It was found that when using the second mode of processing sesame seeds with hot milk at a temperature of 75 ± 2 °C and holding for 25-30 minutes, the microbiological indicators are quite high, so pathogenic microorganisms such as salmonella, Listeria monocytogenes, Enterobacter sakazakii, bacteria of the genus Yersinia have not been identified , the number of molds and yeasts is significantly lower than the regulated values, so at a rate of 50 CFU/g, 15 CFU/g of yeast and 12 CFU/g of mold were detected. Extraneous microflora was not identified. In addition, sesame seeds become softer and acquire a pleasant pronounced taste of pasteurized milk, harmoniously combined with the fermented milk taste of the curd base.

IOP Conf. Series: Earth and Environmental Science **965** (2022) 012042 doi:10.1088/1755-1315/965/1/012042

For the production of whey proteins, we used cottage cheese whey obtained in the production of fat-free cottage cheese, containing 0.6% of albumin and globulin, which amounted to 3.0% of whey proteins from the mass of processed milk to obtain the required fat-free cottage cheese. Thus, in the future, when developing an enriched curd product, we will use exactly this amount of whey proteins.

To establish the dose of sesame seeds introduced in laboratory conditions, curd products were developed with different doses of sesame seeds. The results are shown in Table 5.

Table 5. Quality indicators of the curd product, depending on the dose of introduced sesame seeds.

Dose of sesame seeds	Appearance, consistency	Taste and smell	Color
0.5%	Soft, spreadable, slightly crumbly texture with single sesame seeds	Pure, fermented milk with a mild flavor of sesame seeds	Milky white, with occasional blotches of sesame seeds, evenly distributed throughout the mass
1.0%	Soft, spreadable, slightly crumbly texture with single sesame seeds	Pure, fermented milk with a pronounced flavor of sesame seeds, harmoniously combined with the curd base	Milky white, with occasional blotches of sesame seeds, evenly distributed throughout the mass
1.5%	Soft, smudging, slightly crumbly consistency with the presence of sesame seeds	Pure, fermented milk with a pronounced taste of sesame seeds	Milky white, interspersed with sesame seeds, unevenly distributed throughout the mass

As can be seen from the Table 5, the second sample has the best organoleptic properties, where the dose of sesame seeds was used - 1.0%. Therefore, in further research, this dose will be used - 1.0% of sesame seeds. Considering the above presented research results, it is advisable to introduce prepared sesame seeds and whey proteins at the technological stage - making up a batch, which, according to the developed recipe, includes low-fat cottage cheese, whey proteins, prepared sesame seeds, pasteurized and chilled cream to normalize fat, everything is thoroughly mixed and is sent for packing in consumer containers, after which the obtained samples of the curd product were subjected to an assessment of its consumer properties. The results are shown in Table 6 and Table 7.

As can be seen from the table, due to the introduction of whey proteins and sesame seeds into its composition, the fat content increases by 0.5%, proteins - by 1.0%, incl. 0.8% whey proteins, carbohydrates appeared (1.0%), incl. dietary fiber (0.1%), which indicates the increased nutritional value of the fortified curd product and characterizes it as a functional food product.

Microbiological indicators characterize curd products as living products due to the presence of lactic acid bacteria in the amount of 2.2×10^6 CFU/g and safe due to the absence of pathogenic and opportunistic microorganisms, such as bacteria of the Escherichia coli group, staphylococcus S. aureus and bacteria of the genus Salmonella, which meet the requirements of TR CU 033/2013 for this group of fermented milk products.

It was found that the enriched curd product has higher organoleptic characteristics than the control sample, does not have a pronounced taste and smell, which will allow them to be used by different groups of the population, both children and the elderly.

From the results obtained, it can be seen that the developed curd products have a high nutritional value and safety, which confirms their physiological value.

Table 6. Physicochemical and microbiological indicators of curd products.

No.	Indicator name		Control sample	Experimental sample
	Physical and chemical indicators			
1	Mass fraction of fat, not less %		18.0 ± 0.3	18.5 ± 0.9
2	Protein mass fraction, not less %		9.0 ± 0.2	10.0 ± 0.3
3	Mass fraction of moisture, not more %		75.0 ± 1.5	70.5 ± 1.2
4	Mass fraction of carbohydrates, %		-	1.0 ± 0.01
	incl. mass fraction of dietary fiber, %		-	0.1
5	Titratable acidity, °T		162 ± 0.2	160 ± 0.1
6	Phosphatase		not found	
7	Temperature when leaving the factory, °C		4 ± 2	
	Microbiological indicators			
1	The number of lactic acid bacteria at the end of the shelf life, CFU/g		1×10^6	2.2×10^6
2	Product weight (g) in which it is not allowed	Pathogenic microorganisms, including salmonella in 25 g of product	Not allowed	Not found
		E. coli group bacteria is not allowed in 0.1 g of the product	Not allowed	Not found
		Staphylococcus aureus in 1 cm^3 of product	Not allowed	Not found
3	The number of molds in 1 cm^3 of the product		50	20
4	Yeast, at the end of the shelf life, CFU/g		50	16

Table 7. Organoleptic characteristics of curd products.

Indicator name	Curd products	
	Control sample	Experimental sample
Appearance, consistency	Soft, smudging, slightly crumbly consistency	Soft, spreadable, slightly crumbly texture with single sesame seeds
Taste and smell	Clean, fermented milk, without foreign tastes and odors	Pure, fermented milk with a pronounced aftertaste of harmoniously combined pasteurized milk and sesame seeds
Color	Milky white, evenly throughout the mass	Light cream, with sporadic splashes of sesame seeds and whey proteins, evenly distributed throughout the mass

The biological value of the developed curd product was calculated (Table 8).

Analysis of the Table 8 showed that the limiting amino acid in the enriched curd product is methionine, its amino acid rate was 95.8%. The enriched curd product contains 26.5% more essential amino acids than the reference protein, including higher levels of six essential amino acids, such as: valine (by 9.6%), lysine (by 60.0%), threonine (by 29.7%), isoleucine (by 17.4%), leucine (by 40.6%) and tryptophan - almost twice the values in the standard.

The calculation of the biological value (BV) of the enriched curd product was 65.3%, which characterizes it as a product with a high biological value.

The calculation of the utilitarian coefficient of the amino acid composition (U) indicates that the enriched product had a value of 0.77, and the higher the utility coefficient, the better-balanced amino acids in the protein and the more rationally they can be used by the body. This proves the balance of the amino acid composition of the enriched curd product.

Table 8. Biological value of the developed curd product.

Name	FAO standard, (g/100 g protein)	Curd product (g/100 g protein), ($p \leq 0.05$)	Indicator values					
			Amino acid score, %	Difference of the amino acid scoring of an amino acid	Coefficient of difference of amino acid score	BV	K_i	U
Valine	5.2	5.7	109.6	13.8			0.87	
Lysine	5.5	8.8	160.0	64.2			0.60	
Methionine	2.8	2.7	96.4	0.6			0.99	
Phenylalanine	4.8	4.6	95.8	0			1.0	
Threonine	3.7	4.8	129.7	33.9	34.7	65.3	0.74	0.77
Isoleucine	4.6	5.4	117.4	21.6			0.82	
Tryptophan	1.0	1.95	195.0	99.2			0.49	
Leucine	6.9	9.7	140.6	44.7			0.68	
Total	34.5	43.65		277.9			6.20	

4. Conclusion

The authors studied the composition and properties of sesame seeds and milk whey protein, established their high nutritional and biological value. High functional activity is shown. The method of obtaining whey proteins from whey obtained in the production of curd, preliminary preparation of functional ingredients, the dose of adding whey proteins and sesame seeds and the technological stage of their introduction into the prepared curd base - at the kneading stage, on the basis of which the technology of the enriched curd product was developed. It was found that the organoleptic and physicochemical indicators of the enriched curd product are significantly higher than the control sample - the curd product without additives. Enrichment of a curd product with whey proteins and sesame seeds contributes to an increase in its consumer properties: nutritional, physiological and biological value. The number of essential amino acids exceeds the value of the reference protein by 26.5% for six amino acids, the biological value was 65.3%, and the utility coefficient was 0.77, which proves the balance of the amino acid composition of the protein of the enriched curd product and the more rationally they can be used by the body.

5. Acknowledgments

This work was carried out under the grant of the Russian Science Foundation 21-16 00025 SSI NIIMMP. Grant sponsors were not directly involved in the development, analysis, or writing of this article.

References

[1] Giménez-Bastida J A and Zieliński H 2015 Buckwheat as a functional food and its effects on health *J Agric Food Chem* **63(36)** 7896-7913
[2] Asgary S, Rastqar A and Keshvari M 2018 Functional food and cardiovascular disease prevention and treatment: a review *J. Am. Coll. Nutr.* **37(5)** 429-455
[3] Singh P, Tripathi M K, Yasir M, Khare R, Tripathi M K and Shrivastava R 2020 Potential inhibitors for SARS-CoV-2 and functional food components as nutritional supplement for COVID-19: a review *Plant Foods Hum. Nutr.* **75(4)** 458-466
[4] *National Standard of the Russian Federation State Standard R 52349 2005 Foodstuffs. Functional Food. Terms and Definitions* 2006 (Moscow: Standartinform)
[5] Kontareva V Yu, Belik S N, Morgul E V, Gorlov I F and Slozhenkina M I 2020 Yogurt enriched to correct intestinal microflora in dysbiosis *IOP Conf. Series: Earth and Environmental Science* **548(8)** 82051
[6] Zobkova Z S, Zenina D V and Fursova T P 2016 On cottage cheese as a national product *Dairy Industry* **1** 28-30

[7] Danilova O A and Gulyaeva A Y 2017 Cottage cheese as a national Russian product *Topical Issues of Improving the Technology of Production and Processing of Agricultural Products* **19** 90-92
[8] Kryuchkova V V, Gorlov I F, Korneichuk K M, Belik S N and Mosolova N I 2020 Brine-ripened cheese enriched with vegetable ingredients: technology and quality *IOP Conf. Series: Earth and Environmental Science* **548(8)** 82063
[9] Li C, Miao H, Wei L, Zhang T, Han X and Zhang H 2014 Association mapping of seed oil and protein content in Sesamum indicum L. using SSR markers *PLoS One* **9(8)** 105757
[10] Zhou L, Lin X, Abbasi A M and Zheng B 2016 Phytochemical contents and antioxidant and antiproliferative activities of selected black and white sesame seeds *Biomed. Res. Int.* 8495630
[11] Alipoor B, Haghighian M K, Sadat B E and Asghari M 2012 Effect of sesame seed on lipid profile and redox status in hyperlipidemic patients *Int. J. Food Sci. Nutr.* **63(6)** 674-678
[12] Khadem Haghighian M, Alipoor B, Malek Mahdavi A, Eftekhar Sadat B, Asghari Jafarabadi M and Moghaddam A 2015 Effects of sesame seed supplementation on inflammatory factors and oxidative stress biomarkers in patients with knee osteoarthritis *Acta Med. Iran* **53(4)** 207-213
[13] Deng S, Zhou J L, Fang H S, Nie Z G, Chen S and Peng H 2018 Sesamin protects the femoral head from osteonecrosis by inhibiting ROS-induced osteoblast apoptosis in rat model *Front Physiol.* **9** 1787
[14] Pereira P C 2014 Milk nutritional composition and its role in human health *Nutrition* **30(6)** 619-627
[15] Tokajuk A, Zakrzeska A, Chabielska E and Car H 2019 Whey protein concentrate limits venous thrombosis in rats *Appl. Physiol. Nutr. Metab.* **44(8)** 907-910
[16] Patel S 2015 Emerging trends in nutraceutical applications of whey protein and its derivatives *J. Food Sci. Technol.* **52** 6847-6858
[17] Cheison S C and Kulozik U 2017 Impact of the environmental conditions and substrate pre-treatment on whey protein hydrolysis: a review *Crit. Rev. Food Sci. Nut.* **57** 418-453
[18] Morgul E V, Belik S N, Morgul A R, Gorlov I F, Slozhenkina M I and Mosolov A A 2021 The possibility of using amino acid supplements for stress prevention *IOP Conf. Series: Earth and Environmental Science* **677** 32068
[19] Kryuchkova V V, Gorlov I F, Belik S N and Kamlatsky A S 2020 As vegetable ingredients in functional fermented milk products *IOP Conf. Series: Earth and Environmental Science* **548(8)** 82092
[20] Kontareva V Yu, Belik S N, Morgul E V, Gorlov I F and Slozenkina M I 2020 The effect of prebiotic components on the quality of yogurt *IOP Conf. Series: Earth and Environmental Science* **548(8)** 82054

IOP Conf. Series: Earth and Environmental Science **965** (2022) 012043 doi:10.1088/1755-1315/965/1/012043

Improving the quality characteristics of flour confectionery when adding vegetable puree for the sustainable development of food production

E N Efremova, E S Taranova, E A Zenina, E A Melnikova, A G Melnikov and A A Shershnev

Volgograd State Agrarian University, 26 University Avenue, Volgograd, 400002, Russia

Email: Elenalob@rambler.ru

Abstract. The confectionery industry is an independent production sector in the food processing industry. The purpose of the work is to improve the quality of products and expand the assortment of confectionery - cupcake, which consists in adding vegetable puree (pumpkin). The object of the study is the Dutch cupcake. In the Volgograd State Agrarian University at the Department of Processing Technologies and Food Safety, test baking of a cupcake confectionery with the addition of pumpkin puree was carried out. According to organoleptic and physico-chemical parameters, the cupcakes met the requirements of State Standard 15052-2014 "Cupcakes. General technical conditions". When comparing the test samples, there are no significant differences, it differs in that the product with the addition of pumpkin puree has the smell and taste of pumpkin, and the porosity of the product decreases slightly, all this has a positive effect on the studied product. The profitability of the cupcake products with the addition of pumpkin puree is 1% higher than the control sample. A cupcake with the addition of pumpkin puree can be recommended as a preventive nutrition.

1. Introduction

Confectionery products are considered to be the most demanded and popular in the world in terms of taste and energy value among the products of the food industry [1].

Currently, this industry is one of the fastest growing industries. The annual growth in demand for confectionery products contributes to the development of industry in the Russian market. The development of confectionery industry enterprises is characterized by quick payback, high profitability and low risks [2].

Flour confectionery products occupy a leading place among confectionery products. Thirty percent of the assortment and volume of production of all flour confectionery products is given to the production of cupcakes [2, 3].

Currently, the population strives for a healthy lifestyle, proper nutrition, food industry enterprises are trying to keep up with the desires of consumers. In turn, enterprises, in order to expand the assortment and increase the demand for their products, introduce natural ingredients containing vitamins, macro- and microelements with therapeutic, preventive and functional properties into the formulation.

The use of berries, fruits, and vegetables in confectionery production is currently relevant [3].

The purpose of the work is to improve the quality of products and expand the assortment of confectionery products when adding pumpkin puree. To achieve the goal, the following tasks were solved:

- the recipe of a cupcake with pumpkin puree was studied;
- thc quality of the cupcake was investigated;
- the economic efficiency of the cupcake was calculated.

The object of the study is a confectionery product - cupcake "Dutch".

2. Materials and methods

In our study, pumpkin was used in the form of puree, as a confectionery additive in the manufacture of dough for cupcakes.

In the Volgograd State Agrarian University at the Department of Processing Technologies and Food Safety in 2019-2020, test baking of cupcake confectionery with the addition of pumpkin puree was carried out.

Pumpkin is a perennial, frost-resistant herbaceous plant. For confectionery production, the following varieties are mainly interesting: nutmeg, vitamin, honey.

Pumpkin is a very fortified vegetable. Vitamins such as vitamin D, thiamine (vitamin B1), riboflavin (vitamin B2), tocopherol (vitamin E) are found in pumpkins. Vitamin D accelerates the growth of children, helps to digest coarse food well and enhances the vitality of the body. Pumpkin contains many other useful substances that are of great benefit to the body [4, 5]. Table 1 describes the chemical composition of pumpkin.

From Table 1, it can be concluded that the basis of pumpkin is water. Pumpkin is a low-calorie product. There is a small amount of protein and fat in the pumpkin.

Table 1. Chemical composition of pumpkin.

Indicators	Content, per 100 g
Fats	0.10
Proteins	1.00
Carbohydrates	6.50
Water	91.60
Ash	0.80

Pumpkin carbohydrates are represented by polysaccharides. This means that they are slowly absorbed and the body receives energy from them gradually, so it does not feel hungry for a long time. Since pumpkin has a composition rich in nutrients, it is easily digested and is a low-calorie product, it can be used in the confectionery industry [6].

Organoleptic and physico-chemical studies were carried out. According to organoleptic and physico-chemical parameters, the investigated product must comply with the requirements of State Standard 15052-2014 "Cupcakes. General technical conditions". Organoleptic research in the production of cupcakes studied the following indicators: structure, appearance, consistency, color, shape, smell and taste [7]. The condition of the crumb, the degree of porosity and acidity under certain baking conditions, as well as the elasticity and plasticity of the product were considered.

3. Results and discussion

The improvement of the production technology and the variety of the assortment consists in adding pumpkin puree to the investigated Dutch cupcake to improve the performance of confectionery products. Table 2 describes the recipe of the cupcake per 100 kg.

Comparing the recipes of the control and experimental samples of cupcakes, it should be concluded that the use of chicken eggs in the experimental version decreased by 14.3 kg. The use of granulated sugar decreased by 7.2 kg. The amount of wheat flour, baking powder and vegetable oil remained

IOP Conf. Series: Earth and Environmental Science **965** (2022) 012043 doi:10.1088/1755-1315/965/1/012043

unchanged. Pumpkin puree in the amount of 55.6 kg (29% of the total mass of raw materials) was added to the experimental version at the stage of kneading and knocking down the dough.

Table 2. Cake recipe per 100 kg.

Raw material	Control sample	Exptrimental sample
Chicken eggs, kg	57.1 (36.2%)	42.8 (22.73%)
Granulated sugar, kg	48.5 (30%)	41.3 (21.5%)
Wheat flour, kg	48.5 (30%)	48.5 (25%)
Baking powder, kg	0.14 (0.08%)	0.14 (0.07%)
Vegetable oil, kg	3.35 (2.13%)	3.35 (1.7%)
Pumpkin puree, kg	0	55.6 (29%)

A comparison of the organoleptic parameters of the product "Dutch Cupcake" and cupcake with the addition of pumpkin puree is presented in Table 3.

Comparing the control version and experimental version with the addition of pumpkin puree in Table 3, there are no significant differences, it differs in that the product with the addition of pumpkin puree has the smell and taste of pumpkin, and also the porosity of the product decreases slightly, all this has a positive effect on the studied producty.

Table 3. Organoleptic indicators of the cupcake.

Indicators	"Dutch Cupcake"	Cupcake with the addition of pumpkin puree
Taste and smell	A product with a rich taste and characteristic aroma, without foreign tastes and odors	A product with a rich taste and with a pronounced pumpkin aroma, there is no foreign smell and taste
Surface	The upper surface is convex, without cracks, without finishing, with the presence of a pronounced lateral surface; on the lower and side surfaces there is no presence of emptiness, burnt, irregularities and breaks	The upper surface is convex, without cracks, without finishing, with the presence of a pronounced lateral surface; on the lower and side surfaces there is no presence of emptiness, burnt and r breaks
View in the break	At the break, the baked product is without traces of non-kneading, there are no lumps, voids and hardening, the porosity is uniform	At the break, the baked product is without traces of non-kneading, there are no lumps, voids and hardening, the porosity is uniform
Structure	Soft, bound, porous loosened, having no voids and compaction	Finely porous bound, soft, loosened, having no voids and compaction
Form	The lower and side surfaces are smooth, without voids and cavities, regular with a convexity on the surface	The lower and side surfaces are smooth and without voids and cavities, regular with a convexity on the surface

Figure 1 shows a cupcake Dutch control sample and experimental sample with the addition of pumpkin puree in its whole form.

(a) (b)

Figure 1. The studied Dutch cupcake in its whole form (a - control sample, b – experimental sample).

In Figure 2, you can clearly see the analyzed control and experimental samples in the section.

(a) (b)

Figure 2. The analyzed Dutch cupcake in the section (a - control sample, b - experimental sample).

The physico-chemical parameters of the control and experimental samples are described in Table 4.

Table 4. Physico-chemical parameters of the cupcake.

Indicators	"Dutch Cupcake"	Cupcake with the addition of pumpkin puree
Mass fraction of moisture, %	26.2	28
Mass fraction of total sugar, %	33.3	30
Alkalinity in cupcakes prepared with chemical baking powder, in degrees, no more	1.0	1.0
Mass fraction of fat, %	6.8	6.6

As a result of the analysis of Table 4, it can be concluded that when pumpkin puree is added to the cupcake, the indicators correspond to the standards given in State Standard 15052-2014 "Cupcakes.

General technical conditions". The addition of pumpkin puree significantly improved the quality indicators of the cupcakes studied. The mass fraction of total sugar decreased by 3.3%, as less sugar is used by 15%. The mass fraction of fat decreased slightly by 0.2%. The mass fraction of moisture increased by 1.8%, as pumpkin puree makes the product more moist, which has a positive effect on taste indicators.

Table 5 calculates the nutritional and energy value of the studied samples.

Table 5. Nutritional and energy value per 100 g of cupcake.

Indicators	"Dutch Cupcake"	Cupcake with the addition of pumpkin puree
Caloric content, kcal	324	243
Proteins, g	2.9	3.9
Fats, g	9.1	9.75
Carbohydrates, g	61.6	34.83

Table 5 compares the nutritional and energy values of the studied samples. In the experimental sample, the caloric content is significantly lower than in the control sample. The protein and fat content is higher in the experimental sample, and the carbohydrate content is lower.

Thus, after conducting an organoleptic and physico-chemical assessment of the quality of the product, it was determined that the experimental sample complies with State Standards for the studied type of confectionery and is quite suitable for introducing it for a variety of assortment and for improving product quality.

Increasing the level of profitability is the main task of production [8, 9]. The calculation of the production cost of confectionery of the control and experimental samples was carried out (Table 6).

Table 6. Calculation of the production cost of the cupcake, rubles/kg.

Indicators	"Dutch Cupcake"	Cupcake with the addition of pumpkin puree
Raw and basic materials	56.8	52.5
Auxiliary materials	5.68	5.25
Electricity	3.21	3.21
Salary (with deductions)	19.28	19.28
Depreciation	0.33	0.33
Maintenance and operation of equipment	0.68	0.68
The cost of packaging	0.28	0.26
Transportation costs	3.4	3.2
Other costs	1.14	1.05
Total production cost	90.8	85.73

Calculations of the production cost of the control and experimental samples of cupcakes showed that the cost of a cupcake with the addition of pumpkin puree is 5 rubles/kg less than the control sample. The difference in the production cost of the cake was influenced by the cost of raw materials and basic materials.

Table 7 shows the economic efficiency of the production of "Dutch Cupcake" and cupcake with the addition of pumpkin puree [10].

Based on calculations (Table 7), it can be seen that the cost of cupcakes with pumpkin puree is higher than for "Dutch Cupcakes". But with the help of the production of cupcakes with the addition of pumpkin puree, you can expand the assortment of confectionery products.

Table 7. Economic efficiency of the production of the studied types of cupcakes.

Indicators	Units of measurement	"Dutch Cupcake"	Cupcake with the addition of pumpkin puree
Volume of products sold	kg	75000	75000
Cost of products sold	rubles/kg	90.8	85.73
Product price	rubles/kg	147.12	138.81
Revenue from sales	rubles	9193500	8735250
Profit	rubles	2383500	2308500
Net profit	rubles	1906800	1846800
Profitability of products sold	%	28	29

The following economic indicators also justify the cost of cupcakes with pumpkin puree: the price of products will be lower, which means that demand for the product will increase; the profitability of products sold is 1% higher, which is a good return on investment in the studied production.

4. Conclusion

Cupcake with the addition of vegetable (pumpkin) puree has a reduced nutritional and energy value. Adding pumpkin puree to a flour confection (cupcake) makes the color of the baked product yellowish and adds a pleasant pumpkin flavor. Pumpkin puree improves the taste and digestibility of confectionery. Pectin and dietary fiber contained in pumpkin in large quantities make this product low-calorie. The finished product contains vitamins, iron, magnesium, provitamin A, minerals. Minerals play an important role in the metabolism in the body.

When calculating the economic efficiency of the production of “Dutch cupcake” and cupcake with the addition of pumpkin puree, it can be concluded that the profitability of products sold with the addition of pumpkin puree will have a beneficial effect on economic efficiency.

Cupcake with the addition of pumpkin puree can be recommended as a preventive nutrition for functional disorders of the liver and kidneys, obesity, atherosclerosis, diabetes and other diseases.

References

[1] Fomchenkova K A and Kolyakin E A 2020 Confectionery industry *Student Bulletin* **27-3(125)** 13-14
[2] Efremova E N, Taranova E S and Zenina E A 2019 Application of food supplement in production in order to promote health and prophylactic properties *IOP Conf. Series: Earth and Environmental Science* **341** 012142
[3] Timotijevic L, Khan S S, Raats M and Braun S 2019 Research priority setting in food and health domain: European stakeholder beliefs about legitimacy criteria and processes *Food Policy* **83** 116-124
[4] Lukina S I, Ponomareva E I, Botasheva Kh Yu, Zhirkova E A and Alekseev A E 2019 Substantiation of the rational dosage of pumpkin puree in the production of rusks *Bread Products* **2** 52-53
[5] Tang H-R, Godward J and Hills B 2020 The distribution of water in native starch granules - a multinuclear NMR study *Carbohydrate Polymers* **43** 375-387
[6] Temnikova O E, Bakharev V V and Mustakaeva A K 2018 The possibility of using pumpkin in the technology of bakery products *Bread Products* **10** 45-46
[7] *State Standard 15052-2014. Cupcakes. General Technical Conditions* 2019 (Moscow: Standardinform) p 6
[8] Laukemper R, Jekle M and Becker T 2019 Time-dependent adhesion behavior between dough and contact surfaces in bakeries *J. of Food Engineering* **255** 24-31

[9] Efremova E N 2015 Improving the recipe of wheat bread with additives with functional and technological properties *Proc. of the Nizhnevolzhsky Agrouniversity Complex: Science and Higher Professional Education* **4(40)** 207-213
[10] Gupanova Yu E 2008 Organizational and economic mechanism for managing the quality of products in the bakery industry *Expert. Food industry* **8** 7-9

Ham production using processed apple products

M I Slozhenkina[1, 2], I F Gorlov[1, 2], S E Bozhkova[1], K A Ermolova[1], A E Serkova[1, 2] and V N Khramova[1]

[1] Volgograd State Technical University, 28 Lenin Avenue, Volgograd, 400005, Russia
[2] Volga Region Research Institute of Manufacture and Processing of Meat-and-Milk Production, 6 Rokossovsky Street, Volgograd, 400131, Russia

E-mail: bozhkova@mail.ru

Abstract. The article analyzes the possibility and rationality of using apple processing products as food additives to increase the nutritional value of products, substantiates the expediency of using a vegetable component (apple pomace) in the technology of smoked-boiled ham products. To assess the technological potential of using apple pomace in the production of smoked-cooked sausages, experimental studies were carried out using generally accepted methods to study the basic properties of raw materials and control samples - products without the addition of powdered apple pomace, and prototypes - products with the addition of apple pomace according to the developed recipe. Formulations have been optimized and a method has been developed for the production of a combined composition of protein-fat emulsion, which is recommended to be introduced instead of 22% of the main unsalted meat raw materials. According to the results obtained, a sample containing 6% apple pomace has the best quality indicators.

1. Introduction

The development of the domestic food industry, including the meat processing industry, is urgent. For the growth of competitiveness in the markets of meat products, a necessary and indispensable condition is the substitution of imported raw materials with domestic ones. In this regard, at present, it is required to use all available reserves of animal protein to the maximum and efficiently. One such source is poultry meat, in particular ducks.

The good digestibility of poultry meat (96%) is explained by its chemical composition. Due to the fact that connective tissue is relatively poorly developed in poultry meat, it contains more complete and digestible proteins in comparison with meat of slaughter animals, and proteins are mainly water-soluble. Collagen and elastin are practically absent in proteins of duck meat, which characterizes its good digestibility and nutritional value [1, 2, 3].

In modern meat-processing industry, the use of protein-fat emulsions is widely used, this is justified by the need to bind fatty raw materials, improve its quality, avoid the appearance of fatty edema, reduce its cost in some cases, etc. [4, 5].

Collagen-containing raw materials, in particular chicken skin, are a promising source of food protein in the meat industry. The advantage of using chicken skin is that collagen has a high water-holding and texture-forming ability, which allows it to be used in various food systems. The use of chicken skin in the composition of protein-fat emulsion (PFE) will stabilize the quality of hams, which are emulsified products, expand the possibilities of using secondary protein-containing resources from the processing of poultry meat and the range of finished products [6, 7, 8].

The use of apple pomace, formed in the production of directly squeezed juice from apple varieties growing in large quantities in the Lower Volga region of Russia, in particular, the Volgograd region, in the formulations of ham products, is promising not only from an economic point of view. The chemical composition of apple pomace determines their high nutritional value (they are a source of pectin, fiber, water-soluble vitamins and minerals). In addition, they exhibit antioxidant properties due to the presence in the composition of organic acids, vitamins (ascorbic acid) and water-soluble antioxidants (quercetin), which are synergists of the nitrite salt in stabilizing the color of the finished product.

The aim of this work is to improve the technology of smoked-boiled ham products from poultry meat using PFE of a complex composition. In accordance with the set goal, the following tasks were solved:

– substantiate and select the ingredient composition;

– to optimize the formulation of the protein-fat emulsion and to improve the technology of the restructured ham;

– to develop an original recipe and method of production of the product.

2. Materials and methods

The work was carried out on the basis of the educational and scientific center "Department Technologist" of the Department of Food Production Technologies of the Volgograd State Technical University and the complex analytical laboratory of the State Scientific Research Institute of the Russian Academy of Medical Sciences and consisted of the following stages: selection and preparation of raw materials, development of prototypes of ham products, conducting organoleptic and physico-chemical studies to assess the quality of the products obtained, optimization of the recipe and development of the method of production of the product in accordance with the research scheme (Figure 1).

The objects of research were duck meat (State Standard 31990-2012); broiler chicken meat (State Standard 31962-2013); soy protein isolate (State Standard R 53861-2010); apple pomace, which is a secondary resource of apple processing, formed in the production of direct-pressed juice from apple varieties Golden Delicious, Rennie Simirenko, Granny Smith; as well as samples of smoked-boiled ham, developed on the basis of these components using PFE combined composition.

Sampling and preparation of samples for laboratory studies was carried out according to a single method in accordance with the requirements of State Standard R 51447-99 (ISO 3100-1-91).

In the course of experimental studies, the following indicators were studied: 1 – organoleptic parameters (State Standard 9959-91; State Standard R 53159-2008; State Standard R 53161-2008); 2 – pH value (using a portable measuring device Testo 206 pH2 using a piercing electrode); 3 – dry matter content (State Standard 9793-2016); 4 – fat mass fraction (State Standard 23042-86); 5 – protein mass fraction (State Standard 25011-2017); 6 – sodium chloride mass fraction (State Standard P 51480-99); 7 – the content of residual sodium nitrite (State Standard 29299-92); 8 – moisture binding capacity (according to the method of Grau and Hamm in the modification of Volovinskaya and Kelman); 11 – product yield (by weight method); 12 – organic acid content (State Standard 32771-2014); 13 – pectin content (State Standard 32223-2013); 14 – fiber content (State Standard 31675-2012); 15 – thiobarbitur number (State Standard R 55810-2013); 16 – peroxide number (State Standard 34118-2017); 17 – computer modeling using the Microsoft Office software package on a PC; 18 – energy value (calculated by the actual content of proteins, fat, and carbohydrates in the sample).

The cost of production of experimental samples was calculated on the basis of actual costs and prices at the end of 2020, taking into account shop costs in the "Department Technologist" of the Volgograd State Technical University.

Product safety is ensured by the implementation of the technological process and compliance with the expiration dates (TR CU 034/2013 Technical Regulations of the Customs Union "On the safety of meat and meat products").

The experimental data were processed using mathematical statistics methods using a computer. The repeatability of the experiments is at least 3 times with a 3-to 4-fold repeatability of the analyses. Confidence level = 0.95.

3. Results and discussion

Duck meat with the following functional and technological parameters was used as the main raw material in the preparation of the ham recipe: pH was 6.27; dry matter content – 38.8%; fat content – 17%; protein content – 18%. To optimize the functional and technological properties of duck meat in the production of ham, minced meat was prepared using the meat of broiler chickens. The recipe also included ingredients such as poultry skin, soy protein isolate and apple pomace.

According to the chemical composition and biological value of poultry skin (from the neck, legs), it contains 14-17% protein and 20-25% fat and vitamins (A, B1, B2, B3, PP, C, E), Ca. Due to the fact that there is a lot of fat in the skin, it is prone to rancidity. The connective tissue in the fat can split into gelatin and fat during frying, thereby contributing to the formation of a porous texture. To eliminate this disadvantage and stabilize the fat present in the skin, it must first be converted into an emulsion.

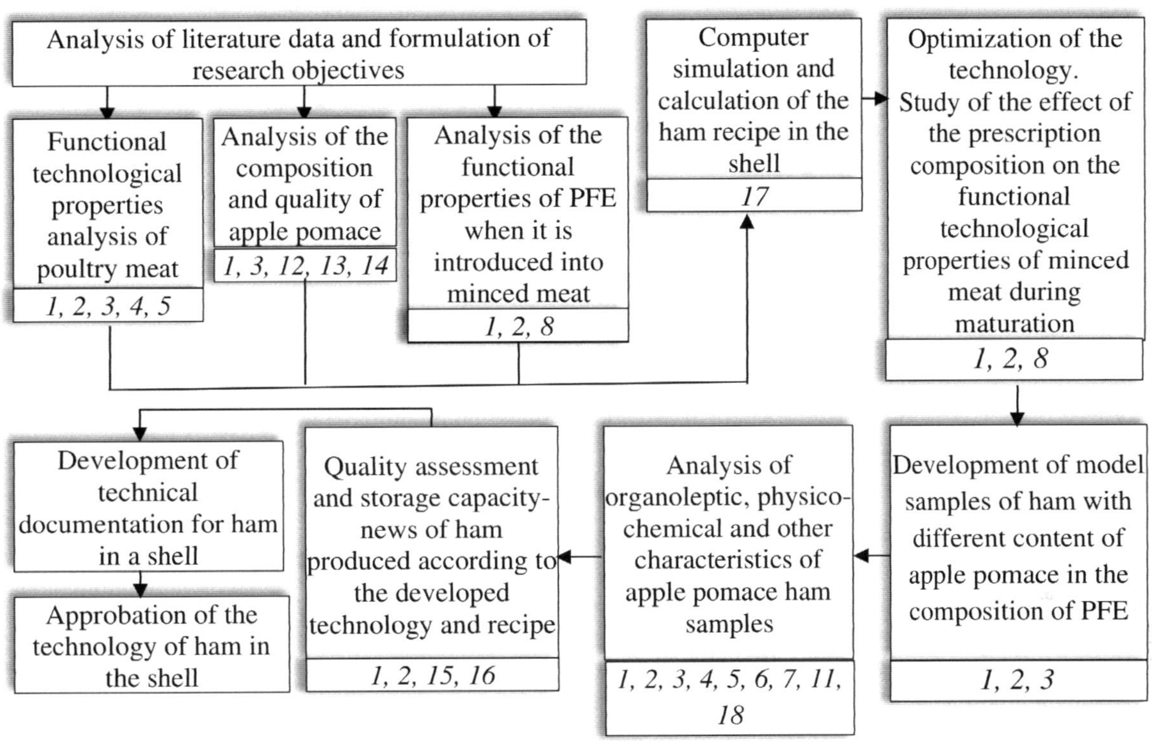

Figure 1. Research scheme.

Also, the developed PFE recipe is enriched with apple pomace, which are secondary resources of apple processing formed during the production of direct-pressed juice from local Volgograd apple varieties (Golden Delicious, Renet Simirenko, Granny Smith). Apple pomace is a fine powder without foreign impurities. It is allowed to have a fibrous fraction of pectin in the form of flakes, slightly acidic taste, without foreign smell, from light gray to cream color. According to the chemical composition and biological value, the pomace of the selected apple varieties contains about 13.5% dry matter, 17-1.9% fiber, and 1.8-4.4% organic acids. The studied varieties are characterized by a high content of carbohydrates, mineral and pectin substances (Figure 2). At the same time, the chemical composition of apple pomace changes depending on the botanical characteristics, morphology and anatomy of plant tissue, the variety and degree of maturity of apples. Pomace is a valuable source of complex antioxidants such as phenolic compounds, contains pectin substances, cellulose and hemicellulose, B vitamins, macro-and microelements. Pectin, in turn, plays the role of a stabilizer and a moisture-retaining component.

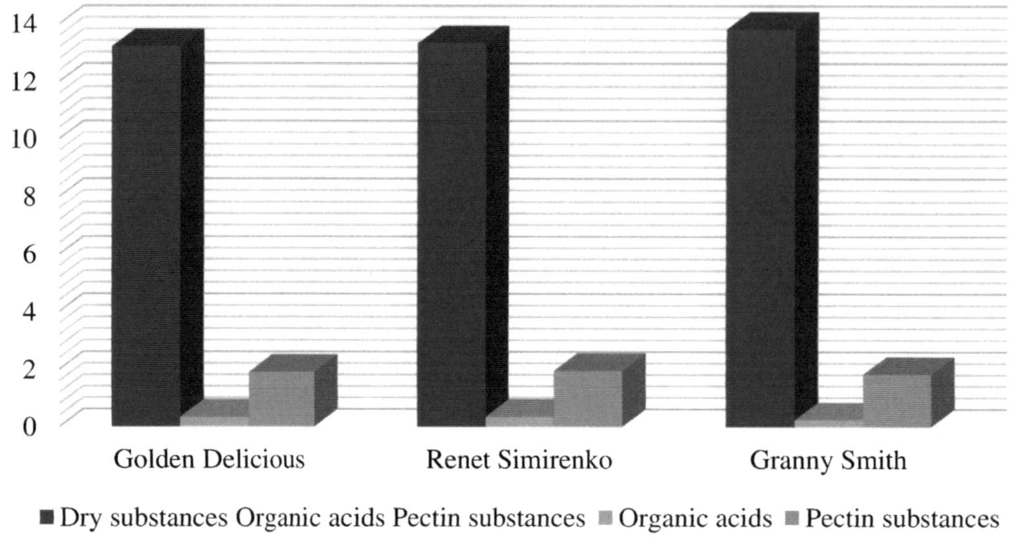

*content in terms of raw weight

Figure 2. Characteristics of apple pomace of different apple varieties.

During the research, a method of ham production was developed, which includes the following main stages: acceptance of raw materials, grinding on a top, mixing, maturation in a refrigerator, molding, heat treatment, cooling, quality control, packaging, labeling and sales. In parallel with the main stages of production, PFE was prepared in the following sequence of stages: grinding of fat raw materials, represented by the skin from the carcass of birds, on a cutter, adding apple pomace and hydrated soy protein, mixing, directing PFE to mix with minced meat. The technological diagram of the production of experimental ham samples is shown in Figure 3.

The recipes of the developed products are presented in Table 1. The use of emulsions is justified by the need to bind fat raw materials, improve their quality, avoid the appearance of fat edema, reduce its cost in some cases, etc., but the main task of using fat emulsions is to improve the quality of the finished product, its appearance, consumer and organoleptic properties [4].

For the production of PFE, low-grade fat raw materials (chicken skin and fat), drinking water and protein are used. The ratio of components can be different (for example, protein: fat: water = 1:5:5) [4].

Studies of the dynamics of the water-binding capacity of meat raw materials during salting have established that the nature of the change depends on the quality of raw materials and the composition of PFE. Figure 4 shows the dynamics of changes in the water-binding capacity of experimental samples during the exposure of minced meat without and in the presence of PFE in the composition of minced meat.

When salting with the use of PFE and without, the maximum values of water-binding capacity are reached by 12 hours of maturation. The air force level stabilizes after 10 hours of exposure.

Further treatment does not lead to noticeable changes in pH, but contributes to a gradual increase in destructive changes in muscle tissue and a decrease in the level of moisture binding.

The developed ham samples have high organoleptic characteristics: straight loaves, with a clean, dry surface, non-uniform structure on the cut (there are pieces of chicken fillet in the minced meat) on the cut, light pink muscle tissue of various intensity, elastic consistency, with a ham smell, a moderately salty pleasant taste, without foreign tastes and odors.

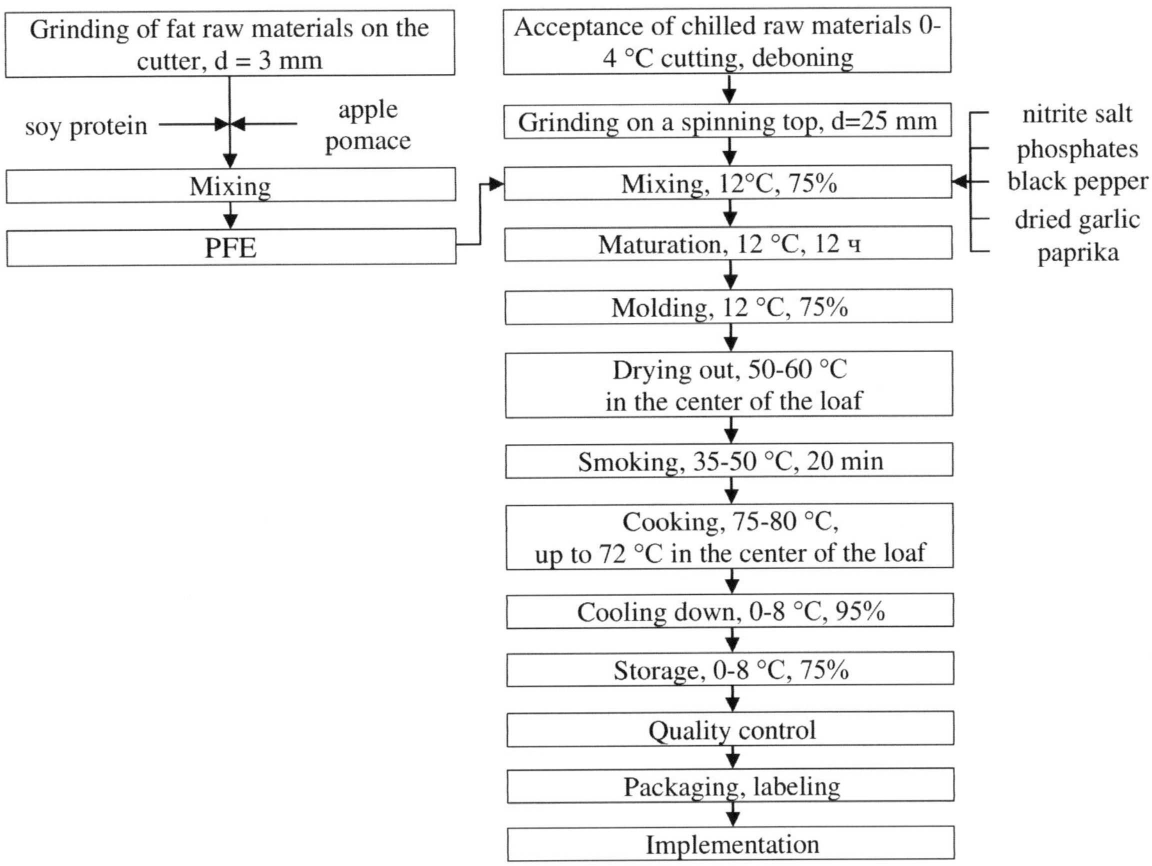

Figure 3. Technological scheme of ham production.

Table 1. Recipes of experimental ham samples.

Name of raw materials, materials, spices	kg, per 100 kg of unsalted raw materials			
	control sample (without pomace)	Experimental No. 1 (4% pomace)	Experimental No. 2 (6% pomace)	Experimental No. 3 (8% pomace)
Duck carcasses	50.0	39.0	39.0	39.0
Meat of broiler chickens	50.0	39.0	39.0	39.0
Protein-fat emulsion, including	0	22	22	22
poultry thigh fillet with skin	0	8	7	6
soy protein isolate	0	2	2	2
apple pomace	0	4	6	8
water	0	8	7	6
TOTAL unsalted raw materials	100.0	100.0	100.0	100.0
Nitrite salt	2	2	2	2
Food phosphates	0.2	0.2	0.2	0.2
Dried garlic	1	1	1	1
Black pepper	0.5	0.5	0.5	0.5
Ground paprika	1	1	1	1
Water	25	25	25	25

According to the comparative assessment of organoleptic parameters presented in Figure 5, it can be noted that the prototypes have a slight advantage due to the slightly more intense color of the minced meat and a more delicate consistency due to the addition of PFE of the combined composition.

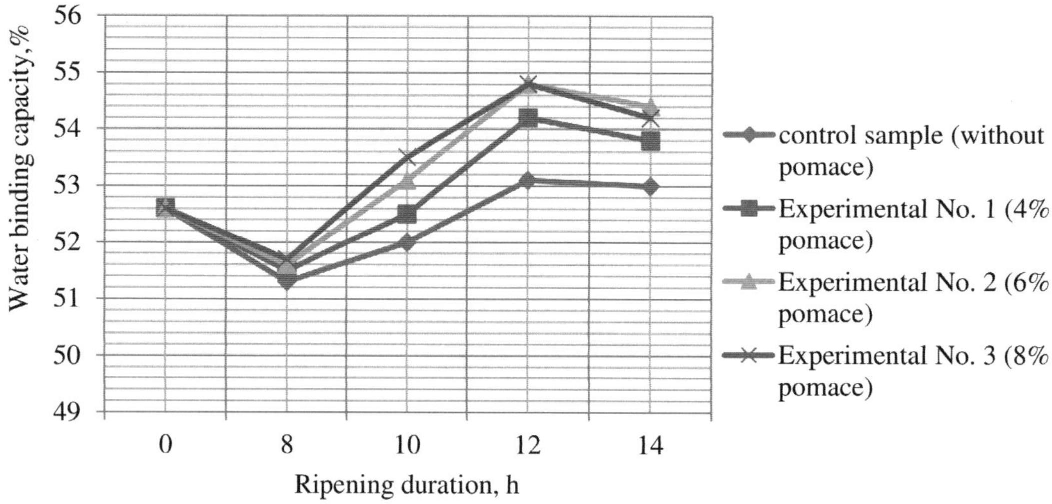

Figure 4. Change in the water-binding capacity of the prototypes of finished minced meat for ham without and in the presence of different compositions in the composition of PFE during ripening.

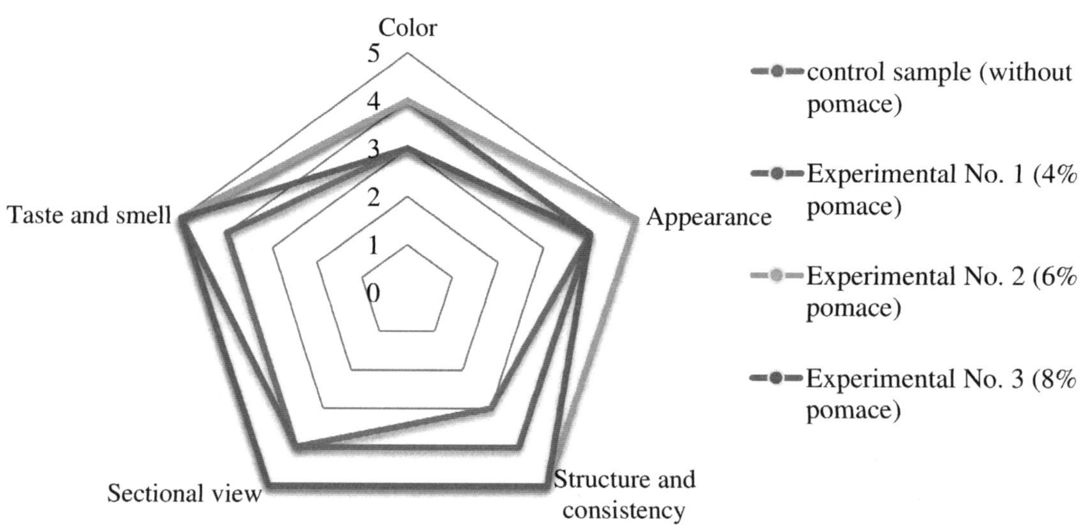

Figure 5. Profilogram of organoleptic indicators prototypes: 1 - unsatisfactory; 2 - satisfactory; 3 - good; 4 - very good; 5 – excellent.

The results of the analysis of the main indicators of the quality of experimental samples are presented in Table 2.

Table 2. Results of the analysis of indicators of quality and yield of finished products.

Indicator	Characteristic			
	control sample (without pomace)	Experimental No. 1 (4% pomace)	Experimental No. 2 (6% pomace)	Experimental No. 3 (8% pomace)
Mass fraction of sodium chloride (table salt),%	2.0	2.0	2.0	2.0
Protein mass fraction, %	19.0	22.6	22.5	22.4
Mass fraction of fat, %	10.0	9.6	9.0	8.8
Mass fraction of sodium nitrite,%, no more	0.005	0.005	0.005	0.005
Caloric content 100 g, kcal	195	180	177	174
Product yield, % not less than	94	96	97	97

On the basis of the carried out, the main indicators of quality were analyzed and it was concluded that the optimal sample, with the best indicators, is the sample with the addition of 6% apple pomace, taking into account the highest yield (97%) and high physical and chemical characteristics (fat content - no more than 9.0%, protein content - not less than 22.5%).

At the next stage, we studied the changes in thiobarbituric and lipid peroxide numbers of control and experimental ham samples during storage. The samples were stored for 30 days at a temperature of 4 ± 2 °C (Figure 6).

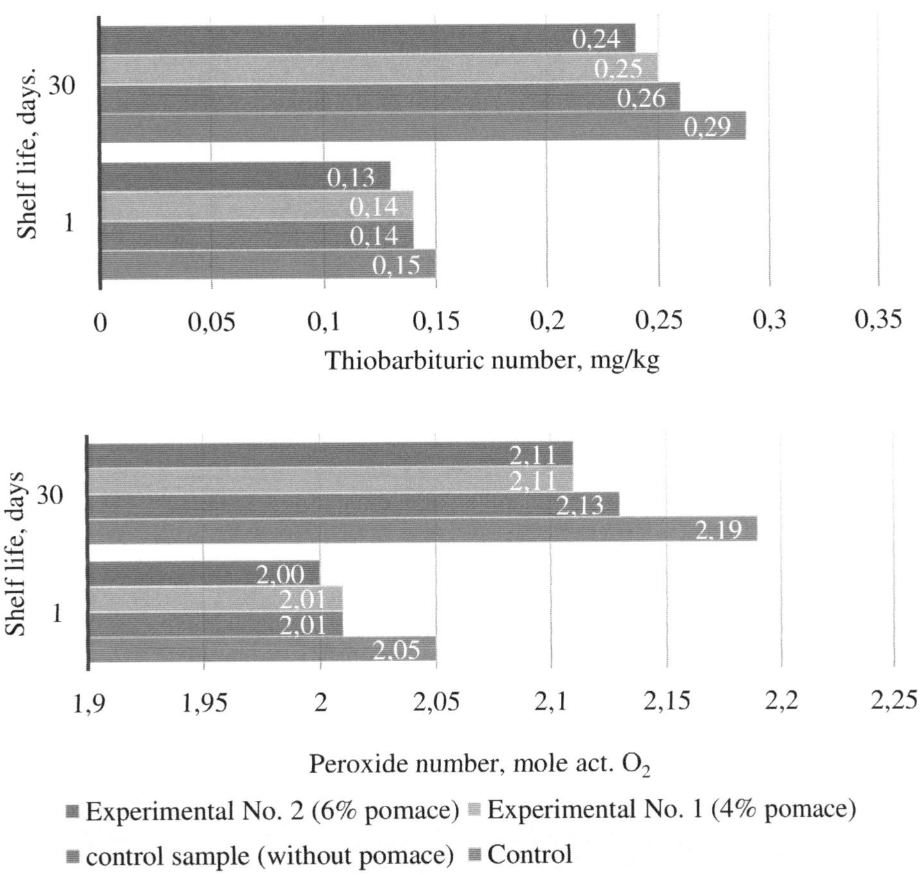

Figure 6. Changes in thiobarbituric and peroxide numbers during storage of samples.

According to the results obtained, it was found that during storage of ham samples developed according to new formulations with the addition of apple pomace, oxidative processes in the lipid fraction proceeded less actively than in the control sample. The best result was noted in samples with the addition of 6% and 8% apple pomace, which is explained by the presence of substances with antioxidant properties (organic acids, vitamins (ascorbic acid) and water-soluble antioxidants (quercetin), etc.).

An assessment of the economic efficiency of ham production was carried out by calculating the cost of the product, presented in Table 3.

The cost of production of a prototype ham from poultry meat, enriched with PFE of a complex composition, amounted to 269 rubles per 1 kg, which at the moment, taking into account other costs of the manufacturer and the distribution network, is the average for the market (the average retail price of ham from poultry meat of well-known manufacturers in St. Volgograd at the end of 2020 amounted to 280-300 rubles per 1 kg).

Table 3. Calculation of the cost of production of an experimental sample.

Raw materials	Price for 1 kg	Quantity, kg	Experimental No. 2 (6% pomace)
Duck carcasses	350	0.42	145.0
Broiler chicken meat	170	0.43	73.1
TOTAL for the main raw materials	520	0,85	218.1
Nitrite salt	200	0.02	4.0
Food phosphates	790	0.002	1.6
Dried garlic	145	0.01	1.45
Black pepper	82	0.005	0.41
Ground paprika	340	0.01	3.4
Soy Protein Isolate	1000	0.02	20
Pomace	30	0.06	1.8
Water (* RUB per 1 m^3)	23*	0.32	0.007
TOTAL for additional raw materials	-	0.45	32.7
Supporting materials	–	–	10
Total	–	–	261
Output, % not less	–	–	97
Cost price, RUB per 1 kg of finished product for raw materials and materials	–	–	269

4. Conclusion

As a result of the conducted experimental studies, it can be concluded that the use of PFE in the formulations of various meat products allows:

- to solve the technological problem of forming the necessary consistency and improving the functional properties of meat products;
- increase the yield of the finished product;
- reduce moisture loss during storage and stabilize the consistency of finished products;
- get a juicy product of a monolithic structure with increased nutritional value;
- reduce the rate of oxidative processes of lipids, thereby ensuring the stability of the quality during storage;
- reduce the cost of raw meat.

The expediency of introducing PFE in an amount of up to 22% instead of the main meat raw materials in order to improve the functional and technological properties of the minced meat system and the quality of the finished product is proved.

The developed products have a pleasant taste and aroma, characteristic of this type of product, elastic consistency, and are characterized by high quality indicators. When storing samples of smoked-boiled ham with apple pomace in the composition of PFE, a decrease in the activity of oxidative processes in the lipid fraction was noted, which will allow maintaining high consumer properties for at least 5 days of the shelf life. The cost of a sample of poultry ham was 269 rubles per 1 kg, which confirms the feasibility of this development.

5. Acknowledgments

This work was carried out as part of a state assignment SSI NIIMMP of the Ministry of Science and Higher Education, No 1021032427412-9.

References

[1] Mashanova N S *et al* 2016 Improving poultry meat by adding functional components *Young Scientist* **25(129)** 56-59
[2] Gargaeva A G and Gurinovich G V 2017 Development of recipes of protein-fat emulsions for pates based on poultry meat *Technique and Technology of Food Production* **4** 33-39
[3] Rogov I A, Zabashta A G and Gutnik B E 1993 *Handbook of the sausage production technologist* (Moscow: Kolos) p 431
[4] Shkhalakhov D S and Nesterenko A A 2016 The use of vegetable proteins in the meat industry by adding a protein-fat emulsion *Young Scientist* **23(127)** 109-111
[5] Pryanishnikov V V 2016 Emulsion from chicken skin in technologies of meat products *Rational Nutrition, Food Additives and Biostimulants* **5** 27-29
[6] Gorlov I F, Bozhkova S E, Danilov Y D, Anisimova E Y, Mosolova N I and Starodubova J V 2020 Analysis of efficiency of production of sausage products using non-traditional vegetable raw materials *IOP Conference Series: Earth and Environmental Science* **547** 82032
[7] Gorlov I F *et al* 2019 Meat and vegetable pate: optimization of functional and processing properties and quality parameters *Indo American J. of Pharmaceutical Sciences* **8** 14998-15005
[8] Shakhray T A, Viktorova E P, Velikanova E V and Kornen N N 2020 Modern research in the field of obtaining food ingredients from the secondary resources of apple processing *New Technologies* **3** 78-86

IOP Conf. Series: Earth and Environmental Science **965** (2022) 012045 doi:10.1088/1755-1315/965/1/012045

Development of resource-saving technology of chopped semi-finished products based on lamb

Y A Kolosov[1], N V Shirokova[1], N G Chamurliev[2], N N Kolosova[1], E S Vorontsova[2, 3] and A S Shperov[2]

[1] Don State Agrarian University, 24 Krivoshlykov Street, Persianovsky settlement, Oktyabrsky district, Rostov region, 346493, Russia
[2] Volgograd State Agrarian University, 26 University Avenue, Volgograd, 400002, Russia
[3] All-Russian Research Institute of Irrigated Agriculture, 9 Timiryazev Street, Volgograd, 400002, Russia

E-mail: zootexnia@volgau.com

Abstract. The aim of the study was to develop the formulation and technology of chopped semi-finished products based on mutton with a multicomponent protein mixture, as well as the characteristic of the nutritional value of the product intended for functional nutrition. When conducting experimental studies, a set of indicators of functional and technological properties of sheep meat of the Edilbaev breed was determined by generally accepted methods: moisture binding ability, moisture-retaining ability, fat-retaining ability, organoleptic indicators of the finished product. Taking into account the high content of collagen and adipose tissue in the middle part of the carcass, the low moisture-retaining capacity of meat, we recommend using it in the production of chopped semi-finished products. Based on the results of the evaluation of the qualitative characteristics of soy, amaranth, artichoke and inulin, the choice of vegetable raw materials for further use in chopped semi-finished products is justified. Recipes of chopped semi-finished products using a protein module have been developed. The organoleptic characteristics of the finished product were studied, the optimal ratio of components was selected and the optimal dosage of the protein module was determined.

1. Introduction

The pulp part of the carcasses of agricultural and wild animals, as well as their processed products, are an integral element of the human diet. At the current stage of human society development, there is a steady trend towards an increase in the production and consumption of meat products [1, 2]. In this regard, the problem of healthy food that meets the expectations of consumers, based on an increase in the content of desirable ingredients, is brought to the fore.

In the Russian Federation, the share of chilled and frozen products prevails in the structure of meat. Its importance in the structure of the population's nutrition is evidenced by the fact that, according to the Federal State Statistics Service, more than 9% of all consumer expenditures of Russian residents are accounted for by meat products [3]. Currently, significant efforts in the food industry are aimed at improving the quality of food products, since nutrition is a key factor affecting the prevention of certain diseases. Therefore, along with general-purpose products, the range of functional nutrition products is expanding every year [4]. When developing the formulation and technology of functional meat products, various types of biologically active additives are used that have a positive effect on human health, for example, in terms of increasing the body's resistance. In the meat industry, there is an increase in the use of various types of antioxidants. Their attraction is connected with the purpose of reducing the fat content in finished products, which prevents cardiovascular diseases in humans and helps to solve the problem of weight loss [5, 6, 7]. From a technological point of view, antioxidants have an effect on increasing the shelf life of products.

No less significant attention is paid to food additives, which are used to enrich diets with dietary fibers, micro- and macroelements, increase shelf life, improve taste characteristics and expand the range of chilled and frozen semi-finished products [8].

In order to expand the range and increase the nutritional value, more and more attention is being paid to fortified foods. These include functional foods obtained by adding one or more functional food ingredients to traditional foods.

In order to create products with the most balanced consumer qualities, a combination of meat and vegetable raw materials is relevant. For this purpose, various vegetable supplements obtained from herbs, fruits, vegetables, nuts are attracted. As a rule, they are used to improve the taste characteristics, expand the range of meat products, including various types of chilled and frozen semi-finished products and increase the shelf life of finished products.

Along with the tasks listed above, there are also acute problems associated with protein deficiency in the human diet, an increase in the number of diseases of the gastrointestinal tract and the cardiovascular system. In this regard, there is a need to create and use functional products to prevent a number of these problems. To do this, it is necessary to choose such components that will combine a number of qualities, such as: antioxidant properties, the presence of dietary fiber and high protein content. An important condition for the use of certain compositions of ingredients is their synergy [9].

From the analysis of literary sources, it was revealed that artichoke has antioxidant properties, which helps to reduce the number of age-related and cardiovascular diseases. Soy and artichoke contain a high protein content, which has a good effect on the brain, immunity, the general condition of the body, contain a large amount of dietary fiber, have a positive effect on the activity of the gastrointestinal tract. The use of these additives reduces the energy value of the product. The developed protein complex has a high nutritional value and has not been previously used in meat products. Therefore, the development of semi-finished meat products from minced meat in chilled and frozen form using a protein complex is relevant.

Lamb, along with beef, is one of the highest quality types of meat. That is why its consumption is so high, especially in rich countries. Modern theories of rational nutrition are developed on the principles of conformity of the food product and the organism consuming it. In this regard, pork is considered useful for a growing organism, as it has an increased biological value for this age category. But pork also has properties that, to some extent, are harmful to the elderly. Lamb, due to its dietary properties, on the contrary, significantly slows down the aging process, therefore, to the greatest extent meets the needs of older people. Beef occupies an intermediate position in terms of its biological effect on the human body.

Considering the above, the aim of our research was to develop a recipe and evaluate the characteristics of the nutritional value of chopped semi-finished products for functional nutrition based on lamb with a multicomponent protein mixture.

2. Materials and methods

Experimental studies were conducted on the basis of the Don State Agrarian University. Based on the analysis of literature sources, chemical composition, technological characteristics and the experience of our previous work, vegetable additives were selected: soy protein, amaranth protein, artichoke and inulin prebiotic. The technological process of manufacturing the finished product - chopped semi-finished product - included the following technological operations: preparation of meat raw materials, preparation of other raw materials, materials and additives, grinding of meat raw materials, preparation of minced meat, molding and packaging.

When conducting experimental studies using generally accepted methods, the following indicators were determined: the complex of functional and technological properties (FTP) of meat systems (moisture-binding ability (MBA), moisture-retaining ability (MRA), fat-retaining ability (FRA), organoleptic indicators of the finished product).

The amino acid composition was determined by chromatography on an amino acid analyzer. Mathematical processing of experimental data based on the results of (3-5) repetitions was carried out using the Microsoft Excel 2010 program.

3. Results and discussion

At the initial stage of research, in order to identify the most desirable parts of the carcass for our type of product, we studied the functional and technological properties of sheep meat of the Edilbaev breed. As a result, it was found that the meat of different cuts differs in functional and technological properties (Table 1).

Table 1. Functional and technological properties of sheep meat of the Edilbaev breed.

Name	Moisture-binding ability, %	Moisture-retaining ability, %	Moisture-retaining ability, %
Scapular cut	63.56 ± 1.16	66.96 ± 1.86	90.66 ± 1.11
Neck cut	54.56 ± 1.13	65.27 ± 1.61	89.24 ± 1.10
Middle part	58.32 ± 1.31	60.21 ± 1.16	88.38 ± 2.14
Hip part	65.38 ± 1.16	64.63 ± 1.45	88.11 ± 1.32

Note: * $P < 0.05$ in relation to other cuts.

The analysis of the experimental data presented in Table 1 showed that the largest MBA and MRA is characterized by the meat of the scapular part, which contains the least amount of connective tissue proteins. Therefore, it is rational to direct the meat of the scapular part to the production of lumpy semi-finished products. Since the yield of pulp in neck cut is small, it can be used in soup sets.

Meat containing a sufficiently large amount of collagen is advisable to use mainly in industries related to the softening of collagen raw materials, for example, when lamb is kept in acidic environments (shish kebab) and high-temperature processing (canned food). Taking into account the high content of collagen and adipose tissue in the middle cut, the low moisture-retaining capacity of meat, we recommend using it in the production of chopped semi-finished products. In this case, you can use such technological methods that will improve the structure and increase the FTP of meat.

When choosing a multicomponent protein mixture, we proceeded from the availability of raw materials and chemical composition. A multicomponent protein mixture balanced by amino acid composition - protein module - contains soy protein, amaranth protein, artichoke and inulin prebiotic. These vegetable components have a positive compatibility, have high quality indicators, contain a large amount of protein, vitamins, minerals.

As a result of the conducted studies, it was found that the introduction of a protein module in the amount of 5% increases the MRA of the minced meat system by 17.1%. With the addition of 10 and 15%, the MRA increases by 21.5 and 30.5%, respectively, and the FRA by 6.1 and 6.7% (Figure 1).

For the final selection of the optimal amount of the introduced protein module, a comparative sensory evaluation of the combined minced meat was carried out after temperature treatment by tasting. Sensory evaluation showed that the introduction of a 15% protein module into minced meat worsens the smell, appearance and consistency.

At the next stage of the study, recipes for chopped semi-finished products (cutlets) were developed using a protein module. In order to develop a new product in the original recipe, wheat bread was replaced with a protein module in a dosage corresponding to the conducted research. According to the developed recipes, samples of semi-finished products were developed, which were evaluated according to organoleptic indicators. The quality assessment was carried out after heat treatment using a profile method and a single unified scale.

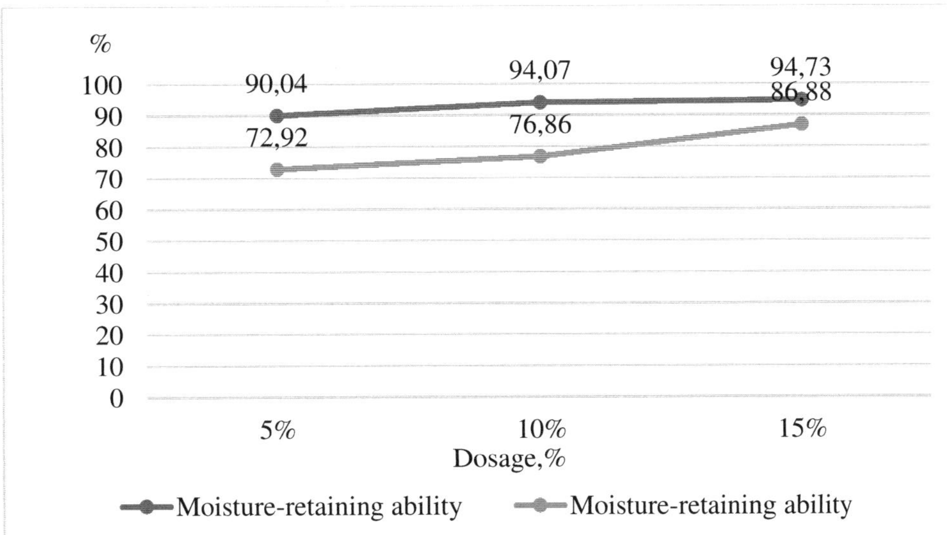

Figure 1. The effect of the dose of the protein module on the level of MRA and FRA of model minced meat, %.

The organoleptic evaluation profilogram shows (Figure 2) that the overall evaluation of the sample developed by us had a positive characteristic against the background of the control sample: all the studied characteristics improved or did not decrease.

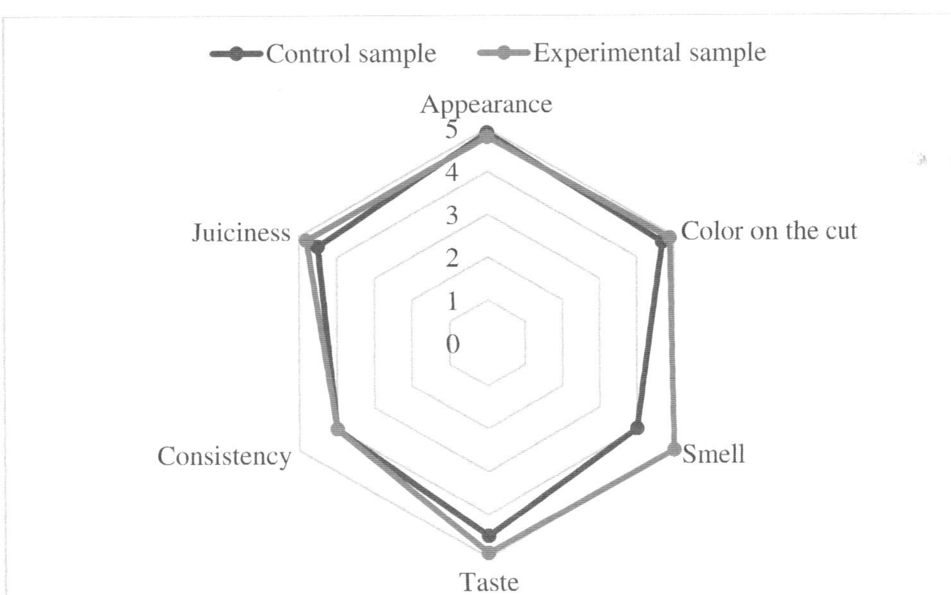

Figure 2. Profilogram of the general organoleptic evaluation of the control and test sample.

When developing the product technology, we were based on a number of postulates given below. Animal proteins are the main qualitative superiority among the structural elements of meat. They are part of the main regulators of metabolism - enzymes, provide the level of immunity of the human body, control the activity of the endocrine glands. Adults who lead a moderate active lifestyle are recommended to consume an average of 0.75 g of protein per kilogram of body weight per day. This is about 55 g for males and 45 g for females. With an increased load, the maximum permissible amount of protein in the daily diet of an adult is about 1.5 g per 1 kg of body weight. The exchange of proteins in the body is carried out with the participation and under the control of enzymes. In humans, depending on a number of factors, including the quantity and quality of incoming food, proteins of the blood, internal organs (liver, intestines, etc.) regenerate on average every 5-7 days. The average protein deficiency in the daily diet of the Russian population, according to researchers and statisticians, is decreasing in relation to the beginning of the 2000s. However, it is still estimated at 30-35%. The deficiency of animal protein is especially acute [10]. According to FAO estimates, out of 7 billion the world's population, about 670 million people do not eat meat. Statistical data for the USA and Europe indicate that in these regions the consumption of this product exceeds dietary norms and amounts to more than 100 kg of meat per inhabitant per year. In Russia, the average per capita consumption is almost twice as low. Meanwhile, the norm recommended by nutritionists is at least 75 kg per year per adult. An adult man should eat up to 220 grams of meat per day to maintain muscle mass. A woman's need for meat is even higher, but, according to statistics, up to 30% of women of childbearing age suffer from anemia, which is especially dangerous during pregnancy.

The category of the most significant meat proteins includes myosin and albumin. The first one is highly soluble in slightly salted water, and insoluble in fresh water, which must be taken into account when processing meat and making meat products. To prevent the loss of myosin during cooking, this process is recommended to be carried out without the use of salt. Albumin is soluble in both salt and fresh water. In connective tissue - in captivity, cartilage and tendons - there are proteins with a lower quality amino acid composition. Depending on their combination, connective tissue can be of two colors - white and yellow. Collagen is the most important component in white connective tissue. It plays an important role in the formation of a jelly-like consistency of the product, which is achieved by cooking meat and converting collagen into gelatin. Thanks to this technological operation, the broth solidifies. More collagen is found in the meat of young animals. Connective tissue of yellow color is formed mainly from elastin protein. Its peculiarity is that it does not dissolve in water and does not undergo transformation during boiling. To achieve softening of the connective tissue, it is recommended to keep the meat in a marinade, i.e. in an acidic environment, before heat treatment.

The main amino acid of collagen and elastin is oxyproline (up to 13%). Therefore, the amount of oxyproline in the muscles is used as an indicator of the level of connective tissue in them, and, consequently, the qualitative characteristics of meat. The higher the specific weight of connective tissue in meat, the lower its nutritional value [11]. The value of sheep muscle tissue proteins is that they contain all the essential amino acids. Their level is in the most favorable ratio for the human body. This category includes: lysine, tryptophan, methionine, leucine, isoleucine, valine, threonine, phenylalanine, histidine, arginine. Especially important – critical – in this category are lysine, tryptophan and methionine. Lysine deficiency in the human diet leads to violations of hematopoietic functions and nitrogen balance, anemia and problems in the reproductive sphere, muscle wasting, changes in the liver and lungs. It is necessary for the normal bone formation and growth of children. Tryptophan is also associated with tissue metabolism and body growth. It is used for the synthesis of serotonin, which is an essential neurotransmitter. It is necessary to stabilize the nervous system and significantly reduces (neutralizes) the harmful effect of nicotine on the body. It plays an important role in the formation of whey proteins, hemoglobin and some vitamins. One of the most important proteins is methionine, which provides normalization of the processes of fat and phosphatide metabolism in the liver, and, as a result, serves as a preventive agent against atherosclerosis [12]. An important role in the human body is played not only by essential amino acid, but also by most of the nonessential amino acids that can be synthesized in the body of animals and humans, including from essential amino acid. The necessary level of nonessential amino acids that come with food ensures normal growth and development of the body, since it does not require the expenditure of nutrients and energy for their formation.

Table 2. Amino acid content of samples.

Essential amino acids	Control sample			Experimental sample		
	mg/g of protein	Amino acid score. %	a_j	mg/g of protein	Amino acid score. %	aj
Valin	54.2	108.4	0.55	61.2	122.4	0.65
Isoleucine	42.2	106.0	0.54	46.2	115.5	0.69
Leucine	79.3	113.3	0.53	89.3	127.6	0.62
Lysine	82.4	149.8	0.40	87.4	158.9	0.50
Methionine + cysteine	36.5	104.3	0.57	38.6	110.3	0.72
Threonine	23.9	60.0*	-	31.7	79.3*	-
Tryptophan	18.4	184.0	0.33	21.8	218.0	0.36
Phenylalanine + tyrosine	72.0	120.0	0.50	79.3	132.2	0.60
Total		408.9			455.5	

Analysis of experimental data and calculation results indicates that samples with the sample of a protein module have a more valuable amino acid composition compared to the control (Table 2). In samples with the addition of a protein module, there is a noticeable improvement in score.

4. Conclusion

A study of the functional and technological properties of sheep meat of the Edilbaev breed was carried out. Recipes of chopped semi-finished products using a protein module have been developed. For the production of chopped semi-finished products, it is recommended to use the middle part of the carcass, characterized by a high content of fat and collagen. Based on the results of the evaluation of the qualitative characteristics of soy, amaranth, artichoke and inulin, the choice of vegetable raw materials for further use in chopped semi-finished products is justified. The introduction of a 10% protein module into the formulation improves organoleptic parameters, functional and technological properties, as well as the nutritional value of the product compared to the control. The organoleptic characteristics of the finished product were studied, the expediency of replacing wheat bread with a protein module in the production of chopped semi-finished products from lamb was substantiated.

References

[1] Antipova L V, Glotova I A and Zharinov A I 2000 *Methods of Research of Meat and Meat Products* (Voronezh: Voronezh State Technological Academy) p 332

[2] Baranikov A I, Kolosov Yu A and Shirokova N V 2013 Creation of new meat products using mutton *Scientific J. of the Kuban State Agrarian University* **05(089)** 0891305052

[3] Kostenko Yu G and Matveev O A 2009 Production control is the basis for obtaining high-quality and safe meat products *Meat Industry* **6** 23-24

[4] Khramova V N, Gorlov I F, Dolgov V A and Khramova Ya I 2013 Development of functional meat products using prebiotics and plant components of regional origin. Khramova, // *Proc. of the Nizhnevolzhsky Agrouniversity Complex: Science and Higher Professional Education* **3(31)** 170-174

[5] Shirokova N V, Kolesnikov A N and Belyaevskaya A V 2019 The use of vegetable raw materials in the production of meat bread *Innovations in Food Production: From Animal Breeding to Food Production Technology* (Persianovsky) 103-106

[6] Gorlov I F *et al* 2016 CAST/MspI gene polymorphism and its impact on growth traits of Soviet Merino and Salsk sheep breeds in the South European part of Russia *Turkish J. of Veterinary and Animal Sciences* **40** 399-405

[7] Gorlov I F, Slozhenkina M I, Knyazhechenko O A, Mosolova N I, Vasilyeva V V, Shinkareva S V, Khramova V N and Mosolova D A 2019 Advances in technological productivity and biological evaluation of sausage products *Research J. of Pharmaceutical, Biological and*

Chemical Sciences **10(1)** 2075-2080

[8] Gorlov I F, Fedotova G V, Sazonov S P, Sergeev V N and Yuldashbaev Yu A 2018 *Cognitive Approach to the Study of Food Security Problems* (Volgograd: Publishing house of Volgograd Institute of Management) p 168

[9] Poznyakovskiy V M, Gorlov I F, Tikhonov S L and Shelepov V G 2015 About the quality of meat with PSE and DFD properties *Foods and Raw Materials* **3** 104-110

[10] Slobodyanik V S, Uspenskaya M E and Ryabykh A G 2015 Vegetable ingredients as sources of biologically active substances in the recipe for semi-finished meat products *Veterinary and Sanitary Aspects of the Quality and Safety of Agricultural Products* pp 324-326

[11] Morozova L A, Mikolaichik I N and Okhokhonina E N 2018 Modern aspects of the production technology of chopped semi-finished products of a functional orientation *Ways of implementing the Federal Scientific and Technical Program for the Development of Agriculture for 2017-2025* pp 583-587

[12] Browning C J, Qiu Z and Yang H 2019 Food, eating, and happy aging: the perceptions of older Chinese people *Frontiers in Public Health* **7** 1-12

IOP Conf. Series: Earth and Environmental Science **965** (2022) 012046 doi:10.1088/1755-1315/965/1/012046

Cleaning of grain sowing material from weed seeds for the sustainable development of agriculture

V V Alekseev[1], L K Mitrukhin[1], I V Fadeev[2], I A Uspensky[3] and I A Yukhin[3]

[1] Chuvash State University named after I N Ulyanov, 15 Moskovsky Avenue, Cheboksary, 428015, Russia
[2] Chuvash State Pedagogical University named after I Ya Yakovlev, 38 Karl Marks Street, Cheboksary, 428003, Russia
[3] Ryazan State Agrotechnological University named after P A Kostychev, 1 Kostychev Street, Ryazan, 390044, Russia

E-mail: ivan-fadeev-2012@mail.ru

Abstract. The article deals with the optimization of cleaning grains and seeds of agricultural crops from weed seeds. The process of cleaning grain material by one-dimensional horizontal vibrations depends on the friction coefficients of the seeds against each other. It is optimized by selecting the amplitude and angle, relative to the vertical, under which the cylindrical container containing the grain mixture is located. Experiments were carried out with almost all types of seeds most widely used in the domestic agricultural industry. The cleaning process was investigated only at amplitudes smaller than the seed size. The seeds of cultivated plants and weeds move upward or downward, depending on the specific gravity. Calculations show that with correctly selected values of the amplitude and angle, the average trajectories of the grains are practically straight lines parallel to the generatrix of the cylinder containing the seeds. Studies have shown that the maximum cleaning speed is achieved under the condition that the angle of inclination of the cylindrical vessel is equal to the natural slope of the seeds with the highest specific gravity.

1. Introduction

One of the most important stages of clearing fields from clogging is cleaning the seeds. Separation of grain mixtures is an integral part of various technological processes, including the preparation of seed material and plant materials [1, 2]. This is a preventive phase that is important in the integrated control of weed. Even though the seeds of weed grasses are not completely released when the sowing material is cleaned, the cleaning is very effective [3]. Many weed plants that have severely clogged crops have almost come to naught with an improvement in the quality of seed purification. The quality of the separation process can be improved by a combination of vibration and additional factors such as rising air flow [4] and electrification of the sorted seeds [5], etc. In various variations during cleaning, the physical and mechanical properties of seeds of cultivated and weed plants are involved - geometric dimensions, windage, specific gravity, degree of surface roughness, etc.

This, however, leads to the fact that technologies and equipment become more complex. Almost all of the currently known cleaning methods have (with various variations) basically a vibrating trier (cellular) surface on which seeds are poured in a thin layer. And since the processes of cleaning seeds entail certain economic costs, the problem of increasing the efficiency and optimizing the methods of vibration separation of the seed remains urgent [6].

The analysis shows that the vibrational component of mechanical energy in the total energy consumption when cleaning grain mixtures is predominant. Therefore, in terms of optimization, it seems most preferable to determine the optimal conditions for vibration impact on the seed. Here, the most significant is the selection of the direction, frequency and amplitude of oscillations for each specific grain mixture with its physical and mechanical characteristics. In this regard, the article discusses aspects of the behavior of grain material in one-dimensional, horizontal, low-frequency oscillations for various angles of inclination of cylindrical vessels containing planting material.

2. Materials and methods

Practically without using any idealizations, the seeds of cultivated plants and weeds can be attributed to a certain kind of bulk materials. And bulk materials under the influence of vibrations and vibrations, under certain conditions can behave both as a solid and as a liquid [7, 8, 9]. In this work, an attempt is made to use the above-mentioned properties to describe aspects of the process of cleaning planting seed material from weed seeds [10]. Let's consider a vibrating cylindrical container of seed material.

Oscillation energy from the cylinder is also transmitted to individual seeds in the mixture. As a result of making up the system with vibration energy, under certain modes, a certain state occurs (analogue of the "phase" transition) that the entire volume of planting seed material starts to move. The grains begin to "flow" through the vessel, as in a liquid. Averaged over the oscillation period, the values of the coordinates of the seeds make it possible to calculate a certain average speed of their "drift" relative to the cylindrical container vessel. It is important that, unlike a liquid, in this case, there is dry friction in the interaction between seeds in the mixture. Collisions are not always elastic and, unlike a liquid, there are no forces of attraction [11]. This can lead to instability with respect to the separation of seeds into fractions or clustering [12]. That is, if a weed seed or a seed of another variety appears next to the grain of a cultivated plant, then the resulting forces acting on them are different.

The seeds move in different directions. Conversely, when the seed gets to itself like this the forces are such that the relative movement becomes minimal but they continue to move relative to the cylindrical vessel. So the seeds of the same type unite and, uniformly filling the space, push out seeds of other types.

The following forces act on each seed in the mixture: gravity, inertial force, reaction force from the surrounding seeds, friction force. In this work we will neglect the effects of seed electrification due to friction. When seeds are poured onto a horizontal surface, they form a cone. The angle of the generatrix of the cone and the normal to the surface depends on the coefficient of friction between the seeds. Let's consider a seed on an inclined surface and the angle of inclination is such as $\mu = \mathrm{tg}\alpha$. Let $\vec{i}$ be the unit vector (ort) in the direction of the greatest descent (gradient), let $\vec{j}$ be the unit vector of the tangent to the trajectory of the seed and let $\vec{k}$ be the unit vector of some averaged force F acting from the surrounding seeds. First, we write down the condition of quasi-static equilibrium, when the seed practically does not move:

$$\vec{F} + mg\sin\alpha\,\vec{i} + \vec{F}_f = 0, \tag{1}$$

where F_f is friction force.

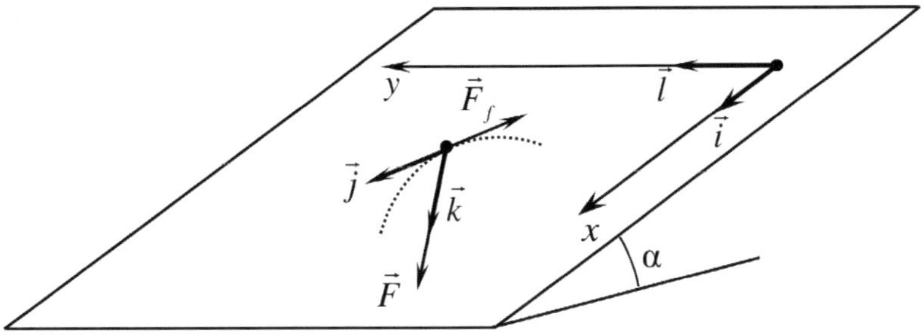

Figure 1. Forces acting on each seed.

As far as $\vec{F}_f = -\mu mg \cos\alpha \vec{j}$, and $\mu = \text{tg}\alpha$ we have

$$\vec{F} + mg \sin\alpha \vec{i} = \text{tg}\alpha \; mg \cos\alpha \vec{j}. \tag{2}$$

Squaring both parts and simplifying:

$$F^2 + m^2 g^2 \sin^2\alpha + 2Fmg \sin\alpha \cos\delta = m^2 g^2 \sin^2\alpha, \tag{3}$$

$$F(F + 2mg \sin\alpha \cos\delta) = 0, \tag{4}$$

here $\cos\delta = (\vec{k}, \vec{i})$.

It follows from equation (4) that at values from the range $90 \le \delta \le 180$ we have $F = -2mg \sin\alpha \cos\delta > 0$. If δ takes values from the range $0 \le \delta \le 90$ then from (4) it follows that F = 0. The seed trajectory determines the resulting force

$$\begin{aligned} \vec{R} &= \vec{F} + mg \sin\alpha \vec{i} = mg \sin\alpha(\vec{i} - 2\cos\delta \vec{k}) = \\ &= mg \sin\alpha[\vec{i} + 2\cos\delta(\sin\delta \vec{l} - \cos\delta \vec{i})] = \\ &= mg \sin\alpha \cos 2\delta(\text{tg}\, 2\delta \vec{l} - \vec{i}), \end{aligned} \tag{5}$$

where $\vec{l}$ is axis eagle Oy.

We got that under certain conditions, the seed moves, sliding down or rising up along the vector $\vec{i}$. If it is necessary to obtain a trajectory for other conditions, for example, for $90 \le \delta \le 180$ then from (5) it follows

$$\frac{dy}{dx} = -\text{tg}\, 2\delta = -\frac{2\sin\delta\cos\delta}{\cos^2\delta - \sin^2\delta} = \frac{2xy}{x^2 - y^2}, \tag{6}$$

that is, the seed moves in an arc of a circle.

Thus, from (4) and (6) there follows a deterministic relationship between the efficiency of the seed cleaning process by one-dimensional horizontal oscillations on the value of the angle of natural friction. Let's consider a two-component blend of peeled and unpeeled millet seeds. When such a mixture is poured onto a flat surface a complex cone is formed. This is a figure consisting of a truncated cone on which another cone is located. The different values of the angles between the generatrices and the bases are due to different natural slope angles. The more the angles of repose differ, the more pronounced the effect. The shape of the cone and kink depend on the ratio of the components. Table 1 shows the dynamics of the angles of the cones depending on the proportion (ν) of the components that make up the mixture. For untreated seeds, the angle is 36.2° and for cleaned millet seeds, is 31.4°. Analysis of the data in the table shows that the most statistically significant for describing the dependence is the dependence f(ν) = 0.14ln(ν) + 36.19, that is, as the proportion of unrefined seeds in the mixture increases, the angle corresponding to the upper cone increases according to the logarithmic law from 31° to 36°.

IOP Conf. Series: Earth and Environmental Science **965** (2022) 012046 doi:10.1088/1755-1315/965/1/012046

Table 1. Changes the corners of the cone generatrix to the horizontal when the percentage of components changes.

The proportion of unrefined seeds in the mixture (ν)	Upper cone, °	Bottom cone, °
0.125	34.4	33.0
0.250	34.5	31.4
0.375	34.7	31.0
0.500	35.1	26.6
0.625	35.6	28.6
0.875	36.1	44.9

In order to study the features of the separation of a two-component mixture of bulk materials, an experimental setup was made. The setup consists of a cylindrical container, the angle of the axis of which relative to the horizontal can be changed by means of a rotary mechanism. The cylinder itself is fixed to the platform. oscillating in a horizontal plane with an adjustable frequency.

The directions of all forces, except for gravity and inertia are constantly changing due to the chaotic movement of seeds relative to each other due to vibration. The resulting gravity and inertia force is constant at a given vibration frequency. Under certain conditions in the oscillatory system, the resulting force allows seeds with a large specific gravity to reduce their potential energy by moving downward relative to seeds with a lower specific gravity.

3. Results and discussion

For an experimental study of the optimal cleaning conditions buckwheat and millet seeds containing impurities and weed seeds were selected. This choice is mainly due to the clarity of the demonstration of the cleaning process - a good color difference in seeds with a small difference in size. In addition, the considered not the smallest seed sizes make it possible to neglect the influence of electrostatic forces arising from electric charging of seeds due to mechanical contact. The characteristics of the components of the studied mixtures are presented in Table 2.

The studied mixtures of seeds were placed in a cylindrical vessel made of transparent material with an adjustable angle of inclination. The vessel was rigidly connected to the vibrator.

The stages of the cleaning process are shown in Figure 2. As mentioned above, the amplitude of fluctuations did not exceed the average linear dimensions of the seeds. This is due to the fact that at which amplitudes and frequencies, instead of the process of cleaning seeds in a mixture into components, on the contrary, their intensive mixing begins. The vibration frequency varied from 5 to 10 Hz. Under these conditions making measurements is not difficult. The ability to use a high frame rate when recording video on modern smartphones (up to 960 frames per second) allows to determine both the average and instantaneous speed of movement of seeds in the cylinder.

Table 2. Characteristics of materials to be separated.

Seeds	Angle of natural slope, °	Density, kg/m^3
Buckwheat	31.1	530
Millet	34.7	760

Figure 2. Mixture separation process.

The results of measurements of the dependence of the average "drift" speed during the cleaning process in the form of graphs of dependence on the angle of inclination of the cylindrical vessel to the horizon are shown in Figure 3.

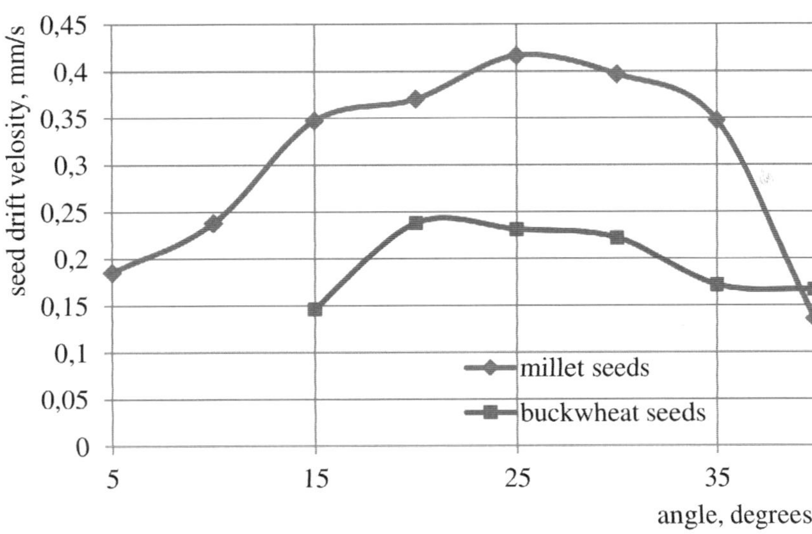

Figure 3. Dependence of the average speed of "drift" of millet and buckwheat seeds on the angle of orientation of the axis of the vertical cylindrical bunker (vibration frequency ν = 10 Hz).

4. Conclusion

The manufacture and use of the experimental setup made it possible to collect sufficient material for statistical verification and confirmation of the theoretical premises for describing the process of cleaning seeds of cultivated plants from weed seeds. The relationship between the average speed of the "drift" of seeds in a mixture poured into a cylindrical container with the frequency and amplitude of oscillations has been investigated. For all the used samples the extremum of the function was found - the maximum value of the average "drift" speed corresponding to the angle of inclination of the cylinder approximately equal to the natural angle of slope of these seeds with a higher specific weight. The maximum scatter of the measurement results relative to the average values turned out to be no more than 8-12%. It is also important that this approach to cleaning among other things allows to separate whole seeds from damaged ones - broken and eaten. As a supplement to the research, it should be noted that it is necessary to improve the experimental setup to eliminate the effect of electrifying seeds. In order to save energy and time further research on this problem should be directed to the study of the critical values of the amplitude and frequency of oscillations.

References

[1] Rosatoa A, Blackmoreb D L, Zhanga N and Lanc Y 2004 *Chem. Eng. Sci.* **57** 265-275
[2] Akhmadiev F, Gizzyatov R and Nazipov I 2020 *IOP Conf. Series: Materials Science and Engineering* **890** 012106
[3] Alekseev V et al 2019 *IOP Conf. Series: Materials Science and Engineering* **560** 012137
[4] Konopka S, Choszcz D and Markowski P 2017 *Sustainability* **9(11)** 2134
[5] Rogovskii I L, Palamarchuk I P, Voinash S A, Troyanovskaya I P, Butenko A F and Sokolova V A 2020 *J. of Physics: Conf. Series* **1679** 042034
[6] Kovalyshyn S and Kovalyshyn O 2018 *Acta Universitatis Agriculturae et Silviculturae Mendelianae Brunensis* **66(5)** 1157–1164
[7] Kovalyshyn S and Shvets O 2018 *Acta Universitatis Agriculturae et Silviculturae Mendelianae Brunensis* **66** 677-683
[8] Bocquet L, Losert W, Schalk D, Lubensky T C and Gollub J P 2002 *Phys. Rev. E Stat. Nonlin. Soft Matter Phys.* **65(1)** 011307
[9] Andreotti B, Forterre Y and Pouliquen O 2013 *Granular Media: Between Fluid and Solid* (Cambridge: Cambridge University Press)
[10] Tupolskikh T, Gucheva N and Kirishiev O 2018 *MATEC Web of Conf.* **224** 05021
[11] Shvets O, Dadak V and Husak Y 2017 The use of an electric field as an additional body of work during additional cleaning of seeds of agricultural *15th International Scientific and Technical Conference Motorization and Energetics in Agriculture MOTROL* (Lublin: Poiska Academy of Sciences)
[12] Shvets O and Duhanets V 2014 The research results trajectories of rape seeds moving on an inclined plane *Bulletin of the Lviv National Agrarian University Agro-Engineering Investigations* **18** 148-152

Problems of safety of gas systems in the places of residence of agricultural workers

D V Belomutenko, M V Istochkina, N A Kolobanova, G V Konovalova and P S Kryukov

Volgograd State Agrarian University, 26 University Avenue, Volgograd, 400002, Russia

E-mail: kolobanova.nina@yandex.ru

Abstract. The article presents the results of the analysis of emergencies that occurred in 2011-2020 in Russia as a result of improper operation of gas stoves in residential houses of rural residents, as well as the results of studying the circumstances that preceded the occurrence of these emergencies. Separate provisions of the relevant regulatory legal acts of Russia and departmental regulatory documents of Russia in relation to the placement of gas stoves and their operation in residential houses of rural residents are given. The ongoing gasification of rural areas of Russia, aimed at improving the quality of life of its population, will significantly change the lives of millions of people in rural areas, primarily those who still had to heat stoves with wood or coal, and at the same time will require solving the problem of preventing emergencies during the placement and operation of gas stoves, which is still relevant and unresolved in full.

1. Introduction

In modern conditions, blue fuel is not only a more environmentally friendly type of fuel, used, among other things, for heating water and heating in places of residence and stay of the population of Russia, but it is also much cheaper in cost than other options. Electric heating will cost for the owner of a residential building about ten times more expensive than gas heating. The installation of the gas tank allows not to depend on the central pipeline, but connecting to the central pipeline subsequently creates less hassle.

Emergencies associated with improper placement and operation of gas stoves in the places of residence and stay of the Russian population, including the rural population, occur with enviable regularity and often entail serious consequences: the death of people, injuries of various kinds by people, partial or complete destruction of buildings, thereby causing a negative public response. At the same time, they also entail additional unforeseen costs for the restoration of damaged and destroyed buildings, for the payment of monetary compensation to injured persons and families of deceased people, for the resettlement of citizens to a new home, etc.

This article presents a study of emergency situations that arose in 2011-2020 in Russia as a result of improper operation of gas stoves in the places of residence and stay of rural residents and the previous circumstances, as well as existing inconsistencies between the individual requirements of the relevant regulatory legal acts of Russia and departmental regulatory documents regarding the placement of gas stoves and their operation in the places of residence and stay of rural residents.

IOP Conf. Series: Earth and Environmental Science **965** (2022) 012047 doi:10.1088/1755-1315/965/1/012047

2. Materials and methods

The following materials were used for the study:

- statistical data on emergencies that occurred in 2011-2020 in Russia as a result of improper placement and operation of gas stoves in the places of residence and stay of rural residents, information about which is taken from open sources [1-7];
- requirements of national standards [8, 9] for combustible gases for industrial and municipal purposes, as well as for household appliances for cooking;
- fire-hazardous properties of combustible gases [10, 11] for municipal purposes used in household appliances for cooking;
- separate provisions of the relevant regulatory legal acts of Russia and departmental regulatory documents [12, 13 ,14, 15] related to the protection of the population and territories of Russia from natural and man-made emergencies, to ensuring safety when using and maintaining of intra-house and intra-apartment gas equipment in Russia, to the safe use of gas to serve municipal needs in Russia.

3. Results and discussion

The study of the causes of emergencies indicated in Table 1 and arising in 2011-2020 in Russia as a result of improper operation of gas stoves in the places of residence and stay of rural residents allowed us to identify the following average patterns:

- each emergency ended with the death of about one person and injury to four more people, including about one child;
- 33% of emergencies were realized in the form of an explosion of a gas-air mixture, which led to a further fire;
- 43.5% of emergencies resulted in the destruction or damage of a building in which villagers lived or were located;
- in eleven cases, emergencies were realized in the form of an explosion of a gas-air mixture due to improper operation of gas-cylinder equipment;
- in six cases, emergencies were realized by the explosion of a gas-air mixture due to improper operation of a gas stove connected to a gas cylinder;
- in two cases, emergencies realized in the form of an explosion of a gas-air mixture arose due to unauthorized connection of gas equipment by rural residents;
- in four cases, emergencies realized in the form of an explosion of a gas-air mixture, arose due to a breach of tightness or damage to the gas pipeline connecting the gas cylinder to the gas stove;
- in 19 cases, emergencies occurred during the daytime, and in 20 cases they occurred at night, that is, during the period of stay of rural residents in their homes;
- in 37 cases, emergencies realized in the form of an explosion of a gas-air mixture took place directly in the places of residence of rural residents and only in two cases they took place in places of temporary residence of rural residents;
- in three cases, emergencies realized in the form of an explosion or a clapping of a gas-air mixture took place in rooms where the ventilated window openings were filled with single glazing and which was destroyed afterwards;
- in 17 cases, emergencies realized in the form of an explosion of a gas-air mixture took place in rooms with unventilated window openings or in which double-glazed windows were installed. The result of these emergencies were more serious consequences, including partial or complete destruction of the building.

In the national standard [8] it is recorded that in Russia the concentration limits of ignition (for methane) of natural combustible gas mixed with air are in the range from 4.4 to 17.0% by volume and its acceptance is carried out during transfer from the supplier to the consumer in accordance with the procedure established by the relevant agreement.

The national standard [9], which applies to household gas appliances for cooking and establishes requirements for freestanding and built-in appliances operating on natural or liquefied petroleum gas, provides that:

- the use of a gas-powered cooking appliance leads to an increase in temperature and humidity in the room and, in this regard, good ventilation should be provided in the kitchen room due to open natural ventilation openings or an installed mechanical ventilation device;
- with intensive and prolonged use of the gas-powered cooking appliance, additional ventilation is required by opening a window and effectively ventilating the kitchen room or turning on the mechanical ventilation device;
- if the delivery set of the device includes parts intended for adjustment to a different gas family or to a different pressure, it is necessary to provide complete clear instructions for replacing parts, cleaning, setting up and testing the device, as well as for sealing after setting up;
- the largest (without specifying its capacity) possible gas cylinder for cooking appliance should be specified in its operating manual.

Table 1. Statistical data on emergencies that occurred in 2011-2020 in Russia as a result of improper placement and operation of gas stoves in the places of residence and stay of rural residents.

Time period	Total number of emergencies	Including the number of explosions that caused a fire	The number of dead people/including children	Number of injured people/including children	Number of destroyed and damaged buildings
2011	4	-	8/1	55/2	2
2012	3	2	3/2	5/2	2
2013	4	2	10/4	9/2	2
2014	3	-	1/-	9/-	1
2015	2	-	-	4/-	-
2016	1	1	2/1	-	1
2017	9	3	4/-	23/-	4
2018	6	1	1/1	22/8	3
2019	6	3	1/1	17/5	2
2020	1	1	-	4/1	-
Total	39	13	30/10	148/20	17

At the same time, this document does not establish requirements for cylinders for liquefied gases and gas pressure regulators for them.

The reference book [10] shows the properties of natural gas, including fire-hazardous: it consists of methane (CH_4) 93.05%, ethane (C_2H_6) 2.73 %, propane (C_3H_8) 1.04 %, butane (C_4H_{10}) 0.22%, hydrogen (H_2), hydrogen sulfide (H_2S), carbon dioxide (CO_2), nitrogen (N_2), helium (Ne) and other inert gases; it has no color and odor; its explosive concentrations of a gas-air mixture are in the range from 4.5 to 13.5% by volume; it is 1.8 times lighter than air, so it rises up when leaking.

The reference book [10] shows the properties of butane (C_4H_{10}), including fire-hazardous: it is a colorless and odorless combustible gas; its concentration limits of flame propagation in the air are in the range from 1.8 to 9.1% by volume; it is twice as heavy as air, so it falls down when leaking.

The reference book [11] shows the properties of methane (CH_4), including fire-hazardous: it is a colorless and odorless combustible gas; its concentration limits of flame propagation in the air are in the range from 5.28 to 14.1% by volume; it is 1.8 times lighter than air, so it rises up when leaking.

The reference book [10] shows the properties of propane (C_3H_8), including fire hazards: it is a colorless and odorless combustible gas; its concentration limits of flame propagation in the air are in the range from 2.3 to 9.4% by volume; it is one and a half times heavier than air, therefore, when leaking, it falls down.

At the same time, propane or butane contained in cylinders for gas appliances, in case of leakage, will accumulate in the lower part of the kitchen room and one and a half to two times faster than methane,

IOP Conf. Series: Earth and Environmental Science **965** (2022) 012047 doi:10.1088/1755-1315/965/1/012047

will create an explosive concentration in a mixture with air. Removing these propane or butane vapors from the lower part of the kitchen room is almost impossible in a natural way, since open natural ventilation openings are located in the upper part of the room and are able to remove only methane vapors when it leaks. And electric motors of compressors of household refrigerators and other kitchen equipment, together with the "lower baseboard wiring" of electrical wiring, as possible sources of ignition, supplement the gas (steam) air mixture with air and create a so-called "fire triangle", creating an actual fire hazard.

According to [12]:

- prevention of emergencies is measures carried out in advance aimed at minimizing the risk of emergencies, as well as at preserving people's health, reducing the amount of environmental damage and material losses in case of their occurrence;
- Russian citizens have the right to protection of life, health and personal property in case of emergencies;
- Russian citizens are obliged to observe safety measures in everyday life, as well as to prevent violations of safety requirements that may lead to emergencies.

According to [13]:

- replacement of the equipment that is part of the inra-house gas equipment is carried out by a specialized organization as part of the execution of an agreement on maintenance and repair of inra-house gas equipment. At the same time, payment for these works is carried out by the owner of this equipment, independent replacement of the specified equipment by its owner without the involvement of a specialized organization is not allowed;
- the proper maintenance of smoke and ventilation ducts is provided by the owner of the household by checking their condition and their functioning or by concluding an agreement for their inspection, as well as, if necessary, for cleaning and (or) repair with the organization carrying out the specified work.

In [14], the minimum requirements for the places of residence and stay of rural residents with the presence of household gas appliances are given:

– at the entrance to single-flat residential buildings, including residential buildings of blocked development, as well as to premises of buildings and structures in which gas cylinders are used, a fire safety warning sign with the inscription “Flammable. Gas cylinders ";

– when using household gas appliances, it is prohibited: operation of household gas appliances in case of gas leakage; connecting parts of gas fittings with the help of a spark-forming tool; checking the tightness of connections using open fire sources.

The fulfillment of the following requirements of the approved [15] "Instructions for the safe use of gas to serve municipal needs", ensuring the safe use of gas by owners (users) of households in relation to intra-house gas equipment, is possible with the following additions:

- paragraph 4.20 with minimum values of the safe distance from furniture and other flammable objects and materials to the gas stove;
- paragraph 4.21 with minimum values of the multiplicity of air exchange and the rate of air inflow, the cross-sectional area of the grate or the height of the gap between the door and the floor, as well as the cross-sectional area of special supply devices in the outer wall (window);
- paragraph 6.29 with minimum values of the safe distance from furniture and other flammable objects and materials to household gas-using equipment.

The joint participation of local government and rural residents themselves in the work on replacing gas stoves with expired service life with modern ones, the use of composite gas cylinders, the installation of gas pollution detectors, the dismantling of plastic windows with double-glazed windows, the maintenance of ventilation ducts of exhaust natural ventilation, the inadmissibility of connecting to them hoods from gas stoves, and other measures will ensure the safe operation of gas stoves both at the enterprises of the agro-industrial complex of Russia and in their residential area.

4. Conclusion

After the conducted research, it can be concluded that:

– only a joint partnership of local governments and rural residents of Russia themselves can ensure the safe installation and operation of gas stoves, both at the enterprises of the agro-industrial complex of Russia and in their residential area;

– it is necessary to make appropriate additions and changes to certain regulatory legal acts and departmental regulatory documents that exclude the above inconsistencies regarding the placement and operation of gas stoves, both at the enterprises of the agro-industrial complex of Russia and in their residential area.

References

[1] Statistics of explosions in residential buildings for 2011 https://chr.aif.ru/incidents/crime/181946
[2] Household gas explosions in residential buildings in Russia in 2012-2013 *RIA NEWS* https://ria.ru/20130420/933630800.html
[3] Explosions and fires of household gas in residential buildings in Russia in 2014-2016 *RIA NEWS* https://ria.ru/20161023/1479808976.html
[4] Household gas explosions in residential buildings in Russia in 2017-2018 *RIA NEWS* https://ria.ru/20180112/1512529694.html
[5] Household gas explosions in residential buildings in Russia in 2018-2019 *RIA NEWS* https://ria.ru/20190214/1550888421.html
[6] Cases of explosions of gas cylinders in residential buildings in the Russian Federation in 2019-2020 *RIA NEWS* https://ria.ru/20200408/1569724718.html
[7] Statistics and comparison of accidents, emergencies, carbon monoxide poisoning in the Russian Federation - 2017, 2018, 2019 *https://гиго.рф/blog/2019/03/11/статистика-и-сравнение-аварий-чс-отра/*
[8] *Natural Combustible Gases for Industrial and Municipal Purposes. Technical Specifications* 2019 State Standard GOST 5542-2014 (Moscow: Standartinform)
[9] *Household Gas Appliances for Cooking. General Technical Requirements and Test Methods* 2006 State Standard GOST R 50696-2006 (Moscow: Standartinform)
[10] Korolchenko A Ya and D A Korolchenko 2004 Fire and Explosion Hazard of Substances and Materials and their Means of Extinguishing: Reference Book vol 1 (Moscow: Association "Pozhnauka) p 713
[11] Korolchenko A Ya and D A Korolchenko 2004 Fire and Explosion Hazard of Substances and Materials and their Means of Extinguishing: Reference Book vol 2 (Moscow: Association "Pozhnauka) p 774
[12] Federal Law of December 21, 1994 No. 68-FZ On the protection of the population and territories from natural and man-made emergencies (as amended on 08.12.2020) 1994 *Collected Legislation of the Russian Federation* **35** 3648
[13] Decision of the Government of the Russian Federation dated 14.05.2013 No. 410 On measures to ensure safety when using and maintaining indoor and indoor gas equipment (as amended on 19.03.2020) 2013 *Collected Legislation of the Russian Federation* **21** 2648
[14] Decision of the Government of the Russian Federation dated 16.09.2020 No. 1479 On the approval of the Rules for the fire regime in the Russian Federation 2020 *Collected Legislation of the Russian Federation* **39** 6056
[15] Order of the Ministry of Construction and Housing and Communal Services of the Russian Federation dated 05.12.2017 No. 1614/pr On approval of the Instructions for the safe use of gas to serve municipal needs 2018 Ministry of Justice of Russia 50945

IOP Conf. Series: Earth and Environmental Science **965** (2022) 012048 doi:10.1088/1755-1315/965/1/012048

Intra-farm transportation of easily damaged agro food products for sustainable development of agricultures

S N Borychev, I A Uspensky, I A Yukhin, A A Golikov and D I Kosorukov

Ryazan State Agrotechnological University named after P A Kostychev, 1 Kostychev Street, Ryazan, 390044, Russia

E-mail: yuival@rambler.ru

Abstract. The article presents some aspects of the issue of damage to easily damaged agricultural products during intrafarm transportation, as well as the results of a patent and license review of the scientific and technical groundwork of the authors of the article in this research area. Based on the results new scientifically grounded technical solutions of devices for increasing the safety of agricultural products during transportation are proposed, in particular containers, to reduce the damage of products during their transportation by increasing the area of contact between fruits and the container, as well as reducing the space for free movement of products.

1. Introduction

One of the main factors characterizing agricultural goods is the variability of their mechanical properties under the influence of humidity, pressure, temperature, and storage duration. As the height of the stored material increases, the risk of caking increases. Many cargoes are easily damaged. Most of all, fruits, roots and vegetables are damaged, especially when transported in bulk [1, 2].

The general structure of transportation of agricultural goods in the Russian Federation is presented as follows (Table 1).

Table 1. The structure of agricultural cargoes by the mode of transportation.

Mode of transportation	To the entire volume of transportation, %
In bulk	76
Including specialized bodies	28
In various containers, packages, bales	14
Including specialized bodies	8
Shot lots (mass of a one-time dispatch up to 2 tons)	4
In tanks	6

At the same time, depending on the agricultural crop, transportation methods can be varied (potatoes are transported both in bulk and in containers) or individual (for example, tomatoes neatly packed in boxes). But even a rationally made choice cannot guarantee the absence of damage to the fruit during transportation.

IOP Conf. Series: Earth and Environmental Science **965** (2022) 012048 doi:10.1088/1755-1315/965/1/012048

If one considers the location of the fruit in the container (Figure 1), the likelihood of their damage will be different. The main areas here will be [3]:

- the upper layer (bruises when the upper layer of the fruit "bounces" and collisions with the underlying ones, due to the movement of the vehicle on an uneven road surface);
- the bottom layer (bruises and cracks on the fruit when interacting with the bottom of the container, due to the movement of the vehicle on an uneven road surface);
- fruits near the side walls of the container (bruises when the container is tilted sharply).

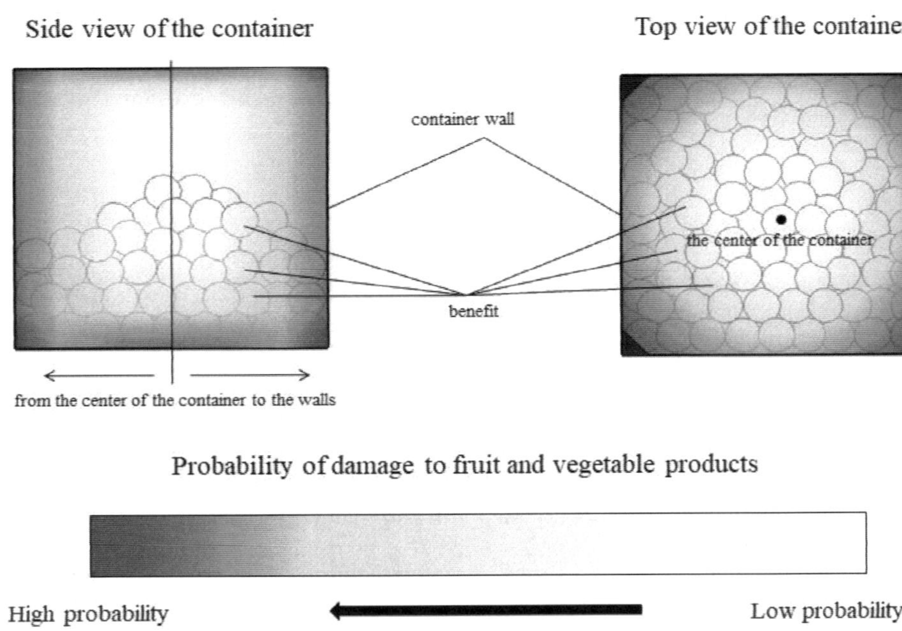

Figure 1. Layout of agricultural cargo in a container and the likelihood of damage.

In addition to the location of individual fruits in the container, the degree of damage will be influenced by a number of factors, such as the speed of the vehicle and its technical condition (which will directly affect the vibration speed of the cargo platform of the vehicle), biological properties of the fruits, their ripeness, etc. [4, 5].

2. Materials and methods

Accelerometers ADXL 320 and ADXL 103 were used to register the maximum amplitudes of the vibration velocity of the cargo platform for each of the generalized coordinates. The quality of the fruits is assessed in accordance with State Standard 34314-2017 "Fresh apples sold in retail. Technical specifications for fresh apples". The varieties of apples "Antonovka ordinary" and "Pepin saffron" were studied.

3. Results and discussion

Studies were carried out, during which the degree of damage to fruits during intrafarm transportation was determined (fruits for transportation were loaded into a container), depending on the vibration velocity of the cargo platform of the vehicle using the example of apples [6].

For this purpose, 200 control apples, marked and clean from damage, were placed in the container. The control fruits were placed on 3 levels (bottom, middle, top), evenly distributed over the entire area. Three containers with experimental fruits were installed in the body of the vehicle, one at a time, respectively, in the front and rear of the platform and in the middle. After that, the body was fully loaded, the fruits were taken to the checkpoint and unloaded. In the process of movement, the data necessary to determine the maximum vibration velocity of the body.

The received data were processed and on their basis the following graphical dependence was built (Figure 2).

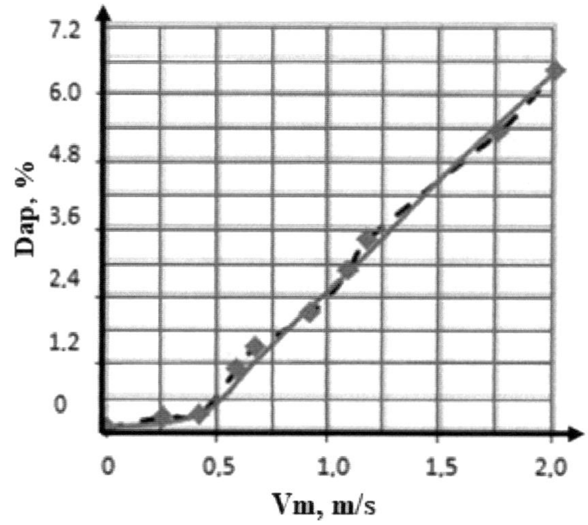

Figure 2. The influence of the vibration velocity of the cargo platform of a vehicle on the degree of damage to agricultural products (apples) transported in a container.

Consequently, at a vibration velocity of the loading platform of a vehicle equal to 0.36 m/s, the damage to the transported agricultural products is insignificant (less than 0.35 %), and at a vibration velocity of 0.7-2.0 m/s it reaches 6.3%.

Based on this, it can be concluded that for the implementation of an effective process of intrafarm transportation of easily damaged agricultural products, a significant change in the design of the containers used is required.

To prevent damage to the top layer of fruits placed in the container for transportation (when the vehicle is driving on an uneven micro-profile of the road), there are locking devices, like the ones in Figure 3 [7].

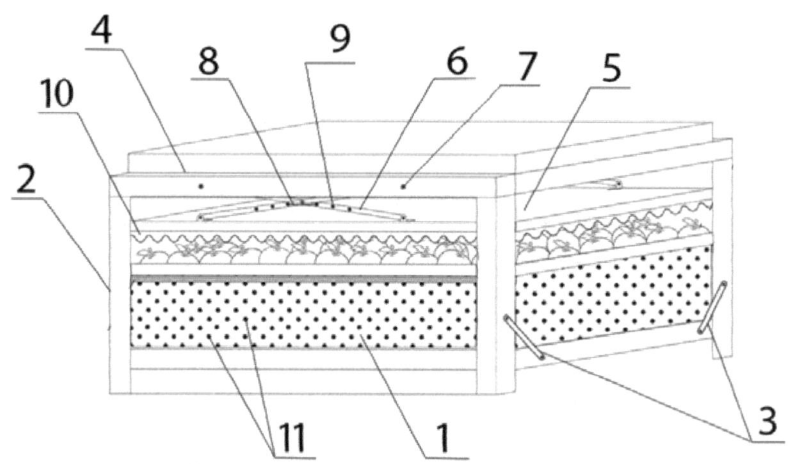

Figure 3. The container for transportation of easily damaged agricultural products (utility model patent No. 176885): 1 - container; 2 - frame; 3 - clamps; 4 - grooves; 5 - pressure cover; 6 - clamping mechanism; 7 - splint pin; 8 - slide bar; 9 - holes for pressure adjustment; 10 - damping material; 11 - through ventilation holes.

In this technical solution, the reduction in the degree of damage to agricultural products during transportation is carried out by using a non-rigid damping material of a wave-like profile installed on the walls, bottom and lid of the container, which, together with a three-position scissor-type clamping mechanism with a locking rod, seal the fruits inside the container, which does not allow them move relative to each other during transportation. At the same time, the grooves made on the upper part of the frame make it possible to stack several such devices, excluding the interaction of fruits in filled containers.

To reduce the degree of damage to easily damaged agricultural products located at the bottom of the container, the following structure has been developed (Figure 4) [8], which includes a bottom and walls, at the bottom of which there are staggered protrusions that form a cell with the bottom plane, while the protrusions have the shape of a hemisphere, are made of a damping material, are hollow and filled with a gas.

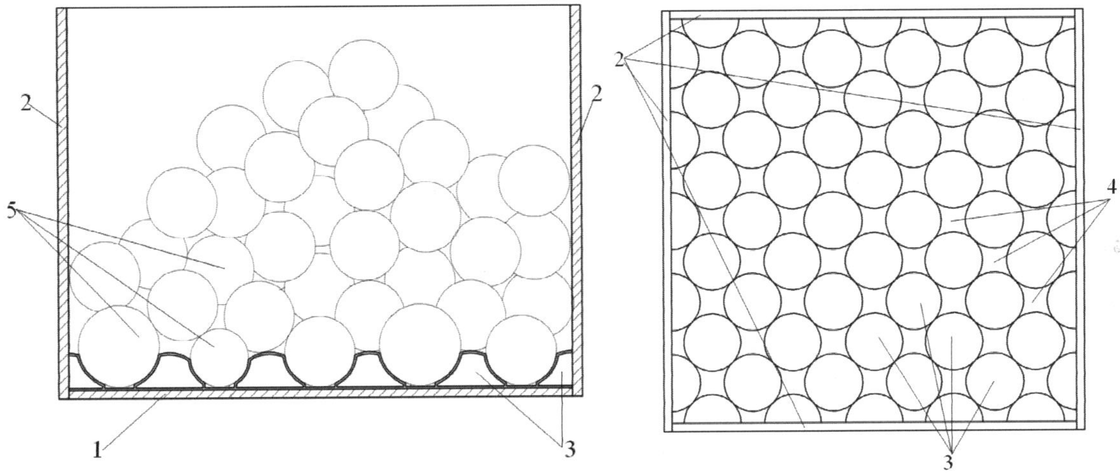

(a) (b)

Figure 4. The container for transportation of easily damaged agricultural products (utility model patent No. 166384): 1 - the bottom of the container; 2 - container walls; 3 - hollow protrusions; 4 - cells; 5 – fruits.

The proposed design is aimed at reducing the degree of damage to transported agricultural products (transportation is carried out in the container). This result is achieved by reducing the dynamic impact on the bottom layer of the fruit in the container.

Consider another device for reducing damage to agricultural products during transportation (the device prevents the transverse movement of cargo inside the container). It (Figure 5) [9] contains a container, including an external fence in the form of rigidly connected frames, to which the enclosing elements are attached. Vertical flexible posts are rigidly fixed on its bottom in a checkerboard pattern, at a distance from each other and from the side surfaces of the container not exceeding 20% of its width. Their height is equal to the height of the container, and the diameter does not exceed 5% of the height.

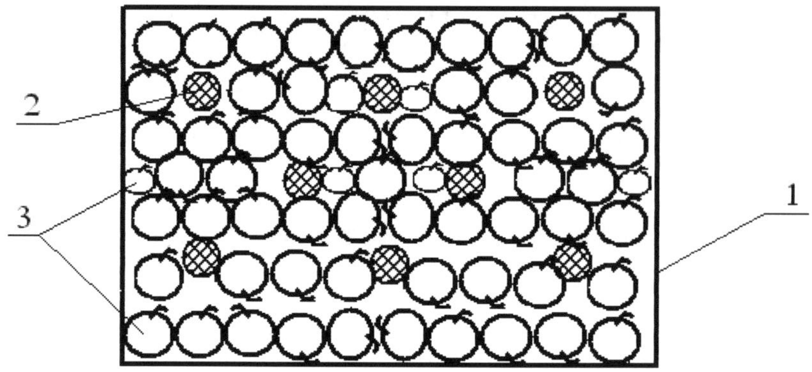

Figure 5. The container for transportation of easily damaged agricultural products (utility model patent No. 2636569): 1 - container; 2 - vertical flexible struts; 3 - fruits.

The degree of damage to agricultural products is reduced by reducing their horizontal movement on uneven roads, braking or accelerating a vehicle, as well as redistributing the load between the vertical layers of fruits located in the area of the struts and the vertical flexible struts themselves.

The technical solutions presented above make it possible to solve local problems [6, 10], and in the aggregate, reduce their number to practically zero.

4. Conclusion

The application of a scientific approach to the organization of intrafarm transportation of easily damaged agricultural products (for example, compliance with the recommended speed limit for a vehicle transporting lightly damaged agricultural products) makes it possible to achieve high efficiency indicators of this process.

The use of the above technical solutions for intrafarm transportation of easily damaged agricultural products allows to preserve its appearance, and, consequently, the cost in the consumer market.

References

[1] James P Yanta and Tong C B 2013 Commercial postharvest handling of potatoes (Solanum Tuberosum) *Regents of the University of Minnesota*

[2] Byshov N V, Borychev S N, Uspensky I A, Yukhin I A, Filyushin O V and Golikov A A 2019 Improving the performance parameters of vehicles for intrafarm transport in the agro-industrial complex *IOP Conf. Series: Earth and Environmental Science* **341(1)** 012145

[3] Byshov N V, Borychev S N, Simdyankin AA, Yukhin I A and Golikov A A 2018 Increasing the safety of agricultural products during its transportation and unloading *ACM Int. Conf. Proc. Series* 176-179

[4] Byshov N V, Borychev S N, Uspensky I A, Rembalovich G K, Kokorev G D, Yukhin I A, Zhukov K A and Bulatov E P 2012 Prospects for improving the operational performance of vehicles for on-farm transportation of fruit and vegetable products *Polythematic Network Electronic Scientific J. of Kuban State Agrarian University* **78** 227-238

[5] Byshov N V, Borychev S N, Uspensky I A, Rembalovich G K, Yukhin I A and Anikin N V 2013 Improving the efficiency of operation of motor vehicles and mobile agricultural machinery in on-farm transportation *Polythematic Network Electronic Scientific J. of Kuban State Agrarian University* **88** 519-529

[6] Yukhin I A *et al* 2014 For the modernization of vehicles for the agro-industrial complex *Int. Conf.: Energy-Efficient and Resource-Saving Technologies and Systems* 181-187

[7] Byshov N V, Borychev S N, Krekov S A, Simdyankin A A, Uspensky I A and Yukhin I A 2018 Device for transporting easily damaged fruit and vegetable products Patent of the Russian Federation 176885

IOP Conf. Series: Earth and Environmental Science **965** (2022) 012048 doi:10.1088/1755-1315/965/1/012048

[8] Shaforostov V A, Yukhin I A, Uspensky I A, Byshov N V, Borychev S N, Anikin N V and Golikov A A 2016 Container for transportation of fruit and vegetable products Patent of the Russian Federation 166384

[9] Yukhin I A *et al* 2017 Device for transporting easily damaged fruit and vegetable products Patent of the Russian Federation 2636569

[10] Rembalovich G K, Byshov N V, Borychev S N, Uspensky I A, Yukhin I A, Golikov A A, Berkasov K S and Pavlov V A *et al* 2013 Improvement of operational and technological indicators of transport and special equipment for potato harvesting *Polythematic Network Electronic Scientific J. of Kuban State Agrarian University* **88** 509-518

IOP Conf. Series: Earth and Environmental Science **965** (2022) 012049 doi:10.1088/1755-1315/965/1/012049

Finite element modeling of the processes of elastic-plastic deformation of reclamation objects of the agro-industrial complex

Yu V Klochkov[1], S D Fomin[2], O V Vakhnina[1], T A Sobolevskaya[1], M Yu Klochkov[3] and A S Andreev[1]

[1] Department of Higher Mathematics, Volgograd State Agrarian University, 26 University Avenue, Volgograd, 400002, Russia
[2] Department of Mechanics, Volgograd State Agrarian University, 26 University Avenue, Volgograd, 400002, Russia
[3] Faculty of Physics, Lomonosov Moscow State University, 1 Leninskie Gory, Moscow, 119991, (City service post office-3), Russia

E-mail: Klotchkov@bk.ru

Annotation. To study the processes of nonlinear deformation of reclamation objects and engineering systems of the agro-industrial complex, taking into account the plastic stage of the used structural material, a finite element model was created based on a volumetric prismatic discretization element with quadrangular bases. The plastic stage of deformation of the applied structural material of the object is taken into account on the basis of the provisions of the deformation theory of plasticity. The plasticity matrix at the (j + 1)-th stage of sequential loading was compiled as a result of applying the operation of differentiating the stress tensor components accumulated over the j previous stages of sequential loading with respect to the strain tensor components. The stiffness matrix and the column of nodal forces of a prismatic discretization element with quadrangular bases at the (j + 1)-th stage of loading were obtained by minimizing the Lagrange functional. Numerical experiments on the calculation of a tubular outlet proved the adequacy of the developed finite element model and the accuracy of calculating the required strength parameters of the studied reclamation objects and other engineering systems of the agro-industrial complex, sufficient for engineering practice.

1. Introduction

Sustainable development of agricultural production in the Volgograd region and in the Lower Volga region as a whole is impossible without the revival of reclamation irrigation systems and the corresponding infrastructure. Achievement of this goal provides for the construction of new and reconstruction of existing reclamation systems and facilities.

To reduce the payback period of such systems, the optimal design of objects and systems for reclamation purposes becomes very important, which is unthinkable without the use of modern numerical methods of calculation [1, 2, 3, 4, 5], which, in particular, includes the finite element method (FEM) [6, 7, 8, 9]. Moreover, the application of this method at the present stage is based on the use as discretization elements, mainly, three-dimensional finite elements [10, 11, 12, 13, 14], as the most versatile, adapted for modeling both thin-walled fragments of reclamation structures (pipelines, tanks,

tanks) and thick-walled massive bodies (dams, retaining walls, joints and others). The most relevant computational algorithms based on the FEM are currently finite element models that take into account the plastic stage of the used structural material [15, 16, 17, 18]. The presented work presents a computational algorithm for studying the strength parameters of reclamation systems and objects, based on the modeling of the latter by three-dimensional prismatic finite elements with quadrangular bases, taking into account the plastic stage of the applied structural material.

2. Materials and methods

2.1. Object geometry

Suppose that the reference surface of the investigated reclamation object can be specified by the radius vector in a rectangular cartesian coordinate system with unit vectors $\vec{i}_k$ $(k = 1, 2, 3)$

$$\vec{R}^0 = x^k(\theta^\rho)\vec{i}_k, \tag{1}$$

where x^k are cartesian coordinates, which are functions of generalized curvilinear coordinates θ^ρ $(\rho = 1, 2)$.

The differentiation procedure (1) by θ^ρ can determine the local basis of point M^0 of the reference surface

$$\vec{a}_\rho = \vec{R},^0_{\theta^\rho}; \qquad \vec{a}_3^0 = \vec{a},^0_1 \times \vec{a},^0_2 / |\vec{a},^0_1 \times \vec{a},^0_2|. \tag{2}$$

For point M^{0t}, spaced from point M at a distance of t along the perpendicular to the reference surface, the radius vector will have the form

$$\vec{R}^{0t} = \vec{R}^0 + t\vec{a}_3^0. \tag{3}$$

To linearize the set task, in the overwhelming majority of cases, the method of successive loading of the studied reclamation object is used. In the process of implementing this method, point M^{0t} will sequentially occupy new positions: after a series of j consecutive loads is M^t and after the $(j+1)$-th loading is M^{*t}. The specified positions can be defined by radius vectors

$$\vec{R}^t = \vec{R}^{0t} + \vec{V}; \quad \vec{R}^{*t} = \vec{R}^t + \vec{W}, \tag{4}$$

where $\vec{V}$ and $\vec{W}$ are the displacement vectors of point M^{0t} after j consecutive loads and after the $(j+1)$-th loading, respectively.

The vectors of bases of points M^{0t}, M^t and M^{*t} can be obtained by calculating the derivatives of (3) and (4) with respect to the generalized curvilinear coordinates θ^ρ and t

$$\vec{g}_k^0 = \vec{R},^{0t}_k; \quad \vec{g}_k = \vec{R},^t_k; \quad \vec{g}_k^* = \vec{R},^{*t}_k \quad (k = 1, 2, 3). \tag{5}$$

Subscripts 1, 2, 3 after the decimal point indicate that differentiation is performed along curvilinear coordinates θ^1, θ^2, and t.

Applying the scalar product operation to (5), one can obtain the covariant components of the metric tensors at points M^{0t}, M^t and M^{*t}, respectively

$$\overset{0}{g}_{mn} = \vec{g}_m^0 \cdot \vec{g}_n^0; \quad g_{mn} = \vec{g}_m \cdot \vec{g}_n; \quad \overset{*}{g}_{mn} = \vec{g}_m^* \cdot \vec{g}_n^*. \tag{6}$$

The deformations of the investigated reclamation object at point M^{0t} after a series of j successive loads and their increments after the $(j+1)$-th loading can be calculated by the formulas proposed by academician Sedov [19]

$$\varepsilon_{mn} = (g_{mn} - \overset{0}{g}_{mn})/2; \quad \Delta\varepsilon_{mn} = (\overset{*}{g}_{mn} - g_{mn})/2. \tag{7}$$

2.2. Physical relationships

When taking into account the plastic stage of deformation of the applied structural material, the deformation theory of plasticity or the theory of plastic flow is usually used. In this work, the deformation theory of plasticity was used, according to the hypotheses of which [20], we can write the following relation

$$\sigma_{mn} = \frac{2}{3}\frac{\sigma_i}{\varepsilon_i}\varepsilon_{mn} - \frac{1}{3}I_1(\varepsilon)g_{mn}\cdot\left(\frac{2}{3}\frac{\sigma_i}{\varepsilon_i} - K\right), \tag{8}$$

where $K = E/(1-2\nu)$; σ_i, ε_i are stress and strain intensities determined by formulas

$$\sigma_i = \sqrt{\frac{3}{2}S_{mn}S^{mn}}; \quad \varepsilon_i = \sqrt{\frac{2}{3}E_{mn}E^{mn}}. \tag{9}$$

The co- and contravariant components of the stress and strain deviators included in (9) are calculated from the following dependences

$$\begin{aligned} S_{mn} &= \sigma_{mn} - I_1(\sigma)g_{mn}/3; \quad S^{mn} = \sigma^{mn} - I_1(\sigma)g^{mn}/3; \\ E_{mn} &= \varepsilon_{mn} - I_1(\varepsilon)g_{mn}/3; \quad E^{mn} = \varepsilon^{mn} - I_1(\varepsilon)g^{mn}/3, \end{aligned} \tag{10}$$

where $I_1(\sigma) = \sigma_{mn}g^{mn} = \sigma^{mn}g_{mn}$ is the first invariant of the stress tensor; $I_1(\varepsilon) = \varepsilon_{mn}g^{mn} = \varepsilon^{mn}g_{mn}$ is the first invariant of the strain tensor.

The increments in stresses at the $(j+1)$-th stage of sequential loading can be expressed in terms of the increments in deformations at the same stage as follows

$$\Delta\sigma_{mn} = \frac{\partial\sigma_{mn}}{\partial\varepsilon_{11}}\Delta\varepsilon_{11} + \frac{\partial\sigma_{mn}}{\partial\varepsilon_{12}}\Delta\varepsilon_{12} + \frac{\partial\sigma_{mn}}{\partial\varepsilon_{13}}\Delta\varepsilon_{13} + \frac{\partial\sigma_{mn}}{\partial\varepsilon_{22}}\Delta\varepsilon_{22} + \frac{\partial\sigma_{mn}}{\partial\varepsilon_{23}}\Delta\varepsilon_{23} + \frac{\partial\sigma_{mn}}{\partial\varepsilon_{33}}\Delta\varepsilon_{33} \tag{11}$$

or in matrix form

$$\{\Delta\sigma_{mn}\} = [C]\{\Delta\varepsilon_{mn}\}; \tag{12}$$

$$\{\Delta\sigma_{mn}\}^T = \{\Delta\sigma_{11}\ \Delta\sigma_{12}\ \Delta\sigma_{13}\ \Delta\sigma_{22}\ \Delta\sigma_{23}\ \Delta\sigma_{33}\}; \quad \{\Delta\varepsilon_{mn}\}^T = \{\Delta\varepsilon_{11}\ 2\Delta\varepsilon_{12}\ 2\Delta\varepsilon_{13}\ \Delta\varepsilon_{22}\ 2\Delta\varepsilon_{23}\ \Delta\varepsilon_{33}\}.$$

The plasticity matrix included in (12) is used to arrange the stiffness matrix of the prismatic discretization element at the $(j+1)$-th stage of loading.

3. Results

3.1. 3D prismatic bin with quadrangular bases

A universal element of discretization of a reclamation object, both thin-walled and in the form of a solid massive body, was a prism with quadrangular bases with nodes located at its vertices. The sought-for nodal unknowns of this bin are the components of the displacement vector of the $(j+1)$-th stage of sequential loading and the first-order partial derivatives of these components. In matrix form, the sought nodal unknowns in the global (superscript G) and local (L) coordinate systems can be represented as follows

$$\underset{1\times 96}{\{W^G\}^T} = \left\{\underset{1\times 32}{\{w^{1G}\}^T}\ \underset{1\times 32}{\{w^{2G}\}^T}\ \underset{1\times 32}{\{w^{3G}\}^T}\right\}; \tag{13}$$

$$\underset{1\times 96}{\{W^L\}^T} = \left\{\underset{1\times 32}{\{w^{1L}\}^T}\ \underset{1\times 32}{\{w^{2L}\}^T}\ \underset{1\times 32}{\{w^{3L}\}^T}\right\}, \tag{14}$$

where

$$\underset{1\times 32}{\{Q^G\}^T} = \{Q^i\ Q^j\ Q^k\ Q^l\ Q^m\ Q^n\ Q^p\ Q^s\ Q^i_{,\theta 1}\dots Q^s_{,\theta 1}\ Q^i_{,\theta 2}\dots Q^s_{,\theta 2}\ Q^i_{,t}\dots Q^s_{,t}\};$$
$$\underset{1\times 32}{\{Q^L\}^T} = \{Q^i\ Q^j\ Q^k\ Q^l\ Q^m\ Q^n\ Q^p\ Q^s\ Q^i_{,\xi}\dots Q^s_{,\xi}\ Q^i_{,\eta}\dots Q^s_{,\eta}\ Q^i_{,\zeta}\dots Q^s_{,\zeta}\}; \quad (15)$$
$$Q = w^1, w^2, w^3; \qquad -1 \le \xi, \eta, \zeta \le 1.$$

Here, superscripts i, j, k, l denote nodes located at the vertices of the lower base, and subscripts m, n, p, s denote nodes located at the vertices of the upper base of the prismatic bin.

The matrix dependence can be established between (13) and (14)

$$\underset{1\times 96}{\{W^L\}} = \underset{96\times 96}{[T]}\ \underset{96\times 1}{\{W^G\}}. \quad (16)$$

Local coordinate system $-1 \le \xi, \eta, \zeta \le 1$ is used to organize the procedure for numerical integration over the volume of a prismatic sampling unit using Gaussian quadrature with a variable number of integration points.

The component of the displacement vector of the $(j+1)$-th stage of the sequential loading of a point located in the inner region of the prismatic sampling unit can be expressed in terms of column (15) by means of the following interpolation dependence

$$Q = \underset{1\times 32}{\{\psi\}^T}\ \underset{32\times 1}{\{Q^L\}}, \quad (17)$$

where $\underset{1\times 32}{\{\psi\}^T}$ is a row matrix whose elements are triadic products of third-order Hermitian polynomials.

Applying the differentiation operation to (17), we can obtain expressions for the derivatives of the components of the displacement vector of the $(j+1)$-th stage of sequential loading

$$Q_{,\theta\rho} = \left\{\{\psi_{,\xi}\}^T \xi_{,\theta\rho} + \{\psi_{,\eta}\}^T \eta_{,\theta\rho} + \{\psi_{,\zeta}\}^T \zeta_{,\theta\rho}\right\}\{Q^L\};$$
$$Q_{,t} = \left\{\{\psi_{,\xi}\}^T \xi_{,t} + \{\psi_{,\eta}\}^T \eta_{,t} + \{\psi_{,\zeta}\}^T \zeta_{,t}\right\}\{Q^L\}. \quad (18)$$

The stiffness matrix and the column of nodal forces of the discretization element in the form of a prism with quadrangular bases at the $(j+1)$-th stage of sequential loading can be arranged using the procedure of minimizing the Lagrange functional

$$\Phi_L = \int_V \{\Delta\varepsilon_{mn}\}^T \left(\{\sigma^{mn}\} + \{\Delta\sigma^{mn}\}\right) dV - \int_F \{W\}^T \left(\{P\} + \{\Delta P\}\right) dF, \quad (19)$$

where $\{\sigma^{mn}\}^T = \{\sigma^{11}\ \sigma^{12}\ \sigma^{13}\ \sigma^{22}\ \sigma^{23}\ \sigma^{33}\}$ are stresses accumulated over j stages of sequential loading; $\{P\}$, $\{\Delta P\}$ is external surface load, also accumulated during j stages of loading and its increment at the $(j+1)$-th stage of sequential loading.

The matrix-row of the components of the displacement vector in (19) at the $(j+1)$-th stage of loading with the help of (17) can be represented by the matrix product

$$\{W\}^T = \{W^L\}^T [A]^T, \quad (20)$$

where $\underset{3\times 96}{[A]} = \begin{bmatrix} \{\psi\}^T & \{0\}^T & \{0\}^T \\ \{0\}^T & \{\psi\}^T & \{0\}^T \\ \{0\}^T & \{0\}^T & \{\psi\}^T \\ 1\times 32 & 1\times 32 & 1\times 32 \end{bmatrix}$.

Expressing the increments of deformations at the $(j+1)$-th stage of loading through the column of unknown unknowns

$$\{\Delta\varepsilon_{mn}\} = [B]\{W^L\}, \quad (21)$$

functional (19) can be represented in a form prepared for the minimization procedure

$$\Phi_L = \{W^G\}^T [T]^T \int_V [B]^T [C][B] \mathrm{d}V [T] \{W^G\} + \{W^G\}^T [T]^T \int_V [B]^T \{\sigma^{mn}\} \mathrm{d}V - \\ - \{W^G\}^T [T]^T \int_F [A]^T \{\Delta P\} \mathrm{d}F - \{W^G\}^T [T]^T \int_F [A]^T \{P\} \mathrm{d}F. \tag{22}$$

Carrying out the procedure of minimization (22) by $\{W^G\}^T$, we can obtain the following matrix relation

$$\underset{96\times96}{[MG]}\underset{96\times1}{\{W^G\}} = \underset{96\times1}{\{f\}} - \underset{96\times1}{\{NR\}}, \tag{23}$$

where $\underset{96\times96}{[MG]} = [T]^T \int_V [B]^T [C][B] \mathrm{d}V [T]$ is stiffness matrix of a prismatic sampling element with quadrangular bases at the $(j+1)$-th stage of sequential loading; $\underset{96\times1}{\{f\}} = [T]^T \int_F [A]^T \{\Delta P\} \mathrm{d}F$ is matrix-column of nodal forces of the discretization element at the $(j+1)$-th stage of loading; $\underset{96\times1}{\{NR\}} = [T]^T \left(\int_V [B]^T \{\sigma^{mn}\} \mathrm{d}V - \int_F [A]^T \{P\} \mathrm{d}F \right)$ is column matrix of the Newton-Raphson correction.

4. Discussion

To assess the accuracy of calculating the required strength parameters of reclamation facilities and engineering systems of the agro-industrial complex, taking into account the plastic deformation of the applied structural material, a test problem was solved to study the stress-strain state of a tubular outlet made of carbon steel, loaded with various types of external surface load.

Various versions of the deformation diagram were used: with linear and nonlinear hardening. In the process of performing numerical experiments, the convergence of the computational process, the correspondence of the obtained values of strength parameters to the values corresponding to the equations of static equilibrium of the outlet were analyzed.

5. Conclusions

The analysis of the obtained numerical values of the sought parameters of the strength of the object under study made it possible to conclude that there is a convergence of the computational process when the grid of prismatic sampling elements with a quadrangular base increases and the number of stages of sequential loading increases.

At the controlled outlet points, the correspondence of the calculated values of the strength parameters to the values of these parameters obtained from the static equilibrium equations was recorded.

On the basis of the performed numerical experiments, the final conclusion was made about the adequacy of the developed finite element model and the accuracy of calculating the required strength parameters of the studied reclamation objects and engineering systems of the agro-industrial complex satisfactory for engineering practice.

6. Acknowledgments

The study was carried out with the financial support of the Russian Foundation for Basic Research and the Administration of the Volgograd Region in the framework of the research project No. 19-41-340005 r_ a.

References

[1] Sultanov L U 2016 Medium kinematics and constitutive equations *Lobachevskii J. of Mathematics* **37(6)** 787-793
[2] Ivanov V N and Shmeleva A A 2020 *Construction Mechanics of Engineering Structures and Structures* **16(1)** 38-44

[3] Zheleznov L P, Kabanov V V and Boiko D V 2018 *Russian Aeronautics* **61(2)** 175-182
[4] Solodovnikov A S and Sheshenin S V 2017 *Moscow University Mechanics Bulletin* **72(4)** 94-100
[5] Mamieva I A 2019 *Construction Mechanics of Engineering Structures and Structures* **15(4)** 299-307
[6] Storozhuk E A, Maksimyuk V A and Chernyshenko I S 2019 *International Applied Mechanics* **55(4)** 504-514
[7] Agapov V 2018 *MATEC Web of Conferences* 04035
[8] Klochkov Y V, Nikolaev A P and Vakhnina O V 2016 *Moscow University Mechanics Bulletin* **71(5)** 114-117
[9] Kirichevsky R V and Skrynnykova A V 2019 *Tomsk State University J. of Mathematics and Mechanics* **57** 26-37
[10] Voronkova G V, Rekunov S S and Dushko O V 2020 *IOP Conf. Series: Materials Science and Engineering* **753** 032088
[11] Klochkov Y V, Nikolaev A P, Kiseleva T A and Marchenko S S 2016 *J. of Machinery Manufacture and Reliability* **45(4)** 328-336
[12] Shlyannikov V N, Zakharov A P and Tumanov A V 2018 *Russian Aeronautics* **61(3)** 340-346
[13] Kayumov R A 2017 *Mechanics of Solids* **52(5)** 575-580
[14] Evdokimov A P and Shikhnabieva T S 2017 *J. of Machinery Manufacture and Reliability* **46(2)** 199-203
[15] Klochkov Y V, Nikolaev A P, Vakhnina O V and Kiseleva T A 2017 Using Langrange multipliers in the triangular element of a nonshallow shell under variable interpolation of displacements *J. of Applied and Industrial Mathematics* **11(4)** 535-544
[16] Klochkov Yu V, Nikolaev A P, Fomin S D, Vakhnina O V and Klochkov M Yu 2019 *Proc. of the Nizhnevolzhsky Agrouniversity Complex: Science and Higher Professional Education* **4(56)** 227-237
[17] Paimushin V N, Badriev I B, Makarov M V and Kholmogorov S A 2019 *J. of Physics: Conf. Series* 032042
[18] Lalin V, Rybakov V and Sergey A 2014 *Applied Mechanics and Materials* **578-579** 858-863
[19] Sedov L I 1976 *Mechanics of a Continuous Medium* (Moscow: Nauka)
[20] Malinin N N 1975 *Applied-Theory of Plasticity and Creep* (Moscow: Mechanical Engineering)

IOP Conf. Series: Earth and Environmental Science **965** (2022) 012050 doi:10.1088/1755-1315/965/1/012050

A dynamic algorithm for stabilization of the working body of a mobile robot weeding for the future of agriculture

A G Ivanov[1], V V Zhoga[2], N S Vorob'yeva[1] and V V Dyashkin-Titov[1]

[1] Volgograd State Agrarian University, 26 University Avenue, Volgograd, 400002, Russia
[2] Volgograd State Technical University, 28 Lenin Avenue, Volgograd, 400005, Russia

E-mail: leha_2106@list.ru

Abstract. The article considers the development of a control algorithm for the actuating links of a manipulator with a parallel-serial structure. The authors suggest a two-stage procedure for resolving the problem of controlling the robot operating body motion into a set neighborhood of the final state for a set time. The first stage covers the process of solving the problem of defining the manipulator generalized coordinates at the set coordinates of the operating body. Then the problem of determining the laws of the formation of reference influences for the actuators is solved, which ensures the movement of the working body in the vicinity of a given point. The technique for the synthesis of dynamic algorithms for stabilizing the effector relative to a given position and implementing the program paths is based on generating the control signals of actuators by solving the inverse dynamic problem using a control signal generation algorithm provided that deviations from the current program path values are the solutions of a second-order differential equation. The laws of change in control voltages are obtained. The results of numerical modeling are presented, confirming the efficiency of the proposed algorithm on the example of a specific manipulator.

1. Introduction

When cultivating vegetables and cucurbits, one of the time-consuming and energy-consuming tillage operations is weeding [1, 2], commonly carried out by cultivation. The effectiveness of mechanical weed control largely depends on the weeding quality. Manual weeding is still practiced for raising vegetables and cucurbits in open ground. This is also explained by the fact that in case of the mechanical destruction of weeds the operating bodies of the cultivator ("arms") process only the aisles. To solve the problem of complete weeding, both in row spacings and in rows, a weeding robot was developed (Figure 1) for the target weeding.

The weeding robot consists of a frame 1, steered wheels 2, a control and navigation system with instrumentation 3, a power supply system 4 and a computer vision sensor 7 [3].

The motion mechanism of the operating body of the weeding robot (Figure 2) is a flat mechanism of a parallel-serial structure with three degrees of freedom [4]. To move the operating body 8 in the horizontal plane, three linear actuators 5 are used, and the linear actuator 6 is used to move the operating body in a vertical plane. The operating body is rotated by means of the electric motor 9.

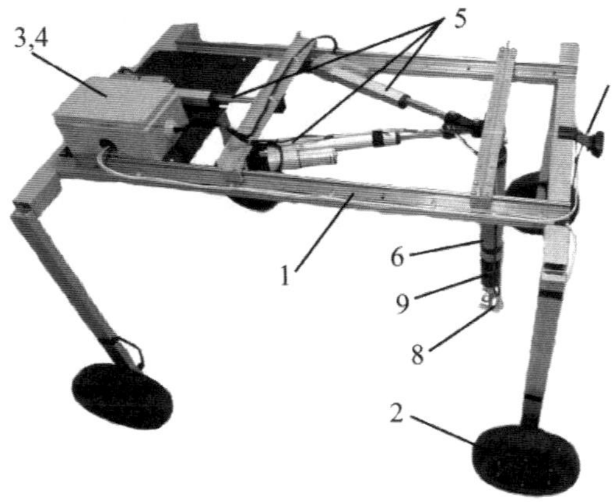

Figure 1. Mobile robot weeding.

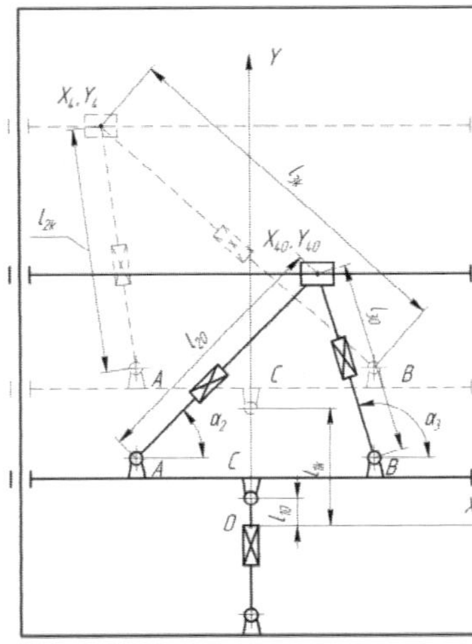

Figure 2. Design diagram of the mechanism for moving the working body of the robotic - weeding machine. View from above.

The robot operates as follows. With the help of the control and navigation system, the weeding robot enters the plant beds. In automatic mode, using a computer vision system, the coordinates of the location of weeds in a row are determined. The cutting tool of the implement moves to the weed. When the implement cutter is above the weed, the linear drive moves the implement vertically downward, which cuts the weed and loosens the soil next to it.

The sensor system consists of two main components: one is rigidly attached to actuators, and the second, which performs most of the functions of global navigation and spatial orientation, as well as computer vision, is connected to the top-level control unit.

The feedback on the running drives implements odometric navigation, and the feedback of the actuators provides accurate positioning of the manipulator. The global navigation system includes a GPS receiver, the data from which are combined with the readings of odometric sensors on wheels.

1.1 Problem statement

For the implementation of the described technology, it is necessary to develop algorithms for the formation of a system to control the motion of the weeding robot operating body. The control system must set the motion of the operating body to a given point in the operating area with a required accuracy. The motion of the manipulator in the horizontal plane is considered. The generalized manipulator coordinates are the lengths of the actuating links $q_1 = l_1(t)$, $q_2 = l_2(t)$, $q_3 = l_3(t)$. The motion of the operating body from a known initial position with coordinates (x_{40}, y_{40}) to a predetermined final position (x_4, y_4) is provided by changing the lengths $l_i(t), (i = 1 \div 3)$ of the manipulator actuating links. The problem is solved in two stages. First, the positioning problem is solved: for the given final coordinates (x_4, y_4) of the attachment point of the vertical actuator (Figure 2), it is required to obtain the generalized coordinates of the manipulator l_i^{op}, $(i = 1 \div 3)$. Since the number of the Cartesian coordinates of the actuator attachment point is two, and the number of generalized coordinates of the manipulator is three, the maneuverability of the manipulator is one [5]. That is, the final position of the actuator with the operating body corresponds to an infinite number of manipulator configurations. Thus, the solution of this problem is of an optimization nature.

At the second stage, it is necessary to determine the laws of change in the control signals $u_i(t)$, $(i = 1 \div 3)$ of DC drive motors when moving the attachment point of the working body from the initial to the vicinity of the designated point. Acceleration of change of generalized coordinates is taken as controlled parameters $\ddot{l}_i(t),\ (i=1\div 3)$.

2. Materials and methods

2.1. Manipulator positioning

The design of the manipulator stipulates for the holonomic constraints between the coordinates of the point $O_4(x_4, y_4)$, the lengths of the actuating links and the manipulator geometric parameters

$$l_2=\sqrt{(y_4-l_1)^2+(AC+x_4)^2},\qquad l_3=\sqrt{(y_4-l_1)^2+(BC-x_4)^2} \tag{1}$$

The configuration of the manipulator is determined from the minimum condition of the quadratic function - the criterion of generalized energy [6]

$$\Phi=\sum_{i=1}^{3} c_i\cdot(l_{ik}-l_{i0})^2$$

with inequality constraints

$$l_{i\min}\le l_i^{op}\le l_{i\max}, \tag{2}$$

where c_1, c_2, c_3 are weighting factors, the values of which are proportional to the loads on the actuating links; $l_{i\min}, l_{i\max}$ are the minimum and maximum permissible values of the lengths of the manipulator actuating links (Figure 2); l_i^{op} is an optimal length.

The Lagrange objective function has the form [7]

$$\Phi^*=\sum_{i=1}^{3} c_i(l_{ik}-l_{i0})^2+\lambda_{2i-1}\left(\varphi_{2i-1}^2+l_i-l_{2i-1,\max}\right)+\lambda_{2i}\left(\varphi_{2i}^2+l_{i\min}-l_i\right), \tag{3}$$

where φ_i - are auxiliary functions; λ_i - are Lagrange multipliers.

The necessary conditions for the extremum of function (3) are written in the form

$$\frac{\partial\Phi^*}{\partial l_1}=c_1\frac{\partial(l_{1i}-l_{10})^2}{\partial l_1}+c_2\frac{\partial(l_{2i}-l_{20})^2}{\partial l_1}+c_3\frac{\partial(l_{3i}-l_{30})^2}{\partial l_1}+\lambda_1-\lambda_2=0, \tag{4}$$

$$\begin{gathered}\lambda_{2i-1}=0,\ if\ l_i^{op}<l_{imax};\ \lambda_{2i}=0,\ if\ l_i^{op}>l_{imin},\\ \lambda_{2i-1}>0,\ if\ l_i^{op}=l_{imax};\ \lambda_{2i}>0,\ if\ l_i^{op}=l_{imin}\end{gathered} \tag{5}$$

Since the Lagrange function (3) is convex, and the multipliers $\lambda_{2i-1}\ge 0,\ \lambda_{2i}\ge 0$, the necessary conditions (4), (5) are sufficient.

2.2. Stabilization of the manipulator given position

Since the manipulator has great rigidity, the Cartesian coordinates of the actuator attachment points with the operating body are determined through the lengths of the actuating links with a high accuracy [8]

$$x_4=\frac{l_3^2-l_2^2}{4AC},\quad y_4=l_1+\sqrt{l_2^2-\left(\frac{l_3^2-l_2^2}{4AC}\right)^2+\frac{l_3^2-l_2^2}{2}-AC^2} \tag{6}$$

Below is a method for synthesizing a control system algorithm for the laws of change in the lengths of executive links.

The requirements for control accuracy are set only to the end point, without imposing restrictions on the path of the characteristic point, which may change during the motion of the operating body due to external disturbances. In this case, the solution of the problem is reduced to determining the laws of formation of the setting actions $u_i(t)$ for the executive drives, ensuring the motion of the operating body in the neighborhood of a given point on the plane to which the lengths of the actuating links $l_i(\tau) = l_i^{op}$, $i = 1-3$ correspond. Assume that on the motion path the deviations from the final state of the actuating links $\Delta l_i(t) = l_i^{op} - l_i(t)$ change in accordance with the solution of the differential equation

$$\Delta \ddot{l}_i(t) + a_{1i}\Delta \dot{l}_i(t) + a_{2i}\Delta l_i(t) = 0, \tag{7}$$

where a_{1i}, a_{2i} - are constant positive numbers.

The following differential equations correspond to the laws of the formation of the setting actions $l_k(t)$ of the actuators determined by equations (7)

$$\ddot{l}_i(t) + a_{i1}\dot{l}_i(t) + a_{i2}l_i(t) = a_{i2}l_i^{op}, \tag{8}$$

under the initial conditions describing the manipulator state at $t_0 = 0$

$$l_i(0) = l_{i0}, \quad \dot{l}_i(0) = \dot{l}_{i0}. \tag{9}$$

A general solution for the equation (8) is a total of a general solution for the homogeneous equation and a partial solution of a heterogeneous equation; it has the form

$$l_i(t) = C_{1i}e^{p_{1i}t} + C_{2i}e^{p_{2i}t} + l_i^{op} \tag{10}$$

where the roots p_{1i}, p_{2i} - are the roots of the corresponding characteristic homogeneous equation.

The integration constants are obtained from (8), taking into account (9)

$$C_{1i} = \frac{\Delta l_i(0)p_{2i} + \dot{l}_{i0}}{(p_{1i} - p_{2i})}, \quad C_{2k} = -\frac{\Delta l_i(0)p_{1i} + \dot{l}_{i0}}{(p_{1i} - p_{2i})}, \quad \Delta l_i(0) = l_i^{op} - l_i(0) \quad , \tag{11}$$

The coefficients of equations (8) are determined using the roots of the characteristic equation

$$a_{1i} = -(p_{1i} + p_{2i}), \; a_{2i} = p_{1i}p_{2i}$$

To satisfy the requirement $l_i(t) \to l_i^{op}$ for $t \to \infty$, the difference of the characteristic equation roots is negative.

The law of changing the lengths of the actuating links (10) is programmed. To construct a feedback control law, it is necessary to calculate the required acceleration from the current values of the generalized coordinates

$$\ddot{l}_i(t) = a_{i2}l_i^{op} - a_{i1}\dot{l}_i(t) - a_{i2}l_i(t), \tag{12}$$

Thus, in the expressions (11), it is necessary to consider the current values of the generalized coordinates as initial.

IOP Conf. Series: Earth and Environmental Science **965** (2022) 012050 doi:10.1088/1755-1315/965/1/012050

3. Results and discussion

3.1. Complete model of manipulator dynamics

The generalized coordinates of the manipulator are the lengths of the executive links $q_1 = l_1$, $q_2 = l_2$, $q_3 = l_3$, and the angles of rotation of the executive links relative to the movable coordinate axes $q_4 = \alpha_2$, $q_5 = \alpha_3$. The three coordinates are independent. The coordinates $q_2 = l_2$, $q_3 = l_3$, $q_4 = \alpha_2$, $q_5 = \alpha_3$ are superimposed with holonomic connections.

$$f_1 = l_2 \cdot \cos\alpha_2 - l_3 \cdot \cos\alpha_3 - AB = 0 \tag{13}$$

$$f_2 = l_2 \cdot \sin\alpha_2 - l_3 \cdot \sin\alpha_3 = 0 \tag{14}$$

We assume that the mechanism of movement of the working body consists of seven masses: the mass of the rod of the first executive link together with the mass of the slider m_1; two bodies of actuators weighing m_{21}, m_{31}; two rods weighing m_{22} and m_{32} ; the mass of the articulated slider m_4 together with the working body and the mass of the slider m_5.

Since the weeding robot is a multi-mass mechanism, the dynamics of its movements is described by a system of nonlinear differential equations, which are formed using the Lagrange equations with indeterminate multipliers with additional holonomic connections [9].

$$\frac{d}{dt}\left(\frac{\partial T}{\partial \dot{q}_i}\right) - \frac{\partial T}{\partial q_i} = Q_i + \lambda_1 \frac{df_1}{dq_i} + \lambda_2 \frac{df_2}{dq_i}, \quad i = 1 \div 5 \tag{15}$$

The total kinetic energy T of the manipulator has the form

$$T = T_1 + T_2 + T_3 + T_4 + T_5$$

where T_1 - kinetic energy of the 1st link; T_2 - total kinetic energy of the 2nd link; T_3 - total kinetic energy of the 3nd link; T_4 - kinetic energy of the 4th link; T_5 - kinetic energy of the 5th link.

The first link of the mechanism performs a rectilinear forward movement. The bodies of the second and third actuators perform complex movements: portable translational and relative rotational around a fixed axis. The rods of these actuators also perform complex movements: portable translational and relative plane-parallel movement. The movement of the articulated slider with a mass of m_4 consists of the sum of translational movements, and the slider with a mass of m_5 performs a translational movement.

Using expressions for kinetic energy and equations of holonomic constraints (13, 14), from (15) we obtain five differential equations describing the dynamics of the robot weeding mechanism

$$\begin{aligned} F_1 = {} & M\ddot{l}_1 + (m_{22}+m_4+m_5)\ddot{l}_2 \cdot \sin\alpha_2 + m_{32}\ddot{l}_3 \cdot \sin\alpha_3 + \ddot{\alpha}_2\left[(m_{22}+m_4+m_5)l_2 + 0.5(m_{21}-m_{22})l_{2\min}\right]\cos\alpha_2 + \\ & + \ddot{\alpha}_3\left[m_{32}l_3 + 0.5(m_{31}-m_{32})l_{3\min}\right]\cos\alpha_3 + (m_{22}+m_4+m_5)\cdot(2\cdot\dot{l}_2\cdot\cos\alpha_2\cdot\dot{\alpha}_2 - l_2\cdot\sin\alpha_2\cdot\dot{\alpha}_2^2) + \\ & + m_{32}\left(2\cdot\dot{l}_3\cdot\cos\alpha_3\cdot\dot{\alpha}_3 - \cdot l_3\sin\alpha_3\cdot\dot{\alpha}_3^2\right) - \\ & - 0.5(m_{21}-m_{22})l_{2\min}\sin\alpha_2\cdot\dot{\alpha}_2^2 - 0.5(m_{31}-m_{32})l_{3\min}\sin\alpha_3\cdot\dot{\alpha}_3^2, \end{aligned} \tag{16}$$

$$M = (m_1 + m_{21} + m_{22} + m_{31} + m_{32} + m_4 + m_5),$$

$$\begin{aligned} & (m_{22}+m_4)\cdot\left(\ddot{l}_2 + \ddot{l}_1\cdot\sin\alpha_2\right) - m_{22}\cdot\dot{\alpha}_2^2\cdot\left(l_2 - \frac{l_{2\min}}{2}\right) - m_4\cdot l_2\cdot\dot{\alpha}_2^2 + m_5\cdot\sin\alpha_2\cdot(2\dot{l}_2\dot{\alpha}_2\cos\alpha_2 - \\ & - l_2\cdot\sin\alpha_2\cdot\dot{\alpha}_2^2 + \ddot{l}_1 + \ddot{l}_2\cdot\sin\alpha_2 + l_2\cdot\cos\alpha_2\cdot\ddot{\alpha}_2) = F_2 + \lambda_1\cdot\cos\alpha_2 + \lambda_2\cdot\sin\alpha_2, \end{aligned} \tag{17}$$

$$m_{32}\left(\ddot{l}_3+\ddot{l}_1\cdot\sin\alpha_3-\dot{\alpha}_3^2\left(l_3-\frac{l_{3\min}}{2}\right)\right)=F_3-\lambda_1\cdot\cos\alpha_3-\lambda_2\cdot\sin\alpha_3, \tag{18}$$

$$\begin{gathered}m_{22}\left(l_2-\frac{l_{2\min}}{2}\right)\left[\ddot{\alpha}_2\cdot\left(l_2-\frac{l_{2\min}}{2}\right)+\ddot{l}_1\cos\alpha_2+2\dot{l}_2\cdot\dot{\alpha}_2\cdot\right]+\ddot{\alpha}_2\cdot\left(I_{z21}+I_{z22}\right)+m_4(l_2^2\cdot\ddot{\alpha}_2+\\+2\cdot l_2\cdot\dot{l}_2\cdot\dot{\alpha}_2+l_2\cdot\ddot{l}_1\cdot\cos\alpha_2)+m_{21}l_{2\min}\left(\frac{l_{2\min}\cdot\ddot{\alpha}_2}{4}+\frac{\ddot{l}_1\cdot\cos\alpha_2}{2}\right)+m_5\cdot l_2\cdot\cos\alpha_2(2\cdot\dot{l}_2\cdot\\\cos\alpha_2\cdot\dot{\alpha}_2-l_2\cdot\sin\alpha_2\cdot\dot{\alpha}_2^2+\ddot{l}_1+\ddot{l}_2\cdot\sin\alpha_2+l_2\cdot\cos\alpha_2\cdot\ddot{\alpha}_2)=-\lambda_1\cdot l_2\cdot\sin\alpha_2+\lambda_2\cdot l_2\cdot\cos\alpha_2,\end{gathered} \tag{19}$$

$$\begin{gathered}m_{32}\cdot\left(l_3-\frac{l_{3\min}}{2}\right)\left[\cdot\ddot{\alpha}_3\cdot\left(l_3-\frac{l_{3\min}}{2}\right)+\ddot{l}_1\cdot\cos\alpha_3+2\cdot\dot{l}_3\cdot\dot{\alpha}_3\right]+\ddot{\alpha}_3\cdot\left(I_{z31}+I_{z32}\right)+\\+m_{31}l_{3\min}\left(\frac{l_{3\min}\cdot\ddot{\alpha}_3}{4}+\frac{\ddot{l}_1\cdot\cos\alpha_3}{2}\right)=\lambda_1\cdot l_3\cdot\sin\alpha_3-\lambda_2\cdot l_3\cdot\cos\alpha_3,\end{gathered} \tag{20}$$

where: α_2,α_3 - angles of rotation of actuators 2 and 3 relative to the mounting axis; $l_{2\min},l_{3\min}$ - minimum actuator lengths; l_2,l_3 - current actuator lengths; I_{z21},I_{z31} - the main central moments of inertia of the masses of the 2 and 3 housing of the actuators; I_{z22},I_{z32} - the main central moments of inertia of masses 2 and 3 of the actuator rods; F_1,F_2,F_3 - forces required to perform programmed movement; λ_1,λ_2 - Lagrange multipliers.

With the known laws of displacement of the working body of the manipulator $l_i(t)$, from expressions (13, 14, 16-20), the driving forces $F_i(t)$, $i=1-3$ are found, which ensure the execution of the programmed movement, as well as the values of dynamic loads in kinematic pairs.

The solution to this problem is associated with the integration of differential equations. However, due to the inertia of the mechanisms of the gearbox and the ball screw pair of actuators, the actual displacements may differ from the specified.

To form the control forces of the executive drives of the weeding robot, we use the acceleration control algorithm. The control circuits of the executive motors are synthesized in the process of constructing the control algorithm. The synthesis of the stabilization algorithm is formed according to the equations of the dynamics of the manipulator (16-20).

Expressing the indefinite factors $\lambda_1(q_s,\dot{q}_s,\ddot{q}_s)$ and $\lambda_2(q_s,\dot{q}_s,\ddot{q}_s)$ from equations (19), (20)

$$\begin{aligned}\lambda_1=&\frac{\ddot{l}_1\cos\alpha_2\cos\alpha_3\left[l_2l_3\left(m_4+m_5+m_{22}+m_{32}\right)+0.5l_2l_{3\min}\left(m_{31}-m_{32}\right)+0.5l_3l_{2\min}\left(m_{21}-m_{22}\right)\right]}{l_2l_3\sin\left(\alpha_3-\alpha_2\right)}+\\&+\frac{\ddot{l}_2l_2l_3m_5\cos\alpha_2\cos\alpha_3\sin\alpha_2+\ddot{\alpha}_3l_2\cos\alpha_2\left(I_{z31}+I_{z32}-l_3l_{3\min}m_{32}+0.25l_{3\min}^2\left(m_{31}+m_{32}\right)+l_3^2m_{32}\right)}{l_2l_3\sin\left(\alpha_3-\alpha_2\right)}+\\&+\frac{\ddot{\alpha}_2\left[l_3\cos\alpha_3\left(I_{z21}+I_{z22}-l_2l_{2\min}m_{22}\right)+l_2^2l_3\cos\alpha_3\left(m_4+m_{22}+m_5\cos^2\alpha_2\right)+0.25l_3l_{2\min}^2\cos\alpha_3\left(m_{21}+m_{22}\right)\right]}{l_2l_3\sin\left(\alpha_3-\alpha_2\right)}+\\&+\frac{\dot{l}_2\dot{\alpha}_2\cos\alpha_3\left[2l_2l_3\left(m_4+m_{22}+m_5\cos^2\alpha_2\right)-l_3l_{2\min}m_{22}\right]+\dot{l}_3\dot{\alpha}_3l_2m_{32}\cos\alpha_2\left(2l_3-l_{3\min}\right)-\dot{\alpha}_2^2l_2^2l_3m_5\cos\alpha_2\cos\alpha_3\sin\alpha_2}{l_2l_3\sin\left(\alpha_3-\alpha_2\right)},\end{aligned} \tag{21}$$

IOP Conf. Series: Earth and Environmental Science **965** (2022) 012050 doi:10.1088/1755-1315/965/1/012050

$$\lambda_2 = \frac{\ddot{l}_1\left[l_3\cos\alpha_2\sin^2\alpha_3\left(2l_2(m_4+m_5+m_{22})+l_{2\min}\left(m_{21}-m_{22}\right)\right)+0.5l_2\sin 2\alpha_3\sin\alpha_2(l_{3\min}\left(m_{31}-m_{32}\right)+2l_3m_{32})\right]}{l_2l_3\left[\cos\alpha_2-\cos\left(\alpha_2-2\alpha_3\right)\right]}+$$
$$+\frac{\ddot{l}_2l_2l_3m_5\sin 2\alpha_2\sin^2\alpha_3+2\ddot{\alpha}_2l_3\sin^2\alpha_3\left[I_{z21}+I_{z22}+l_2^2\left(m_4+m_{22}+m_5\cos^2\alpha_2\right)+0.25l_{2\min}^2\left(m_{21}+m_{22}\right)-l_2l_{2\min}m_{22}\right]}{l_2l_3\left[\cos\alpha_2-\cos\left(\alpha_2-2\alpha_3\right)\right]}+$$
$$+\frac{2\ddot{\alpha}_3l_2\sin\alpha_2\sin\alpha_3\left[I_{z31}+I_{z32}+l_3m_{32}\left(l_3-l_{3\min}\right)+0.25l_{3\min}^2\left(m_{31}+m_{32}\right)\right]}{l_2l_3\left[\cos\alpha_2-\cos\left(\alpha_2-2\alpha_3\right)\right]}+ \tag{22}$$
$$+\frac{4\dot{l}_2\dot{\alpha}_2l_3\sin^2\alpha_3\left[l_2\left(m_4+m_{22}+m_5\cos^2\alpha_2\right)-0.5l_{2\min}m_{22}\right]+2\dot{l}_3\dot{\alpha}_3l_2m_{32}\sin\alpha_2\sin\alpha_3\left(2l_3-l_{3\min}\right)-\dot{\alpha}_2^2l_2^2l_3m_5\sin 2\alpha_2\sin^2\alpha_3}{l_2l_3\left[\cos\alpha_2-\cos\left(\alpha_2-2\alpha_3\right)\right]}$$

and substituting in (18, 19), we find expressions for the efforts F_i depending on the kinematic variables of the manipulator q_s and their speeds $\dot{q}_s$.

To calculate the control forces (21), it is necessary to calculate the current values of the angles of rotation of the actuators, their speed and acceleration

$$\cos\alpha_2 = \frac{l_2^2-l_3^2+AB^2}{2l_2AB},\quad \cos\alpha_3 = \frac{l_2^2-l_3^2-AB^2}{2l_3AB},\quad \sin\alpha_2\cdot\dot{\alpha}_2 = \left(\frac{\cos\alpha_2}{l_2}-\frac{1}{AB}\right)\dot{l}_2+\frac{l_3\dot{l}_3}{l_2AB},$$

$$\sin\alpha_2\cdot\ddot{\alpha}_2 = \left(\frac{\cos\alpha_2}{l_2}-\frac{1}{AB}\right)\ddot{l}_2+\frac{\ddot{l}_3l_3}{l_2AB}-\frac{\dot{l}_2^2\cos\alpha_2}{l_2^2}+\frac{\dot{l}_3^2}{l_2AB}-\frac{\dot{l}_2\dot{l}_3l_3}{l_2^2AB}-\frac{\dot{\alpha}_2\dot{l}_2}{l_2}\sin\alpha_2-\cos\alpha_2\dot{\alpha}_2^2 \tag{23}$$

$$\sin\alpha_3\cdot\dot{\alpha}_3 = -\frac{l_2\dot{l}_2}{l_3AB}+\left(\frac{\cos\alpha_3}{l_3}+\frac{1}{AB}\right)\dot{l}_3,$$

$$\sin\alpha_3\cdot\ddot{\alpha}_3 = -\frac{\ddot{l}_2l_2}{l_3AB}+\left(\frac{\cos\alpha_3}{l_3}+\frac{1}{AB}\right)\ddot{l}_3-\frac{\dot{l}_2^2}{l_3AB}-\frac{\dot{l}_3^2\cos\alpha_3}{l_3^2}+\frac{\dot{l}_2\dot{l}_3l_2}{l_3^2AB}-\frac{\dot{\alpha}_3\dot{l}_3}{l_3}\sin\alpha_3-\cos\alpha_3\dot{\alpha}_3^2.$$

The required laws of change in the control efforts are found after the substitution of expressions (12) into the formulas for the control efforts F_i.

The static characteristic of a DC drive motor with independent excitation has the form

$$F_i = ru_i - s\dot{l}_i, (i=1-3), \tag{24}$$

where r, s – coefficients depending on the parameters of the engine and the mechanical transfer; $u_i(t)$ – control voltage.

The dynamic algorithm for stabilization of the grip of the manipulator is based on the implementation of the expression ratios for the control voltages

$$u_i = \frac{F_i(q_s,\dot{q}_s)}{r}+\frac{s}{r}\dot{l}_i, (i=1-3). \tag{25}$$

Figure 3 shows a block diagram of the stabilization algorithm.

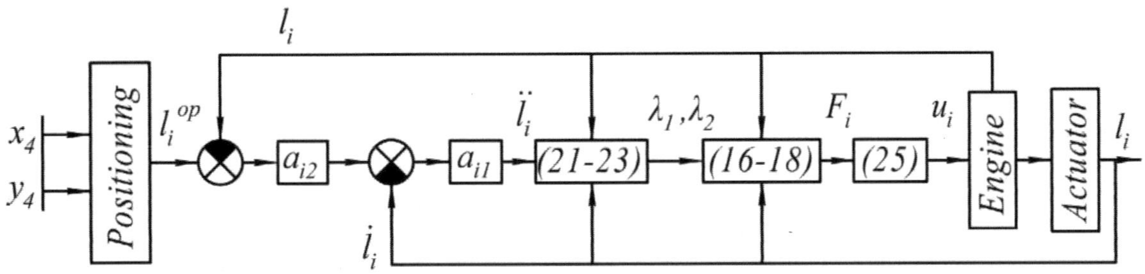

Figure 3. Block diagram of the dynamic algorithm for stabilizing the grip of the manipulator in a given position.

3.2. Numerical example of the control algorithm implementation

At the initial moment of time, the working body of the manipulator was located at the point with the coordinates $x_{40} = 60mm,\ y_{40} = 300mm$. Distance between the axes of the attachment points of the actuators $AC = BC = 140mm$. These coordinates correspond to the lengths of the executive links $l_{10} = 0mm,\ l_{20} = 360mm,\ l_{30} = 310mm$. By setting the coordinates of the end point $x_4 = -120mm,\ y_4 = 420mm$, as a result of applying the positioning algorithm, we obtained $l_1^{op} = l_1^0 = 93.5mm,\ l_2^{op} = l_2^0 = 327mm,\ l_3^{op} = l_3^0 = 417.4mm$.

In Figure 4 and Figure 5 show the results of calculations when moving the working body from a position determined by coordinates $x_{40} = 60mm,\ y_{40} = 300mm$ to a point with coordinates $x_4 = -120mm,\ y_4 = 420mm$ according to the laws (10): initial speed $\dot{l}_{i0} = 0$;

$$l_i(t) = l_i^0 + \frac{\Delta l_{i0}}{p_{1i} - p_{2i}}\left(p_{2i}e^{p_{1i}t} - p_{1i}e^{p_{2i}t}\right),\quad \Delta l_i(t) = l_i^0 - l_{i0},\ p_{1i} = -12,\ p_{2i} = -20$$

$\Delta l_{10} = -93.5mm,\ \Delta l_{20} = -33mm,\ \Delta l_{30} = 107.4mm.$

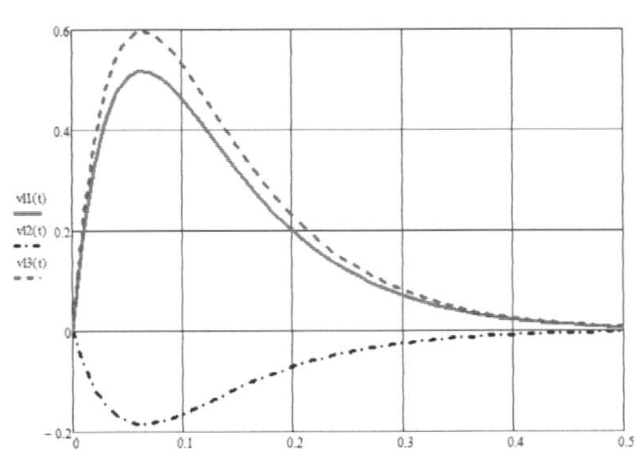

Figure 4. Laws of change of speeds of executive drives.

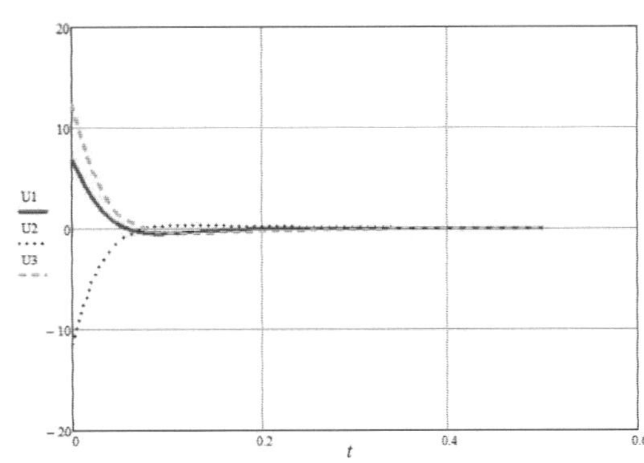

Figure 5. Laws of change of control voltages of executive drives.

From the graphs (Figure 4, 5), it can be seen that within $\tau = 0.5s$ the gripper device is stabilized in the specified position. Deviations from the specified position are not more than 0.1%, and the speeds and accelerations are almost zero. The trajectory of the point differs slightly from the straight line. If the generalized coordinates of the manipulator deviate from the program trajectory, it does not make sense to stabilize them on the original program trajectory, since at each current point a different program trajectory is constructed that satisfies the same final conditions [10].

4. Conclusion

The problem of positioning the manipulator and the problem of weak terminal control is solved. The program trajectory is corrected in such a way that at each moment of time it passes through the current position of the working body. Over time, the working body falls into the final area of the target point. The algorithm for generating control voltages is formed taking into account the kinematic parameters of the manipulator for all degrees of freedom. Mathematical modeling has shown that the proposed algorithm allows you to implement the movement of the gripper along the program trajectory with an error of no more than 1%. The algorithm also functions well in the presence of initial perturbations.

References

[1] McAllister W, Osipychev D, Davis A and Chowdhary G 2019 Agbots: Weeding a field with a team of autonomous robots *Computers and Electronics in Agriculture* **163** 104827

[2] Fawakherji M, Youssef A, Bloisi D, Pretto A and Nardi D 2019 Crop and weeds classification for precision agriculture using context-independent pixel-wise segmentation *3rd IEEE Int. Conf. on Robotic Computing* 8675654 pp 146-152

[3] Ovchinnikov A S, Bocharnikov V S, Vorob`yeva N S, Dyashkin A V, Ivanov A G, Nesmiyanov I A, Zhoga V V and Dyashkin-Titov V V 2019 Robotic weeding with the function of fertigation Patent of the Russian Federation 2694588 MPK A01V 39/18 Bulletin **20**

[4] Bushuev V V and Hol'shev I G 2001 The mechanisms of parallel structure in mechanical engineering *STIN* **1** 3-8

[5] Kolovskii M Z and Sloushch A V 1998 *Foundations of Industrial Robot Dynamics* (Moscow: Nauka) p 240

[6] Kozlov V V, Makary`chev V P, Timofeev A V and Yurevich E I 1984 *The Dynamics of Robot Control* (Moscow: Nauka) p 336

[7] Banday B D 1988 Basic Optimisation Methods (London: Edward Arnold (Publishers) Limited) p 128

[8] Korendesev A I, Salamandra B L and Tyves L I 2006 Theoretical Foundations of Robotics (Moscow: Nauka) vol 1 p 383

[9] Ivanov A G, Zhoga V V, Pavlovskij V E and Vorob`yeva N S 2019 Dynamic model of end-effector actuator used for mobile robotic weeder *IOP Conf. Series: Materials Science and Engineering* **747** 012084

[10] Pshikhopov V Kh 1995 Analytical design of nonlinear terminal control systems *New Concepts of The General Theory of Management* (Moscow) 125-141

IOP Conf. Series: Earth and Environmental Science **965** (2022) 012051 doi:10.1088/1755-1315/965/1/012051

Algorithm for calculating flat-loaded shells of environmental structures

A P Nikolaev, A P Kiselev, V N Yushkin and S S Marchenko

Federal State Educational Institution of Higher Education "Volgograd State Agrarian University", 26 University Avenue, Volgograd, 400002, Russia

E-mail: marchenkosergey1978@yandex.ru

Abstract. Structures made of thin-walled shells are widely distributed elements in the environmental protection systems. In this connection, it is necessary to develop their effective and refined calculation. The point of the shell of the environmental protection structures is considered with a plane load in the initial position, deformed after j loading steps (displacement vector) and adjacent after (j + 1) loading step. The displacement increments and their derivatives are taken as nodal unknowns. The displacement vectors of the inner point of a finite element are represented in the initial basis, and their components are approximated through nodal unknowns using Hermite polynomials of the 3rd degree. To determine the deformed state of the shell, the algorithm of the method of discrete continuation with respect to the parameter in the vicinity of a singular point is used, in which the increment of loading at the step is the desired parameter. Based on the obtained dependencies, a step-by-step calculation procedure is organized, which allows you to get correct results if there is a special point.

1. Introduction

There is a range of problems in the mechanics of a rigid deformable body that can be reduced to solving algebraic, differential or integral equations that contain an explicit parameter. This can be a temperature parameter, a load parameter, a geometric parameter, a design parameter, etc. [1]

The problem that occurs in geometrically nonlinear systems is already found in the simplest two-rod Mises truss, and consists in the fact that the analytical form of the solution can be obtained only in individual cases, if the ideal rods are considered and their longitudinal bending is not taken into account. If you complicate the problem by adding an additional condition, then it is not possible to solve the resulting system of equations analytically, and numerical methods are used to obtain the result.

The complexity of the problem increases when solving problems for determining the stress-strain state of structures in the form of arches and shells. In solving such problems, the finite element method shows itself well [2, 3, 4, 5], which is shown in a number of papers [6, 7, 8] regarding the determination of the stress-strain state of shells.

The method of continuation of the solution by the parameter is actively used together with numerical methods in scientific research in determining the stress-strain state of geometrically nonlinear systems.

In [9], the method of solving the continuous parameter solution in combination with the original variation of the finite-difference method and the nonlinear spectral problem is used to solve the geometrically nonlinear stability problem of a composite multilayer plate.

Flat shells are also proposed to be calculated using an algorithm based on the Bubnov-Galerkin method and the method of continuation of the solution with respect to the parameter [10], while the results show good convergence, including with the solutions of other authors.

The results of studies of the stress-strain state of shells based on the Ritz method and the method of continuation by the best parameter are given in [11], where the authors use examples in comparison with the results of other researchers to prove the workability of the proposed method, and analyze the deformation of a square shell with a double curvature.

This article is devoted to the development of a geometrically nonlinear prismatic finite element of a shell with a cross-section in the form of a quadrangle and the algorithm of its use in the presence of a special point of deformation of plane-loaded shells widely used in environmental protection structures.

2. Materials and methods

Shell geometry. The position of an arbitrary point M^0 on the median surface of the shell is determined by the radius-vector $\vec{R}^0$

$$\vec{R}^0 = x\vec{\imath} + z\vec{k}, \tag{1}$$

where x, z are coordinates of the Cartesian system; $\vec{\imath}$, $\vec{k}$ are orthoses of the Cartesian system.

The unit vectors of the local basis of an arbitrary point on the median surface are defined by the expressions

$$\begin{aligned} \vec{g}_1^0 &= \vec{R}_{,s} = x_{,s}\vec{\imath} + z_{,s}\vec{k}, \\ \vec{g}_3^0 &= \vec{g}_1^0 \times \vec{\jmath} = -\vec{\imath}z_{,s} + \vec{k}x_{,s}. \end{aligned} \tag{2}$$

The derivatives of the vectors of the local basis are determined by the differentiation (2)

$$\begin{aligned} \vec{g}_{1,s}^0 &= \vec{R}_{,ss} = x_{,ss}\vec{\imath} + z_{,ss}\vec{k}, \\ \vec{g}_{3,s}^0 &= -\vec{\imath}z_{,ss} + \vec{k}x_{,ss}. \end{aligned} \tag{3}$$

Using (2), the derivatives of vectors of a local basis can be represented by a decomposition over vectors of the same basis

$$\underbrace{\{\vec{g}_{,s}^0\}}_{2\times1} = \underbrace{[l]}_{2\times2}\underbrace{\{\vec{g}^0\}}_{2\times1}, \tag{4}$$

where $\{\vec{g}_{,s}^0\}^T = \{\vec{g}_{1,s}^0\, \vec{g}_{3,s}^0\}$; $\{\vec{g}^0\}^T = \{\vec{g}_1^0\, \vec{g}_3^0\}$;

$$[l] = \begin{bmatrix} x_{,ss}x_{,s} + z_{,ss}z_{,s} & -x_{,ss}z_{,s} + z_{,ss}x_{,s} \\ -z_{,ss}x_{,s} + x_{,ss}z_{,s} & z_{,ss}z_{,s} + x_{,ss}x_{,s} \end{bmatrix}.$$

The position of an arbitrary point of the shell M^{0t} located at a distance t from the median surface is determined by the radius vector

$$\vec{R}^{0t} = \vec{R}^0 + t\vec{g}_3^0. \tag{5}$$

The vectors of the local basis of a point M^{0t} are determined by the differentiation (5)

$$\begin{aligned} \vec{G}_1^0 &= \vec{R}_{,s}^{0t} = \vec{R}_{,s}^0 + t\vec{g}_{3,s}^0 = \vec{g}_1^0(1 + tl_{21}) + tl_{22}\vec{g}_3^0, \\ \vec{G}_3^0 &= \vec{R}_{,t}^{0t} = \vec{g}_3^0. \end{aligned} \tag{6}$$

In step loading, three positions of an arbitrary point of the shell are distinguished: M^{0t} – initial, M^t- after j loading steps, M^{t^*} - after the $(j + 1)$ loading step.

The displacement vector $\vec{W}$ of the point M^{0t} after j loading steps and the displacement vector $\vec{w}$ after the $(j + 1)$ loading step are determined by the components in the basis of the point M^0

$$\begin{aligned} \vec{W} &= W^1\vec{g}_1^0 + W\vec{g}_3^0, \\ \vec{w} &= w^1\vec{g}_1^0 + w\vec{g}_3^0. \end{aligned} \tag{7}$$

Their derivatives in the basis of the point M^0 are determined by differentiating (7) taking into account (4)

$$\begin{aligned} \vec{W}_{,s} &= \vec{g}_1^0(W_{,s}^1 + W^1 l_{11} + Wl_{21}) + \vec{g}_3^0(W^1 l_{12} + W_{,s} + Wl_{22}) = f_1^1\vec{g}_1^0 + f_1\vec{g}_3^1; \\ \vec{W}_{,t} &= W_{,t}^1\vec{g}_1^0 + W_{,t}\vec{g}_3^0; \end{aligned} \tag{8}$$

$$\begin{aligned} \vec{w}_{,s} &= \vec{g}_1^0(w_{,s}^1 + w^1 l_{11} + wl_{21}) + \vec{g}_3^0(w^1 l_{12} + w_{,s} + wl_{22}) = \alpha_1^1\vec{g}_1^0 + \alpha_1\vec{g}_3^0; \\ \vec{w}_{,t} &= w_{,t}^1\vec{g}_1^0 + w_{,t}\vec{g}_3^0. \end{aligned} \tag{9}$$

IOP Conf. Series: Earth and Environmental Science **965** (2022) 012051 doi:10.1088/1755-1315/965/1/012051

The positions of points M^t and M^{t*} are determined by the radius vectors

$$\vec{R}^t = \vec{R}^{0t} + \vec{W};\quad (10)$$
$$\vec{R}^{t*} = \vec{R}^t + \vec{w}.$$

The vectors of their local bases are equal to the derivatives (10)

$$\vec{G}_1^0 = \vec{R}_{,s}^t = \vec{g}_1^0 + \vec{W}_{,s};\quad (11)$$
$$\vec{G}_3^0 = \vec{R}_{,t}^t = \vec{g}_3^0 + \vec{W}_{,t}.$$
$$\vec{G}_1^* = \vec{R}_{,s}^{t*} = \vec{G}_1 + \vec{w}_{,s};\quad (12)$$
$$\vec{G}_3^* = \vec{R}_{,t}^{t*} = \vec{G}_3 + \vec{w}_{,t}.$$

Deformations and stresses. Total deformations after j loading steps are determined by the relations of continuum mechanics [12]

$$\varepsilon_{\alpha\beta} = \frac{1}{2}\left(G_{\alpha\beta} - G_{\alpha\beta}^0\right) = \frac{1}{2}\left[\vec{G}_\alpha^0\vec{W}_{,\beta} + \vec{G}_\beta^0\vec{W}_{,\alpha} + \vec{W}_{,\alpha}\vec{W}_{,\beta}\right],\quad (13)$$

where the Greek indices α, β take the values 1 and 3 (for a value of 1 $\alpha \equiv s$, and for a value of 3 – $\alpha \equiv t$).

Given (11-12) and (8) from (13), we can obtain the expressions

$$\varepsilon_{\alpha\beta} = \frac{1}{2}\left(G_{\alpha\beta} - G_{\alpha\beta}^0\right) = \frac{1}{2}\left[\vec{G}_\alpha^0\vec{W}_{,\beta} + \vec{G}_\beta^0\vec{W}_{,\alpha} + \vec{W}_{,\alpha}\vec{W}_{,\beta}\right],\quad (14)$$

Deformations at the $(j + 1)$ loading step are determined by the expressions

$$\Delta\varepsilon_{\alpha\beta} = \frac{1}{2}\left(G_{ij}^* - G_{ij}\right) = \frac{1}{2}\left[\vec{G}_\alpha\vec{w}_\beta + \vec{G}_\beta\vec{w}_\alpha + \vec{w}_\alpha\vec{w}_\beta\right].\quad (15)$$

Using expressions (11 – 12) and (9), we obtain

$$\Delta\varepsilon_{11} = c_1 w_{,s}^1 + c_2 w_{,s} + w(c_1 l_{11} + c_2 l_{12}) + w(c_1 l_{21} + c_2 l_{22}) + $$
$$+\frac{1}{2}\vec{w}_{,s}\vec{w}_{,s};$$
$$\Delta\varepsilon_{33} = w_{,t}^1 W_{,t}^1 + w_{,t}(1 + W^1) + \frac{1}{2}\vec{w}_{,t}\vec{w}_{,t};\quad (16)$$
$$2\Delta\varepsilon_{13} = w_{,t}^1 c_1 + w_{,t}c_2 + w_{,s}^1 W_{,t}^1 + w_{,s}(1 + W_{,t}) +$$
$$+w^1\left[l_{11}W_{,t}^1 + l_{12}(1 + W_{,t})\right] + w\left[l_{21}W_{,t}^1 + l_{22}(1 + W_{,t})\right] +$$
$$+\vec{w}_{,s}\vec{w}_{,t}.$$

Nonlinear terms at the loading step can be ignored, and relations (16) can be represented in matrix form

$$\underbrace{\{\Delta\varepsilon\}}_{3\times1} = \underbrace{[L]}_{3\times2}\underbrace{\{w\}}_{2\times1},\qquad \{\Delta\varepsilon\} = [L]\{w\},\quad (17)$$

where $\{\Delta\varepsilon\ \varepsilon\}^T = \{\Delta\varepsilon_{11}\Delta\varepsilon_{33}2\Delta\varepsilon_{13}\}$; $\{w\}^T = \{w^1 w\}$; $[L]$ is a matrix of algebraic and differential operators.

The relations between stresses and deformations are determined by the expressions [12, 13]

$$\sigma^{ij} = \lambda I_1(\varepsilon)G^{ij} + 2\mu G^{im}G^{jn}\varepsilon_{mn},\quad (18)$$

where λ, μ are the Lame parameters.

For an flat stress state, the following relations are obtained from (18):

$$\varepsilon_{11} = \sigma^{11}\frac{1}{E}G_{11}G_{11} + \sigma_{33}\left[\frac{1+\nu}{E}G_{13}G_{13} - \frac{\nu}{E}G_{33}G_{11}\right] +$$
$$+\sigma^{13}\frac{2}{E}G_{11}G_{13};$$
$$\varepsilon_{33} = \sigma^{11}\left[\frac{1+\nu}{E}G_{31}G_{31} - \frac{\nu}{E}G_{11}G_{33}\right] + \sigma^{33}\frac{1}{E}G_{33}G_{33} +\quad (19)$$
$$+\sigma^{13}\frac{2}{E}G_{31}G_{33};$$
$$2\varepsilon_{13} = \sigma^{11}\frac{1}{E}G_{11}G_{31} + \sigma^{33}\frac{1}{E}G_{13}G_{33} +$$

$$+2\sigma^{13}\left[\frac{1+\nu}{E}G_{11}G_{33}-\frac{\nu}{E}G_{13}G_{13}\right],$$

where E is the elastic modulus of the material, and v is the Poisson's ratio.

Relations (19) can be represented as

$$\underbrace{\{\varepsilon\}}_{3\times1}=\underbrace{[D]}_{3\times3}\underbrace{\{\sigma\}}_{3\times1}, \qquad \{\varepsilon\}=[D]\{\sigma\}, \tag{20}$$

where $\{\varepsilon\}^T=\{\varepsilon_{11}\varepsilon_{33}2\varepsilon_{13}\}$; $\{\sigma\}^T=\{\sigma^{11}\sigma^{33}\sigma^{13}\}$.

The stresses for j loading steps and the stress increments at the $(j+1)$ loading step are determined by the expressions

$$\underbrace{\{\sigma\}}_{3\times1}=\underbrace{[C]}_{3\times3}\underbrace{\{\varepsilon\}}_{3\times1};\ \underbrace{\{\Delta\sigma\}}_{3\times1}=\underbrace{[C]}_{3\times3}\underbrace{\{\Delta\varepsilon\}}_{3\times1}. \qquad \{\sigma\}=[C]\{\varepsilon\};\ \{\Delta\sigma\}=[C]\{\Delta\varepsilon\}. \tag{21}$$

where $[C]=[D]^{-1}$.

Finite element stiffness matrix at the $(j+1)$-th loading step. We consider a three-dimensional finite element that has a XOZ cross-section in the XOZ plane in the form of an arbitrary quadrilateral with nodes i, j, k, and l. The displacements and their first derivatives with respect to the S and t coordinates are chosen as nodal unknowns.

To perform numerical integration, we use a square with local coordinates ξ_1,ξ_2, varying within $-1\le\xi_1,\xi_2\le1$. Global coordinates S, t are related to their nodal values by bilinear functions of coordinates ξ_1,ξ_2

$$\lambda=\underbrace{\{\varphi\}^T}_{1\times4}\{\xi_1,\xi_2\}^T\underbrace{\{\lambda_y\}}_{4\times1}, \qquad \lambda=\{\varphi\}\{\xi_1,\xi_2\}^T\{\lambda_y\}, \tag{22}$$

where λ is coordinates S, t; $\{\lambda_y\}=\{\lambda^i\lambda^j\lambda^k\lambda^l\}$ - a string of node coordinate values λ.

Differentiation of (22) are determined by the derivatives of the global coordinates in the local system $S_{\xi_1},S_{\xi_2},t_{\xi_1},t_{\xi_2}$ and the local coordinates in the global system $\xi_{1,s},\xi_{2,s},\xi_{1,t},\xi_{2,t}$.

The vectors of nodal unknowns of a finite element in the local and global coordinate systems have the form

$$\underbrace{\{v_y^{loc}\}^T}_{1\times24}=\left\{\underbrace{\{w_y^{1l}\}^T}_{1\times12}\ \underbrace{\{w_y^{l}\}^T}_{1\times12}\right\};$$
$$\underbrace{\{v_y^{glob}\}^T}_{1\times24}=\left\{\underbrace{\{w_y^{1g}\}^T}_{1\times12}\ \underbrace{\{w_y^{r}\}^T}_{1\times12}\right\}, \tag{23}$$

where

$$\underbrace{\{w_y^{1loc}\}^T}_{1\times12}=\{w^{1i}\,w^{1j}\,w^{1k}\,w^{1l}\ w^{1j}_{,\xi_1}\,w^{1k}_{,\xi_1}\,w^{1l}_{,\xi_1}\,w^{1i}_{,\xi_2}\,w^{1j}_{,\xi_2},w^{1k}_{,\xi_2},w^{1l}_{,\xi_2}\};$$
$$\underbrace{\{w_y^{1glob}\}^T}_{1\times12}=\{w^{1i}\,w^{1j}\,w^{1k}\,w^{1l}\,w^{1i}_{,s}\,w^{1j}_{,s}\,w^{1k}_{,s}\,w^{1l}_{,s}\,w^{1i}_{,t}\,w^{1j}_{,t}\,w^{1k}_{,t}\,w^{1l}_{,t}\}. \tag{24}$$

The vectors $\underbrace{\{w_y^{loc}\}^T}_{1\times12}$ and $\underbrace{\{w_y^{glob}\}^T}_{1\times12}$ are written similarly to (24).

Based on (22), the matrix relation can be written between the vectors (23)

$$\underbrace{\{v_y^{loc}\}}_{24\times1}=\underbrace{[R]}_{24\times24}\underbrace{\{v_y^{glob}\}}_{24\times1}. \tag{25}$$

Movements of the internal point of a finite element areюapproximated through nodal values by the relations

$$w^1=\underbrace{\{\psi(\xi_1,\xi_2)\}^T}_{1\times12}\underbrace{\{w_y^{1л}\}}_{12\times1};\ w=\underbrace{\{\psi(\xi_1,\xi_2)\}^T}_{1\times12}\{w_y^{л}\}, \qquad \begin{aligned}w^1&=\{\psi(\xi_1,\xi_2)\}^T\{w_y^{1loc}\};\\ w&=\{\psi(\xi_1,\xi_2)\}^T\{w_y^{loc}\},\end{aligned} \tag{26}$$

where the elements of a function of the form ψ are Hermite polynomials of the 3rd degree.

IOP Conf. Series: Earth and Environmental Science **965** (2022) 012051 doi:10.1088/1755-1315/965/1/012051

The relation (26) can be represented in matrix form

$$\underbrace{\{v\}}_{2\times 1} = \begin{Bmatrix} w^1 \\ w \end{Bmatrix} = \underbrace{[A]}_{2\times 24} \underbrace{\{v_y^{loc}\}}_{24\times 1}; \tag{27}$$

Using (26) of strain (17) at the loading step, we write in matrix form

$$\underbrace{\{\Delta\varepsilon\}}_{3\times 1} = \underbrace{[L]}_{3\times 2} \underbrace{[A]}_{2\times 24} \underbrace{\{v_y^l\}}_{24\times 1} = \underbrace{[B]}_{3\times 24} \underbrace{\{v_y^{loc}\}}_{24\times 1}. \tag{28}$$

To form the stiffness matrix at the loading step, the Lagrange functional is used, which expresses the equality of the work of external and internal forces at the loading step

$$\Phi_L = \int_V \left[\underbrace{\{\sigma\}}_{1\times 3}^T + \underbrace{\{\Delta\sigma\}}_{1\times 3}^T\right] \underbrace{\{\Delta\varepsilon\}}_{3\times 1} dV = \int_S \underbrace{\{v\}}_{1\times 2}^T \left[\underbrace{\{q\}}_{2\times 1} + \underbrace{\{\Delta q\}}_{2\times 1}^T\right] dS = 0,$$

$$\int_V \left[\underbrace{\{\sigma\}}_{1\times 3}^T + \underbrace{\{\Delta\sigma\}}_{1\times 3}^T\right] \underbrace{\{\Delta\varepsilon\}}_{3\times 1} dV = \int_S \underbrace{\{v\}}_{1\times 2}^T \left[\underbrace{\{q\}}_{2\times 1} + \underbrace{\{\Delta q\}}_{2\times 1}^T\right] dS, \tag{29}$$

where $\{q\} = \{q \ =_{q1} q_{q2}\}$; $\{\Delta q\} = \{\Delta q_1 \Delta q_2\}$ – vectors of full and step loads.

Using relations (20), (25), and (28), equality (29) can be represented as

$$\underbrace{[K]}_{24\times 24} \underbrace{\{v_y^{glob}\}}_{24\times 1} = \underbrace{\{\Delta f\}}_{24\times 1} + \underbrace{\{R\}}_{24\times 1}, \tag{30}$$

where $[K] = \int_V [B]^T [C][B] dV$ – stiffness matrix of the finite element in the $(j + 1)$-th step of loading; $\{\Delta f\} = \int_S [A]^T [\Delta q] dS$ is the vector of nodal loads; $\{R\} = \int_S [A]^T [q] dS \int_V [B]^T \{\sigma\}^j dV$ – residual errors in the $(j + 1)$ step of loading.

3. Results and discussion

At the loading step, the stiffness matrix of a separate finite element and the vector of its nodal forces are formed. If a special point appears in the process of loading the structure during deformation, then the developed algorithm can be used in this sequence. Based on (30), the stiffness matrix of the calculated shell is formed at the loading step. Since in the calculations in the presence of a singular point, the increment of the external load parameter is an unknown quantity, the system of equations of the calculated shell will consist of m equations with $(m + 1)$ the number of unknowns and can be represented as

$$\underbrace{[K_0]^j}_{m\times m} \underbrace{\{\Delta M_y\}^{j+1}}_{m\times 1} = \Delta\lambda^{j+1} \underbrace{\{F\}}_{m\times 1} + \underbrace{\{R^*\}^j}_{m\times 1}, \tag{31}$$

where $\{\Delta M_y\}^{j+1}$ – increments of nodal unknowns at the $(j + 1)$ loading step; $\{F\}$ – nodal force vector; $\Delta\lambda^{j+1}$ – increment of the external load parameter at the $(j + 1)$ loading step; $\{R^*\}$ - the discrepancy in the values of the external and internal forces for j loading steps.

The vectors of nodal unknowns in the space R_{m+1} at the j-th and $(j + 1)$ loading steps can be written by the expressions

$$\underbrace{\{\Delta X^j\}^T}_{1\times(m+1)} = \left\{\underbrace{\{\Delta M_y^j\}^T}_{1\times m} \Delta\lambda^j\right\};$$

$$\underbrace{\{\Delta X^{j+1}\}^T}_{1\times(m+1)} = \left\{\underbrace{\{\Delta M_y^{j+1}\}^T}_{1\times m} \Delta\lambda^{j+1}\right\}. \tag{32}$$

The components $\{\Delta M_y^j\}$ are deferred in the m-directions, and the components $\Delta\lambda^j$ are deferred in the $(m + 1)$-th direction [1]

At the $(j + 1)$ step of loading, the vector of increments of unknowns $\{\Delta\vec{X}^{j+1}\}$ is represented in the following form

$$\underbrace{\{\Delta X^{j+1}\}^T}_{1\times(m+1)} = \left\{ t \underbrace{\{\Delta X^j\}}_{1\times(m+1)}{}^T + \underbrace{\{\Delta Z^{j+1}\}^T}_{1\times(m+1)} \right\}, \quad (33)$$

where the new vector of increments of unknowns has the form

$$\underbrace{\{\Delta Z^{j+1}\}^T}_{1\times(m+1)} = \left\{ \underbrace{\{\Delta Z_w^{j+1}\}^T}_{1\times m}, \Delta Z_\lambda^{j+1} \right\}, \quad (34)$$

T is the assigned numeric parameter.

The components $\{\Delta Z_w^{j+1}\}^T$ are shifted in the m-directions of the $Rm + 1$ space R_{m+1}, and the component ΔZ_λ^{j+1} in the $(m + 1)$ direction.

From the comparison of (32) and (33), we can write the expressions

$$\underbrace{\{\Delta M_y^{j+1}\}}_{m\times 1} = t\underbrace{\{\Delta M_y^j\}}_{m\times 1} + \underbrace{\{\Delta Z_w^{j+1}\}}_{m\times 1};$$
$$\Delta\lambda^{j+1} = t\Delta\lambda^j + \Delta Z_\lambda^{j+1}, \quad (35)$$

where the unknown quantities at the loading step are the vector $\{\Delta Z_Z^{wj+1}\}$ and the parameter $\Delta Z_{\lambda j}^{+1}$.

Substituting (35) into (31) we can obtain a system of m-equations with $(m + 1)$ unknowns

$$\underbrace{[K_0]^j}_{m\times m} \left\{ t\underbrace{\{\Delta M_y^j\}}_{m\times 1} + \underbrace{\{\Delta Z_w^{j+1}\}}_{m\times 1} \right\} = \underbrace{\{F\}}_{m\times 1} \left(t\Delta\lambda^j + \Delta Z_\lambda^{j+1}\right) - \underbrace{\{R^*\}^j}_{m\times 1}. \quad (36)$$

To get an additional equation to the system (36) the condition of orthogonality of vectors is used

$$t \underbrace{\{\Delta X^j\}}_{1\times(m+1)}{}^T \underbrace{\{\Delta Z^{j+1}\}}_{(m+1)\times 1} = 0, \quad (37)$$

which, taking into account (33) and (34), will take the form

$$\underbrace{\{\Delta M_y^j\}^T}_{1\times m} \underbrace{\{\Delta Z_w^{j+1}\}}_{m\times 1} + \Delta\lambda^j \Delta Z_\lambda^{j+1} = 0. \quad (38)$$

Equations (36) and (38) are resolved jointly.

From (38), the value of the load parameter at the $(j + 1)$ loading step is determined

$$\Delta Z_\lambda^{j+1} = -\frac{1}{\Delta\lambda^j} \underbrace{\{\Delta M_y^j\}^T}_{1\times m} \underbrace{\{\Delta Z_w^{j+1}\}}_{m\times 1}. \quad (39)$$

Substituting (39) into (36) yields m equations with m unknowns

$$\underbrace{[K]}_{m\times m}{}^j \underbrace{\{\Delta Z_w^{j+1}\}}_{m\times 1} = \underbrace{\{f\}}_{m\times 1}, \quad (40)$$

where $[K]^j = [K_0]^j + \frac{1}{\Delta\lambda^j}\{R^*\}\{\Delta M_y^j\}^T$ is the structural rigidity matrix; $\{f\} = \{F\}t\Delta\lambda^j - \{R^*\}^j - [K_0]^j t\{\Delta M_y^j\}$ is the vector of nodal forces in the vicinity of the singular point.

After determining the increment vector $\{\Delta Z_w^{j+1}\}$, the value ΔZ_λ^{j+1} is determined from relations (39), and by the formulas (35) we find the required values $\{M_y^{j+1}\}$ and $\Delta\lambda^{j+1}$, which are necessary for forming the structural stiffness matrix at the next loading step.

4. Conclusion

The developed algorithm makes it possible to calculate the displacements in the vicinity of a singular point using the method of parameter propagation and on the basis of the finite element method in the form of the displacement method when solving geometrically nonlinear problems for determining the stress-strain state of axially symmetric shell structures.

5. Acknowledgments
This work was supported by the Russian Foundation for Basic Research and the Administration of the Volgograd Region No. 19-41-340004 r_a.

References
[1] Grigolyuk E I and Shalashilin V I 1980 *Problems of Nonlinear Deformation* (Moscow: Science) p 230
[2] Zenkevich O and Morgan K 1986 *Finite Elements and Approximation* (Moscow: World) p 318
[3] Oden J 1976 Finite Elements in the Nonlinear Mechanics of Continuous Media (Moscow) p 464
[4] Postnov V A 1977 *Numerical Methods for Calculating Ship Structures* (Leningrad: Shipbuilding) p 280
[5] Rikards R B 1988 The Finite Element Method in the Theory of Shells and Plates (Riga: Zinatne) p 284
[6] Gureeva N A, Nikolaev A P and Yushkin V N 2020 Comparative analysis of finite element formulations at plane loading of an elastic body *Structural Mechanics of Engineering Constructions and Buildings* **16(2)** 139–145
[7] Gureeva N A, Klochkov Yu V, Nikolaev A P and Yushkin V N 2020 Stress-strain state of shell of revolution analysis by using various formulations of three-dimensional finite elements *Structural Mechanics of Engineering Constructions and Building* **16(5)** 361–379
[8] Gureeva N A, Nikolaev A P and Yushkin V N 2020 Elastic deformation of a solid based on various variants of finite element formulations *Bulletin of Volgograd State University of Architecture and Civil Engineering. Civil Engineering and Architecture* **2** 5-13
[9] Badriev I B, Buyanov V Yu and Makarov M V 2020 Solution of the geometrically nonlinear stability problem for a three-layer plate *Trends in the Development of Science and Education* **4(58)** 65-74
[10] Alaeva D R, Semenov A A and Karpov V V 2020 Methodology for the study of subcritical and transcritical deformation of shells in the hinge-movable fixing of the contour *Engineering and Construction Bulletin of the Caspian Sea* **32** 49-53
[11] Semenov A A and Leonov S S 2019 Method of continuous continuation of the solution for the best parameter in the calculation of shell structures *Scientists of the Kazan University. Physical and Mathematical Sciences* **161(2)** 230-249
[12] Sedov L I 1976 *Continuum Mechanics* vol 1 (Moscow: Science) p 535
[13] Sedov L I 1994 *Continuum Mechanics* vol 2 (Moscow: Science) p 560

IOP Conf. Series: Earth and Environmental Science **965** (2022) 012052 doi:10.1088/1755-1315/965/1/012052

Melons and gourds sowing quality improving in conditions of risky farming

M N Shaprov, I S Martynov, A A Mikhalyenok, M A Sadovnikov, E Yu Guzenko and T S Ivanova

Volgograd State Agrarian University, 26 University Avenue, Volgograd, 400002, Russia

E-mail: m.shaprov@yandex.ru

Abstract. The article describes the conditions in the agricultural crops cultivation, which greatly affect their productivity. One of the main problems is sparse seedlings. In this regard, it is necessary to use innovative technologies for the cultivation of melons and gourds, which reduce risks and increase their productivity. To solve this problem, the authors developed a technology for multi-depth row crops seedbed planting, which field tests were carried out in the fields of the Volgograd state agrarian university. In the course of the study, the distribution of seeds in the seedbed and the distance between the bunches were studied when sowing them into the soil under various operating modes of the seeding machine. At the same time, the dependence of the distance between seeds in the bunch and between the bunches on the speed of the seeding machine was established. The use of the bunch planting technology at different depths makes it possible to ensure a favorable temperature and humidity regime for seeds in compliance with all agrotechnical requirements.

1. Introduction

In 2020, the Volgograd region in terms of the agricultural crops sown area size was entrenched in the leading positions in the country, it entered the top three among the regions in the Southern Federal Authority (SFA) and amounted to more than 3 million hectares. It can be noted that the Volgograd region among the subjects of the Russian Federation took the second place for melons, the third place for vegetables, when cultivated in the open field, the fourth one for oilseeds (sunflower). The region's peculiarities of the climate and the properties of the soils have a significant impact on the yield of the cultivated crops [1, 2, 3]. First of all, this applies to the row crops cultivation, which seeds germination largely depends on the temperature and soil moisture. The most responsive crops to these parameters are melons and gourds. But their yield is low despite the increase in areas. The main reasons for this are the lack of good predecessors and the imperfection of existing technologies for the agricultural crops cultivation. Sowing is a technological operation in which it is required to correlate several indicators at the same time (temperature and soil moisture, weather conditions, etc.). One of the main problems is sparse seedlings. The reason is the likelihood of late spring frosts, hail and strong winds. Therefore, the seeds field germination and the further plants development, and, consequently, the row crops yield will depend on the quality of sowing [4]. The existing technologies for the row crops cultivation are not able to create optimal conditions for seeds with their subsequent trouble-free growth and development [5-

10]. In this regard, it is necessary to use innovative technologies for the melons and gourds cultivation that reduce the above risks and increase their productivity.

2. Materials and methods

To solve the set problem, the technology for row crops multi-depth bunch planting was developed. To ensure sowing by using the specified technology, a sowing machine was designed (Figure 1, 2) [11].

Its sections consist of bracket 1 (Figure 1), suspension 2, apparatus 3, seed box (hopper) 4, device 5 for distributing seeds, opener 6, covering knife 7, wheel 8 and smoothing harrow 9.

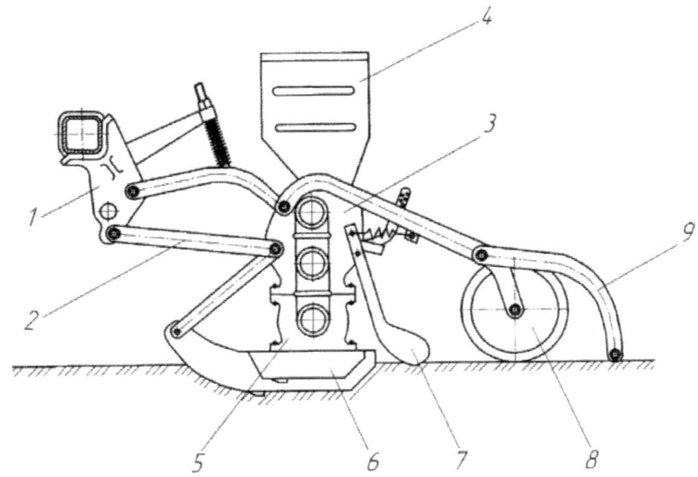

Figure 1. Seeding machine diagram.

The structural differences of the seeding machine relative to the serial row-crop seeding machine include the seeding disc 4, the distributing device (roller) 11 and the opener 8 (Figure 2). Disc 4 with three holes allows seeds bunch planting in a row. The modernized opener 6 creates conditions for them at different levels. Moreover, the right 21 and left 19 compartments form seed furrows in two and four centimeters less than the main one 20.

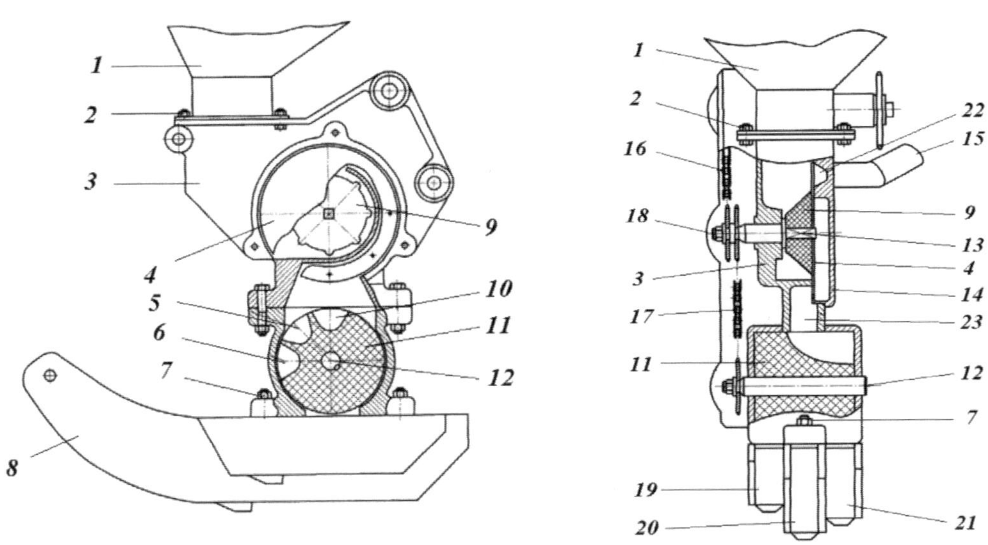

Figure 2. Working bodies of the seeding machine section.

IOP Conf. Series: Earth and Environmental Science **965** (2022) 012052 doi:10.1088/1755-1315/965/1/012052

The roller 11 (Figure 2) is installed between the device 3 and the opener 8 by pins 7. The roller is fixed on the drive shaft 12. This device is necessary to supply seeds to the desired section of the opener with cells 5,6,10. Moreover, cells 1 and 3 (Figure 3) have a curved shape, necessary to displace the seeds from the middle and right or left parts of the roller, respectively. Cell 2 is made in the form of a cone. The disc and roller rotation is provided by a chain drive 17.

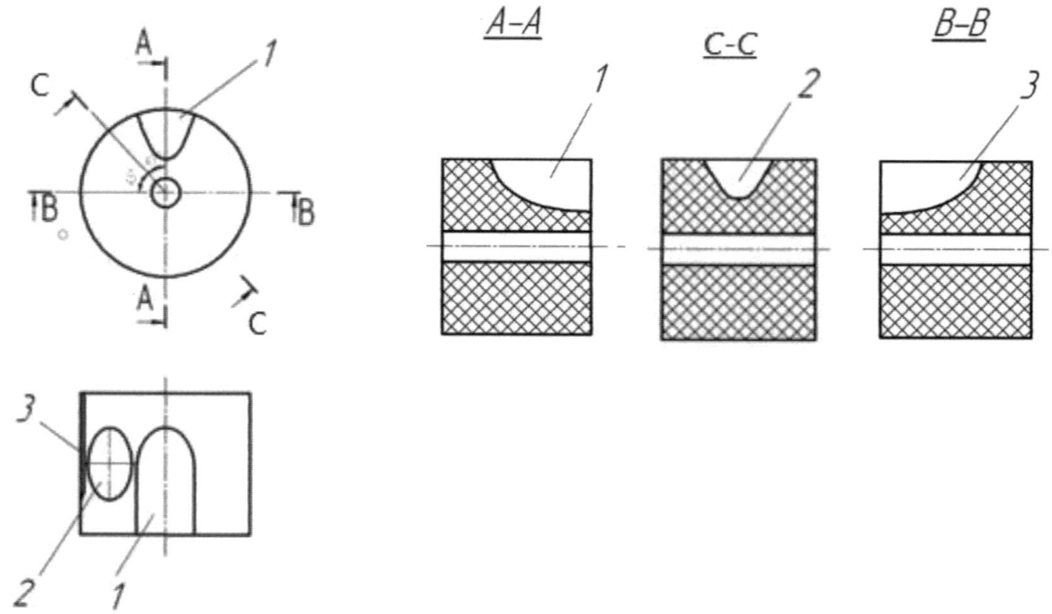

Figure 3. Roller.

Optimization of the seeding machine modernized section design parameters was carried out in the soil box of the Volgograd State Agrarian University [12]. Based on their results, a mock-up test sample was designed, which field tests were carried out in the fields of the Volgograd State Agrarian University on the example of melons and gourds sowing. The sowing was carried out with the seeds of a watermelon variety "Holodok" by two technologies. Under the same conditions, the authors compared the technology of multi-depth bunch planting with the distribution of three seeds in the bunch at the distance of 15...17 cm and the distance between the bunches of 1.18...1.2 m and the technology of row-single grain sowing with the inter-seed distance of 15...17 cm.

Figure 4. Measurement of sowing depth.

In the field conditions, such parameters as the seeding quality and the bunch formation in a row were evaluated. All investigated parameters were assessed depending on the sowing section speed (1.12, 1.5, 1.87 m/s). The actual sowing depth was determined by the plants (Figure 4) by extracting them from the soil. Seedlings were noted at the field surface level and removed from the soil. Direct measurement of the stem length was taken as the sowing depth. The emerging seedlings (Figure 5) determined the distance between the plants in the bunch.

In addition to the carried-out measurements, a comparative assessment of the field germination of watermelons seeds sown according to the proposed technology with the existing one was carried out. The seed germination was determined under the same conditions for the modernized and serial seeding machine with the determination of the specified distances corresponding to the feeding area.

(a) (b)

Figure 5. Plants development: a - day 28, b - day 49.

Taking into account the conditions of the harsh climate and the impact of dry winds, the authors measured humidity, since it is one of the most important indicators for seed germination. This indicator is especially important for row crops, melons and gourds, since its lack leads to blindless in seedlings.

3. Results and discussion

As the result of the data processing (Figure 6), it was revealed that over time, the moisture content in the ten-centimeter soil layer decreased. In just three days, soil moisture dropped from 21 to 18%.

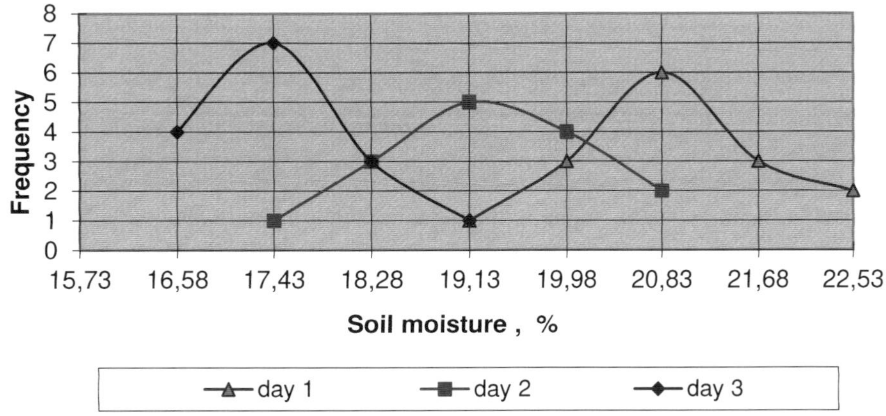

Figure 6. Variational curves of soil moisture.

IOP Conf. Series: Earth and Environmental Science **965** (2022) 012052 doi:10.1088/1755-1315/965/1/012052

When investigating the sowing depth with the modernized seeding machine, the results were obtained (Figure 7), reflecting its performance at the set depth in accordance with the agricultural requirements. The sowing depths are as follows: the left compartment is 4 cm (Figure 7 a), the right compartment is 6 cm (Figure 7 c), and the central compartment is 8 cm (Figure 7 b). Deviation from the set depth does not exceed 15%.

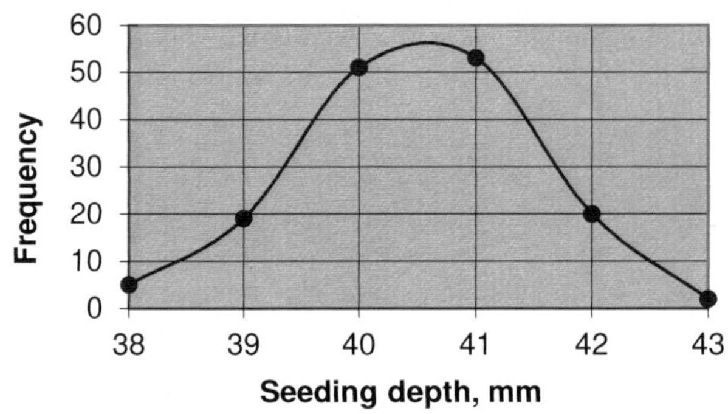

(a)

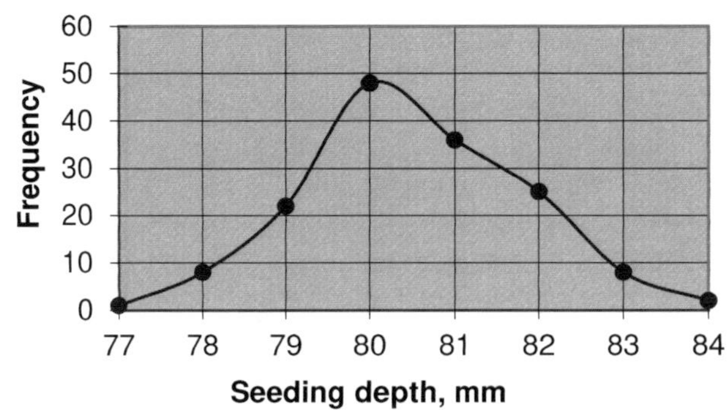

(b)

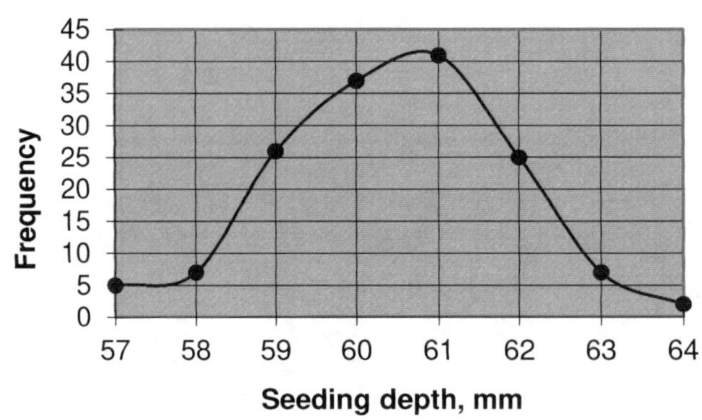

(c)

Figure 7. Sowing seeds by a modernized opener.

When measuring the distances between seeds in the bunch and between the bunches, it was revealed that the values are insignificant, but change depending on the speed of the sowing unit (Figure 8, 9).

IOP Conf. Series: Earth and Environmental Science **965** (2022) 012052 doi:10.1088/1755-1315/965/1/012052

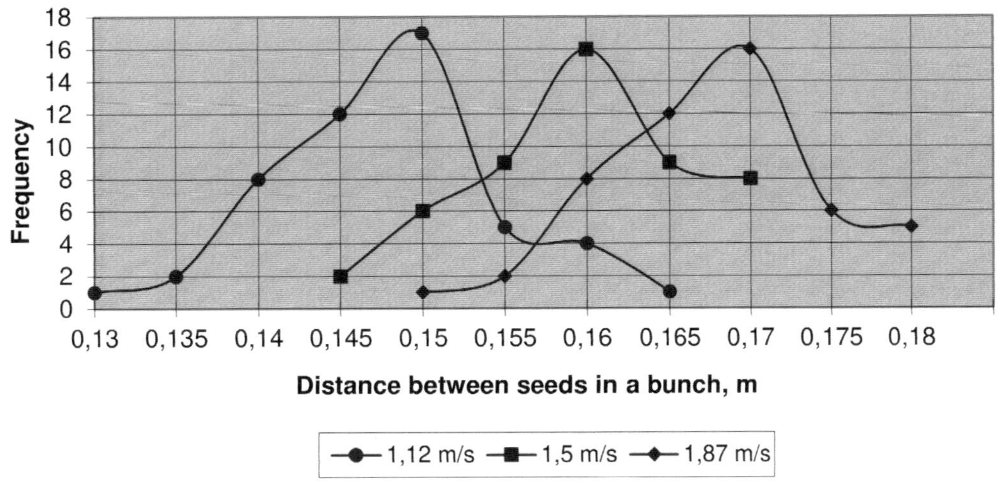

Figure 8. Variational curves of seeds sowing in the bunch.

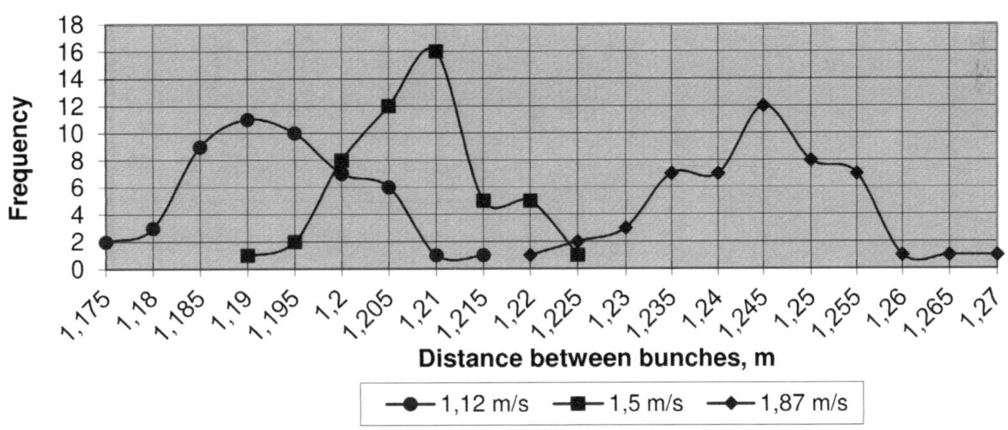

Figure 9. Variational curves of the distance between the bunches.

At the same time, the dependences of the distance between seeds in the bunch (Figure 10) and the size of the inter-bunch (Figure 11) on the seeding machine speed were built, which are described by the equations $y = -0.006x^2 + 0.045x + 0.1$ and $y = 0.113x^2 - 0.269x + 1.35$ respectively. The reliability coefficient of the approximation R^2 in these cases is equal to one.

When analyzing the seeds germination, it was revealed that when sowing by a modernized seeding machine, seedlings in the bunch were observed in all plots (one, two or three plants) with 100% provision of the specified distance between the bunches with the required feeding area (1.18...1.27 m), the provision of the given distance between bunches was 10% while sowing by a serial seeding machine, and 90% provided the distance within 0.9...1.17 m and 1.28...1.8 m. The proposed sowing technology provides the best soil parameters in the "temperature-humidity" ratio, which is confirmed by observations of plant development (Figure 5).

In addition, it was revealed that the yield in the plots sown by using the technology with a different depth of seeding in the bunches is 23% higher compared to the plots in which the sowing was carried out by a serial seeding machine.

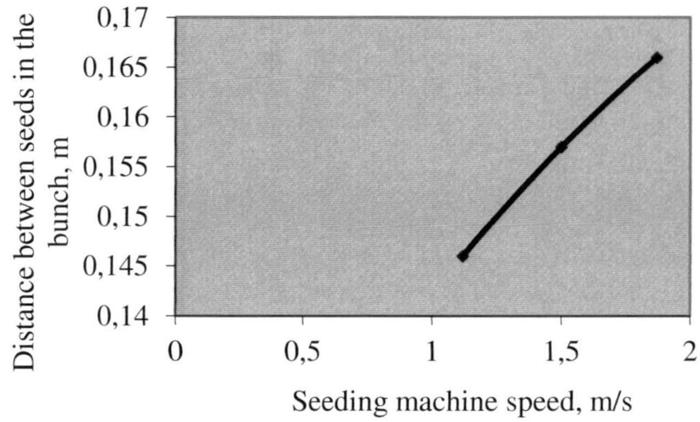

Figure 10. Dependence of the seeds distribution in the bunch.

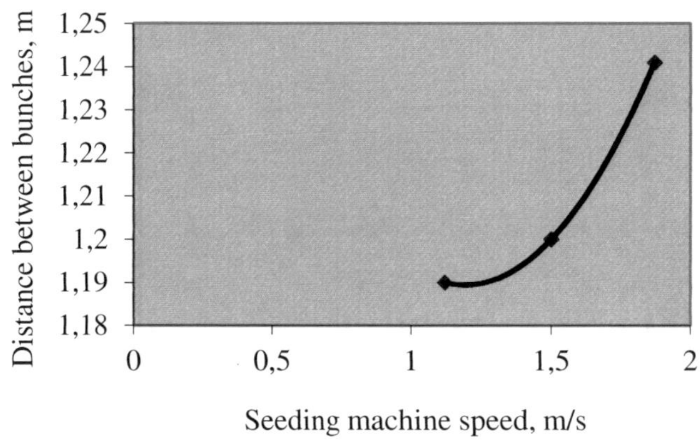

Figure 11. Dependence of seed distribution between bunches.

4. Conclusion

The carried-out studies made it possible to obtain results confirming the modernized seeding machine use effectiveness. The use of the bunch planting technology at different depths makes it possible to ensure a favorable temperature and humidity regime for seeds in compliance with all agrotechnical requirements. Therefore, the improved seeding machine provides optimal conditions for seed germination. High-quality seedlings contribute to the good development of plants, as the result of which their productivity increases in the harsh continental climate of the Volgograd region.

References

[1] Van Mansvelt Ya D and Temirbekova S K 2017 Organic agriculture *Agricultural Biology* **52(3)** 478-486

[2] Tseplyaev A N, Kharlashin A V and Tseplyaev V A 2018 Resource-saving soil-protective technology of row crops sowing seeds in severely arid zones of the Southern Federal District *Proc. of the Nizhnevolzhsky Agrouniversity Complex: Science and Higher Professional Education* **2(50)** 331-338

[3] Milyutkin V A and Dolgorukov N V 2014 Soil-protective agricultural technologies and equipment for the agricultural crops cultivation *Bulletin of the Samara State Agricultural Academy* **3** 37-44

[4] Tseplyaev A N, Shaprov M N and Martynov I S 2016 Soil-saving mechanized technology of row crops sowing *Agrarian Scientific J.* **3** 69-72

[5] Nesmiyan A Yu, Dolzhnikov V V, Cheremisin Yu M and Asaturyan S V 2015 Increasing the efficiency of the seeds capture by the metering elements of the precision seeding machine sowing apparatus *Scientific J. of the KubSAU* **106(02)** 1024-1039

[6] Dorozhko G R, Shabaldas O G, Zaitsev V K and Borodin D Yu 2013 Direct sowing of field crops in the Stavropol territory *Agriculture* **8** 20-23

[7] Kem A A, Miklashevitch V L and Chekushev M S 2017 Opener for two-line sowing of grain crops with multilevel application of mineral fertilizers *Bulletin of the OmSAU* **2(26)** 105-111

[8] Kurdyumov V I and Zykin E S 2018 Energy assessment of a bedder seeder *Bulletin of the Nizhny Novgorod State Engineering and Economic Institute* **5(84)** 66-75

[9] Laryushin N P, Kuvaitsev V N, Buchma A V and Shumaev V V 2014 Theoretical studies of the combined opener for simultaneous seeds multilevel fertilization and sowing *Niva Povolzhya* **1(30)** 82-88

[10] Meinel T 2015 Sowing technology *Jahrbuch Agrartechnik 2014* ed L Frerichs (Braunschweig: Institut für mobile Maschinen und Nutzfahrzeuge) pp 1-9

[11] Martynov I S and Shaprov M N 2019 Multilevel row crops bunch planting *Rural Mechanic* **8** 10-11

[12] Martynov I S, Shaprov M N and Mikhalyenok A A 2020 The results of laboratory studies of opener for mid-depth seed placement work *Proc. of the Nizhnevolzhsky Agrouniversity Complex: Science and Higher Professional Education* **1(57)** 316-324

IOP Conf. Series: Earth and Environmental Science **965** (2022) 012053 doi:10.1088/1755-1315/965/1/012053

Fighting forest and landscape fires in forestry

D V Belomutenko[1], M V Istochkina[1], S V Belomutenko[1], A E Peskov[2], D A Golatov[1], M V Medvedev[1] and E A Klenina[2]

[1] Volgograd State Agrarian University, 26 University Avenue, Volgograd, 400002, Russia
[2] Volgograd State Technical University, 28 V I Lenin Avenue, Volgograd, 400005, Russia

E-mail: Istochkina@yandex.ru

Abstract. Wildfire is a serious disastrous phenomenon for nature and the population of our country and the whole world. Wildfires, which result in enormous economic losses, take a toll in human life and affect ecosystems, are a fundamental problem. The article assesses the fire hazards of Leshchevskoe forestry in the Volgograd region. The total area of the forest fund of the forestry is 17004 ha. The area covered by forests is 9313 ha. According to the forest zoning the territory of Leshchevskoe forestry attributed to the steppe region of the European part of Russia of the steppe forest zone. The article contains natural and climatic characteristics of the forestry territories, the indicators of the characteristics of forest and non-forest zones of the forestry fund, the fire hazard classes, the tree species composition and the major forest types. The main causes of forest fires and factors that increase the fire risks of forestry have been identified. The list and terms of measures for the organization of monitoring and control of fire hazards in forests and forest fires, as well as measures for the organization of fire detection, inventory and monitoring systems on the territory of Leshchevsky forestry are set out.

1. Introduction

Wildfire is a serious disastrous phenomenon for nature and the population of our country and the whole world. Wildfires, which result in enormous economic losses, take a toll in human life and affect ecosystems, are a fundamental problem. Fire causes damage to forestry, agriculture and human settlement [1].

In January 2021, Federal Act No. 454 of 22 December 2020 “On Amendments on Certain Legislative Acts of the Russian Federation in terms of improving activities in the field of fire safety” came into force. The Law introduced a new term - “landscape fire”, and gave a new definition of wildfire. Now these concepts are presented in the Federal Act “On Fire Safety” as follows: “landscape(natural) fire - uncontrolled combustion process, that spontaneously emerging and spreading to nearby surroundings, covering various landscape components”; “wildfire is a type of landscape (natural) fire, spreading in forest” [2].

The causes of wildfires are in many ways connected with geographical and meteorological conditions. There is also a distinction in the frequency of wildfires: long-term, one-year, seasonal, daily. It is connected with alternation of dry and wet periods, with seasonal and daily changes in weather conditions, with a forest type and so on.

This article presents the research of fire hazards on the territory of Leshchevskoe forestry of the

Volgograd region and analyzes the measures on organization of monitoring of fire hazards in forests and on organization of wildfire detection systems.

2. Materials and methods

Research material included:

- fire threat monitoring data in the forests of the Volgograd region;
- the forestry regulation of Leshchevskoe forestry [3];
- particular provisions of the appropriate legal and regulatory instruments of Russia and departmental normative documents [4, 5, 6, 7, 8], related to Forestry;
- requirements of the legislative and regulatory documents in the field of ensuring Fire Safety [2, 9, 10].

3. Results and discussion

Leshchevskoe forestry is located in the south-eastern part of the Volgograd region in the territory of the Leninsky and partially Svetloyarsky district (Table 1). Forest fund of the forestry is located in close proximity to the town Leninsk. Leshchevskoe forestry is bordered to the North by Sredneahtubinskoye, to the South and West by Svetloyarskoe forestry and to the East by the Astrakhan region.

Table 1. Structure of Leshchevskoe forestry.

No.	Forest district name	Administrative district (Municipality)	Total area, ha
1	Leshchevskoe	Leninsky	4741
		Svetloyarsky	5323
2	Karshevitskoe	Leninsky	6940
Total for the forestry			17004

The climate on the territory of the forestry is continental and extremely dry. The maximum temperatures in some years ranging from 40-45 °C. The cold period lasts from the middle of November until the end of March, but the real winter with the steady snow cover comes in the middle of December. The annual amplitude of air temperature reaches 33-35°, the average rainfall is just over 300 mm, while the rate of evaporation is above 800 mm [11].

While the average temperature of the coldest month is about -9 °C, in some winters with about 20% coverage heavy frosts may happen with the temperature falling down below -25 °C...-35 °C.

The total area of the forest fund of the forestry as of January 2021, is 17004 ha (Table 2). The area covered by forests is 9313 ha. According to the forest zoning the territory of Leshchevskoe forestry attributed to the steppe region of the European part of Russia of the steppe forest zone [3].

The characteristics of the forest fund lands of the forestry given in this table show that, the area of forest lands is 60.1% of the total forestry area. While 54.7% of the land is covered with forest vegetation, including 8.8% of forest crops.

Table 2. Characteristics of forest and non-forest zones of the forest fund on the territory of Leshchevskoe forestry

Indicators of land characteristics	Total for the forestry	
	Area, ha	%
1	2	3
The total area of the lands	17004	100.0

IOP Conf. Series: Earth and Environmental Science **965** (2022) 012053 doi:10.1088/1755-1315/965/1/012053

Indicators of land characteristics	Total for the forestry	
	Area, ha	%
1	2	3
Forest lands, total	10211	60.1
Lands covered with forest vegetation, total	9290	54.7
Non-forest lands, total	921	5.4
Including:		
forest zones which had not reached the normative level of canopy cover	72	0.4
forest nurseries, plantations	-	-
natural glades	-	-
burned-out forests	220	1.3
deforestation	106	0.6
wastelands	523	3.1
Non-forest lands, total	6793	39.9
Including:		
biofields	-	-
arable lands	9	-
hayfields	435	2.5
pasture	2239	13.2
waters	1912	11.2
gardens	-	-
roads, clearings	126	0.7
estates	114	0.7
marshes	691	4.1
sands	626	3.7
other lands	641	3.8

And an important point, that characterizes the potential inflammability of woodland and defines the conditions for the emergence and development of wildfires is the existence and ratio between the stands of different forest types and species composition of the forest (Table 3).

In accordance with "Form 3.2 GLR. Characteristics of fire-fighting equipment of forests" of the sectoral statistical reporting for 01.01.2021 in the Volgograd region, the average fire hazard class for the forestry is 4.2 and is distributed as follows:

class I – 238 ha;
class II – 481 ha;
class III – 1791 ha;
class IV– 8144 ha;
class V – 6350 ha.

IOP Conf. Series: Earth and Environmental Science **965** (2022) 012053 doi:10.1088/1755-1315/965/1/012053

Table 3. Information about the distribution of forests of Leshchevskoe forestry, according to the species composition of the forest.

No.	Names of tree species	Area, ha
1	Pine	61
2	Oak	5036
3	Ash	2665
4	Elm	56
5	Acacia	6
6	Poplar	1240
7	Willow	53
8	Apple tree	6
9	Marsh elder	25
10	Currant	94
11	Hawthorn	1
12	Other shrubs	31
13	Other tree species	39
	Total	9313

A significant part of the forestry fund lands (2510 ha or 15%) relates to the I, II and III classes of natural fire hazard, the most dangerous in relation to fire.

Special protection of forests from fires is responsible for the forest areas, which are classified into the following categories of protected forests:

- forests, located in water protection zones;
- forests, performing protection functions of natural or other objects;
- young growths of coniferous and economically valuable species;
- ripe stands with undergrowth of valuable species.

Factors that increase fire hazards of Leshchevskoe forestry are:

- rugged relief;
- frequent droughts with strong and dry winds;
- significant littering of forests;
- high transport accessibility for inhabitants of towns and human settlements in the surrounding area;
- recreational and other use of forests;
- highway of Srednyaya Akhtuba-Leshchev;
- highway of Leninsk-Leshchev;
- highway of Leninsk-Karshevitoe.

In accordance with the Order of Forestry Agency of 05.08.2020 No.753 "On the establishment of forest fire zoning of Forestry Fund lands and recognition invalid the Order of the Federal Forestry Agency of 07.06.2018 No.468" forests of the Volgograd region are classified as a ground detection and extinguishing zone.

The agents of the specialized state budgetary institution of the Volgograd region "Leshchevskoe forestry" monitor fire hazards in forests and wildfires.

The list and terms of measures on organization of monitoring and control of fire hazards in forests and wildfires [9]:

1. Organization of monitoring of the forest fire situation through the Information system for remote monitoring of forest Fires of the Federal Forestry Agency – every day, during the fire season.
2. Organization of monitoring while carrying out ground patrol of forests on approved routes and raiding by the State Forestry Inspectors, and also by the Forest fire station forces - during the fire season on the developed routes in accordance with the patrol schedule.
3. Organization of monitoring from observation points - during the fire season in accordance with the Order of the Russian Ministry of Natural Resources of 23.06.2014 No. 276.
4. Identification and accounting of fire hazards classes depending on the weather conditions - every day, during the fire season.

The list and terms of measures on organization of fire detection systems, accounting of fires, and the monitoring system of their development:

1. Organization of monitoring and wildfire detection through the Information system for remote monitoring of forest Fires of the Federal Forestry Agency - every day, during the fire season.
2. Forest inspection in order to detect wildfires is carried out during the fire season in forests depending on the weather conditions (fire hazard class):
   - at II fire hazard class in 10, 13, 16, 19 hours;
   - at III fire hazard class - from 10 to 20 hours at least once every two hours;
   - at IV fire hazard class - from 9 to 21 hours at least once an hour;
   - at V fire hazard class - from 6 to 24 hours at least once an hour.
3. Organization of monitoring and wildfire detection while carrying out ground patrol of forests by the Forest fire station forces.
4. Organization and implementation of ground patrols aimed to monitor fire hazards in forests and wildfires by the State Forestry Inspectors - during the fire season on the developed and approved routes in accordance with the patrol schedule.
5. Accounting of wildfires by the specialized dispatch service - during the fire season.

Ground patrol of forests that aimed to monitor fire hazards in forests and wildfires is carried out by the Specialized State Budgetary Institution of the Volgograd region "Leshchevskoe forestry" on routes No. 1a, 1b, 1v, 1g.

The period of patrolling, its length, frequency and time depend on the fire hazard class in forests according to the weather conditions and the natural fire hazard class of forests.

Ground patrol of forests aimed to monitor fire hazards in forests and wildfires can be carried out by the state forest inspectors of the forestry according to the approved routes and schedules.

4. Conclusion

The actual fire-dangerous season in the forestry lasts from April till October. The duration of the fire season depends on melting and formation of snow cover, as well as on anthropogenic pressures on forest lands, which rise from year to year, that significantly increases fire hazards of the forestry area. The main cause of wildfires on the territory of Leshchevskoe forestry is unidentified causes. That is why it is necessary to plan the works on firefighting management and carry out fire prevention activities to reduce forest burning of the territory of Leshchevskoe forestry.

References

[1] Medvedev V N, Belomutenko D V and Istochkina M V 2021 Analysis of methods of extinguishing landscape fires used on the territory of the Volgograd region *Innovative Technologies in the Agro-Industrial Complex in Modern Economic Conditions* (Volgograd)

[2] Federal Law of 21.12.1994 No. 69-FZ "On Fire Safety" (as amended on 11.06.2021) 1994 *Collection of Legislation of the Russian Federation* **35** 3649

[3] Forestry Regulations of Leshchevsky Forestry of the Volgograd Region (as amended by Orders of the Committee of Natural Resources, Forestry and Ecology of the Volgograd Region No. 920 of 20.03.2019, No. 2787 of 01.11.2019, No. 525-OD of 16.03.2020)

[4] Federal Law of 04.12.2006 No. 200-FZ "The Forest Code of the Russian Federation" (as amended on 09.03.2021) 1997 *Collection of Legislation of the Russian Federation* **5** 610
[5] Resolution of the Government of the Russian Federation of April 16, 2011 No. 281 " On measures of fire-fighting arrangement of forests" https://base.garant.ru/12184962/
[6] Resolution of the Government of the Russian Federation of May 17, 2011 No. 377 "On approval of the Rules for the development and approval of the Plan for extinguishing forest fires and its Form" (with amendments and additions) https://base.garant.ru/12185983/
[7] Resolution of the Government of the Russian Federation of June 22, 2007 No. 394 "On Approval of the Regulations on the Implementation of the Federal State Forest Supervision (Forest Protection)" (with amendments and additions) https://base.garant.ru/12154199/
[8] Order of the Ministry of Natural Resources of the Russian Federation of 23.06.2014 No. 276 (as amended on 01.06.2016) "On Approval of the Procedure for Monitoring Fire Hazards in Forests and Forest Fires" 2014 Ministry of Justice of the Russian Federation 33144
[9] Resolution of the Government of the Russian Federation of October 7, 2020 No. 1614 "On Approval of Fire Safety Rules in Forests" https://base.garant.ru/74739511/
[10] Federal Law of 22.07.2008 N 123-FZ "Technical Regulations on Fire Safety requirements" 2008 *Collection of Legislation of the Russian Federation* **30** 3579
[11] Kablov V F, Kostin V E and Sokolova N A 2015 *Volga-Akhtuba Floodplain Ecological Situation: Problems and Solutions for Its Improvement* (Volgograd: Volgograd State Technical University) p 241

IOP Conf. Series: Earth and Environmental Science **965** (2022) 012054 doi:10.1088/1755-1315/965/1/012054

Methodology for assessing the energy efficiency of technical means for harvesting Sudanese grass

A I Ryadnov, O A Fedorova and S Yu Fandeev

Volgograd State Agrarian University, 26 University Avenue, Volgograd, 400002, Russia

E-mail: alex.rjadnov@mail.ru

Abstract. The article discusses the design features of the developed technical means for hay bales loading, transporting and unloading, which makes it possible to exclude its downtime when loading bales. The technical means use effectiveness comprehensive assessment method that combine several operations for Sudan grass harvesting for seeds and fodder for cattle is proposed. The methodology includes the following stages: the choice of factors affecting the technical means use efficiency; the substantiation of particular performance indicators, selected from the set of possible ones according to their relative importance; the selection of the mathematical dependence of the effectiveness complex criterion; the effectiveness complex criterion evaluation and the analysis of the compared technical means effectiveness.

1. Introduction

The population's life quality in the Russian Federation is inextricably linked with the food provision, including meat products. The number of meat animals, including cattle, as well as their productivity, determines the increase in the production of meat products.

Regions of the Russian Federation with insufficient moisture content, which include, in particular, the Volgograd, Astrakhan, Orenburg regions, the Republic of Kalmykia, are unfavorable for the production of most cereals. Therefore, in such regions, there is an acute issue of the fodder base for cattle.

A valuable feed for cattle can be Sudan grass [1, 2], which belongs to the agricultural crop "sorghum". Sorghum is capable of growing and producing a high yield in arid conditions [3, 4].

Hay from Sudan grass is harvested in a loose and pressed form. Moreover, at present, hay is more often pressed into bales and rolls, since it is better stored in this form, it loses 1.5-2.0 times less nutrients compared to storage in loose form, and takes up less space. When harvesting hay, it is necessary to carry out technological operations within the time limits established by agrotechnical requirements, with high quality and minimal funds and labor expenditure.

Serially produced machines for Sudan grass harvesting for seeds do not provide the desired seed quality, and their use in harvesting for cattle feed is very costly. Along with serial machines for Sudan grass harvesting, some farms use machines of their own production.

Technological operations for the agricultural crops cultivation and harvesting require significant labor, energy and money expenditures. However, it should be noted that a significant share of these costs falls on transport and handling operations. For example, transport work expenditures are 15-40% of the agricultural products production and harvesting total costs [5]. A special role is assigned to

IOP Conf. Series: Earth and Environmental Science **965** (2022) 012054 doi:10.1088/1755-1315/965/1/012054

transport operations during the agricultural crops harvesting [6], as well as loading and unloading operations [7].

In agricultural enterprises, various trucks are used to transport hay from Sudan grass bales and rolls, for example, GAZ-SAZ 3307, KAMAZ 55102, GAZ-3102 Gazelle, machine-tractor units as the part of a tractor and a tractor cart, for example, K-744R2 + ZPTS-12, Belarus 82.1 + 2PTS-4, as well as various modifications of roll transporters, for example, TP-10. For loading bales and rolls into vehicles, for example, loaders SNU-550 and PF-0.5, equipped with a device PT-F-500, are used. At the same time, according to our research, a significant proportion of the shift time is lost to perform this operation, which reduces the vehicles productivity.

It is possible to reduce the transport and handling operations expenditures by the most effective technical means application into production that ensure the performance of individual technological operations, selected according to the efficiency comprehensive criterion, or designed new machines that combine some technological operations.

Until now, the scientific literature does not represent a unified method of choosing the most effective technical means use in Sudan grass harvesting for seeds and feed for cattle.

The purpose of the article is to develop a methodology for the technical means use effectiveness comprehensive assessment, based on the analysis of technological operations performed by machines for hay bales loading, transporting and unloading, combining some operations on Sudan grass harvesting for seeds and feed for cattle.

2. Materials and methods

In this work, information materials from open sources of sites of research and educational institutions, scientific articles, as well as the results of patent research are used.

The object of the study was a vehicle for hay bales loading, transporting and unloading, designed at the Volgograd State Agrarian University.

A special attention is paid to the particular indicators set selection of the developed vehicle for hay from Sudan grass bales loading, transporting and unloading use, which form an efficiency comprehensive criterion.

3. Results and discussion

For harvesting bales formed from hay materials, various machines that combine several technological operations designs were developed.

In the work [8] a hay bales loader-transporter is proposed. This loader-transporter is simple in design, but it has a significant drawback - when using it, a significant proportion of the shift time is lost while the stops before loading each bale.

The design solution for a vehicle that reduces the time spent on hay bales loading and unloading was proposed in the work [9] to increase the productivity of transport means for the bales transportation.

Its design simplicity and high productivity are the main advantages of this vehicle, and the disadvantage is a significant loss of time for lifting the vehicle from the working position to the transport one after loading the bale and lowering it to the working position when approaching the next loaded bale.

The transport mean design that combines three technological operations: hay bales loading, transportation and unloading [10] was designed in the Volgograd State Agrarian University (Figure 1).

IOP Conf. Series: Earth and Environmental Science **965** (2022) 012054 doi:10.1088/1755-1315/965/1/012054

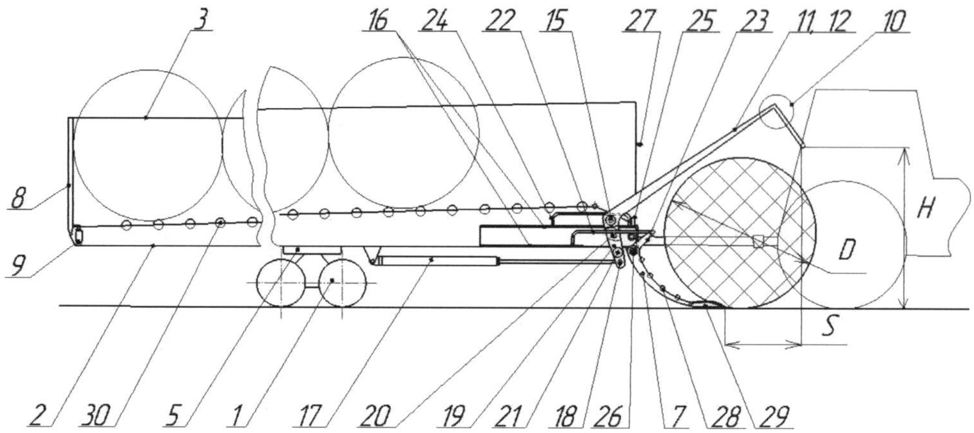

(a)

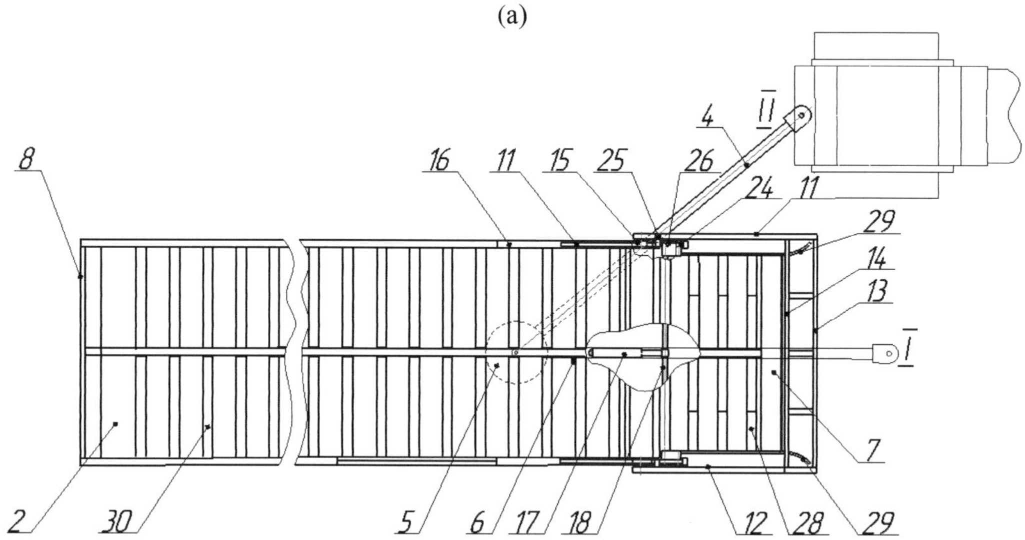

(b)

Figure 1. Vehicle for hay bales loading, transporting and unloading: a - side view, b - top view; 1 - chassis; 2 - platform; 3 - barriers; 4 - trailer; 5 - steering circle; 6 - retainer; 7 - loader; 8 - tailgate; 9 - hydraulic cylinder; 10 - clamp; 11, 12 - longitudinal beams; 13, 14 - cross beams; 15 - rollers; 16 - guideways; 17 - hydraulic cylinder; 18 - axis; 19 - pressure levers; 20, 21, 28, 30 - rollers; 22 - traction; 23 - loader lever; 24 - limiter; 25 - stopper; 26 - retainer; 27 - device for fixing the front side; 29 – guideways.

The developed vehicle works as follows.

Before starting operating, the following preparatory work is performed: the trailer 4 of the vehicle is transferred from the transport position *I* to the working position *II* and fix it in the position *II*.

After the vehicle enters the field with rows of hay bales located on it, the loader 7 is lowered until it stops against the surface of the field, which is then directed to the bale in such a way that the bale is between the guideways 29. Then, with the help of the energy device hydraulic system, the levers 19 are turned around the axes of the upper rollers 15 to the stop of the lower rollers 21 in the guideways 16. In this case, the clamp 10 is lowered to the stop in the bale. With further actuation of the hydraulic cylinder 17 rod on the clamp 10, the hay bale is pulled onto the rollers 28 of the loader 7. Next, the middle rollers 20 move the tractions 22, pivotally connected to the levers 23, actuating on the loader 7, which then feeds the hay bale to the deck of the platform 2. Due to the fact that the deck of the platform has an inclination towards the tailgate 8, the hay bale moves along the rollers 30 until it stops against the tailgate. At the moment the loader 7 touches the front barriers 3, the device 27 for fixing the front side in the closed position is triggered.

When loading the next bale, the handle of the hydraulic valve is moved to the "floating position", the loader 7 is disconnected from the front barriers 3 by using the device 27 for fixing the loader 7. The loader 7 and the clamp 10 are lowered under their own weight. The loader 7 is lowered until it touches the surface of the field, and the clamp 10 - until it stops against the stoppers 25. In this case, each lever 23 moves the corresponding traction 22, actuating upon the middle roller 20, which, in turn, moves the clamp 10 forward to the stop of the upper roller 15 into the limiter 24. The height and place of installation of the stopper 25 provide the lower position, and the limiter 24 - the extreme front position of the clamp 10. The valve handle is moved to the "lift" position. In this case, the rod of the hydro-cylinder extends. The lower roller 21 rotates around the axis of the upper roller 15, the clamp 10 is set to the position in which the first cross beam 13 is above the bale.

The process of loading the vehicle with bales of mainly hay-straw materials is repeated.

When the vehicle is fully loaded with bales, the trailer 4 is transferred from the working position to the transport position. The energy device transports the vehicle to the bale storage area.

When unloading the bales with the help of the hydraulic cylinder 9, the tailgate 8 opens, the vehicle jerkily begins to move and the bales slide down the rollers 30 of the platform 2 reclined deck to the bale storage area.

The proposed vehicle design allows loading hay bales into a vehicle without stopping it, transporting and unloading the bales, which significantly increases the productivity of Sudan grass harvesting compressed into bales, reduces labor and costs expenditures.

It should be noted that when developing new technical means designs for agricultural crops harvesting, it is necessary to take into account the characteristics of the crop, which harvesting is supposed to use the new technical means [11, 12].

The considered machines, which carry out hay-straw materials bales loading, transportation and unloading, allow a greater or lesser degree to improve the operational and economic indicators of their use in comparison with the serially produced by the industry. However, it is impossible to determine even the best of the three technical means considered above, since each of them has its own advantages and disadvantages. In this case, the technical means are chosen by the decision-maker for one or two indicators, which is not always the right decision.

The authors consider that a technical mean must be selected based on a set of use particular indicators included in the efficiency comprehensive criterion.

It is known that the comprehensive performance criterion is developed on the basis of one of the following rational decision-making concepts: suitability, optimization, adaptability [13].

When implementing the concept of suitability, it is necessary that the values of the particular indicators of the technical means use W_i were not lower than the values of the corresponding indicators W_i^{TP} required level.

The analysis of scientific works showed that one of the first papers reflecting the concept of suitability in solving the problem related to the agricultural crops harvesting is the study [14], which presents this work's author long-term research. The work [14] also developed the methodology for the formation of the efficiency complex criterion, which was then used when choosing agricultural machinery, including grain harvesters [15].

In order to select the most efficient vehicle for Sudan grass hay bales loading, transporting and unloading, the basic principle that reflects a general approach to achieving the set goal was determined first - the vehicles using effectiveness assessment should be carried out in an integrated manner, taking into account the results of assessing a set of the most important principle indicators.

Taking into account the formulated principle, the scheme for choosing technical means for Sudan grass harvesting was developed according to the efficiency complex criterion (Figure 2).

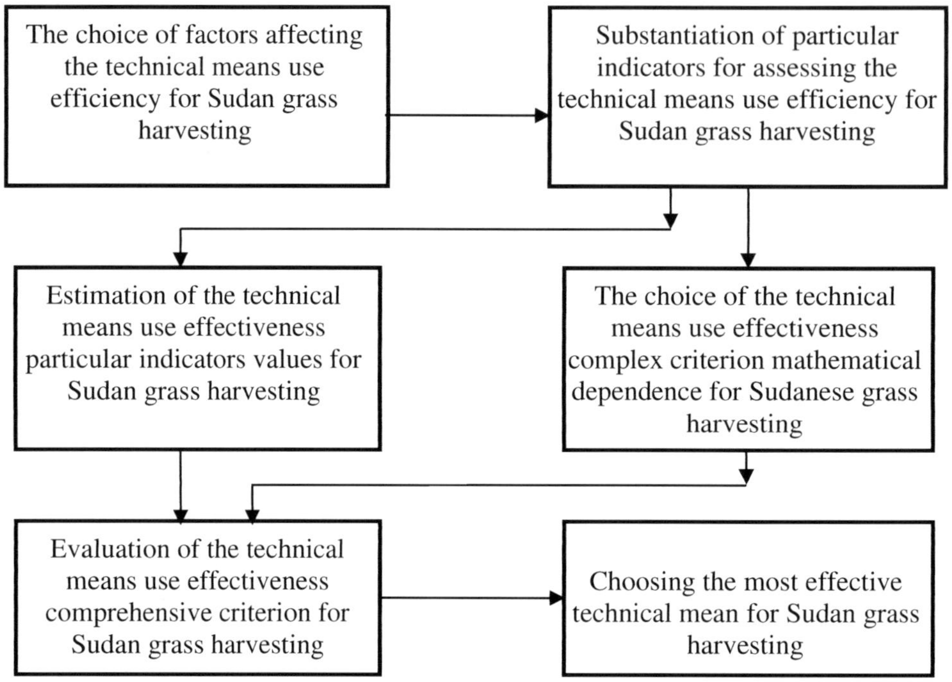

Figure 2. Scheme of technical means selection for Sudan grass harvesting according to the effectiveness complex criterion.

The choice of a technical mean for Sudan grass harvesting according to the efficiency complex criterion involves the implementation of all the stages shown in Figure 2.

In paper [16], the substantiation of the factors influencing the technical means use effectiveness for Sudan grass harvesting is given in detail. All factors are conveniently classified into three large groups: the design and technological factors of machines that press Sudan grass hay into bales, machines that load bales into a vehicle, as well as machines that transport bales from the field to their storage areas used for harvesting; harvesting conditions and technological factors.

Particular indicators for assessing the technical means use effectiveness for Sudan grass harvesting are selected on the basis of the scientific works and practical experience analysis.

Several particular indicators can be selected to assess the efficiency of using vehicles. But some of them may not be significant. To determine the significance of particular indicators, the authors used the method of expert assessment [13].

As the result of the research works analysis and the selected indicators importance assessment, it was found that one of the most important indicators of the technical means use for harvesting crops is their productivity.

For technical means carrying out bales loading, transportation and unloading, the hourly throughput (W_h, kg/h) is determined by the total mass of hay bales delivered per unit of time to the storage site:

$$W_h = \sum_{j=1}^{j=N} m_j N_{bj} / T_{sh}, \tag{1}$$

where m_j is the mass of one bale compressed by the j-th pick-up baler brand, kg; N_{bj} is the number of bales formed by the j-th type of the pick-up baler and transported per shift by a vehicle to the storage site, pcs; T_{sh} is the shift duration, h; $j = 1\ldots N$ is the number of pick-up baler types.

The second most important particular indicator is the total labor intensity of work on Sudan grass hay bales loading, transporting and unloading, since during each of the technological operations, one or several machines can work. Moreover, one or several operators can service each machine. Currently, one machine is serviced, as a rule, by one operator due to the use of automation and robotics. Consequently, the total labor intensity of work on Sudan grass hay bales loading, transporting and unloading T (man-hours) will be determined by the number of machines used at each operation and the number of technological operations.

$$T = \sum_{i=1}^{i=M} t_i N_{opi}, \tag{2}$$

where t_i is the complexity of performing the i-th technological operation, man-hours; N_{opi} is the number of operators employed in the i-th technological operation, people; $I = 1 \ldots M$ is the number of technological operations, pcs.

The third most important particular indicator is the total fuel consumption G (l) by engines of machines used for Sudan grass hay bales loading, transporting and unloading:

$$G = \sum_{k=1}^{k=l} g_{ik} t_{ik}, \tag{3}$$

where g_{ik} is the hourly fuel consumption for the i-th technological operation by the machine of the k-th type, l/h; t_{ik} is the operating time of the machine with the k-th type on the i-th technological operation, h; $k = 1 \ldots l$ is the number of car types, pcs.

In addition to the above particular indicators of the machines use that carry out Sudan grass hay bales loading, transportation and unloading, the specific cost of one hay bale loading, transportation and unloading (C, rubles) and damageability and hay bales possible losses during the transportation (L, %).

Thus, the following indicators should be included in the mathematical dependence of the technical means use effectiveness complex criterion for Sudan grass hay bales loading, transporting and unloading: W_h, T, G, C and L, as well as their relative importance.

The next stage in choosing a technical mean for Sudan grass harvesting is to substantiate the effectiveness complex criterion mathematical dependence. In this case, the possible options for the aggregation function [13] are considered, from which the one that is acceptable for the given case is selected. In this case, the influence of the selected particular indicators on the using technical means efficiency is taken into account, comparing them with the desired or required values. All particular indicators are reduced to dimensionless and scalar values.

After determining the efficiency complex criterion mathematical dependence, its values for all options for technical means are calculated. The technical mean, for which the complex criterion is maximum, is recommended for use.

4. Conclusion

Thus, on the basis of the scientific-research works analysis, the technical means use effectiveness comprehensive assessment method that combine several operations on Sudan grass harvesting for seeds and feed for cattle is proposed. The technique includes several stages. The first stage is the selection of factors influencing the technical means use efficiency. The second stage is the substantiation of particular performance indicators, taking into account their relative importance. For the proposed technical means, the following particular indicators were selected: productivity of the technical means, total labor intensity of work, total fuel consumption by energy means, specific cost of loading, one hay bale transporting and unloading and hay bales damageability. The next stages: the choice of the effectiveness complex criterion mathematical dependence, its assessment for the compared technical means and the choice of the most effective technical means.

5. Acknowledgments

Funding: The reported study was funded by RFBR, project number 20-38-90255.

References

[1] Zhukova M P *et al* 2015 Prospects of using annual spring forage crops in forage production *Bulletin of the Stavropol Agro-Industrial Complex* **3(19)** 149-153

[2] Tagarakis A C 2017 Proximal sensing to estimate yield of brown midrib forage sorghum *Agronomy J.* **109(1)** 107-114

[3] Berdyshev V E 2011 Methodology for assessing the quality of functioning of grain harvesters *Bulletin of the Altai State Agrarian University* **2(76)** 85-89

[4] Biktimirov R A and Lukmanova F H 2010 Results and prospects of selection of sorghum crops in the Bashkir Research Institute *Achievements of Science and Technology of the Agro-Industrial Complex* **1** 44-45

[5] Gamayunov P P and Alekseev S A 2018 Model of the movement of the tractor train *Scientific Review* **4** 8-11

[6] Byshov N V, Borychev S N, Uspensky I A, Yukhin I A, Kokorev G D, Kostenko M Yu, Rembalovich G K, Kolotov A S and Kolupaev S V 2017 Peculiarities of application of the modern tractor transport processes in the cultivation of crops *Scientific J. of Kuban State Agrarian University* **126(02)** 180-198

[7] Guskov Yu A, Tikhonkin I V and Blynsky Yu N 2018 The use of specialized vehicles in technologies for collecting and transporting plant mass pressed into rolls in the conditions of Siberia *Far Eastern Agrarian Bulletin* **2(46)** 146-152

[8] Shaprov M N, Abezin V G, Kharlashin A V and Tseplyaev A N 2016 Patent of the Russian Federation No. 2580163 Loader-conveyor of rolls of hay-straw materials *Bulletin* **10**

[9] Guskov Yu A, Blynsky Yu N, Golub S A and Tikhonkin I V 1998 Patent of the Russian Federation No. 2108022 Vehicle for loading, transporting and unloading rolls *Bulletin* **10**

[10] Ryadnov A I, Fedorova O A, Sharipov R V and Fandeev S Yu 2020 Patent of the Russian Federation No. 2728961 Vehicle for loading, transporting and unloading rolls mainly made of hay-straw materials *Bulletin* **3**

[11] Kim S K *et al* 2018 Changes in physicochemical characteristics of sorghum among different varieties and at different harvest stages after heading *Korean J. of Food Science and Technology* **50(3)** 260-266

[12] Mathanker S K and Hansen A C 2014 Harvesting system design and performance *Engineering and Science of Biomass Feedstock Production and Provision* pp 85-139

[13] Abduyevsky V S *et al* 1986 *Reliability and Efficiency in Engineering* vol 3 ed V F Utkin and Yu V Kryuchkov (Moscow: Mechanical Engineering) p 328

[14] Ryadnov A I 2008 *Methods for Evaluating the Efficiency of Harvesting Agricultural Crops* (Volgograd: Volgograd State Agricultural Academy) p 108

[15] Berdyshev V E 2010 Theoretical definition of the complex indicator of the efficiency of grain harvesters *Proc. of the Nizhnevolzhsky Agrouniversity Complex: Science and Higher Professional Education* **3(19)** 168-172

[16] Ryadnov A I and Fandeev S Yu 2020 Factors affecting the efficiency of using machines for forming and transporting hay rolls *Proc. of the Nizhnevolzhsky Agrouniversity Complex: Science and Higher Professional Education* **4(60)** 460-468

IOP Conf. Series: Earth and Environmental Science **965** (2022) 012055 doi:10.1088/1755-1315/965/1/012055

The use of thin-walled structures for the sustainable development of agro-industrial systems and facilities

Yu V Klochkov[1], A P Nikolaev[1], T R Ishchanov[1], A Sh Dzhabrailov[1], A S Andreev[1] and M Yu Klochkov[2]

[1] Department of Higher Mathematics, Volgograd State Agrarian University, 26 University Avenue, Volgograd, 400002, Russia
[2] Faculty of Physics, Lomonosov Moscow State University, 1 Leninskie Gory, Moscow, 119991, (City service post office-3), Russia

E-mail: Klotchkov@bk.ru

Abstract. The issues of optimal use of strength resources of materials from which elements of thin-walled objects of the agro-industrial complex are made for the purpose of sustainable development of agricultural production are currently coming to the fore. In this regard, the development of modern algorithms for the numerical analysis of the processes of physically nonlinear deformation of thin-walled objects of agro-industrial complex of various sizes and shapes is becoming a rather urgent topic. The article proposes an algorithm for the finite element analysis of physically nonlinear deformation of thin-walled objects of the agro-industrial complex, taking into account the transverse shear, based on the hypotheses of the theory of plastic flow The element of discretization of the thin-walled object of the agricultural and industrial complex was a four-node fragment of its middle surface. The stiffness matrix of this sampling unit was obtained in the process of minimizing the Lagrange functional with respect to the sought nodal parameters, which were the components of the step vector of displacement and their first-order partial derivatives, as well as the components of the step vector of the angles of rotation of the normal.

1. Introduction

As the experience of recent years shows, in the modern world there is a steady demand for food, and with a pronounced tendency to an increase in its cost. Russia, with its huge potential of land resources and a variety of climatic zones, is becoming one of the leading exporters of food and agricultural raw materials. Sustainable development of the agricultural sector and the growth of agricultural production volumes require adequate development of engineering systems of the agro-industrial complex, such as irrigation and drainage networks, bunkers, granaries and others. The overwhelming majority of such systems and objects of the agro-industrial complex belong to the class of thin-walled structures. In this regard, the task of developing modern computing technologies for the nonlinear analysis of the stress-strain state of such objects and systems with optimal design and reconstruction of the latter becomes quite urgent. Taking into account the current trend in resource saving, the most in demand are computational algorithms that take into account the possibility of the used material of the thin-walled structure of the agro-industrial complex operating beyond the elastic limit, which ultimately ensure the sustainable development of the agricultural sector.

The presented study presents a computational algorithm for determining the stress-strain state of engineering objects of the agro-industrial complex, which can be classified as thin-walled (water conduits, hangars, bunker structures, tanks, etc.) taking into account their deformation beyond the elastic limit. The research tool is the finite element method (FEM) [1-26]. When constructing a finite element model, transverse shear deformations are taken into account [27]. The provisions of the theory of plastic flow [28] are used as the theory of plasticity.

2. Materials and methods

2.1. Geometric relationships

The surface equidistant from the outer and inner boundaries of the investigated thin-walled agricultural and industrial complex (AIC) object is specified by the radius vector

$$\vec{R}^0 = x(\alpha^1,\alpha^2)\vec{i} + y(\alpha^1,\alpha^2)\vec{j} + z(\alpha^1,\alpha^2)\vec{k}, \tag{1}$$

where α^1, α^2 curvilinear coordinates of the above surface.

Using the formulas of differential geometry [29], one can obtain the basis vectors of the point M^0 of the surface

$$\vec{a}_1^0 = \vec{R}^0_{,\alpha^1};\ \vec{a}_2^0 = \vec{R}^0_{,\alpha^2}; \tag{2}$$

$$\vec{a}_3^0 = \vec{a}_1^0 \times \vec{a}_2^0 / \left|\vec{a}_1^0 \times \vec{a}_2^0\right|. \tag{3}$$

If, at the point M^0, restore the normal to the contacting plane and fix the point $M^{0\zeta}$ on this normal at a distance ζ from the point M^0, then the radius vector of the point $M^{0\zeta}$ can be represented as the sum

$$\vec{R}^{0\zeta} = \vec{R}^0 + \zeta\vec{a}_3^0. \tag{4}$$

When analyzing the stress-strain state of thin-walled objects of the agro-industrial complex, both geometrically and physically nonlinear formulation, a stepwise application of an external load is usually used. After sequential loading in n steps, point $M^{0\zeta}$ will move to point M^{ζ}, and after the next (n + 1)-th step of loading - to point $M^{*\zeta}$. The positions of the points M^{ζ} and $M^{*\zeta}$ can be fixed by the corresponding radius vectors

$$\vec{R}^{\zeta} = \vec{R}^{0\zeta} + \vec{V};\ \vec{R}^{*\zeta} = \vec{R}^{\zeta} + \vec{W}. \tag{5}$$

The vectors $\vec{V}$ and $\vec{W}$ characterizing the displacements of the point $M^{0\zeta}$ from the initial position to the deformed ones after the n and (n + 1)-th loading steps can be represented by vector sums

$$\vec{V} = \vec{v} + \zeta\vec{\gamma};\quad \vec{W} = \vec{w} + \zeta\Delta\vec{\gamma}, \tag{6}$$

where $\vec{v} = v^m\vec{a}_m^0;\ \vec{w} = w^m\vec{a}_m^0$, (m = 1, 2, 3) - displacement vectors of point M^0 of the surface (1); $\vec{\gamma} = \gamma^\rho\vec{a}_\rho^0;\ \ \Delta\vec{\gamma} = \Delta\gamma^\rho\vec{a}_\rho^0\ (\rho = 1,2)$ - vectors of slope angles of the normal after n loading steps and increment of this vector at the (n + 1)-th loading step.

The basis vectors at the points $M^{0\zeta}$, M^0 and $M^{*\zeta}$ can be obtained by applying the operation of differentiation with respect to the curvilinear coordinates of the surface (1) to the formulas of radius vectors (4), (5)

$$\vec{g}_\rho^0 = \vec{R}^{0\zeta}_{,\rho};\ \vec{g}_3^0 = \vec{R}^{0\zeta}_{,\zeta};\ \vec{g}_\rho = \vec{R}^{\zeta}_{,\rho};\ \vec{g}_3 = \vec{R}^{\zeta}_{,\zeta};\ \vec{g}_\rho^* = \vec{R}^{*\zeta}_{,\rho};\ \vec{g}_3^* = \vec{R}^{*\zeta}_{,\zeta}, \tag{7}$$

Using the fundamental formula of continuum mechanics [25], one can obtain expressions for the covariant components of the strain increment tensor at the point $M^{0\zeta}$ of a thin-walled AIC object after the (n + 1)-th loading step

$$\Delta\varepsilon_{mn}^{\zeta} = (g_{mn}^* - g_{mn})/2, (m,n = 1,2,3) \tag{8}$$

or in matrix form

$$\left\{\Delta\varepsilon_{mn}^{\zeta}\right\} = [D]\{W\}, \tag{9}$$

where $\{W\}^T = \{w^1w^2w^3\}$ a column matrix of the components of the vector of the (n + 1)-th step movement; $[D]$ is a matrix that includes algebraic and differential operators in its structure.

The covariant components of the metric tensors in (8) at the points M^{ζ} and $M^{*\zeta}$ are calculated by performing the scalar products (7)

$$g_{mn} = \vec{g}_m \cdot \vec{g}_n;\ g_{mn}^* = \vec{g}_m^* \cdot \vec{g}_n^*. \tag{10}$$

2.2. Physical relationships

According to the hypothesis of the theory of plastic flow, the increments of plastic deformations are proportional to the stress deviator [28]

$$\Delta\varepsilon_{mn}^{\zeta\rho} = \frac{3}{2}\frac{\varepsilon_i^{\Delta\rho}}{\sigma_i}\left(\sigma_{mn} - \frac{1}{3}g_{mn}I_1(\sigma)\right), \tag{10a}$$

where $I_1(\sigma) = \sigma_{mn}g^{mn}$ is the first invariant of the stress tensor; σ_i - stress intensity; $\varepsilon_i^{\Delta\rho}$ is the intensity of the increments of plastic deformations, which, as a rule, is replaced by an increment in the intensity of plastic deformations $\varepsilon_i^{\Delta\rho} \approx \Delta\varepsilon_i^{\rho}$, which in turn is determined by the difference between the increment in the intensity of deformations and the increment in the intensity of elastic deformations

$$\Delta\varepsilon_i^{\rho} = \Delta\varepsilon_i - \varepsilon_i^{e}. \tag{11}$$

Using the deformation diagram of the material of a thin-walled AIC object, relation (11) at the (n + 1)-th loading step can be represented in the following form

$$\Delta\varepsilon_i^{\rho} = \frac{\Delta\sigma_i}{E_k} - \frac{\Delta\sigma_i}{E_o} = \Delta\sigma_i\left(\frac{1}{E_k} - \frac{1}{E_o}\right) = \Delta\sigma_i\psi, \tag{12}$$

where $\Delta\sigma_i$ is the increment in stress intensity; E_o, E_k - initial and tangent modules of the deformation diagram.

Representing the increments of deformations at a point in the form of the sum of increments of elastic and plastic deformations when using the orthonormal local basis of the point of the surface of a thin-walled AIC object, we can write the following physical relations at the (n + 1)-th loading step

$$\begin{aligned}
\Delta\varepsilon_{11}^{\zeta} &= \frac{1}{E}(\Delta\sigma_{11} - \mu\Delta\sigma_{22}) + \frac{3}{2}\frac{\Delta\sigma_i}{\sigma_i}\psi(\sigma_{11} - \sigma_0);\\
\Delta\varepsilon_{22}^{\zeta} &= \frac{1}{E}(\Delta\sigma_{22} - \mu\Delta\sigma_{11}) + \frac{3}{2}\frac{\Delta\sigma_i}{\sigma_i}\psi(\sigma_{22} - \sigma_0);\\
\Delta\varepsilon_{33}^{\zeta} &= -\frac{\mu}{E}(\Delta\sigma_{11} + \Delta\sigma_{22}) - \frac{3}{2}\frac{\Delta\sigma_i}{\sigma_i}\psi\sigma_0;\\
2\Delta\varepsilon_{12}^{\zeta} &= \frac{\Delta\sigma_{12}}{G} + \frac{3\Delta\sigma_i}{\sigma_i}\psi\sigma_{12};\\
2\Delta\varepsilon_{13}^{\zeta} &= \frac{\Delta\sigma_{13}}{G} + \frac{3\Delta\sigma_i}{\sigma_i}\psi\sigma_{13};\\
2\Delta\varepsilon_{23}^{\zeta} &= \frac{\Delta\sigma_{23}}{G} + \frac{3\Delta\sigma_i}{\sigma_i}\psi\sigma_{23},
\end{aligned} \tag{13}$$

where E, G - elastic moduli of the first and second kind, μ - Poisson's ratio; $\sigma_0 = (\sigma_{11} + \sigma_{22})/3$.

When assembling (13), the hypothesis of the equality of normal stresses and their increments in the direction normal to the surface of the AIC object, generally accepted in the calculations of thin-walled structures, was taken into account.

$$\sigma_{33} = 0;\ \Delta\sigma_{33} = 0. \tag{14}$$

The increment in stress intensity included in (13) can be expressed in terms of the increments in stresses using the formula for the total differential

$$\Delta\sigma_i = \frac{\partial\sigma_i}{\partial\sigma_{11}}\Delta\sigma_{11} + \frac{\partial\sigma_i}{\partial\sigma_{22}}\Delta\sigma_{22} + \frac{\partial\sigma_i}{\partial\sigma_{12}}\Delta\sigma_{12} + \frac{\partial\sigma_i}{\partial\sigma_{13}}\Delta\sigma_{13} + \frac{\partial\sigma_i}{\partial\sigma_{23}}\Delta\sigma_{23}. \tag{15}$$

Partial derivatives of stress intensity with respect to stresses $\partial\sigma_i/\partial\sigma_{mn}$ can be represented by functions of total stresses accumulated over n previous loading stages

$$\begin{aligned}
&\frac{\partial\sigma_i}{\partial\sigma_{11}} = \frac{1}{2\sigma_i}(2\sigma_{11} - \sigma_{22});\quad \frac{\partial\sigma_i}{\partial\sigma_{22}} = \frac{1}{2\sigma_i}(2\sigma_{22} - \sigma_{11});\\
&\frac{\partial\sigma_i}{\partial\sigma_{12}} = \frac{3\sigma_{12}}{\sigma_i};\quad \frac{\partial\sigma_i}{\partial\sigma_{13}} = \frac{3\sigma_{13}}{\sigma_i};\quad \frac{\partial\sigma_i}{\partial\sigma_{23}} = \frac{3\sigma_{23}}{\sigma_i}.
\end{aligned} \tag{16}$$

Physical relations (13), taking into account (16), can be formalized in the form of matrix dependencies

$$\underset{5\times1}{\left\{\Delta\varepsilon_{mn}^{\zeta}\right\}} = \underset{5\times5}{[N]}\underset{5\times1}{\{\sigma_{mn}\}}; \tag{17}$$

$$\Delta\varepsilon_{33}^{\zeta} = \underset{1\times5}{\{G\}^{\mathrm{T}}}\underset{5\times1}{\{\Delta\sigma_{mn}\}}, \tag{18}$$

where $\{\Delta\varepsilon_{mn}\}^T = \left\{\Delta\varepsilon_{11}^{\zeta}\Delta\varepsilon_{22}^{\zeta}2\Delta\varepsilon_{12}^{\zeta}2\Delta\varepsilon_{13}^{\zeta}2\Delta\varepsilon_{23}^{\zeta}\right\}$; $\{\Delta\sigma_{mn}\}^T = \{\Delta\sigma_{11}\Delta\sigma_{22}\Delta\sigma_{12}\Delta\sigma_{13}\Delta\sigma_{23}\}$.

Applying the operation of referring to (17), one can obtain the required plasticity matrix at the (n + 1)-th stage of step loading

$$\{\Delta\sigma_{mn}\} = [\mathrm{C}_{\Pi}]\left\{\Delta\varepsilon_{mn}^{\zeta}\right\}, \tag{19}$$

where $[\mathrm{C}_{\Pi}] = [N]^{-1}$.

When implementing the computational algorithm, it should be noted that, using the strength parameters of the previous (n - 1)-th loading step, the stress increments are first determined by means of (19), and then, using (18), $\Delta\varepsilon_{33}^{\zeta\,(n)}$ is calculated and, by summation, the total deformations in a direction perpendicular to the median surface

$$\varepsilon_{33}^{\zeta\,(n)} = \Delta\varepsilon_{33}^{\zeta\,(n-1)} + \Delta\varepsilon_{33}^{\zeta\,(n)}. \tag{20}$$

The value (20) is used to calculate the intensity of deformations [28]

$$\varepsilon_i = \frac{\sqrt{2}}{3}\sqrt{\left(\varepsilon_{11}^{\zeta}-\varepsilon_{22}^{\zeta}\right)^2+\left(\varepsilon_{22}^{\zeta}-\varepsilon_{33}^{\zeta}\right)^2+\left(\varepsilon_{33}^{\zeta}-\varepsilon_{11}^{\zeta}\right)^2+\frac{3}{2}\left(\left(2\varepsilon_{12}^{\zeta}\right)^2+\left(2\varepsilon_{13}^{\zeta}\right)^2+\left(2\varepsilon_{23}^{\zeta}\right)^2\right)}. \tag{21}$$

Having the values $\varepsilon_i^{(n)}$ (21), we can determine σ_i, E_k from the deformation diagram and again, using formula (19), compose the plasticity matrix $[\mathrm{C}_{\Pi}]$ of the joint venture corresponding to the (n + 1) step of step loading.

3. Results and discussion

The stiffness matrix of a four-node discretization element at the (n + 1)-th stage of step loading is composed based on the Lagrange functional [27]

$$L = \int_V \left\{\Delta\varepsilon_{mn}^{\zeta}\right\}^T (\{\sigma_{mn}\} + \{\Delta\sigma_{mn}\})dV - \int_F \{W\}^T (\{P\} + \{\Delta P\})dF, \tag{22}$$

where $\{P\}$ and $\{\Delta P\}$ is the total surface load accumulated over n stages of sequential loading and the increment of this load at the (n + 1)-th loading step.

The strain increments at the point M^{ζ} at the (n + 1)-th stage of step loading can be represented as a matrix product

$$\left\{\Delta\varepsilon_{mn}^{\zeta}\right\} = [\Gamma]\{\Delta\varepsilon_{mn}\}, \tag{23}$$

where $\{\Delta\varepsilon_{mn}\}^T$ are the increments of deformations at the point of the median surface M of the thin-walled APC object, and the matrix $[\Gamma]$ has the form

$$\underset{5\times10}{[\Gamma]} = \begin{bmatrix} 1 & 0 & 0 & 0 & 0 & \zeta & 0 & 0 & 0 & 0 \\ 0 & 1 & 0 & 0 & 0 & 0 & \zeta & 0 & 0 & 0 \\ 0 & 0 & 1 & 0 & 0 & 0 & 0 & \zeta & 0 & 0 \\ 0 & 0 & 0 & 1 & 0 & 0 & 0 & 0 & \zeta & 0 \\ 0 & 0 & 0 & 0 & 1 & 0 & 0 & 0 & 0 & \zeta \end{bmatrix}.$$

Using the Cauchy relations at the (n + 1)-th stage of step loading (8), the increments of deformations at the point of the median surface M can be expressed in terms of the column $\{W_y\}$ of the sought quantities at the nodes of the four-node sampling element, which are the components of the step vector of displacement $\vec{w} = w^m\vec{a}_m^0$, their first-order partial derivatives and the components of the step vector of the slope of the normal $\Delta\vec{\gamma} = \Delta\gamma^{\rho}\vec{a}_{\rho}^0$

$$\{\Delta\varepsilon_{mn}\} = [B]\{W_y\}. \tag{24}$$

Functional (22), taking into account (19), (23), and (24), can be transformed to the form

$$L = \{W_y\}^T \int_V [B]^T[\Gamma]^T \left(\{\sigma_{mn}\} + [C_\Pi][\Gamma][B]\{W_y\}\right)dV - \{W_y\}^T \int_F [A]^T \left(\{P\} + \{\Delta P\}\right)dF, \tag{25}$$

where [A] is a matrix containing polynomial functions of the form.

Applying to (25) the procedure of minimization with respect to $\{W_y\}^T$, we can obtain the following matrix relation

$$[M]\{W_y\} = \{Q\} - \{NR\}, \tag{26}$$

where $[M] = \int_V [B]^T[\Gamma]^T [C_\Pi][\Gamma][B]dV$ – stiffness matrix of a four-node sampling element at the (n + 1)-th step of step loading; $\{Q\} = \int_F [A]^T \{\Delta P\}dF$ – column of nodal forces at this stage of loading; $\{NR\} = \int_V [B]^T[\Gamma]^T \{\sigma_{mn}\}dV - \int_F [A]^T \{P\}dF$ – Newton-Raphson correction at the (n + 1) th stage of step loading.

4. Conclusions

It should be noted that the algorithm described above for determining the strength parameters of thin-walled AIC objects, taking into account the plastic deformation of the material from which the objects under study are made, provides for the calculation of the total stresses accumulated over n stages of step loading by summing the stress increments at each step

$$\sigma_{mn} = \sum_{i=1}^{n} \Delta\sigma_{mn}. \tag{27}$$

Despite the absence of physical relationships between total stresses and total strains, which are present in the deformation theory of plasticity, the stresses calculated from the relations of the theory of plastic flow implemented in the above algorithm are in better agreement with the experimental results.

5. Acknowledgements

The study was carried out with the financial support of the Russian Foundation for Basic Research and the Administration of the Volgograd Region in the framework of the research project No. 19-41-343003 r_mol_a.

References

[1] Agapov V 2018 *MATEC Web of Conf.* 04035
[2] Golovanov A I 2010 *Russian Aeronautics* **53(2)** 161-166
[3] Paimushin V N, Badriev I B, Makarov M V and Kholmogorov S A 2019 *J. of Physics: Conf. Series* 032042
[4] Kayumov R A, Muhamedova I Z, Shakirzjanov F R and Tazyukov B F 2018 *J. of Phys.: Conf. Series Current Probl.* 012006
[5] Vinogradov Yu I 2021 *Bulletin of the Russian Academy of Sciences. Rigid body mechanics* **4** 110-123
[6] Khairullin F S and Mingaliev D D 2017 *Bulletin of the Kazan Technological University* **20(14)** 102-104
[7] Sultanov L U 2016 Medium kinematics and constitutive equations *Lobachevskii J. of Mathematics* **37(6)** 787-793
[8] Lalin V V and Zdanchuk E V 2017 *Mat. Phys. and Mech.* **31(1-2)** 28-31
[9] Zheleznov L P, Kabanov V V and Boiko D V 2018 *Russian Aeronautics* **61(2)** 175-182
[10] Yakupov N M, Kiyamov H G and Mukhamedova I Z 2020 *Lobachevskii J. of Mathematics* **41(7)** 1310-1314
[11] Tyukalov Yu Ya 2019 *Engineering and Construction J.* **7(91)** 80-85
[12] Nguyen N 2016 *ZAMP. Z. Angew. Math. and Phys* **67(9)** 35/1-35/24
[13] Lei Zh, Gillot F and Jezeguel L 2015 *Int. J. Mech.* **54** 105-119

[14] Paznanova S L, Vasilev G P, Dineva P S and Manolis G D 2016 *Int. J. Solids and Struct.* **80** 1-18
[15] Leonetti L, Magisano D, Madeo A, Garcea G, Kiendl J and Reali A 2019 *Computer Methods in Applied Mechanics and Engineering* **354** 369-396
[16] Garcea G, Liguori F S, Leonetti L, Magisano D and Madeo A 2017 *Int. J. Numer. Methods Eng.* **112(9)** 1154-1174
[17] Antonietti P F, Beirao da Veiga L, Scacchi S and Verani M 2016 *SIAM J. Numer. Anal.* **54(1)** 34-56
[18] Zucco G, Groh R M J, Madeo A and Weaver P M 2016 *Compos. Struct.* **152 (C)** 324-338
[19] Neto M A, Amaro A, Roseiro L, Cirne J and Leal R 2015 Finite element method for plates/shells *Engineering Computation of Structures: The Finite Element Method* (Cham: Springer)
[20] Chi H, Talischi C, Lopez-Pamies O and Paulino G H 2016 *Comput. Methods Appl. Mech. Engrg.* **306** 216-251
[21] Klochkov Y V, Nikolaev A P, Ishchanov T R and Andreev A S 2020 *J. of Machinery Manufacture and Reliability* **49(4)** 301-307
[22] Klochkov Y V, Nikolaev A P and Ishchanov T R 2018 *J. of Machinery Manufacture and Reliability* **47(4)** 349-355
[23] Klochkov Y V, Nikolaev A P, Kiseleva T A and Marchenko S S 2016 *J. of Machinery Manufacture and Reliability* **45(4)** 328-336
[24] Klochkov Y V, Nikolaev A P, Vakhnina O V and Kiseleva T A 2017 Using Langrange multipliers in the triangular element of a nonshallow shell under variable interpolation of displacements *J. of Applied and Industrial Mathematics* **11(4)** 535-544
[25] Klochkov Y V, Nikolaev A P and Vakhnina O V 2016 *Moscow University Mechanics Bulletin* **71(5)** 114-117
[26] Sedov L I 1976 *Mechanics of a Continuous Medium* (Moscow: Nauka)
[27] Rikards R B 1988 *Finite Elements Method in the Theory of Shells and Plates* (Riga: Zinatne)
[28] Malinin N N 1975 *Applied - Theory of Plasticity and Creep* (Moscow: Mechanical Engineering)
[29] Pogorelov A V 1974 *Differential Geometry* (Moscow: Nauka)

IOP Conf. Series: Earth and Environmental Science **965** (2022) 012056 doi:10.1088/1755-1315/965/1/012056

Improving the design of a root crop harvester in order to increase the sustainability of agriculture

M V Ulyanov[1], D V Skripkin[1], A V Kharlashin[1], A V Ulyanova[1], N P Ayugin[2] and R Sh Khalimov[2]

[1] Volgograd State Agrarian University, 26 University Avenue, Volgograd, 400002, Russia
[2] Ulyanovsk State Agrarian University named after P A Stolypin, 1 Novy Venets Boulevard, Ulyanovsk, 432017, Russia

E-mail: ulyanovmv@mail.ru

Abstract. Cultivating agricultural crops assumes the use of mechanical means to accelerate the performance of technical operations in a reasonable agrotechnical amount of time. Vegetable growing is one of the industries where it is necessary to use modern automated and robotized machinery. Technologies and technical means have been considered which are used in enterprises engaged in cultivating tuberous roots at the moment and their disadvantages have been revealed. As a result, a technology of harvesting tuberous roots has been proposed, which is based on the use of the combine-harvester developed. The design solutions of the combine-harvester have been considered, which allow to solve the problem of the vegetable growers, and the technical process of harvesting tuberous roots has been described. The substantiation of a drum cutter bar has been presented and the coefficient of chaff feed to the transporter, the traction force of the conveyor motion and the power at the drive of the chain slatted conveyors have been determined. As a result of the researches carried out, it was determined that the use of the combine-harvester for tuberous roots shall allow to increase the level of mechanization and the productivity at enterprises, which are engaged in cultivating vegetable crops.

1. Introduction

Harvesting tuberous roots contunues remaining the most labour-intensive operation in the cultivation technology nowadays, as tuberous roots are in the soil. A combine-harvester must extract tuberous roots out of a soil layer by simultaneaously performing the destruction and separation of the soil from the harvest and using technical solutions. Because the technical process of separation of tuberous roots from a soil layer is energy-consuming due to their low content volume of 1…3% [1], it follows thence that a combine-harvester must undercut, separate and extract up to 200 kg of soil per one second in order to extract 4…6 kg of tuberous roots from the soil. One should herewith take into account that the capabilities of the separating mechanisms of combine-harvesters are limited by the dimensions and strength properties of tuberous roots [2]. The separation of tubers from the soil is negatively influenced by the tops, which must be cut before digging out the tuberous roots and removed from the separating workign tools. Large tubers (with the weight exceeding 200 g) and those having a prolonged shape are damaged more when being separated.

IOP Conf. Series: Earth and Environmental Science **965** (2022) 012056 doi:10.1088/1755-1315/965/1/012056

The process of harvesting tuberous roots comprises the following operations: removal of the tops a few weeks before harvesting, extraction of the tuberous roots from the soil and their separation from it, and loading into a means of transport to move the harvest to the place of its sorting and storage.

In dependance to a technology used to cultivate tuberous roots, harvesting is finished by loading the harvest into storage facilities (the in-line technology) or placing the harvest into temporary storage clamps and then to a sorting station with its dispatch for storage [2].

The time when one must start harvesting is determined by the ripeness of a tuber, the destination of the potatoes, and the condition of the culture. The start of combine-harvesting is the moment of the top necrosis. The air temperature during harvesting must be at least +5 °C and, simultaneously, it must be approximately +10 °C in the soil at the level where the tuberous roots are [3]. The humidity of the soil is of a great importance in the course of harvesting as it must be in the range of 10…15% in the sandy loam soils and 16…22% in the clay loam soils. The tops must be therewith removed before the beginning of harvesting down to the height of the stalks of 6…8 cm for combine-harvesting and, when the harvesting is performed by lifters, it must not exceed 20 cm in height. On studying the literature sources, it was found out that the maximum depth of location of tuberous roots is 22 cm and the width of their location in a row is at most 40 cm. [3].

By using the technology of combine-harvesting of tuberous roots, it is necessary to observe the agricultural requirements, which correspond with machine-harvesting: the losses after a combine-harvester must not exceed 2% on the surface and 1% of tuberous roots in the soil, the volume of admixtures, which get into the bunker, must not exceed the following parameters: the total number – 20%, the remains of the tops – 0.5%, and no more than 0.1% of other admixtures, including soil clods and remains of the root system, with mechanical damage to tuberous roots not exceeding the following values: up to 4% in medium and light soils and not more than 10% in those heavy and stony.

Notwithstanding a method of harvesting, the preharvest tops destruction is an obligatory operation. The underripe and developed green tops are exposed to desiccant treatment two weeks before the beginning of harvesting and, after their decay, they are destroyed mechanically [4]. The conventional technical means and technologies for harvesting potatoes do not meet the agricultural requirements presented and need modernizing and improvement.

2. Materials and methods

As a result of the analysis of modern technologies and technical means for harvesting tuberous roots carried out, a design of a combine-harvester for tuberous roots has been developed, which is set out in Figure 1. The design of the combine-harvester for tuberous roots consists of basic carrying frame 1 with disc knives 2 installed vertically at the front, which have the external dimensions corresponding to the condition of cutting through the superior layer down to the depth of location of tuberous roots. The disc knives are located between the rows of tuberous roots and fastened to bearing units. Shovels 3, which are connected to lower conveyor 4, are installed in the rear part of knives 2. The spaces in the row of the slats of conveyor 4 were selected based upon the requirement to prevent losses of tuberous roots and spillage of minor soil fractions. Conveyor 5 is placed upon chain slatted conveyor 4, the distance between its slats being based upon the criterion of avoiding losses of tuberous roots between the slats. Lower chain slatted conveyor 4 is mounted on frame 1. It is possible to change its inclination angle and horizontal plane through the use of degree axis 6. Upper chain slatted conveyor 5 is mounted on frame 5 on degree axis 7. Lower chain slatted conveyor 4 is mounted on frame 8, and upper chain slatted conveyor 5 is mounted on frame 9. Upper chain slatted conveyor 5 is provided with clod destructors 10, which are made of elastic material preventing injuries to roots. The slats of conveyors 4 and 5 are covered with elastic material. The drive of lower 4 and upper 5 chain slatted conveyors is made from reducer 11 through cardan drive 12 and power takeoff shaft of the tractor. Top mounted conveyor 5 is set in motion through reducer 11 from chain gearing 1, and conveyor 4, which is installed at the bottom after the shovel, is set in motion by the same reducer 11 by means of chain 14. The motion speed of conveyor 4 has lower velocity V_1 than that of conveyor 5 (V_2), which is mounted at the top. The upper and lower conveyors finish in the internal plane of drum type loading belt 15, which consists of circular

IOP Conf. Series: Earth and Environmental Science **965** (2022) 012056 doi:10.1088/1755-1315/965/1/012056

rods 16 and blades 17, which are installed in its internal part and grip and hold tuberous roots. The distance between circular rods 16 is preset therewith based upon the condition of avoiding creeping of tuberous roots into the gap between them.

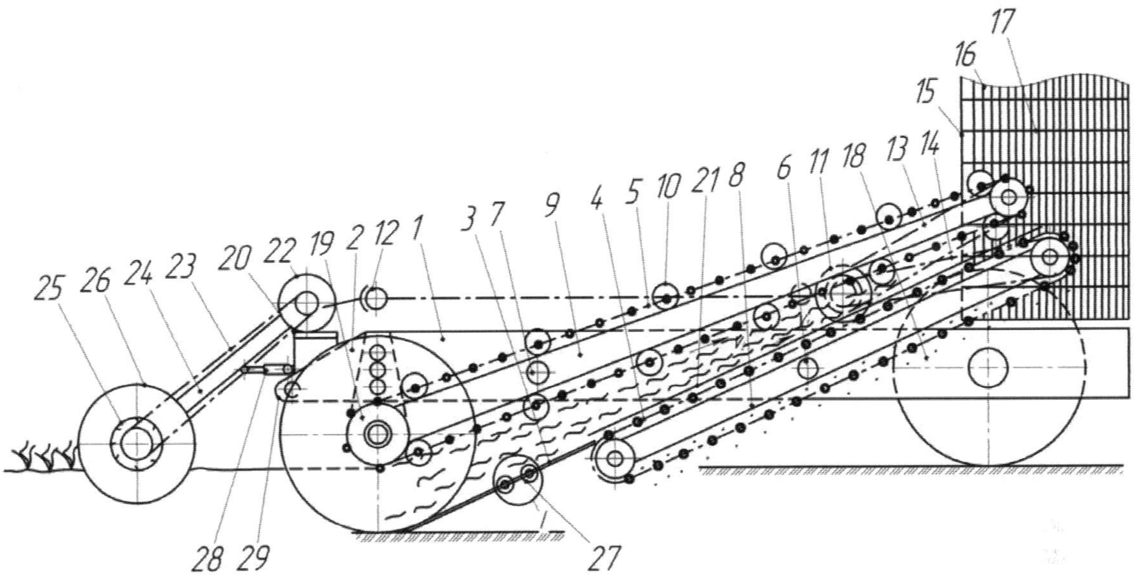

Figure 1. The diagram of the design and technical operation process of the combine-harvester for tuberous roots.

Frame 1 of the combine-harvester for tuberous roots is installed on axes with pneumatic wheels 18.; a guide groove is mounted at the top, in drum type loading belt 15. The drive shaft of upper conveyor 5 is installed in the bearings of disc knives 2. Carrier 20 with the drum of the top cutter bar, which is fastened to it, is installed before frame 1.

Side shields 21 are installed on the sides of lower conveyor 4 with the purpose of reducing the descend of tuberous roots aside off it. Reducer 22 of the drive of the cutter bar is installed on carrier 20 though chain gearing 23. Cutter bar 25 with gauge wheels 26 is mounted on drawbars 24. Soil destroying rollers 27 with teeth, which are located in multiple-thread spiral lines, with the left and right threads, are installed in the upper part of shovel 3 before the coupling with the lower conveyor. Top pulverizer 25 is set in motion by means of hydraulic cylinder 28. Lug 29 of the hitch is used to connect the combine-harvester with the tractor.

Soil destructing rollers 27 are installed on carriers 30 by means of plain bearings 31 in hole 12 of cutting shovel 3.

The technical process of operation of the combine-harvester for tuberous roots runs as outlined below.

By travelling along a field lot where the process of harvesting takes place, top cutter bar 25 chops and scatters about the tops, which are cut along the field lot, the tops having been exposed to desiccant treatment two weeks before the beginning of harvesting.

Disc knives 2, which are installed in the spaces between the rows, cut through the surface of the soils below the depth of location of the tuberous roots, and allow to reduce the traction force of shovel 3, owing to the absence of resistance to the separation of a soil layer from the side walls. Shovel 3 separates a soil layer with tuberous roots in the horizontal plane and passes it to the soil destroying rollers 27, under the impact of which the soil crumbles, with tuberous roots being separated. It is gripped by upper chain slatted conveyor 5 and moved along cutting shovel 3 to lower chain slatted conveyor 4, which also considerably reduces the drawbar resistance of the combine-harvester. Soil destroying rollers 27

IOP Conf. Series: Earth and Environmental Science **965** (2022) 012056 doi:10.1088/1755-1315/965/1/012056

grind the layer and separate tuberous roots from it due to the different direction of the spiral lines of the teeth. When soil with tuberous roots moves along the lower chain slatted conveyor 4, upper chain slatted conveyor 5 provides grinding of the soil and separation of the tuberous roots from it owing to the difference between the motion speeds of the conveyor slats.

Clod destroyer 10 impacts the soil clods, which are difficult to destroy, in the incoming layer, thus improving the separation of the combine-harvester. The separated tuberous roots go to drum type loading belt 15 where, owing to its rotation, they get under the impact of holding blades 17 and get into the guide groove, by means of which loading of containers or means of transport is performed.

The use of top cutter bar 25 in the design allows to reduce the labor intensity of harvesting and improves the conditions for separating the soil and tuberous roots. The use of disc knives in the design proposed allows to considerably decrease the traction force of the combine-harvester for tuberous roots. The soil layer, which is undercut therewith by shovel 3, is then carried by upper conveyor 5 and this also contributes to a reduction in the traction force.

The use of upper conveyor 4 and lower 5 with different motion speeds of their runs improves the process of destruction of a soil layer and increases the quality of separation of tuberous roots from the incoming layer, which enhances the output of the combine harvester and its productivity.

3. Results and discussion

When substantiating the parameters of the cutter bar, angles of setting the knife blade γ, cutter angle φ of the knife, and angle of cutting θ were considered, which are in the plane, which is perpendicular to the cutting plane, being therewith in the following relationship [5]

$$\theta = \varphi + \beta = 90 - \gamma, \tag{1}$$

γ means the angle of setting the blade; it is the angle between its rear surface and the direction of cutting.

The knife of the cutter bar has one sided sharpening and it is set edge forward. It allows to support the constant clearance between the knife blades and the cross bar, which facilitates sharpening of the knives on the surface of the edges without removing them from the drum. [6, 7].

The normal course of the technical process of cutting requires deflection of the blade edge for angle γ from the direction of cutting (Figure 2). This deflection is oriented to the side of movement of the material fed and prevents it from excessive friction on the blade edge.

$$tg\,\gamma = \frac{V_{\text{mat}}}{V_{knife}}, \quad , \tag{2}$$

where V_{mat} means the velocity of the tops feed; V_{knife} means the knife speed.

The reaction of the cutting force of the tops of tuberous roots, which is applied to the knife blade edge, is deflected from the line of the angle of cutting for angle ψ [8]

$$\psi = \gamma + \frac{\beta}{2}, \tag{3}$$

where β means the angle of blade sharpening.

Top cutting force P_{cut} can be distributed into the following elements: normal P_{n}, which is directed along the line of angle of sharpening θ, and bending force P_{k}, which is perpendicular to it [9]

$$P_n = P_{cut} \cos\left(\gamma + \frac{\beta}{2}\right), \tag{4}$$

$$P_k = P_{cut} \sin\left(\gamma + \frac{\beta}{2}\right). \tag{5}$$

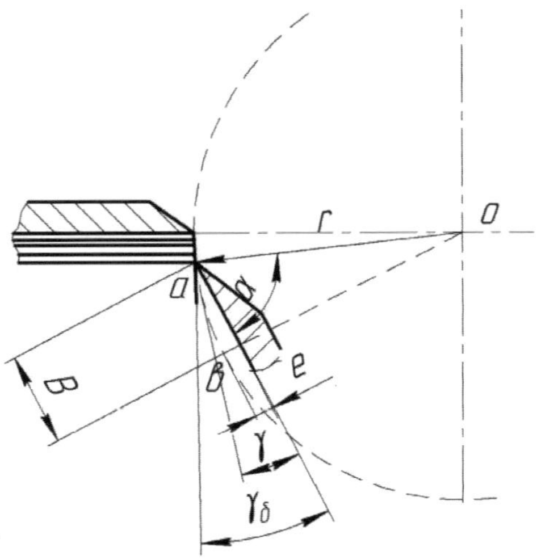

Figure 2. The diagram to substantiate the angles of fastening the knife in the top cutter bar.

The apex of the blade bends under the impact of bending stress P_k and breaks off in a certain section thereof at distance ℓ from the cutting edge of the blade [10]. Bending moment M of stress P_k in the section shall be

$$M = P_k \cdot \ell. \tag{6}$$

The bending strain of the cutting edge can be written in this section as:

$$\sigma_I = \frac{P_k \ell}{\frac{b\,h^2}{6}} = \frac{6\,P_k \cdot \ell}{h^2}, \tag{7}$$

where b means the blade length, m. h means the blade thickness in the section, m.

Value ℓ defines by itself the place of the point of fracture of the knife tip, it is proportional to bending stress P_k and inversely proportional to allowable stress σ_u.

The blade thickness is determined by the formula:

$$\delta = \frac{3\,P_k}{\sigma_I}\left(1 + \frac{1}{\sin\frac{\beta}{2}}\right). \tag{8}$$

It is necessary to use value e of the return of the rear knife edge $e = tg\gamma B_o$ for the drum cutter bar, where $B_o = \frac{B}{\cos\tau}$ means the section of the knife width in the plane, which is perpendicular to the axis of the top chopping drum;

Then B means the knife width, m.; τ - means the gliding angle of the knife.

It is necessary to prevent the knife, which moves in the plane, from bumping so that its rear plane should be at a certain distance from the cylindric generatrix, which is described by the blades. It requires fastening of the knife blade at angle α

$$\cos\alpha = \frac{r^2 + B_o^2 - (r + B_o tg\gamma)^2}{2\,rB_o}. \tag{9}$$

The angle of fastening of chopping knife γ_δ for the cutter bar of the combine-harvester for tuberous roots can be found from the difference between the angles of the rear plane of the knife and the tangent line, which forms a cylinder in the apex of the blade. [11].

$$\gamma_\delta = 90 - \alpha. \tag{10}$$

It has been experimentally established that the value of clearance between the blades of the cutting couple influences the process energetics and the quality of cutting off and chopping.

An empirical dependence has been established between specific work of cutting A_{sp} and clearance Δ.

$$A_{sp} = k\Delta^m + c, \tag{11}$$

where k, m and c mean the coefficients for potato tops: $k = 0.0735$, $m = 1.46$, and $c = 3$.

The soil destroyed is gripped from the tuberous roots by the chain slatted conveyor and moved to digging shovel 3.

The productivity per hour can be determined by formula [12]

$$\Upsilon = \frac{3600\, q_{tot} v}{1000} = 3.6\, q_{tot} \cdot v\ (m^3/h), \tag{12}$$

where q_{tot} means the average quantity of load per unit of length of a load-bearing element; v means the motion speed.

Or by the adjusted formula:

$$v = h\, b \rho v k, \tag{13}$$

where h means the size of the soil chaff to move, m; b means the width of the moving conveyor, m; ρ means the thickness of the soil chaff, kg/m^3; v means velocity of the conveyor, m/s; k means the feed coefficient of chaff soil to the conveyor.

In order to determine the coefficient of chaff feed to the conveyor, let us use the following formula:

$$k = k_1 \cdot k_2 \cdot k_3 \cdot k_4 \cdot k_5, \tag{14}$$

where k_1 means the filling factor ($k_1 = 0.5$); k_2 means the coefficient, which considers thickening of the chaff when it is moved ($k_2 = 1.13$); k_3 means the velocity coefficient ($k_3 = 0.9...0.95$); k_4 means the coefficient, which considers the volume, which is taken up by the chain and the slats ($k_4 = 0.97$); k_5 means the coefficient, which considers the angle of inclination of the inclined conveyor ($k_5 = 0.8...0.9$).

The traction force, N, of the conveyor motion can be determined by the following formula:

$$P = P_1 + P_2 + P_3 + P_4 + P_5, \tag{15}$$

where P_1 means the resistance from friction of the chaff on the soil layer with tuberous roots, which is moved, N; P_2 means the resistance from friction of the chaff on the side walls; P_3 means the resistance from lifting the chaff by the inclined conveyor; P_4 means the resistance from the motion of the conveyor chain, H; P_5 means the resistance from the motion of the chaff in the direction towards the unloading to the drum conveyor, N [13].

The resistance from friction of the chaff on the soil layer of with tuberous roots, which is moved, N

$$P_1 = h \cdot b \cdot \ell \cdot \rho \cdot g \cdot \cos\beta, \tag{16}$$

where ℓ means the way made by the chaff, m; g means the free fall acceleration ($g = 9.81$ m/s^2); f means the coefficient of soil-on-soil friction; β - means the angle of installation of the inclined conveyor, degree.

The resistance from chaff friction on the side walls of the fencing bords:

$$P_2 = h_2 \cdot \rho \cdot g \cdot \ell \cdot f \cdot \zeta \cdot \cos\beta, \tag{17}$$

where ζ means the coefficient of lateral pressure, which is equal to

$$\zeta = \psi \left(1 + f_m^2 - \sqrt{(1 + f_m^2)(f_m + f^2)} - f_m \left(\sqrt{(1 + f_m^2)} - \sqrt{f_m^2 - f}\right)\right), \tag{18}$$

where ψ means the dynamic coefficient (1,5...1,8); f_{in} means the coefficient of internal friction in the chaff.

The effort spent on lifting the chaff by the inclined conveyor, N,

$$P_3 = h \cdot b \cdot \ell \cdot \rho \cdot g \cdot \sin\beta. \tag{19}$$

The resistance to the motion of the chains of the empty conveyor, N,

$$P_4 = 2 \cdot g_c \cdot l_0 \cdot \cos\beta, \tag{20}$$

where g_c means the specific force of gravity of 1 m of the chains with slats, N/m; ℓ_0 means the distance between the axes of the chain-wheels, m.

The resistance of the chaff, which is moved by the drum conveyor, N,

$$P_5 = 0.25 P_4\,. \tag{21}$$

The power at the drive of the chain slatted conveyors, kW,

$$N = kPv/(102\, \eta_T), \tag{22}$$

where k means the coefficient, which provides for the tensile force of the chain of the drive chain-wheel ($k = 1.1$); η_T - means the efficiency of the gearing.

4. Conclusion

The use of the technology and design of the combine-harvester for tuberous roots developed will allow to increase the level of mechanization in enterprises, which are engaged in cultivating vegetable crops, up to 80%, reduce the energy consumption of harvesting down to 20% and enhance the labor productivity up to 50%.

References

[1] Ovchinnikov A S *et al* 2019 The choice of technology of agricultural crops cultivation on the basis of the manual labor costs indicator optimization (for the example of vegetable and melon crops cultivation) *ARPN J. of Engineering and Applied Sciences* **14(22)** 33897-33905

[2] Uspensky I A, Yukhin I A and Zhukov K A 2014 Ways to reduce the traumativeness of fruit and vegetable products during on-farm transportation *Polythematic Network Electronic Scientific J. of the Kuban State Agrarian University* **96(02)** 26

[3] Tseplyaev A N, Ulyanov M V, Tseplyaev A V, Klimov S A and Tseplyaev V A 2019 Determination of the optimal ratio of the speeds of working elements of machines when harvesting watermelons and cabbage *Agrarian Scientific J.* **5** 89-94

[4] Bulgakov V, Pascuzzi S, Ivanovs S, Fedosiy I and Santoro F 2020 A new spiral potato cleaner to enhance the removal of impurities and soil clods in potato harvesting *Sustainability (Switzerland)* **12(23)** 1-19

[5] Bayboboev N G, Goyipov U G, Hamzayev A X, Akbarov S B, Tursunov A A 2021 Substantiation and calculation of gaps of the separating working bodies of machines for cleaning the tubers *IOP Conf. Series: Earth and Environmental Science* **659(1)** 012022

[6] Khamaletdinov R, Martynov V, Mudarisov S, Khasanov E and Pervushin A 2020 Substantiation of rational parameters of the root crops separator with a rotating inner separation surface *J. of Agricultural Engineering* **51(1)** 15-20

[7] Hrushetsky S M, Yaropud V M, Duganets V I, Pryshliak V M and Kurylo V L 2019 Research of constructive and regulatory parameters of the assembly working parts for potato harvesting machines *INMATEH - Agricultural Engineering* **59(3)** 101-110

[8] Yang W, Yang R, Li J, Wei L, Yang J 2018 Optimized tuber-lifting velocity model for cassava harvester design *Advances in Mechanical Engineering* **10(9)**

[9] Bulgakov V, Nikolaenko S, Adamchuk V, Ruzhylo Z, Olt J 2018 Theory of retaining potato bodies during operation of spiral separator *Agronomy Research* **16(1)** 41-51

[10] Yang R and Shang S 2012 Design and test of multifunctional curved-surface shovel in digging harvest *Nongye Gongcheng Xuebao/Transactions of the Chinese Society of Agricultural Engineering* **28(2)** 47-52

[11] Wilczek U, Kulig B, Koch H-J, Kälberloh R and Hensel O 2020 Sensor System development for low-damage sugar beet harvesting - State and perspectives *Zuckerindustrie* **145(5)** 299-306

[12] Dorokhov A S, Aksenov A G, Sibirev A V and Sazonov N V 2020 Justification of design and technological parameters of the onion harvester bed-shaping roller spiral drum *INMATEH - Agricultural Engineering* **60(1)** 107-114

[13] Hevko R B, Tkachenko I G, Rogatynsky R M, Flonts I V, Pohrishchuk B V 2019 Impact of parameters of an after-cleaning conveyor of a root crop harvester on its performance *INMATEH - Agricultural Engineering* **59(3)** 41-48

IOP Conf. Series: Earth and Environmental Science **965** (2022) 012057 doi:10.1088/1755-1315/965/1/012057

Reducing damage of melons and gourds during automated harvesting

M N Shaprov, A V Sedov, O P Sedova, M V Ulyanov and A V Gurba

Volgograd State Agrarian University, 26 University Avenue, Volgograd, 400002, Russia

E-mail: sedov7020@yandex.ru

Abstract. Agricultural enterprises where harvesting is manual contribute significantly to the production of cucurbit crops in modern economic conditions. Their low automation level of harvesting is explained by the absence of modern highly productive machine-harvesters. The goal of the present study is to model the process, wherein cucurbit crops roll in on the floating skid of the guide of a pickup platform without being damaged. An analytical model was composed in order to model the process of interaction between the crops and the floating skid, which allowed to carry out the theoretical substantiation of the technical process. The study carried out resulted in modeling the technical process, wherein crops roll on the floating skid of the guide of a platform, as well as in a theoretical substantiation of the condition, at which the force of impact interaction between a crop and the floating skid must not exceed the ultimate pressure on a crop, which would result in its destruction. The value of this force depends above all on the strength properties and the size and weight characteristics of crops, as well as on the kinematic and structural parameters of the rolling-in device.

1. Introduction

Volgograd Region is amongst the key regions-producers of cucurbit crop cultures in Russia. Market sowing of cucurbit crops is localized to agricultural enterprises, farming companies, and major specialized farms for growing cucurbit crops in the Region. Each of them has its own specific production conditions of harvesting and storing the harvest, as well as different material and technical resources, and production experience. The square occupied by cucurbit crops is therewith at most 20 ha in most agricultural enterprises. Harvesting in such minor squares is carried out manually nearly everywhere [1]. The costs for manual labour for harvesting take up 40 to 60% of all the costs for growing these cultures in the structure of the prime cost of cucurbit crops [2]. The low automation level of the harvesting process is explained by the absence of modern highly productive machine-harvesters, the use of which would allow to decrease considerably the prime cost of the products obtained [3, 4, 5]. Previously developed melon field machine-harvesters are not virtually manufactured, but even these did not comply with the agrotechnical requirements under the quality parameters [6].

Therefore, a necessity to create a rotor machine-harvester has arisen, which would guarantee the obtaining of cucurbit crops products of high quality at the least costs [6]. The goal of our studies is to model the process, wherein cucurbit crops roll in on the floating skid of the guide of a pickup platform without being damaged.

IOP Conf. Series: Earth and Environmental Science **965** (2022) 012057 doi:10.1088/1755-1315/965/1/012057

2. Materials and methods

The analysis of the techniques and methods used nowadays for harvesting cucurbit crops, which we carried out, has demonstrated that total automated harvesting, which is performed along the field for a single pass, with the quality control of the crops at a stationary point, has the most efficiency in terms of quality and productivity of the technical process [7]. In order to introduce this prospective technology, a towed rotor pickup was developed, which carries out harvesting of ripe cucurbit crops in an aggregate with a tractor in a quality manner and guarantees the conformity with all the technical requirements specified for this type of machines [8]. The operating tool is designed in this machine as two honeycomb rotors, which rotate towards each other and seize and displace crops into the middle part of the machine, and then lift them on the platform and move them on the loading conveyor. The flow diagram of the rotor pickup proposed differs essentially from the conventional designs of picking-up devices so that picking up and loading crops are carried out without prior making rolls, which decreases significantly the costs and damage to the crops [9, 10].

The object of study we have considered is the technical process of picking up cucurbit crops by a harvesting rotor. When modeling the process, wherein cucurbit crops roll in on the floating skid of the guide of a pickup platform, the following assumptions were admitted: the rotor pickup moves evenly and rectilinearly; the harvesting rotors rotate at a constant angular velocity; the high of lifting crops on the platform does not vary in the course of harvesting; we admit the diameter and weight of the crops equal to the maximum value, and the shape of the crops as spherical.

3. Results and discussion

The theoretical researches carried out before have demonstrated that the technical process of picking up crops by the rotor pickup developed represents a sequence of operations, which are followed by each other, during which interaction between a blade of the harvesting rotor and a crop takes place: seizing of a crop by the blade, rolling it into a crop-receiving cell and moving along the field, rolling it in on the floating skid of the crop guide, lifting it on the platform, and releasing from the cell.

As a result of the theoretical substantiation of the interaction between cucurbit crops cultures and the operating tool of the pickup, it was found out that, after a crop is seized by a blade of the harvesting rotor and rolls into its crop-receiving cell, it is moved under the impact of impelling force P_I along the surface of the field at speed V_C in the direction towards the platform and spins therewith with angular velocity ω_C [11, 12].

The model of the motion of a crop, which performs a flat movement when this operation is carried out, can be presented as a rolling motion of a pseudo free ball-shaped body along a horizontal surface without gliding, with an instant speed center in point P, and the crop itself can be presented as a homogenous ball with weight m and radius r_C.

The next operation, which is carried out when cucurbit crops cultures are picked up, is rolling in crops on the guide, which is installed on the front bar of a platform and consists of separate floating skids (Figure 1) assembled so that it is possible for them to move freely in the vertical plane. Floating skids 1 are designed in the form of a horizontal section with thickness h_{FS}, which goes by the radius into the sloping section, which is installed at the angle coinciding with angle α of the slope of the platform to the horizon. Regulated support 2 is installed under each floating skid, which limits its sinking when the ground profile is copied or the platform is lifted into the transport position. The front bar and the regulated support of the platform are covered with delivery tray 3 at the top, in which slots are made in order to move the bars of the floating skid.

When the blades of the harvesting rotor rotate, the crops are moved to the platform and their contact with the floating skid takes place in point A (Figure 2). At the moment of this contact, the center of the linear velocities of the crop points instantly moves from point P to point A. In this case, the contact between the crop and the rolling motion surface disappears in point P and the crop can be considered as a free body as the impact impulse coincides with the impulse of the supporting force in contact point A.

IOP Conf. Series: Earth and Environmental Science **965** (2022) 012057 doi:10.1088/1755-1315/965/1/012057

Let us admit in terms of reducing the damage rate of the crops that a perfectly inelastic collision without a bounce, with coefficient of restitution $K = 0$, occurs at the moment of the contact between a crop and the floating skid.

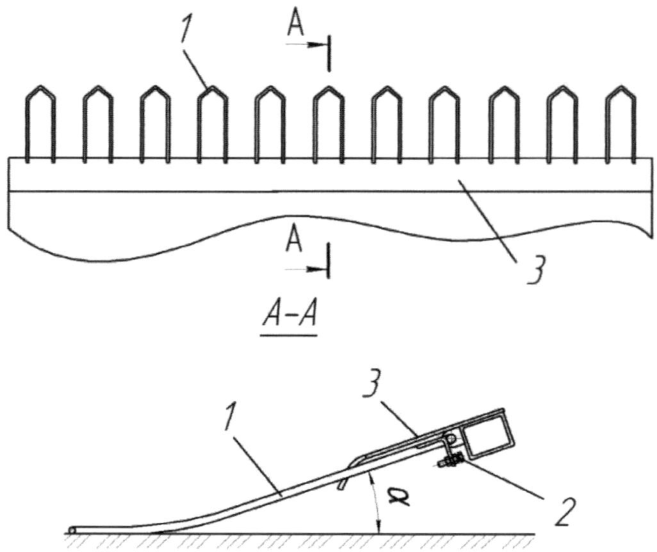

Figure 1. The crop guide design: 1 – floating skid; 2 – regulated support; 3 – delivery tray.

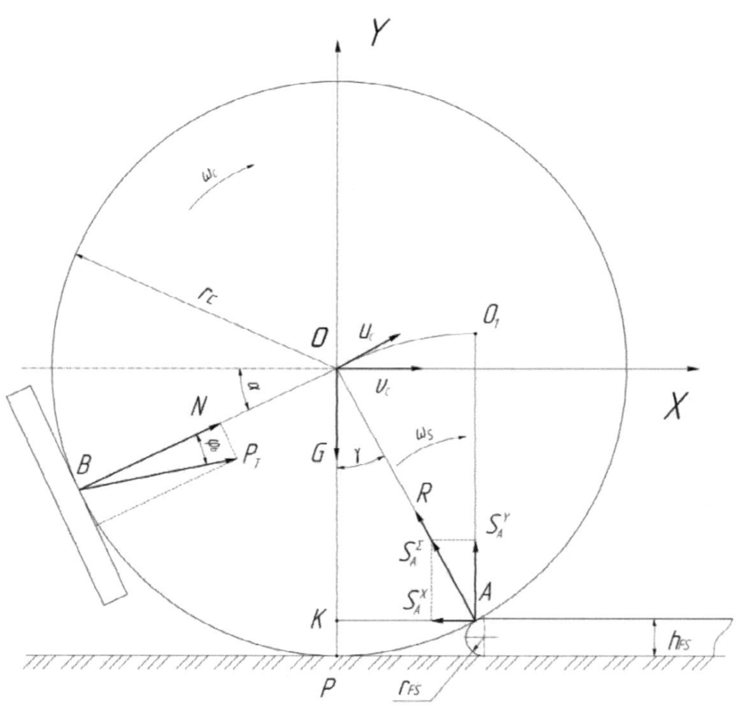

Figure 2. The diagram of interaction between a crop and the floating skid of the guide.

The only support of the crop therewith shall be the edge of the floating skid (contact point A), which creates reaction R preventing the crop from spinning. The line of action of this reaction will pass through center of the crop O at angle γ between the vertical axis of the crop and the radius, which joins contact point A to center of the crop O.

This angle γ can be determined from right-angled triangle OAK, hypothenuse OA of which equals crop radius r_C and cathetus $OK = r_C - h_{FS}$, from the expression:

$$\gamma = arccos\frac{r_c - h_{FS}}{r_c}, \quad (1)$$

where h_{FS} means the thickness of the floating skid, m.

The interaction between the crop and the floating skid results in the transformation of its flat motion into a spinning motion around contact point A, which will have an impact character since the interaction time is little at this transformation. The horizontal trajectory of the motion of crop center O, which was before the moment where the crop met the floating skid, goes therewith to the new trajectory of motion along the radius in relation to contact point A when the impact takes place. The trajectory of the motion of crop center O will be a horizontal line again only after point O_1 goes on vertical line O_1A.

Hence, it follows that the rotational energy of a crop in relation to its own geometrical axis O is completely lost at the moment of impact of a free crop on the floating skid and the linear velocity of crop center of mass V_C will change its direction as a result of the contact and it will become equal to U_C after the impact and angular rolling velocity of the crop ω_C transforms into angular spin rate ω_S of the crop around point A after the impact.

Let us determine a change in the kinetic momentum for the mechanical system "crop – floating skid" regardless of the proper rotation of the crop when it rolls in on the floating skid. Arbitrarily, let the axis, in relation to which a change in the kinetic momentum is formulated, pass through contact point A perpendicularly to the drawing plane.

Kinetic momentum $K^1{}_M$ before the impact is determined from the expression:

$$K_M^1 = m\, v_c\, (r_c - h_{FS}) \quad (2)$$

After the impact, kinetic momentum $K^2{}_M$ is written as:

$$K_M^2 = m\, u_c\, r_c \quad (3)$$

The kinetic momentum of the mechanical system "crop–floating skid" before and after the impact will not change as the impact pulses will not create a momentum in relation to contact point A between the crop and the floating skid, i. e., we shall obtain the equality of expressions (2) and (3):

$$m\, v_c\, (r_c - h_{FS}) = m\, u_c\, r_c \quad (4)$$

Considering that the linear velocity of the crop equals $V_C = \omega_C\, r_C$, and $U_C = \omega_S\, r_C$, expression (4) will be as follows:

$$m\omega_c\, r_c\, (r_c - h_{FS}) = m\, \omega_S\, r_c^2 \quad (5)$$

After the transformation of this equality, we shall obtain angular spin rate ω_S of the crop around contact point A:

$$\omega_S = \omega_c \left(1 - \frac{h_{FS}}{r_c}\right) \quad (6)$$

The model, wherein a free crop rolls in on the floating skid of a guide, which is presented in expression (6), demonstrates that if the thickness of the floating skid and the radius of a crop are equal, $h_{FS} = r_C$, a direct impact occurs, wherein the kinetic energy of the motion of the crop is completely absorbed, which results in the crop stopping to spin in relation to contact point A.

If the thickness of the floating skid is much less than the radius of a crop, $h_{FS} << r_C$, an insignificant loss in the kinetic energy occurs, and the force interaction between the crop and the floating skid will be in this case characterized by impact pulse $S^{\Sigma}{}_A$, which will allow us to obtain horizontal $S^X{}_A$ and vertical $S^Y{}_A$ projections of this pulse, after being expanded along axes X and Y [13].

By using the momentum theorem, we shall find the projection of impact pulse $S^X{}_A$ on axis OX by the expression:

$$-S_A^X = m\, u_c^X \; - m\, v_c^X \quad (7)$$

and the projection of impact pulse $S^Y{}_A$ on axis OY:

$$S_A^Y = m\, u_c^Y \; - m\, v_c^Y \quad (8)$$

By using the diagram, which is set out in figure 2, we shall find the projections of velocity vectors U_C and V_C on axes OX and OY and use them in expressions (7) and (8):

$$-S_A^X = m\,u_c\cos\gamma\ - m\,v_c \tag{9}$$

$$S_A^Y = m\,u_c\ \sin\gamma - 0 \tag{10}$$

By using them in expressions (9) and (10) $V_C = \omega_C\ r_C$ and $U_C = \omega_S\ r_C$, we shall obtain the constituents of the impact pulse:

$$S_A^X = m\,r_c(\omega_c - \omega_S\cos\gamma) \tag{11}$$

$$S_A^Y = m\omega_S\,r_c\ \sin\gamma \tag{12}$$

The modulus of resulting impact pulse S^{Σ}_A, which effects on the crop on the side of the floating skid, is determined by the expression:

$$S_A^{\Sigma} = \sqrt{(S_A^X)^2 + (S_A^Y)^2} \tag{13}$$

With regard to expressions (11) and (12), we shall obtain:

$$S_A^{\Sigma} = \sqrt{m^2 r_c^2(\omega_c - \omega_S\cos\gamma)^2 + m^2\omega_S^2 r_c^2\sin^2\gamma} \tag{14}$$

By performing the conversion of this expression, after the extraction of the root, we shall obtain the modulus of the resulting impact pulse:

$$S_A^{\Sigma} = m\,r_c\omega_c\ \sin\gamma \tag{15}$$

In order to determine the force of impact interaction of a crop, which results in reaction R of the floating skid, let us use H. Hertz formula, which determines the relation between the impact force and the deformation of bodies coming into collision in the contact area:

$$R = Ka^{3/2}, \tag{16}$$

where K means the coefficient, which depends on the properties of the material and the curvature of their surfaces in the contact area; a means the value of the resulting deformation of the crop and the floating skid.

As the material of the floating skid is considerably harder than the material of a crop, the deformation of the floating skid can be neglected and coefficient K will be determined by the following expression:

$$K = \frac{4E\sqrt{R^*}}{3(1-\mu^2)}, \tag{17}$$

where E means the elastic modulus of the crop; μ means Poisson coefficient; R^* means the equivalent radius.

Given that a crop on the contact surface has a sphere with radius r_C, and the floating skid has a sphere with radius r_{FS}, which equals $h_{FS}/2$ on the contact surface, the expression to determine the equivalent radius R^* is given by:

$$R^* = \sqrt{\frac{r_c\frac{h_{FS}}{2}}{r_c+\frac{h_{FS}}{2}}} \tag{18}$$

A deformation of a crop and a displacement of the center of gravity takes place in the process, wherein the crop hits on the floating skid when the relative velocity has its initial value v_o. If counting these displacements from the moment of the contact between the crop and the floating skid, maximum deformation of the crop a_{max} will be determined by the expression:

$$a_{max} = \left(\frac{5m}{4K}v_o^2\right)^{2/5} \tag{19}$$

By using expressions (17) and (19) in expression (16) and performing conversions, we shall obtain maximum impact force of the crop on the floating skid R_{max}:

$$R_{max} = \left(\frac{4E\sqrt{R^*}}{3(1-\mu^2)}\right)^{2/5}\left(\frac{5m}{4}v_o^2\right)^{3/5} \tag{20}$$

By substituting the product of mass m and relative velocity v_o with the value of resulting impact pulse S^{Σ}_A, which is determined by expression (15) in this expression, we shall obtain:

$$R_{max} = \left(\frac{4E\sqrt{R^*}}{3(1-\mu^2)}\right)^{2/5}\left(\frac{5\left(S_A^{\Sigma}\right)^2}{4m}\right)^{3/5} = \left(\frac{4E\sqrt{R^*}}{3(1-\mu^2)}\right)^{2/5}\left(\frac{5m\,r_c^2\omega_c^2\,\sin^2\gamma}{4}\right)^{3/5} \quad (21)$$

The value of this force is limited to the strength properties of crops [14, 15] and must not result in damage thereto, consequently:

$$R_{max} = \left(\frac{4E}{3(1-\mu^2)}\sqrt{\frac{r_c\frac{h_{FS}}{2}}{r_c+\frac{h_{FS}}{2}}}\right)^{2/5}\left(\frac{5m\,r_c^2\omega_c^2\,\sin^2\gamma}{4}\right)^{3/5} \leq P_{CP} \quad (22)$$

where P_{CP} means the maximum pressure on a crop, which does not result in its destruction, N.

4. Conclusion

The study carried out resulted in a model of the technical process, wherein crops roll in on the floating skid of the guide of a platform, and in the theoretical substantiation of the condition, at which the force of impact interaction between a crop and the floating skid must not exceed the maximum pressure on the crop, which would result in its destruction. The value of this force depends above all on the strength properties and the size and weight characteristics of crops, as well as the kinematic and structural parameters of the rolling-in device.

References

[1] Giwa S O and Akanbi T O 2020 Mechanization of melon processing and novel extraction technologies: A short review *Scientific African* **9** e00478

[2] Ovchinnikov A S *et al* 2019 The choice of technology of agricultural crops cultivation on the basis of the manual labor costs indicator optimization (for the example of vegetable and melon crops cultivation) *ARPN J. of Engineering and Applied Sciences* **14(22)** 33897-33905

[3] Roshanianfard A, Noguchi N and Kamata T 2019 Design and performance of a robotic arm for farm use *Int. J. of Agricultural and Biological Engineering* **12(1)** 146-158

[4] Xing X, Liu F and Gao J 2017 Design and implementation of watermelon traceable identification algorithm based on biometric texture information *Transactions of the Chinese Society of Agricultural Engineering* **33(18)** 298-305

[5] Kim Y Y, Hwang H and Cho S I 2008 A hybrid robotic system for harvesting heavy produce *Engineering in Agriculture, Environment and Food* **1(1)** 18-23

[6] Tseplyaev A N, Tseplyaev V A, Ulyanov M V and Klimov S V 2019 Analytical determination of technological parameters of the developed combine harvester for harvesting melons *Proc. of the Nizhnevolzhsky Agrouniversity Complex: Science and Higher Professional Education* **1(53)** 321-327

[7] Shaprov M N, Sedov A V and Gurba A V 2020 Analysis of technologies and methods of harvesting melon crops *Proc. of the Nizhnevolzhsky Agrouniversity Complex: Science and Higher Professional Education* **2(58)** 363-374

[8] Shaprov M N and Sedov AV 2013 Justification of the design of a rotary picker for harvesting melon fruits *Proc. of the Nizhnevolzhsky Agrouniversity Complex: Science and Higher Professional Education* **4(32)** 214-217

[9] Byshov N V *et al* 2018 Theoretical studies of the damage process of easily damaged products in transport vehicle body during the on-farm transportation *ARPN J. of Engineering and Applied Sciences* **13(10)** 3502-3508

[10] Uspensky I A, Yukhin I A and Zhukov K A 2014 Ways to reduce the traumativeness of fruit and vegetable products during on-farm transportation *Polythematic Network Electronic Scientific J. of the Kuban State Agrarian University* **96(02)** 26

[11] Tseplyaev A N, Ulyanov M V, Klimov S V and Tseplyaev A V 2019 Theoretical studies on the study of rolling spherical and toroidal bodies on the example of melon fruits *Bulletin of the Kurgan State Agricultural Academy* **1(29)** 62-65

[12] Tseplyaev A N, Ulyanov M V, Tseplyaev A V, Klimov S A and Tseplyaev V A 2019

Determination of the optimal ratio of the speeds of working elements of machines when harvesting watermelons and cabbage *Agrarian Scientific J.* **5** 89-94

[13] Balakin P D, Kuznetsov E A and Zvezdin D S 2014 Modeling of the impact interaction of the support roller of the multi-purpose crawler engine with a single road obstacle *National priorities of Russia* **4(14)** 103-109

[14] Medvedkov Y, Nazymbekova A, Tlevlessova D, Shaprov M and Kairbayeva A 2021 Development of the juice extraction equipment: physico-mathematical model of the processes *Eastern-European J.l of Enterprise Technologies* **1(11-109)** 14-24

[15] Tseplyaev A N, Ulyanov M V, Lebed N I, Tseplyaev V A and Averina M B 2019 Theoretical justification of impact speeds of working tools on water-melons *IOP Conf. Series: Earth and Environmental Science* **341(1)** 012130

IOP Conf. Series: Earth and Environmental Science **965** (2022) 012058 doi:10.1088/1755-1315/965/1/012058

Modeling potato yield statistics by using information technology for sustainable development

A P Darmanyan and S I Bogdanov

Volgograd State Agrarian University, 26 University Avenue, 400002, Volgograd, Russia

E-mail: adarma@inbox.ru

Abstract. The potato yield statistics analysis and modeling in the Russian Federation according to the Federal State Statistics Service data for the period 2000-2020 was carried out. Based on the analysis of the autocorrelation function (Acf) as a mathematical model for modeling potato yield, the choice of a linear autoregressive model of the first order is justified. By using the multiple regression and correlation analysis methods, the parameters of the model were found, their statistical significance was proved, and the potato yield calculations proved the found model correspondence to real data for the period of 2002-2020.

1. Introduction

The successful application of digital technologies in agriculture is possible only on the basis of the processes and phenomena mathematical models that are formed by using modern IT technologies and software products. An important place among such technologies is occupied by statistical methods, which make it possible to create processes adequate mathematical models, which help to perform scientifically, based forecasting for future periods of time.

However, the use of statistical methods in scientific research, in our opinion, currently in the Russian Federation does not meet the needs of the digital economy. Thus, the study carried out by the Institute for Statistical Studies and Economics of Knowledge of the National Research University Higher School of Economics in 2019 showed that "over a third of Russian scientists do not own the tools of statistical analysis" [1]. Similar results were obtained by the authors earlier [2], when, based on the analysis of scientific publications in agriculture, it was revealed that out of 52 statistical methods used for the analysis of experimental data and mathematical modeling, the scientific studies' authors most often use only descriptive statistics (19%), regression and correlation analysis (2.8%), methods of planning experiments (1.4%). Many other effective statistical methods are not used at all [3, 4].

To analyze the statistics of agricultural production yields, time series analysis methods are often used with the identification of linear trends, as, for example, in the research of grain crops productivity [5]. However, such methods, as a rule, lead to large approximation errors, which do not allow forecasting for future periods of time with the accuracy acceptable for practice.

At the same time, such effective statistical models as autoregressive models of various orders are known, which help to simulate time series with low volatility, i.e. small changes in the series values [6]. Autoregressive models are successfully used in econometrics [7], but they are not used at all in agricultural productivity statistics studies, including potato productivity statistics studies.

The aim of this study was the potato yield statistics mathematical modeling in the Russian Federation by using the autoregressive model (ARM).

2. Materials and methods

To perform mathematical modeling, the Federal State Statistics Service statistics on potato yield in the Russian Federation for 1990-2020 was used [8], which is shown in Figure 1 for the period of 2002-2020.

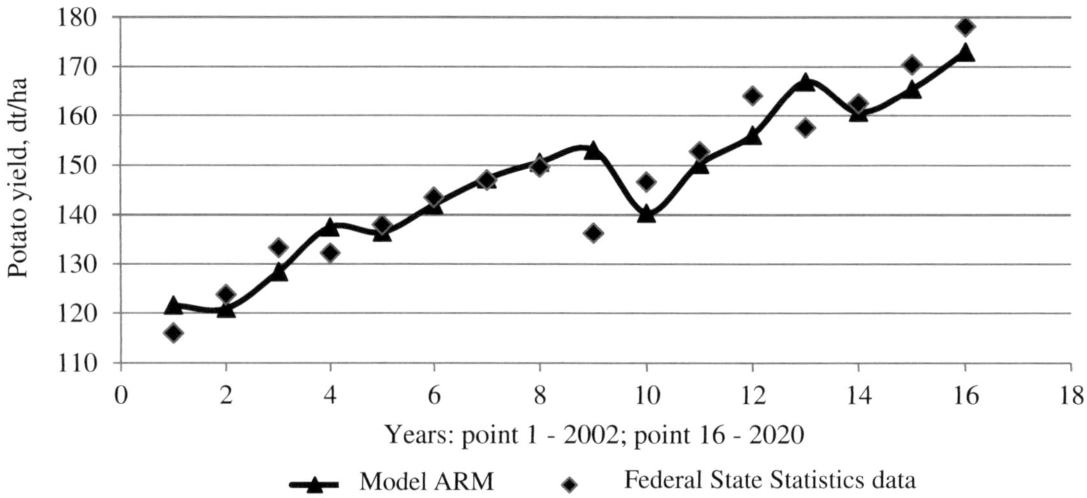

Figure 1. Statistics of potato yield in the Russian Federation for the period of 2002-2020.

As seen in Figure 1, potato yield is a volatile process with a clear positive trend. To simulate such processes, the use of mathematical models with a linear trend allows only to reveal the model of the trend itself, but, as noted above, forecasting by using such a model for future periods of time leads to large errors.

If we do not consider the years with obvious natural anomalies (for example, 2012, point 9 in Figure 1), one of the agricultural crops yield obvious feature is that the value of this indicator in the current year is influenced by the same crop yield in previous years. Therefore, at the research first stage, it is necessary to analyze the presence (or absence) of the relationship between the yield value of the current year and the yield values in previous years (time lags). As it is known [6], this can be done based on the first autocorrelation function (Acf) calculation and analysis.

For the potato productivity statistics in the Russian Federation for the period of 1990-2020 in the MS Excel environment, the authors calculated the Acf, which is shown in the form of the correlogram in Figure 2. As can be seen from Figure 2, the correlation coefficients values on the lags 1 and 2 (1 lag = 1 year) significantly exceed the level of "white noise" shown in Figure 2 by the dashed line. This indicates a strong relationship between current potato yields and potato yields over the previous two years. Therefore, for modeling the statistics of such statistics, it is advisable to choose mathematical models that take into account this dependence. Since autoregressive models of various orders have such properties, this type of model was chosen for potato yield statistics mathematical modeling in the Russian Federation.

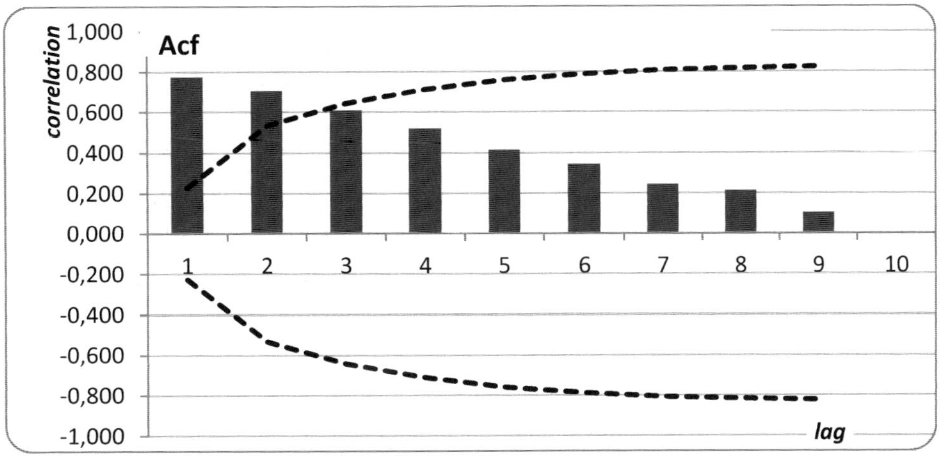

Figure 2. Correlogram for statistics of potato yield in the Russian Federation for the period of 2000-2020 according to the Federal State Statistics Service data [8].

3. Results and discussion

A linear ARM model was chosen in the following form [6] to simulate the potato yield:

$$Y_t = a_0 + \sum_{i=1}^{n} a_i \cdot Y_{t-i}, \qquad (1)$$

where t is the number of the year; n is the order of the model; a_0 and a_i are model parameters.

According to the conclusions made earlier as a result of calculating the autocorrelation function (Acf), the second order or $n = 2$ was chosen for the APM model (1). To find the model parameters a_0, a_1 and a_2, multiple regression analysis was performed further. As the result, the multiple correlation coefficient value turned out to be equal to $r = 0.931$, which indicates that the current year potatoes yield is strongly influenced by both the yield of the last year (t - 1) and the yield that was two years ago (t - 2). This fact confirms the conclusions that were made earlier, based on the autocorrelation function calculation (Figure 1). The values obtained as the result of multiple regression analysis are presented in Table 1 to assess the statistical significance of the ARM model parameters.

Table 1. Calculation results of the second order ARM model parameters.

Parameter	Value	Standard error	t-test	p-value
a_0	10.657	5.103	2.71	0.053
a_1	0.686	0.266	2.58	0.023
a_2	0.275	0.248	1.11	0.287

From the data in Table 1, it follows that in the two-parameter ARM model, only two coefficients a_0 and a_1 are statistically significant, which is confirmed at the 95% significance level of p-values, which are less than 0.05. At the same time, the parameter a_2 turned out to be statistically insignificant, since the p-value for it is greater than 0.05. Therefore, this parameter is excluded from the model. Thus, as the multiple regression analysis result, the two-parameter ARM model was reduced to a one-parameter model with the parameters a_0 and a_1.

Further, for the ARM model, a pairwise regression analysis was performed, which showed that the pairwise correlation coefficient is $r = 0.924$. This indicates that 82% of changes in the current year yield depend on the changes in the previous year yield (coefficient of determination $r^2 = 0.82$), and 18% is on the influence of other factors and random influences. Since the previous year yield influence on the current yield value is quite large, and then the ARM model parameters statistical significance was assessed, which were calculated by using the paired regression analysis methods and are presented in Table 2.

Table 2. The results of calculating the first order ARM model parameters.

Parameter	Value	Standard error	t-test	p-value
a_0	10.076	5.217	3.71	0.051
a_1	0.956	0.106	9.06	0.000

From the data in Table 2, it can be seen that the parameters of the ARM model a_0 and a_1 are statistically significant, which is confirmed by the calculated p-values, which at the 95% significance level are less than 0.05. Finally, the first-order autoregressive model for potato yield statistics in the Russian Federation is as follows:

$$Y_t = 10.076 + 0.956 \cdot Y_{t-1}, \text{ (centners per hectare)} \qquad (2)$$

The potatoes yield calculating results according to the model (2) for the period of 2011-2019 are shown in Table 3. In Figure 1 they are shown in comparison with real values for the period of 2002-2020, which shows a good correlation. The error calculation between the calculated potato yield according to the ARM model and the real values for the period of 2002-2020 averaged 3%, which is acceptable accuracy for practice and forecasting.

Table 3. Comparison of potato yield (centners per hectare) in the Russian Federation according to the Federal State Statistics Service data [8] with potato yield calculated by the model (2).

Years	2011	2012	2013	2014	2015	2016	2017	2018	2019
The Federal State Statistics Service	150	136	147	153	164	158	163	170	178
ARM (2)	151	153	140	150	156	167	161	165	173

4. Conclusion

The potato yield statistics analysis and mathematical modeling in the Russian Federation according to the Federal State Statistics Service data for the period of 2000-2020 was carried out. Based on the calculation and analysis of the first autocorrelation function (Acf) in MS Excel as a mathematical model for modeling potato yield statistics, the choice of the second order linear ARM model was justified. By using the correlation analysis methods, the multiple correlation coefficient value was found and a close relationship between the current year potato yield and the previous two years potato yield was proved, and thus the choice of a two-parameter ARM model was substantiated. By using the multiple regression analysis methods, the values of ARM model two parameters were obtained, their statistical significance was assessed and it was shown that the second parameter of the model is statistically insignificant and therefore it is excluded from the model, and the two-parameter ARM model became a linear paired autoregressive model. By using the paired correlation and regression analysis methods, the ARM model parameters were found and their statistical significance was proved.

Thus, the potato yield statistics mathematical model in the Russian Federation for the period of 2002-2020 can be described by a first-order linear autoregressive model of the following form: $Y_t = 10.076 + 0.956 \times Y_{t-1}$ (centners per hectare). The errors in calculating potato yield according to this model for the period of 2002-2020 averaged 3%, which is acceptable accuracy for practice and forecasting.

References

[1] Volkova G L and Shmatko N A 2020 *Express-Information Science, Technology, Innovation of the Institute of Statistical Research and Knowledge Economics of the National Research University "Higher School of Economics"* **154** 1-4

IOP Conf. Series: Earth and Environmental Science **965** (2022) 012058 doi:10.1088/1755-1315/965/1/012058

[2] Darmanyan A P and Veselova N M 2018 Scientific significance of statistical methods for analyzing experimental data in the system of postgraduate agricultural education *Current Issues of Professional Education* **3(12)** 33-37
[3] Rogachev A F *et al* 2020 A set of data on retrospective grain yield for neural network modeling *IOP Conf. Ser.: Earth Environ. Sci.* **577** 012006
[4] Ryabtsev V G and Utkina T Yu 2016 Information technologies for designing automated process control systems *Control Systems, Communications and Security* **1** 207-239
[5] Kruzhilin I P, Ovchinnikov A S and Melikhov V V 2014 Proc. of the Nizhnevolzhsky Agrouniversity Complex: Science and Higher Professional Education **4(36)** 8-21
[6] Ayvazyan S A 2001 *Applied Statistics. Fundamentals of Econometrics* (Moscow: Unity-Dana) vol 2 pp 234-430
[7] Darmanyan A P and Filippov M V 2013 Forecasting the cost of housing construction in Russia *Business. Education. Right. Bulletin of the Volgograd Institute of business* **2(23)** 120-123
[8] Rosstat URL: https://rosstat.gov.ru/folder/11186 (accessed: 07.05.2021)

IOP Conf. Series: Earth and Environmental Science **965** (2022) 012059 doi:10.1088/1755-1315/965/1/012059

Resource saving in the design of multiple robots control systems

S I Bogdanov[1], V G Ryabtsev[1] and K V Evseev[2]

[1] Volgograd State Agrarian University, 26 University Avenue, Volgograd, 400002, Russia
[2] Moscow Power Engineering Institute, 14 Krasnokazarmennaya Street, Moscow, 111250, Russia

E-mail: bsi@mail.ru

Abstract. The research objective is design of the information technology for automated control system model synthesis for reducing the labour intensity of the digital machine model design by improving the control system information conversion tools. The structural-functional digital machine models are proposed for project development of robotic manipulator control devices. Application of the structural-functional digital machine models allows reducing labour intensity and project timing of robotic center control system design due to the cycle scheme conversion to intermediate representation, which is convenient for digital machine synthesis using the modern development tools. When applying the proposed structural-functional model a process engineer can rely on keeping all important technological process details and a programmer can avoid a large amount of errors when developing the device algorithms. The developed algorithm can be represented as state diagrams, which can be adapted for integrated scheme crystal as a digital machine using the specific tools of computer-aid design.

1. Introduction

In the context of applying robotic centers in producing complex mechanisms and equipment the problem of assembly operations, executed by the robotic manipulators, synchronizing arises. Pooling the working areas into assembly lines makes the interaction between robotic manipulators, which execute the technological process operations, more complicated [1, 2, 3, 4]. In that case the device synchronizing requirements when assembling or processing a detail are raised. The main requirements to robotic manipulators are positional precision and action timing and sequence of the technological operations.

At the present day robotic manipulator control movement is based on electrical engine and pneumatic or hydraulic actuators, control commands generated by motion controllers, synchronizing the technological processes [5]. The motion control algorithms should synchronize multiple parallel tasks. The programming languages allow synchronizing multiple applications, but along with this have time interval precision limitations of actuator active state [6]. When running a robotic manipulator control system on real-time operating system, where multiple tasks are synchronized by the system tools, the control and interface tasks are not separated, what increases the risk of hardware and software errors [7, 8, 9].

According to the mentioned above it follows that the modern control system should consist of the

IOP Conf. Series: Earth and Environmental Science **965** (2022) 012059 doi:10.1088/1755-1315/965/1/012059

hardware tools, which provide robotic manipulator control for complex technological processes and real-time operating with high precision.

The research objective is design of the information technology for automated control system model synthesis for reducing the labour intensity of the digital machine model design by improving the control system information conversion tools.

To reach the objective the following tasks should be solved:

- the structural-functional finite-state machine models design for synthesis of automate technological process control algorithms;
- applying the obtained model to generate the control signals for vehicle body assembly robotic center control system.

2. Materials and methods

To increase the performance developers, when designing the robotic manipulator control algorithms, try to synchronize the actuators. Therefore, each robotic manipulator should meet the higher precision and flow of execution requirements for the following operations:

- actuator response monitoring of robotic manipulator at the specific instant of time;
- monitoring of robotic manipulator reaching the first end point;
- robotic manipulator state period monitoring, when shut down;
- monitoring of robotic manipulator reaching the second end point.

If at least one of the operations is not completed withing the target time interval, set by the process engineer, it could lead to the robotic manipulator fault and/or defective product output.

To manage the manufacturing of the complex mechanisms and equipment in most cases digital machines, which have a finite amount of states, are applied. The digital machine state graph development for automated control of complex processes is labour intensive, for it requires to follow the order of robotic manipulator activation, defined by the process engineer as cycle scheme, when synthesizing algorithms. Besides, when executing the robotic manipulator operations it is needed to control not only the activation events, but the ones, when the robotic manipulator reaches the end points. Along with that it decreases the robotic manipulator dynamic properties, depreciates the mechanisms and causes supply system faults.

The structural-functional model of robotic manipulator control digital machine is:

$$SFM = (S_w, S_c, U, T_d, T_w, T_m, T_s, W),$$

where S_w is technological process operation states; S_c is checking operation states; U is transition conditions; T_d is robotic manipulator end point events; T_w is robotic manipulator activation period; T_m is simulation time; T_s is total operations duration; W is technological process control operations.

It is convenient to represent the model as directed graph, where the robotic manipulator position supervision states are marked as symbol $\oplus$ [10]. The states of working and checking operations, corresponding to the events, are not placed on the plane stochastic, but on the simulation time axis.

The transition conditions U and actions W are implicated $U \rightarrow W$. The working operations W execute for the control signal set $Z = \{Z_1, Z_2, \dots, Z\}$ and are the logical operations, which, when executed, the selected control signal is set $Z_i < '1'$ or reset $Z_i < '0'$. For the control states S_c the current simulation time T_m is compared with the target time, at which the certain robotic manipulator should transit to the finite state. The mechanism conditions are monitored with the set of sensors $D = \{d_1, d_2, \dots, d_k\}$. To reduce the labour intensity of digital machine algorithm development the structural-functional model is proposed, which represents the directed state graph $G = (S, A)$. The graph nodes are placed on time axis, whereas position of each nodeS_i with respect to the initial node S_1 is determined by time interval $T_i, i = \overline{2, k}$, where$k$ is number of graph nodes. The adjacent nodes of graph are connected with edges directed along with the time axis.

3. Results and discussion

Let us consider a case, where five robotic manipulators assembly a vehicle body. The robotic manipulators states under the body assembling process are presented in Figure 1.

The structural-functional model of vehicle body assembly control is presented in Figure 2. The essential operations of body assembly are executed with five robotic manipulators, which start the process on the control signals $Z = \{Z_1 \ldots Z_5\}$. The robotic manipulators activation period is defined by the parameters $T = \{T_{\_z1} \ldots Z_{\_z5}\}$, time lag after the robotic manipulator switching on with respect to the initial point of simulation time is defined by the parameters $Tz = \{T_{-Z1} \ldots Z_{-Z5}\}$, the sensors activation events are defined by $T_d = \{T_{-D2\_1} \ldots T_{-D5}\}$ Total simulation time is defined by variable S_k. When digital machine is operating the current simulation time increases at the predefined in the project time gradient. If the digital machine clock signal frequency is high and the time gradient is small, then it is possible to reach the desired precision of technological operation duration. Therewith, the error will be equal to half of the defined simulation time gradient.

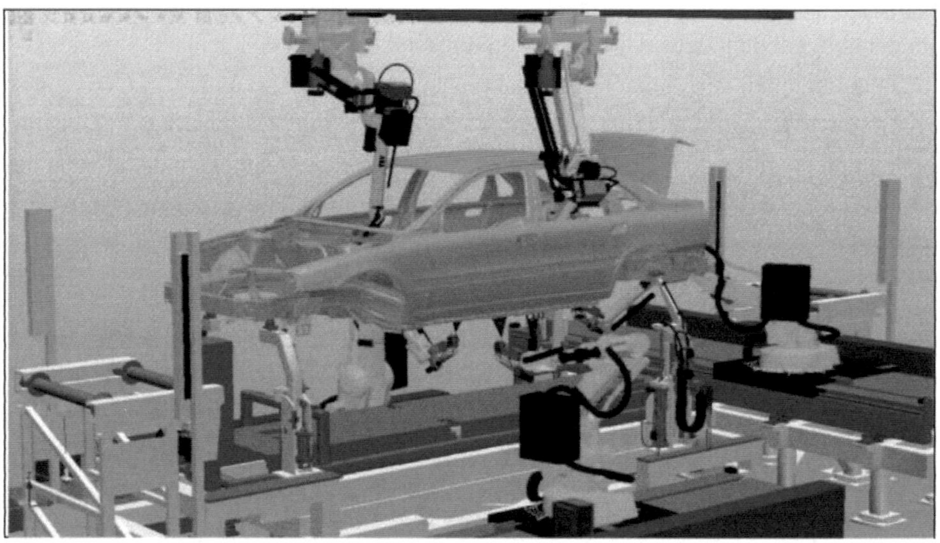

Figure 1. Vehicle body assembly with robotic manipulators.

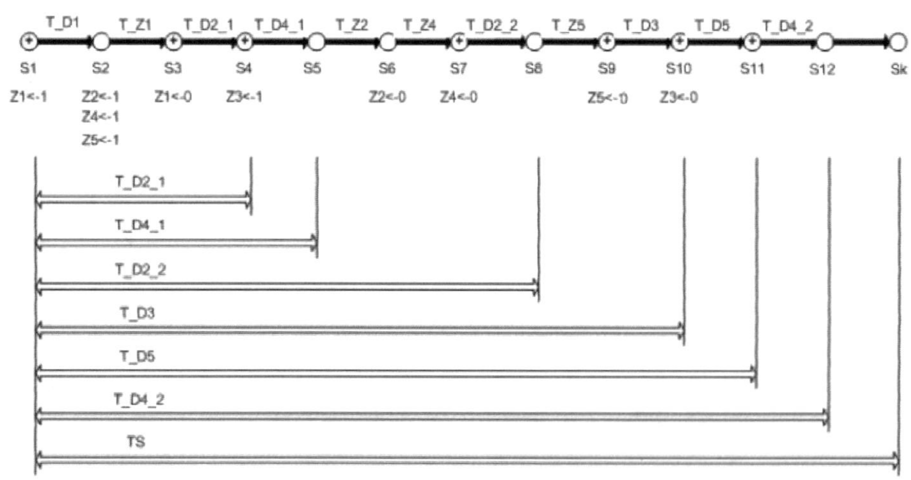

Figure 2. Structural-functional model of robotic manipulator control.

With increase of the executed technological operations the number of cause-effects between the robotic manipulators and sensors, controlling their position, increase, therefore the digital machine graph scheme synthesis becomes more complex, what proves the efficiency of the structural-functional model for control system design.

The developed information technology providing digital machine design for technological process automated control algorithm synthesis is a design system, which includes:

- functional cycle schemes, developed by process engineer;
- robotic manipulator structural-functional models;
- combined function tables of the digital machine transitions and outputs;
- hierarchical digital machine system model.

The time intervals in cycle scheme correspond to the order and duration of the forward and no-load strokes within a single kinematic cycle.

To represent all the technological process cause-effects for further design it is convenient to apply the structural-functional model of robotic manipulator control, which allows defining the set of states for each robotic manipulator according to the target scenario of optimal technological process control, separating the control and checking operations, representing order of the technological process operations on time axis, what in their turn reduce the risk of errors when synthesizing the digital machine project using the most common tools.

The next step is building the combined state transition and output tables of the developed control system digital machine. Wherein, the combined tables are combination of transition functions and outputs.

It is reasonable to synthesize the technological process automated control algorithms, which represent the control and checking operation order to generate the corresponding signals, using the modern digital device design systems, such as Active-HDL, Riviera, MAX + Plus II and others.

It is suggested to use Active-HDL to simulate and verify the automated control algorithms of the technological process, where it is possible to generate digital machine flowgraph from the combined tables of function transitions and outputs.

The hierarchical representation is convenient to use when designing complex digital machines, for such a representation allows determining the common system principles and then analyzing the certain components. Thus, the complex system is being described from top to bottom according to the hierarchy. First the major components are presented at the highest level and then the details are shown for each of the components at the lower level in the hierarchy.

Therefore, a directed graph, where its nodes correspond to the states and edges to the transitions between them, is used to represent the finite state digital machine projects. Such an approach of setting finite state digital machines is supported in FSM (Finite State Machine) Editor in Active-HDL design system.

As a result of simulating the algorithms of the technological process automate control based on the built tables with Active-HDL we can obtain the hierarchical system model of digital machine. The low er level machine graphs are merged into the common hierarchical higher level graph and represent the overall technological process control for the certain scenario.

4. Conclusion

When applying the proposed structural-functional model a process engineer can rely on keeping all important technological process details and a programmer can avoid a large amount of errors when developing the device algorithms. The developed algorithm can be represented as state diagrams, which can be adapted for integrated scheme crystal as a digital machine using the specific tools of computer-aid design.

References

[1] Abd K, Abdhary K, Marian R 2011 A scheduling framework for robotic flexible assembly cells *Int. J. Appl. Sci. Technol.* **4(1)** 31-38

[2] Simon B, Mads H, Morten K and Ole M 2011 Autonomous industrial mobile manipulation (AIMM): from research to industry *Proc. of the 42nd International Symposium on Robotics* 1-9

[3] Bonert M, Shu L H and Benhabib B 2010 Motion planning for multi-robot assembly systems *Int. J. Comput. Integr. Manufact.* **13(4)** 301-310

[4] Abbyasov V M and Buhteeva I V 2014 Optimization of assembly equipment layout solutions at the design stage *Technology of Mechanical Engineering* **5** 27-31

[5] Abbyasov V M, Buhteeva I V and Elkhov P E 2013 Directional selection of the layout of a flexible automated transmission shaft assembly line *Izvestiya of Moscow State University of Mechanical Engineering (MAMI)* **1(15)** 136-140

[6] Krasilnikyants E V, Varkov A A and Tyutikov V V 2011 Manipulative robot control system *Automation in Industry* **5** 38-44

[7] Nikolaev M E, Nesmianov I A and Zakharov E N 2020 Definition of service area of agricultural loading robot with manipulator of parallel-serial structure *IOP Conf. Series: Materials Science and Engineering* **747** 1-7

[8] Krasilnikyants E V, Varkov A A and Tyutikov V V 2012 Software for the control system of an internal manipulative robot *Mechanics. Automation. Management* **3** 31-36

[9] Tyutikov V V, Krasilnikyants E V and Varkov A A 2011 Components of the manipulation robot software *Vestnik IGEU* **4** 40-43

[10] Ryabtsev V G and Utkina T Yu 2016 Information technologies for designing automated process control systems *Control Systems, Communications and Security* **1** 207-239

IOP Conf. Series: Earth and Environmental Science **965** (2022) 012060 doi:10.1088/1755-1315/965/1/012060

Modern digital technologies in accounting and internal audit in agricultural enterprises

S A Vardanyan[1], N N Balashova[1], V I Khoruzhy[2], A S Gorbacheva[1] and T A Chekrygina[1]

[1] Volgograd State Agrarian University, 26 University Avenue, Volgograd, 400002, Russia
[2] Financial University under the Government of the Russian Federation, 6 Krasnostudencheskiy Proezd, Moscow, 127434, Russia

E-mail: sv@volgau.com

Abstract. The paper discusses about the issues of the current state of the using of modern IT technologies in agriculture, about the prospects of the using of information technologies in accounting and internal audit. An objectives of the project "Digital Agriculture" of The Ministry of Agriculture of the Russian Federation was considered. The main advantages for accounting and internal audit by the implementation of IT technologies are briefly presented, a possible obstacle to the large-scale implementation of modern technologies in agricultural organizations are analyzed.

1. Introduction

The relevance of the research is reasoned by the fact that today digital technologies cover most areas of economy in modern developed countries. Agriculture, as a strategic industry for Russia, isn't exception. From gathering to the cultivation of fields and raising of plants, an invention of fertilizers, using of technic for mechanization and automatization of production - each revolutionary innovation raised agriculture to a new stage of development. The modern agrarian revolution implies the introduction of advanced information technology (IT), that will reduce manual labor and costs, at the same time increasing productivity and prolificness.

In international editions a lot of different scientific papers about using of IT in accounting and auditing are published. But the most significant articles, among which are the papers of Carl Benedikt Frey and Michael A. Osborne, Michael Chui, James Manyika, and Mehdi Miremadi and etc, consider the issue of changes in trading, IT-companys, financial areas etc. And it is very difficult to find the results of research about the using of IT technologies in agriculture.

The purpose of this paper is to show how IT technologies are used today in agriculture, what are the most promising areas for the introduction of IT technologies, how it can be used in accounting and internal audit, what positive economic effect can this bring for agrarian companies.

2. Materials and methods

In the process of the research of the prospects of the using of IT technologies in accounting and internal audit in agricultural companies, various general scientific methods were used. The main ones are analysis and synthesis, induction and deduction, comparison, abstraction, observation, modeling, etc.

IOP Conf. Series: Earth and Environmental Science **965** (2022) 012060 doi:10.1088/1755-1315/965/1/012060

3. Results and discussion

About 70% of farms in the US, Canada and Europe are using smart technologies in agriculture. Russian farmers are far from such indicators, but the request for digital technologies is growing. According to experts, digitalization will help the Russian agro-industrial complex make a powerful leap forward.

The Ministry of Agriculture of the Russian Federation is actively working in this direction. The departmental project "Digital Agriculture" was developed last year with a deadline for implementation until 2024. Its main goal is a digital transformation of agriculture through the introduction of digital technologies and platform solutions to ensure a technological leap in the agro-industrial complex and for achieving of productivity growth in «digital» agricultural enterprises [1]. The first stage of this project was a creation of the national platform "Digital Agriculture" - an order to start its formation was issued in the February of 2020. The implementation of the project was not stopped by the Covid-19, at this stage a personal account for the agricultural producer is being created.

There is a clear focus on the efficiency of each operation in the digital agriculture. Digitalized agricultural technique, sensors, drones and other digital elements help to achieve this. At the same time, the management of production processes is automated. Manufacturing demonstrates a relatively high productivity of land at a relatively low cost [2].

A large number of tech companies that had grown out from startups are responding to the main request of agricultural producers - data collecting, aggregating and analyzing. The decisions of American agro-startups declare using of new and very effective technologies - an artificial intelligence, computer vision and machine learning. Some Russian companies are also trying to use these technologies, especially on an industrial scope - for analyzing NDVI images (a map of indicators of the amount of photo synthetically active biomass) of tens of thousands of hectares of arable land, counting the number of apples in thousands of hectares of gardens etc. This makes it possible timely and point wise to solve the problems of poor germination, insufficient of green mass, to identify hearth of the spread of diseases and pests, to predict the harvest with a sufficiently high degree of accuracy and, as a result, to more accurately plan of harvesting, transporting, storage facilities work.

Among the technologies that can be implemented in the Russian agro-industrial complex are software systems for farm management, robotic systems, drones for monitoring of agricultural objects, technologies of precision farming based on the Internet of Things. In particular, drones produced by the Zala Aero Company of the Kalashnikov concern are actively used for aerial photography of agricultural land. Using of that technologies makes it possible to estimate the condition of the soil and plants, increase the yield of land, optimize the cost of fertilizers and plant protection products, and identify areas that need an additional irrigation.

An example of another system that allows agricultural enterprises and farmers to switch to the rational using of fertilizers, based on the needs of a particular area of the field, is the product of JSC NIIIT (Tver). Based on the data of the chemical analysis of the soil, the digital map of farmland is compiled. Taking in attention the condition of the soil, farmers are given recommendations on the optimal planting of crops, the amount and type of fertilizers and plant protection products. Then «smart» agricultural machinery - seeders, sprayers, spreaders - receive the task cards.

There are certain specifics in the field of accounting and internal audit in agricultural organizations. It is due to the presence of accounting objects typical only for agriculture. These objects include biological assets, animals, plants. Such a feature as the seasonality of production also strongly affects on accounting, on the formation of costs and production costs, the procedure of creating of reserves.

Modern IT technologies can greatly facilitate the process of accounting of specific objects in agriculture. For example, some agricultural holdings for accounting of their suitable for agriculture lands, forests, as well as to assess the surface area of their own water resources, use aerial photography using drones that transmit images online and allow to assess the area of arable land, forests and water resources. This makes it possible to carry out an inventory of the company's land and water objects much faster than with traditional geodesy methods. In addition, this method, supplemented with the appropriate software for data analysis, makes it possible to estimate the density of the growth of green biomass, the density of forest stands, the saturation of meadows with grass, etc.

GPS trackers and agronavigators are a useful technology for accounting of the work of agricultural machinery, in particular for accounting of the hours worked of drivers and fuel consumed. Using of these devices allows you to see in real time where the agricultural machinery is located, in what mode is it (work or rest), what fuel consumption, what area of cultivated land, etc. Also high-quality internal control of expenditure of resources is provided.

Modern IT technologies make it possible to improve the accounting of labor and its payment. This issue is especially relevant for agriculture, where the labor of seasonal workers is used, where the places of work are geographically distant from the central office or accounting office. In this case, such a tool as mobile card readers, connected to the special program and the Internet, allows you to quickly register the moment of going to work, all breaks and time of leaving from work for employees who are pre-issued with ID cards.

One of the most important objects of accounting in agriculture is a finished product. Depending on the type of activity that the agricultural organization is engaged in, the finished product can be a harvest of crop products, vegetables or fruits, meat, milk and dairy products, etc. In all of these types of activities in modern organizations, they are equipped with a new technologies, for example, new combines with a harvest accounting system, equipped with measuring devices, which are available to transmit information via the Internet to a central office or warehouse, a new robots for milking milk, immediately transmitting information about the volume of received milk to the accounting department, a personal ID cards of employees, making it easy to record how much vegetables or fruits a person has harvested, etc.

Today, the use of IT in agriculture is not only about the use of computers. Different digital technologies allow you to control the full cycle of crop or livestock production - “smart” devices measure and transmit the parameters of soil, plants, microclimate, etc. All this data from sensors, drones and other equipment is analyzed by special programs. Mobile or online applications can help farmers and agronomists to determine the right time for planting or harvesting, to calculate the fertilization scheme, predict the harvest, and much more [3].

The study conducted by McKinsey in 2016, predicts that 86% of tasks of accountants and auditors can potentially be automated [4]. In Oxford University study [5], accountants and auditors are listed among professionals who may suffer from computerization. Deloitte claims that the UK could lose about half a million of jobs in the financial sector due to automation.

Smacc, Shoeboxed, Wave Accounting and many others (including on the Russian market) state that they are developing artificial intelligence applications for automating accounting and accounting tasks. As a rule, in these applications, artificial intelligence recognizes photographs and scans of documents, translates them into a machine-readable format, classifies them, makes basic checks. The task of recognition and classification is certainly important from the point of view of accounting, and now an accountant spends a lot of time on stuffing data from a paper into a computer, but there is no threat to the accountant as a specialist.

Of course, these technologies can bring a lot of specialists to the labor market, who previously engaged in auditing and analyzing contracts. However, there is no information anywhere, what percentage of errors can artificial intelligence make when it is analyzing a text compared to a man - and this directly affects the responsibility of the auditor [6].

If we talk about the theoretical aspects of the automation of auditing, here, as in accounting, machines can facilitate the work, but the main place will still remain for a person capable of professional judgment.

AI is generally very useful and able to help in a number of situations with manual data processing - and this is a very significant part of traditional audit work. Data retrieval, collation, validation are examples of where this is possible. AI significantly speeds up digitization of the traditionally manual manual data entry or extraction processes, which significantly reduces the time required for their preliminary preparation for verification.

IOP Conf. Series: Earth and Environmental Science **965** (2022) 012060 doi:10.1088/1755-1315/965/1/012060

Using of the modern IT technologies can bring great benefit in internal audit or internal control in agricultural organizations. With the usual organization of these services, when modern technologies are used in limited extent, the main disadvantage is a long time for getting of information and, as a result, a longer analysis and a slower management decision-making. IT technologies make it posible for internal auditors to monitor ongoing business processes in real time or receive complete information from them via the Internet.

For example, using of cloud services or blockchain technology allows you to install a stable connection between the accounting department, warehouses, production departments and internal auditors. The need for an internal auditor to go to the accounting department or to a warehouse for the checking of the accuracy or correctness of a document is no longer so high. To do this, it is enough to place this document in a secure digital service, access to which is limited. The internal auditor will check this document much faster than if he had to come to the place of creating or receiving of this document.

One of the most important tasks which is set for internal auditors is to implement of measures for ensure of the safety of existing resources [7, 8]. With the usual organization of internal control without the use of IT technologies, the most common way to accomplish of this task is to take an inventory. Since its implementation takes a long time, and the organization of this process is associated with a large amount of paperwork, then taking an inventory too often is a problematic. And this increases the risk of violations in the warehouse.

Modern systems such as cameras of video surveillance in real time, measuring equipment, connected to the Internet and capable to transmit information instantly, various programs with artificial intelligence, that allow quick measurements, calculations and analyzes of available resources, etc., make it posible for internal auditors to carry out constant control over the safety of existing resources without the need of regular inventory.

A similar picture is observed in relation to other areas of accounting. For example, accounting of settlements with suppliers, with banks, with customers, with personnel for remuneration, etc. The correctness of these calculations becomes much easier to control for internal auditors who have access to accounting information systems and are able to use various programs to analyze large amounts of data.

As can be seen from the results of the research, the introduction of IT technologies opens up wide opportunities for accounting and internal audit in agricultural organizations to increase the efficiency of their work. But at the same time, there are certain obstacles for the implementation of such systems. The main obstacle is the high cost. This applies to equipment for automating of various processes, measuring equipment, systems for accounting of manufactured products, labor accounting, smart video surveillance and analysis systems, etc. In addition, the installation of such systems is associated with payment for the services of companies involved in installation and subscription services, personnel training. The ability to cover such a high cost is available only for fairly large agricultural organizations; small and medium-sized enterprises generally cannot afford such costs.

The second obstacle is the insufficiently high level of qualifications and skills to work with modern IT technologies among employees of agricultural organizations. Usually, such organizations work in rural areas, a large number of employees do not have higher education and skills in working with modern technologies. The process of retraining workers is time-consuming and expensive.

4. Conclusion

The Ministry of Agriculture of the Russian Federation is implementing the Digital Agriculture project, which sets itself ambitious goals - digital technologies should help to double the productivity of agricultural enterprises by 2024. For the digital transformation of agriculture, specialists with new knowledge are needed, as well as new “smart” solutions that will come to their aid.

Digital solutions are increasingly penetrating at all segments of agriculture. To achieve growth in profitability, it is extremely important for agribusiness to make the most of innovative technologies.

The same companies that in the near future will be able to combine their business into a single system based on a digital platform will become the undisputed market leaders.

References

[1] Alekseev A A, Rushchitskaya O A, Lavrov V N and Yurchenko N A 2020 Improvement of state support for small agricultural business due to Covid-19 coronavirus pandemic *Agrarian Bulletin of the Urals* **4(195)** 86-90

[2] Kurbatova S M, Aisner L Yu and Naumkina V V 2019 Some aspects of the essence and legal regulation of agriculture digitalization as one of the priorities of modern state policy of agriculture development *IOP Conf. Series: Earth and Environmental Science* **315** 032021

[3] *The Field of Opportunity: Digital Solutions for Agriculture* 2020 https://rostec.ru/news/pole-vozmozhnostey-tsifrovye-resheniya-dlya-selskogo-khozyaystva/

[4] Michael Chui, James Manyika and Mehdi Miremadi 2016 Where they could replace humans - and where they can't (yet) *McKinsey Quarterly* https://www.mckinsey.com/business-functions/digital-mckinsey/our-insights/Where-machines-could-replace-humans-and-where-they-cant-yet

[5] Frey C B and Osborne M A 2013 The future of employment: how susceptible are jobs to computerisation? http://www.oxfordmartin.ox.ac.uk/downloads/academic/The_Future_of_Employment.pdf

[6] Vardanyan S A 2017 Basic vectors of accounting and audit development based on blockchain technology in the conditions of the digital economy *Scientific Review: Theory and Practice* **11** 23-27

[7] Balashova N N and Vardanyan S A 2015 Internal audit in the agricultural sector: condition, problems and prospects of development *Proc. of the Nizhnevolzhsky Agrouniversity Complex: Science and Higher Professional Education* **3(39)** 246-250

[8] Ostaev G Ya, Gogolev I M and Zlobina O O 2020 Concept of implementation of internal audit in agricultural organizations *Economics of agriculture in Russia* **8** 61-64

IOP Conf. Series: Earth and Environmental Science **965** (2022) 012061 doi:10.1088/1755-1315/965/1/012061

Formation of investment resources of the grain subcomplex Volgograd region and its sustainability

A S Gorbacheva, V V Ivanov, N N Balashova, T A Chekrygina, N V Chernovanova, E V Yagupova and V A Melikhov

Volgograd State Agrarian University, 26 University Avenue, Volgograd, 400002, Russia

E-mail: annasemenovna26@mail.ru

Abstract. The article discusses the positive experience of the operation of the agricultural holding of the Volgograd region, engaged in the production, processing and sale of grain and grain products. The authors identified and analyzed the factors that both hinder the formation of integration ties in the grain sub-complex of the regional agro-industrial complex, and contribute to their dynamic development. The article describes methodological approaches to the organization of integration mechanisms that take into account the economic interests of all their participants: from producers of grain products, enterprises for the processing and storage of grain, subjects of the processing industry of the agro-industrial complex, transport companies, trade organizations. A competent pricing policy for products, works and services of all economic entities of the agricultural holding has been determined as a guarantee of the ability to compete and maintain its market niche. Prospects for integrating the activities of an agricultural holding with a research organization are considered in order to increase its investment activity in terms of introducing new technological, breeding, marketing and other innovations that contribute to ensuring a high level of development of all subjects of the agro-industrial business and their financial stability.

1. Introduction

The grain industry is one of the most important leading sub-complexes of the agro-industrial complex of Russia, including the Volgograd region. The share of grain and products of its processing is very significant in the structure of the gross product of the Volgograd region [1]. Their share in the food market is not just high, it prevails and thereby ensures regional food security and food independence [2]. Consequently, the search for promising directions for the development of the grain product sub-complex of the Volgograd region, including through the formation of integrated companies, is an urgent problem of the regional economy [3]. The interpretation of integrative growth as a general strategy for the development of the modern agro-industrial complex is based on the fact that, based on the achievements of scientific and technological progress, it reflects a promising global strategy for changing the food market as a whole, necessary to overcome the existing problems of its participants [4, 5, 6].

Based on the foregoing, the aim of the study was to analyze the experience of the effective functioning of integrated formations in the sphere of the agro-industrial complex, as a way to achieve their expanded reproduction, by increasing their investment competitiveness and innovative activity.

2. Materials and methods

The methodological basis for writing the article was the dialectical method of cognition and the systematic approach, which led to an integrated approach to the study of the state of integration and the formation of investment resources of the grain sub-complex of the Volgograd region. In the process of writing the article, monographs and articles of modern economists were studied, devoted to the problems of cooperation and integration of agricultural business entities, as well as statistical materials and accounting and reporting data of the Open Joint Stock Company "Novoanninsky Combine of Bread Products". When writing the article, the methods of expert assessments and the economic and statistical method of scientific knowledge were used.

3. Results and discussion

The article presents objective evidence of the effectiveness of the formation of integration ties between the subjects of agribusiness of the grain sub-complex of the Volgograd region as a way to increase their investment competitiveness and innovative activity.

Despite the dynamic growth in the production of grain and products of its industrial processing, the noted stabilization of the financial condition of individual agribusiness entities in the Volgograd region, its general situation remains rather difficult [7]. Table 1 presents data on the size of the sown area, productivity and gross harvest of grain and leguminous crops.

Table 1. Grain production in the Volgograd region in 2019.

Types of cereals and legumes	Sowing area, thousand hectares	Gross harvest, thousand centner	Crop yield, centner / hectare
Whole grains and pulses	1290.77	28783.02	16.16
Including:			
- winter wheat	906.00	22360.08	24.68
- winter rye	24.30	327.81	13.49
- winter barley	1.09	16.99	15.59
- winter triticale	5.00	85.95	17.19
- spring wheat	65.39	588.51	9.00
- spring barley	130.39	1131.79	8.68
- spring triticale	0.98	5.88	6.00
- oats	4.49	449.00	10.00
- corn	51.19	2548.75	49.79
- millet	21.79	311.82	14.31
- buckwheat	1.68	25.18	14.99
- sorogo	10.47	156.74	14.97
- pulses (total)	68.0	774.52	11.39

The reasons that reduce economic sustainability and investment competitiveness include the following:

- the lack of a balanced mechanism for the formation of the existing infrastructure of the food market for grain products, seed and feed grain, agricultural machines and mechanisms, information and consulting services;
- the presence of numerous intermediaries in the promotion of grain commodity flows from agribusiness and processing industries to the end consumer;
- low solvency of wholesalers - buyers of grain, the use of surrogate forms of mutual settlements by subjects of the agro-food complex;

IOP Conf. Series: Earth and Environmental Science **965** (2022) 012061 doi:10.1088/1755-1315/965/1/012061

- deterioration of all components of the natural potential of grain growing (fertility of agricultural land, reduction of varietal crops, reduction of quality parameters of grain) caused by a shortage of material and production resources and a lack of investment opportunities for its representatives;
- the absence of a mutually beneficial basis in the formation of partnerships of all counterparties of the grain-product half-complex.

Insufficient development of integration ties between the production and processing of grain predetermines: a growing price disparity between agricultural and industrial products, as well as between the cost of grain products and the selling price; a widespread reduction in the capital and power supply of agribusiness entities; reduces their ability to use leasing of agricultural machinery [8].

All of the above reduces the investment competitiveness of economic entities of the regional agro-industrial complex, which results in a drop in their investment activity, deepening of the investment crisis, which interferes with their expanded reproduction.

One of the promising ways out of this situation is the formation of a more profitable production and marketing chain of the grain sub-complex through the cooperation of their activities, which will allow the fullest use of their production potential [1, 2, 9]. The elements of the structure of the grain sub-complex of the Volgograd region, with the proposed integration and the formation of on-farm mutually beneficial relations, are shown in Figure 1.

The above diagram shows that the grain market of the Volgograd region includes all stages of production and marketing chains. It includes four areas.

The first sphere is formed by representatives of the production of means of production, which should include the engineering industry, which produces machinery, equipment and vehicles for the subjects of the grain-food sub-complex of the agro-industrial complex, chemical industry enterprises that provide the above entities with fertilizers and plant protection products, as well as oil refineries that supply fuel and other fuels and lubricants.

Representatives of the second sphere are direct grain producers - these are agricultural enterprises, which have their own individual organizational-production and financial structure and specific intra-farm ties.

The third sphere is formed by processing and service enterprises, which should include grain elevators and warehouses, transport services, flour and cereal plants, feed producers, and bakery enterprises.

IOP Conf. Series: Earth and Environmental Science **965** (2022) 012061 doi:10.1088/1755-1315/965/1/012061

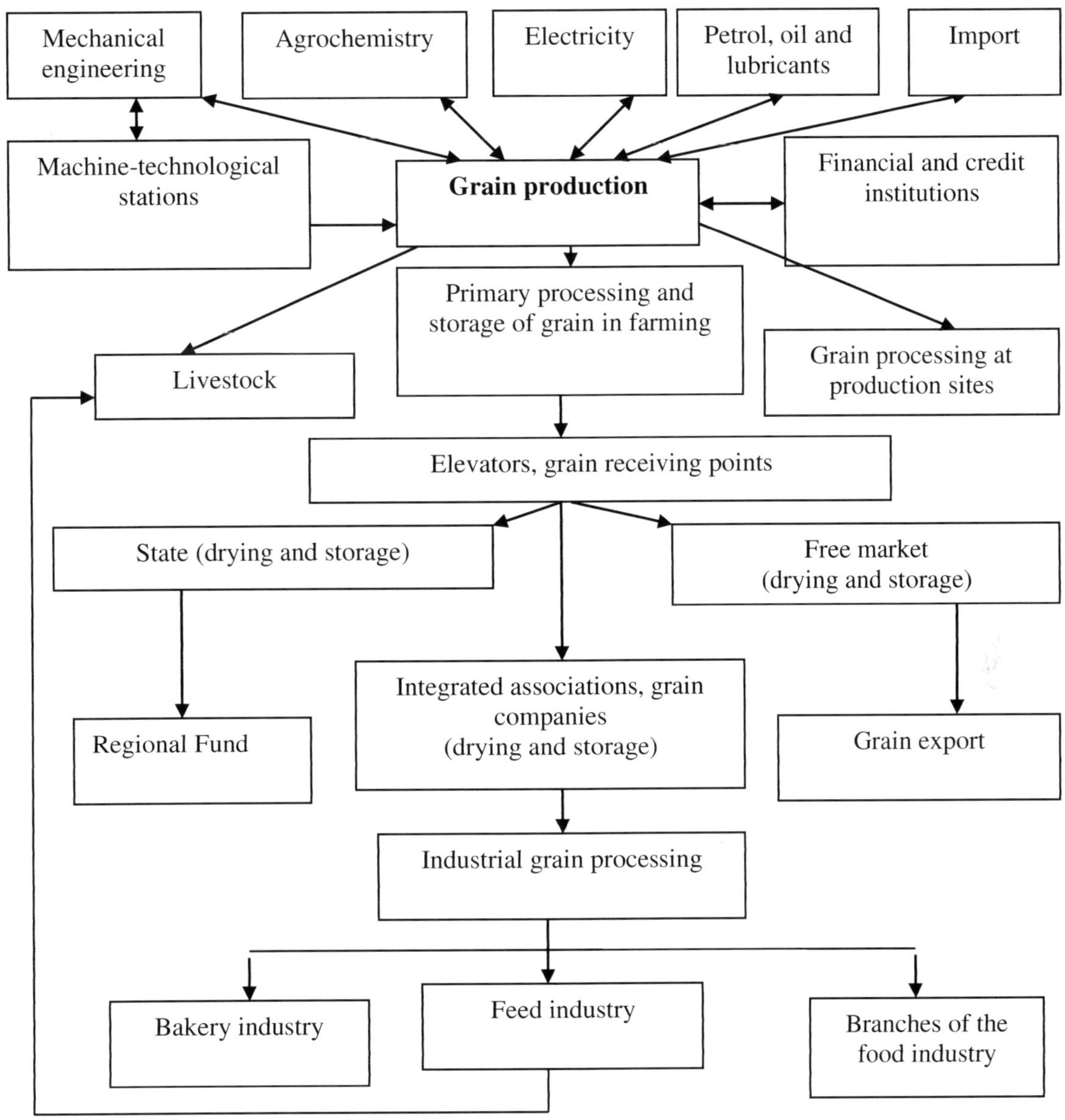

Figure 1. The structure of the grain product sub-complex and the movement of commodity flows.

An important function of promoting goods made from grain is performed by the fourth - the marketing sphere, which includes enterprises and the organization of wholesale and retail trade, their logistics and service (Figure 2).

The existing system of production and marketing of grain and its processed products should be rationalized through the creation and development of horizontal and vertical integrated formations, which makes it possible to stabilize all stages of the production and marketing chain of the grain subcomplex [3, 10]. Practical experience proves that in modern conditions the most stable form of cooperation are vertically integrated organizations. They make it possible to more effectively

IOP Conf. Series: Earth and Environmental Science **965** (2022) 012061 doi:10.1088/1755-1315/965/1/012061

coordinate the actions of all participants in the grain market, using a mutually beneficial pricing policy.

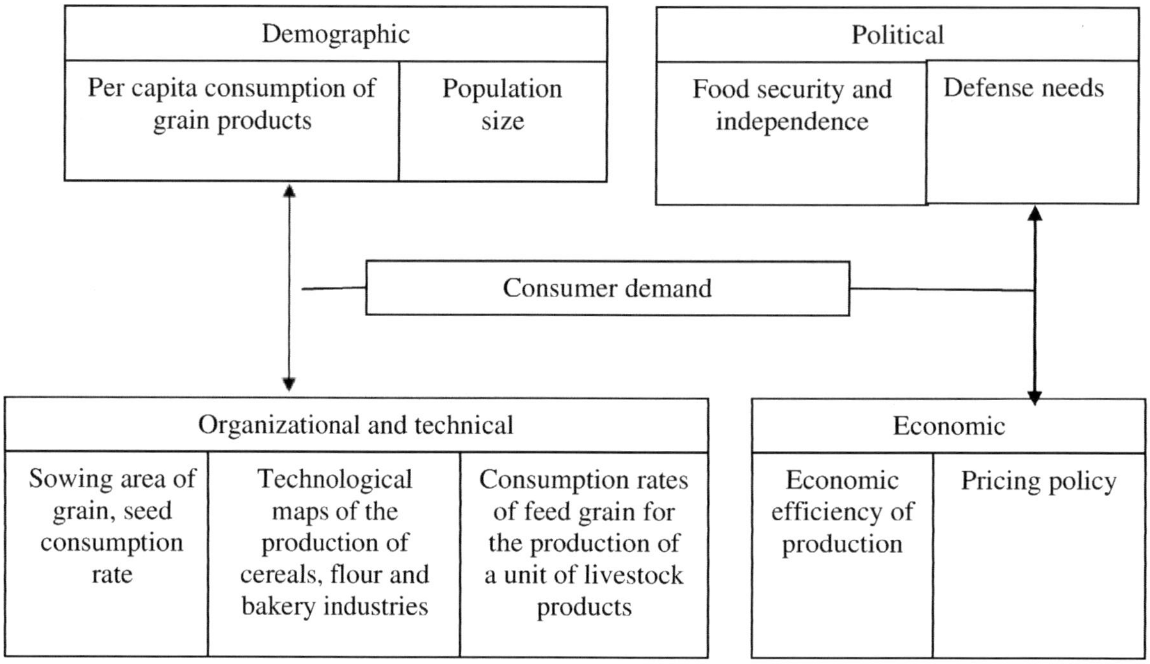

Figure 2. Factors shaping consumer demand for the products of the grain product sub-complex [1].

It is necessary to consider in more detail the factors influencing the development of agro-industrial cooperation and integration. They can be classified in two directions: factors of integration assistance and factors restraining integration processes. The characteristics of these factors, whose synergistic effect affects the formation of cooperative formations in the Volgograd region, is presented in Table 2.

Table 2. Contributing and constraining factors for the development of integration and cooperation in the grain-food sub-complex of the Volgograd region.

Facilitating factors of cooperation	Restraining factors of cooperation
1. Equality and equivalence of participants united by integration ties	1. Decrease in the volume of grain production and its quality parameters
2. Improving the mechanism for regulating the grain market on a mutually beneficial basis	2. The financial crisis of the subjects of individual agricultural producers
3. The interest of grain producers in the development of integration and cooperation in order to balance the price positions for grain and its processed products	3. Counteraction to the development of integration of trade and intermediary commercial structures - owners of elevators and processing enterprises
4. The interest of all subjects of integration ties in increasing their investment competitiveness and investment security	4. Lack of targeted government support - unfinished regulation of integrated business schemes and financial and credit measures
5. The importance of grain products for ensuring regional food independence	5. Insufficient level of qualifications of administrative personnel and owners of enterprises

Despite the influence of constraining factors, an example of the successful formation and development of integration processes in the production and processing of grain in the Volgograd region is the practical experience of the operation of the agricultural holding "Novoanninsky", which unites a number of modern, high-tech, diversified enterprises. It should be noted that over the past few years, thanks to the implementation of a competent investment strategy, this enterprise has been actively working on launching new industries and creating conditions for the creation of new enterprises. In addition, the enterprises that are part of the agricultural holding are constantly updating and expanding their material and technical base - the methodological approaches underlying the formation of this agricultural holding are shown in Figure 3.

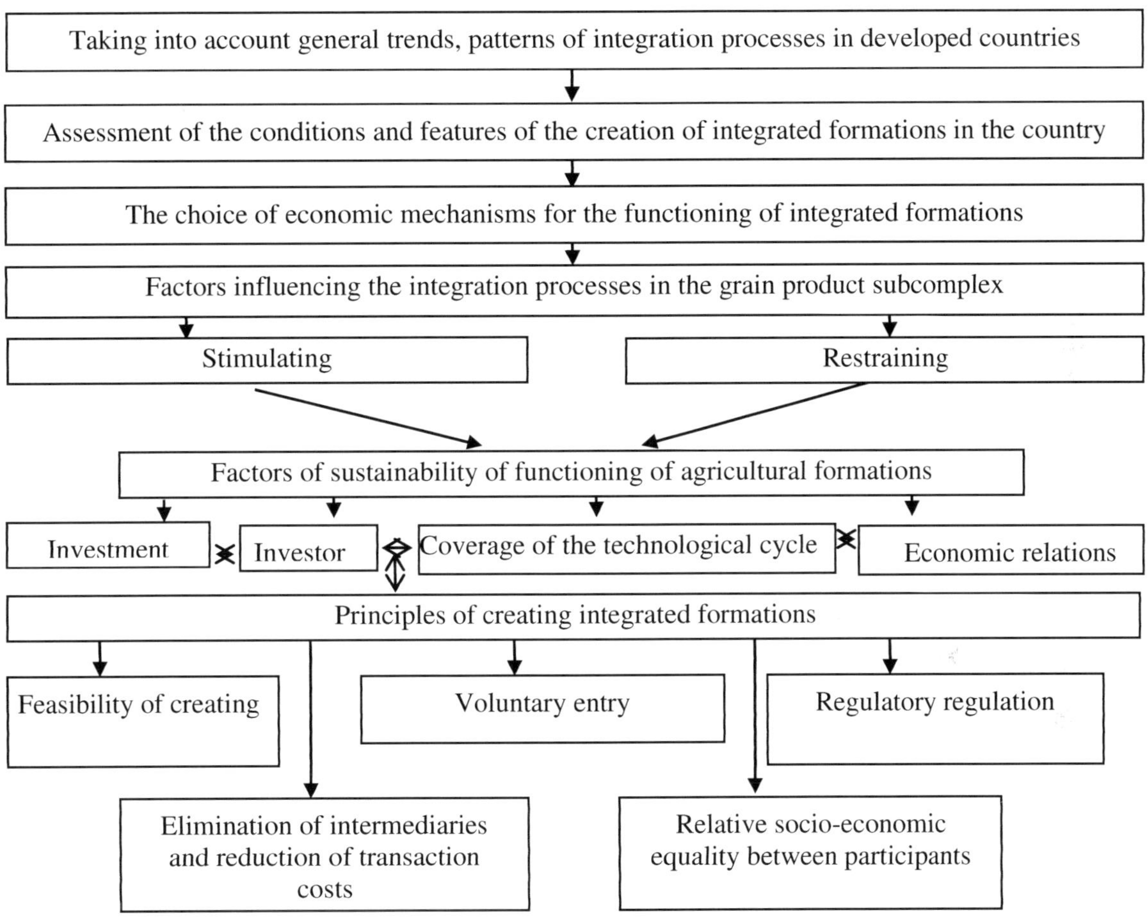

Figure 3. Methodological approaches to the development of integration processes in the grain product sub-complex.

The head enterprise of the agricultural holding in question is "Novoanninsky Combine of Bread Products", whose sphere of activity includes control and management of the production and financial activities of all participants in the integrated formation, namely:

- determination of the target mission of the agricultural holding's functioning;
- formation of a strategy for investment and innovative development of the subjects of the agricultural holding;
- formation of coordination on-farm ties between the participants of the holding system;
- strategic and tactical financial business management;

IOP Conf. Series: Earth and Environmental Science **965** (2022) 012061 doi:10.1088/1755-1315/965/1/012061

- optimization of the formation and use of the material and technical base;
- formation of a pricing policy for products, works and services of the subjects of the agricultural holding;
- choosing a marketing strategy for an agricultural holding and solving problems of promoting products to sales markets;
- solving problematic issues and possible conflicts between agribusiness subjects;
- targeted support of a specific participant in case of crisis situations.

The organizational structure of the agricultural holding in question is represented by enterprises of various fields of activity: direct grain producers (Limited Liability Company "Solnechnaya Strana", Limited Liability Company – Agrofirm "Belorechinskaya"), enterprises providing services for the processing and storage of grain (Open Joint Stock Company "Novoanninsky Combine of Bread Products"), representatives of the bakery industry (Limited Liability Company Novoanninsky Baker; Limited Liability Company "Novoanninsky Bread"), as well as service organizations (Open Joint Stock Company Trading House "Novoanninsky Combine of Bread Products", Open Joint Stock Company "Volgograd Regional Machine-Building Technological Station").

Each of the enterprises included in the agricultural holding is the owner of its property complex, land and independently pays its taxes [8].

A well-thought-out pricing policy is an effective tool for integrated production associations that allows them to maintain competitive advantages, firmly gain a foothold in the market and maintain a market niche. It is based on the principle: "an agricultural holding should be able to sell products, works and services at lower prices than those of competitors. All participants in the agricultural holding's trade and food chain focus their work not on maximizing profit at any cost, but on optimizing it with the rational use of all types of resources: from material and technical to labor. At the same time, the bulk of the profit should be obtained from the expansion of production capacity and the increase in sales, and not due to the irrepressible growth of the selling price.

Let's consider the situation in the market of flour and bakery products and their average retail price in 2019 - 2020 using the example of the Volgograd region.

During 2019, the level of average regional prices for flour was more than 20% higher than the prices of the agricultural holding, which only by the beginning of 2020 increased the price by 4%, while the profit from core activities for 2019 from core activities amounted to more than two million rubles (Table 3).

Table 3. Price range for flour (premium grade), rubles / kg. including value added tax.

Time intervals	Average price for the region	Agricultural holding "Novoanninsky"
As at January 01, 2019	23.50	19.51
As at May 01, 2019	24.91	19.51
As at September 01, 2019	24.91	19.51
As at December 01, 2019	24.91	19.51
As at January 01, 2020	26.48	20.30
As at March 01, 2020	27.30	20.30

A similar situation was observed at the market of bread and bakery products of the Volgograd region at the same time. As an example, consider the dynamics of price fluctuations for a bakery product – "Wheat bread (yeast)" (weight 500 grams). This product enjoys stable consumer demand, its competitive advantages are based on good taste and affordable price.

Among the analogs produced in the municipal districts of the Volgograd region, "Wheat bread (yeast)" of the "Novoanninsky Combine of Bread Products" is sold at the lowest price, the excess of the price of competitors for a similar product range from 11 to 38% (Table 4). A competent policy for managing the cost of production of the structures of the agricultural holding, reducing transfer and sales costs made it possible to achieve such results.

Table 4. Price range for the product "Wheat bread (yeast)" including value added tax.

Municipal areas	Average regional price of a product, ruble	Percentage % to the price of the agricultural holding
Volgograd	30.00	112
Gorodischensky district	30.00	112
Zhirnovsky district	27.90	106
Kikvidzensky district	29.60	112
Elansky district	29.45	111
Olkhovsky district	28.30	107
Rudnyansky district	28.30	107
Novoanninsky district	26.90	102
Frolovsky district	29.80	113
Bykovskiy district	29.80	113
Regional average	29.00	110
Open Joint Stock Company "Novoanninsky Combine of Bread Products"	26.40	100

Due to the established on-farm relations, the enterprises of the agricultural holding have the opportunity to pay for services related to the promotion of the finished product along the technological chain, using more favorable tariffs compared to commodity producers that are not part of its structure.

From the data presented in Table 5, it follows that the economic entities of the agricultural holding cost acceptance services by 9%, and storage services - 24% cheaper than third-party enterprises. The exception is varietal wheat for the production of premium flour, which is purchased by members of the agricultural holding at a single market price of 9600 rubles for one ton.

Table 5. The cost of services for the acceptance and storage of grain of the agricultural holding.

Name of product, service	Members of the agricultural holding					Subjects not included in the agricultural holding
	Open Joint Stock Company "Novoan-ninsky Combine of Bread Products"	Limited Liability Company "Solnech-naya Strana"	Limited Liability Company "Agricult ural enterprise Belo-rechinska ya"	Limited Liability Company "Novoan-ninsky baker"	Limited Liability Com-pany "Novo-anninsky bread"	
Services:						
- storage, rubles/t	*	*	177.65	177.65	177.65	123.30
- acceptance, rubles/t	*	*	164.83	164.83	164.83	171.32
Recycling:						
- grain, rubles/t	*	*	118.94	118.94	118.94	150.83
- sunflower, rubles/t	*	*	120.15	120.15	120.15	122.16
Drying:						
- grain, rubles/t	*	*	221.21	221.21	221.21	223.33
- sunflower, rubles/t	*	*	224.24	224.24	224.24	226.67
Grain processing services, rubles/t	*	*	530.3	530.3	530.3	583.33
Sales:						

Name of product, service	Members of the agricultural holding					Subjects not included in the agricultural holding
	Open Joint Stock Company "Novoan-ninsky Combine of Bread Products"	Limited Liability Company "Solnech-naya Strana"	Limited Liability Company "Agricult ural enterprise Belo-rechinska ya"	Limited Liability Company "Novoan-ninsky baker"	Limited Liability Com-pany "Novo-anninsky bread"	
- high-quality wheat grain for flour, rubles/t	9600	*	9600	*	*	9600
- flour, rubles/t	*	*	*	27300	27300	20300
- finished products, rubles/kg (rubles 26.40 × 2)	*	*	*	*	*	52.80

The relatively low level of tariffs for each link of the production and marketing chain ensures profitability of production and marketing activities for the subjects of the agricultural holding "Novoanninsky", allowing them to carry out their expanded reproduction. Let us consider the mechanism of financing real investments of enterprises of the agricultural holding "Novoanninsky" (Table 6).

Table 6. Distribution of financial resources between the subjects of the agricultural holding for 2019 million rubles.

Financial resources	Open Joint Stock Company "Novoanni nsky Combine of Bread Products"	Limited Liability Company "Solnech-naya Strana"	Limited Liability Company "Agricult ural enterprise Belorechi nskaya"	Limited Liability Company "Novoan-ninsky baker"	Limited Liability Company "Novoanni nsky bread"	Total	
						Million rubles	%
Sources of everything	678.3	36.3	21.4	17.6	20.0	773.6	100.0
Profit	55.7	36.3	8.3	6.3	1.1	107.7	13.9
Loans	222.6	*	13.1	11.3	18.9	265.9	34.4
Investments	400.0	*	*	*	*	400.0	51.7

The data in Table 6 indicate that the largest percentage in the structure of funding sources belongs to borrowed funds: loans and attracted private investment (more than 85%). Only about 14% of investment projects were implemented at the expense of our own sources. Thus, the largest share in financing belongs to private capital in the amount of 400 million rubles, which indicates investment competition and investment security of the enterprise in question. The private investment mechanism is implemented through the head enterprise - Open Joint Stock Company "Novoanninsky Combine of Bread Products", which, in accordance with the investment policy of the enterprise, distributes funds between the structures of the agricultural holding.

Thus, the successfully functioning agricultural holding of the grain sub-complex of the Volgograd region testifies to the viability and competitive and investment advantages of its structures. Financial and production stability of all structures of the agricultural holding "Novoanninsky" is planned to be achieved through their innovative development, the formation of a unified pricing policy, planning and budgeting of income and expenses, an increase in the functionality of the marketing service, and the formation of strong inter-farm ties. As proposals for the future, it is necessary to recommend the introduction of a research structure into its production and marketing chain, which allows for expanded reproduction and ensure a stable raw material base through the use of new varieties of grain crops, to carry out seed production and breeding services [3, 4, 6, 11, 12].

4. Conclusion

All of the above allows us to draw the following main conclusions:

- the considered model of integration ensures the operational independence of all structures of the agricultural holding in production and financial activities;
- all business entities of the agricultural holding, with the help of thoughtful actions and competent investment policy, coordinate their economic interests on a relatively mutually beneficial basis;
- by combining their finances, the subjects of the agricultural holding not only get the opportunity to implement priority and strategic projects, but also provide, if necessary, targeted assistance to its members who are in a difficult economic situation;
- as a production and financial integrated structure, the agricultural holding has investment competitiveness and the ability to carry out innovative development;
- the integration of the agricultural holding's activities with a research organization is a promising direction for increasing its investment activity in terms of introducing new technological, breeding, marketing and other innovations that contribute to ensuring a high level of investment potential development in the interests of social and economic stabilization of its subjects.

References

[1] Donskova O A and Nemkina E A 2019 New approaches to monitoring food security in the digital economy *Int. Scientific and Practical Conf. held within the framework of the Int. Scientific and Practical Forum dedicated to the 75th Anniversary of the Formation of the Volgograd State Agrarian University* vol 4 pp 57-63

[2] Ovchinnikov A S, Balashova N N, Ivanova N V and Koshkarev I A 2018 *Development Strategy for the Regional Market of Leguminous Crops: Implementation Opportunities and Cluster-Network Organization* (Volgograd: Volgograd State Agrarian University) p 200

[3] Shemet E S, Tikhonovich E A, Ruskova E G and Samedova E N 2018 Analysis of trade globalization of integration groups (for example, the Eurasian Economic Union) *European Social Science J.* **12(1)** 14-20

[4] Balashova N N and Ivanova N V 2019 Research of the innovative potential of small-scale farming in the digital economy *Development of the Agro-Industrial Complex Based on the Principles of Rational Nature Management and the Use of Convergent Technologies* pp 147-153

[5] Melikhov V A 2016 Organizational and technical regulations for the formation of integrated reports of agro-industrial holdings *Strategic Guidelines for the Innovative Development of the Agro-Industrial Complex in Modern Economic Conditions* pp 340-344

[6] Yagupova E V and Belousov A S 2019 Strategies for innovative development of the agro-industrial complex *Innovative Technologies and Technical Means for the Agro-Industrial Complex* pp 82-85

[7] Smotrova E E 2019 Efficiency of state regulation of the agro-industrial complex and key problems of its development *Russian Economic Bulletin* **2(5)** 172-177

[8] Vardanyan S A and Terekhina N V 2018 Assessment of financial risks in the formation of the credit policy of the organization *Science and Youth: New Ideas and Solutions* pp 277-280

[9] Popova S A, Rudkova T A and Kolpakova E A 2013 Sustainable development of rural areas of the Volgograd region: problems and solutions *Proc. of the Nizhnevolzhsky Agrouniversity Complex: Science and Higher Professional Education* **3(31)** 235-239

[10] Chernovanova N V and Gorbachev A S 2017 Automated procurement system in agricultural consumer cooperatives *Scientific Review: Theory and Practice* **12** 73-77

[11] Egorova E M 2019 Innovative processes in the agro-industrial complex: problems and prospects for implementation; development of methodological support *Priority Research and Innovative Technologies in the Agro-Iindustrial Complex: Science-Production* pp 103-110

[12] Gorbacheva A S, Balashova N N, Chekrygina T A, Chernovanova N V, Yagupova E V and Vorontsova E S 2020 Organization of registration ensuring innovative development of branch of sheep breeding IOP Conf. Series: Earth and Environmental Science **422(1)** 012073

IOP Conf. Series: Earth and Environmental Science **965** (2022) 012062 doi:10.1088/1755-1315/965/1/012062

Perspective architecture of dairy farming enterprises, using modern digital technologies for sustainable development

O V Kochetkova and E V Shiryaeva

Volgograd State Agrarian University, 26 University Avenue, Volgograd, 400002, Russia

E-mail: ovk555@bk.ru

Abstract. The subject of dairy farming digitalization is relevant due to global factors such as population growth, climate change and food traceability needs. Negative impact of these factors on dairy farming efficiency and safety can be neutralized or significantly reduced by means of modern digital technologies introduction which are aimed at improving of enterprise environmental and economic sustainability, the effectiveness of planning, monitoring and control of animal habitat and health. Main reason for the low rate of dairy farming digitalization is that enterprise top executives are lacking clear understanding of how their organization should be transformed using modern information technologies. Problem could be solved by creation of the perspective enterprise architecture model "To Be" which relies heavily on modern digital technologies usage. System analysis and system engineering techniques, TOGAF architecture management methodology and ArchiMate architectural language are used to develop dairy enterprise architecture. Presented models allow to see digital enterprise of dairy production entirely, to demonstrate interaction of its individual architectural elements from strategy to infrastructure of the enterprise. New business opportunities, namely digital management of dairy farm and various electronic services provide real time monitoring of dairy livestock conditions, traceability, sustainability and efficiency of production processes.

1. Introduction

Dairy farming plays significant role in ensuring safe food production and healthy nutrition of human population. Nowadays due to increasing environmental risks and amount of wastes as well as significant climate change coupled with urgent need to ensure food security of an ever-growing population, there are the actual problems for dairy farming such as significant productivity gains, lower costs, safety and traceability of food products. The biggest portion of dairy farming expenses is cost of labor so automation of dairy farms processes will reduce it. Dairy enterprise is a complex system with its prominent feature being tracking and supporting optimal condition and development of both biological entities and their habitat [1]. It is known [2] that in dairy farming individual identification and eating behavior analysis have a great impact on animal productivity. Besides, individual animal monitoring, environmental microclimate management, development and rationing of feed effectiveness all play significant role in ensuring said productivity [3]. It should be mentioned that due to livestock on farms grows bigger in numbers [4] these factors as well as amount of large farms are now increasing too which imposes new, higher management requirements [5]. Digital technologies used to control the habitat allow to measure and record not only temperature and room

IOP Conf. Series: Earth and Environmental Science **965** (2022) 012062 doi:10.1088/1755-1315/965/1/012062

humidity but also heat stress of milking cows [6] and give ability to investigate food behavior and water consumption by animals [2].

Monitoring of each animal's health is very important. It uses digital technologies to automatically evaluate an animal's body conditions [7], predict calving [8], detect lameness [9] etc. This allows us to determine patterns of cow development and product formation depending on composition and quantity of nutrients and energy, parameters of habitat, animal conditions and to ensure making of informed management decisions [1].

Thus, digital technologies in dairy farming provide not only effective planning, monitoring and control processes [10] but also count as important factor in reducing labor costs, increasing animal productivity and decreasing disease probability [11]. Besides, usage of digital technologies lessens the number of animals needed to produce the same amount of products which leads to environmental and economic sustainability of dairy farming [3].

However, farmers are in no hurry to put digital technologies to use despite their awareness of accurate livestock digital management availability. There are several reasons for this. Major ones for slow digital technologies adaptation to activities of dairy enterprises include not just goods and services markets dynamics and financial instability but also the lack of understanding how rebuilding of production business processes must be done, what specific digital technologies should be applied, how payback of innovative solutions is ensured. For example, the studies [12, 13] mention that it took quite a long time to learn how to make the most of digital technology benefits. Introduction of digital technologies requiring change in the business model which in turn changes all business processes of dairy enterprises is the reason for this [14]. And thus significant changes affect all levels of management. Therefore, it is necessary to identify and develop models "To Be" for improved business processes of dairy enterprise. Also executives of dairy enterprises must always know the current status of both implemented and operated information systems and understand the trends of their further development, taking into account business development strategy. To do this systematic model of enterprise architecture that can control system complexity, manage it in a changing environment, and provide shared vision of structure and functions of business systems and IT systems throughout the organization must be created.

The architecture of enterprise optimizes many separate documented and automated processes inside an integrated environment, balancing IT efficiency with business innovation. This enables safe introduction of innovations, ensures that organization's needs for an integrated IT strategy are met and aligns unified strategy of its development with business requirements. Architecture allows to formalize IT problems and requirements from business point of view and shows how specific tasks are being addressed and how business requirements in IT systems are being implemented. In other words, development of the dairy enterprise perspective architecture is urgent and important task as it provides a way to achieve the organization mission by coordination and efficient functioning of its key business processes within an effective IT environment.

2. Materials and methods

The architecture of enterprise describes a complex system related to the whole organization including strategy, positioning, operating and business model, financial and internal policies, information, digital technologies and assets. Therefore this study is based on system analysis and system engineering methods. These methods define architecture as the main entity of the system represented in its components and their relationships with each other and environment as well as set of guidelines for design and development of the system. Main aspects of these approaches are being studied on ISO/IEC 15288:2008, ISO/IEC 15288:2008, ISO 15704:2000 standards basis. The theoretical basis of this study presented by its architectural approach to management of the organization, functional and logical analysis includes both generalization of domestic and foreign theory and practices of creating architecture of the enterprise and the results that are outlined in works of the world's leading experts in subject area. Carried out to determine applicability to dairy farming and the maturity of the company are studies of the most well-known and common architecture management methodologies. To build architecture of the enterprise this study uses universal TOGAF platform for IT architecture construction and ArchiMate language [15] which contains set of rules and concepts to describe enterprise architecture. ArchiMate usage provides the possibility of abstract composition of multi-layered enterprise architecture with simple and understandable architectural description of various subject areas.

3. Results and discussion

Authors of this study have performed analyses of business processes and architecture of typical dairy enterprises [1, 14] which showed that production processes and animal content mode are not sufficiently supported by IT tools while implementation of the maximum number of factors of production technological equipment and effective management methods based on expert decision-making support is necessary to achieve high results [16]. The majority of dairy enterprises have inefficient business processes and do not comply with standard requirements for their organization which reduces productivity and efficiency of production [14].

Usage of digital systems is necessary to perform business processes that support required parameters of technological procedures implementation and animal conditions in their indoor residence including regulation of internal climate parameters, automatic milking and choice of machine operational modes. Operations of balanced feed mixtures preparations, individual feed dosage, milk deep cooling, sorting and packaging of finished products are practically not automated at all [1]. There is demand for improved cow health management processes because identification of animals through ear tags does not allow to determine their condition and send signal when it changes so urgent actions could be taken.

Thus developed in [1] "As Is" model of typical dairy farms reveals insufficient support and compliance with business and IT goals as well as need to create a model for a perspective "To Be" model.

To develop architecture of the digital dairy enterprise a conceptual model for animal content mode management has been created that encapsulates usage of modern digital technologies in dairy farming enterprises in the form of "To Be" precedents diagram (Figure 1).

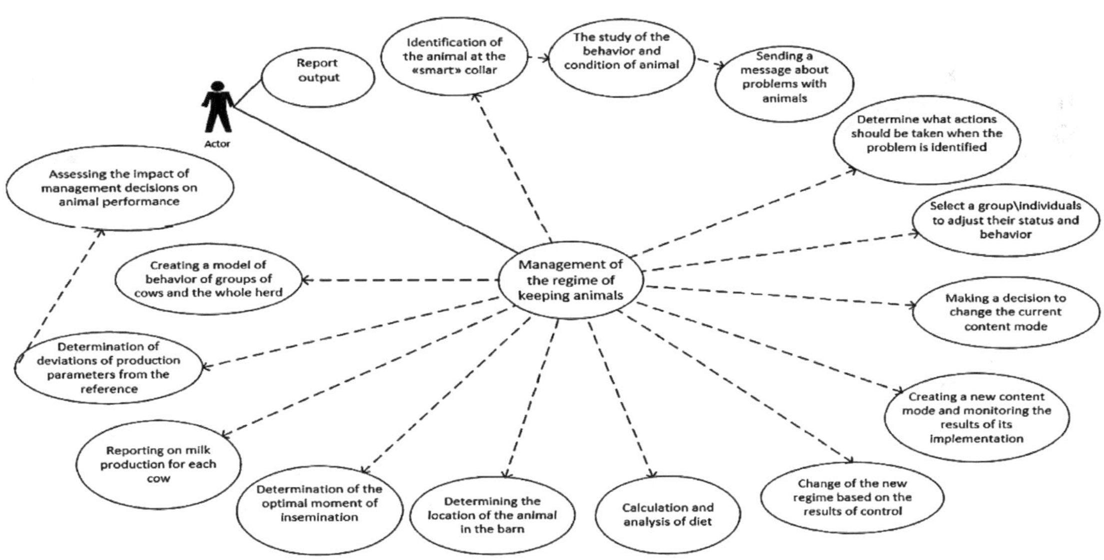

Figure 1. Diagram of animal content mode management precedents in form of IT "To Be".

Modern digital technologies and identification devices that are already being used in foreign dairy cattle breeding practice in form of equipped with various sensors "smart" collars make real time monitoring of each animal's behavior and condition possible. This allows to determine optimal time of cows insemination accurately and automatically send a distress signal in case of conditional or behavioral deviation from desired parameters.

The system proposes to take concrete measures for each pinpointed problem to prevent making incorrect management decisions. It is also possible to select a group of animals or particular specimens for adjusting their condition and behavior, to create patterns of behavior of cow groups or entire herd, to determine location of an animal in the barn and to change cow content mode automatically. Based on results of new animal content mode introduction monitoring decisions can be made to adjust mentioned mode.

System also allows to evaluate the impact of management decisions on animal productivity, to determine deviations of production parameters from reference ones, to form different types of reports for both individual cow and entire herd.

Thus usage of digital technologies in the animal management on dairy farm allows to detect deviations in cow behavior due to disease or malnutrition timely and to take measures to eliminate said deviations, to reduce risks of making incorrect management decisions, to perform competent planning and monitoring of production activities, which leads to higher enterprise efficiency in the end.

The new typical dairy enterprise architecture "To Be" relies on Internet of Things and blockchain technologies. Nowadays these technologies are much more accessible and cheaper than they used to be and their capabilities have also expanded significantly thanks to communication technology and technological development advances, also quantity of sensors and devices that can be implemented in agricultural solutions has increased enormously.

Global technological progress has contributed to emergence of Internet of Things and blockchain technologies and in return has led to appearance of various technology-related concepts such as "smart farm" which uses information technology to perform more comprehensive system analysis. The "Internet of Things" (IoT) term refers to the connection of several diverse objects such as cars, vehicles or buildings with electronic devices such as sensors and executive mechanisms by using different communication protocols to collect and extract data [17]. IoT technology provides basis for research and development of "smart farm" solutions.

The main advantages of Internet of Things usage are reducing costs, increasing productivity, improving traceability, monitoring production and more effective interaction with partners and customers.

The main benefit of blockchain technology is, first of all, its security. Blockchain is a secured digital registry, a network of equal nodes where transactions are stored to transfer just objects ownership not databases of property. Transaction verification and storage functions are distributed among many nodes without participation of regulators [18].

Introduction of blockchain based infrastructure will significantly reduce the costs of maintaining it and negate numerous security risks associated with it. In addition, blockchain allows to replace various models of data reconciliation and thus significantly speed up any processes. At the very least blockchain technology increases the efficiency of already occurring processes [18].

The architecture of the dairy enterprise modernization aims to introduce digital technologies which presents a new agro-industrial platform that facilitates Internet of Things and blockchain technologies usage in "smart farm" environments. The purpose of this platform reflected in the architecture is real time monitoring of livestock conditions while ensuring traceability and sustainability of various processes involved in production.

These applied to "smart farm" technologies allow to coordinate and manage all resources involved in IoT environment as well as to obtain data from disparate sources. IoT multi-layered data solutions offer decision for effective management of diverse data through homogenization and higher level of general management. This approach enables to receive large data sets from multiple scenarios which leads to creation of big data stores where data analysis and machine learning techniques can be applied to provide additional value for the work environment through real time analysis and responsiveness, image recognition, prognosis, etc.

Figure 2 shows the perspective architecture "To Be" of typical dairy enterprise.

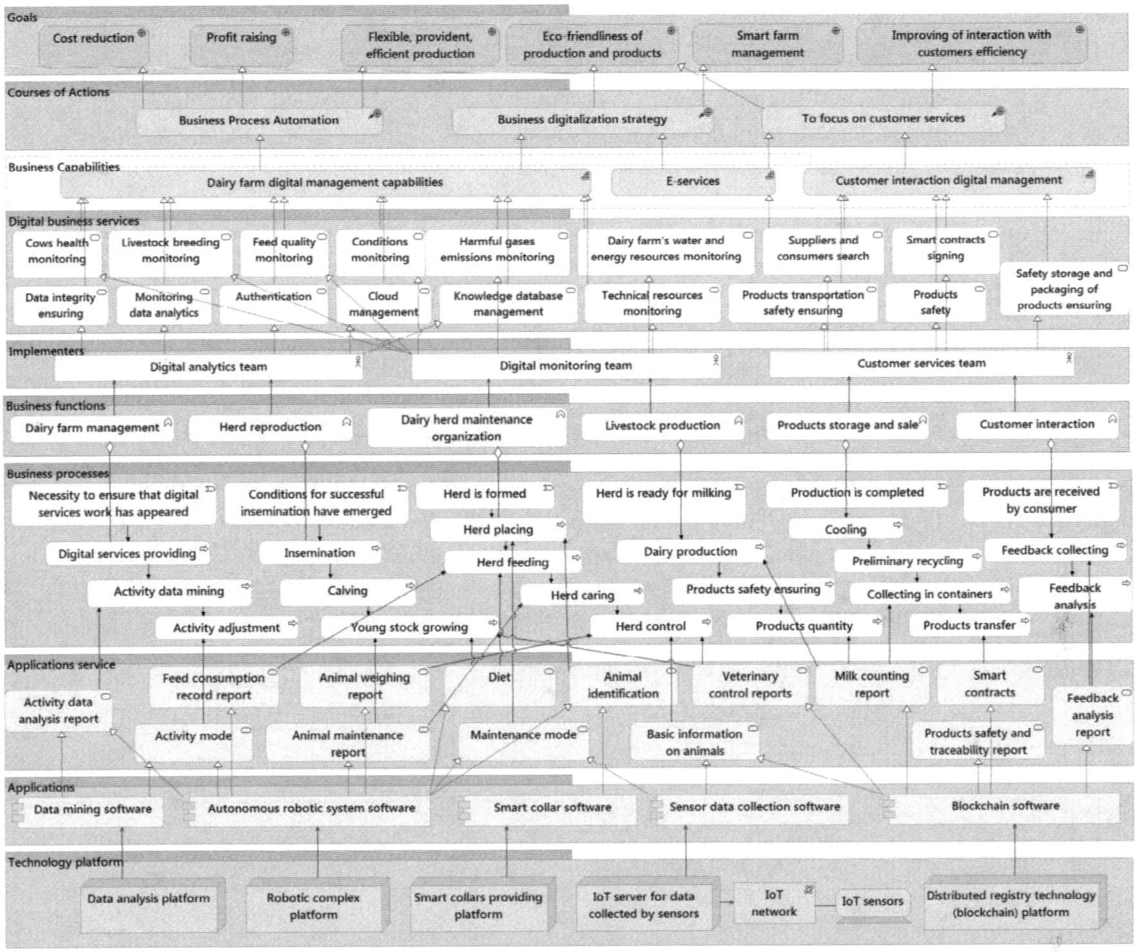

Figure 2. Typical dairy enterprise perspective architecture "To Be".

The main difference between "To Be" and "As Is" architectures is that it focuses on automation and business digitization, the occurrence of new business opportunities such as possibilities of dairy farm digital management, electronic services, customer relationships digital management. As a result many new digital business services have emerged such as cow health monitoring, livestock reproduction monitoring, feed quality monitoring, cow conditions monitoring, harmful gases monitoring, dairy farm water and energy resources monitoring, suppliers and consumers search, smart contracts search, data integrity ensuring, data analytics, authentication, cloud management, knowledge base management, technical resources monitoring, product transportation security, production safety, products storage and packaging safety.

Maturity level of the developed perspective architecture of dairy enterprise must be evaluated to analyze results. Enterprise architecture maturity model [19] includes architectural characteristics, organization maturity levels, maturity parameters system and distributes maturity of an enterprise architecture to one of five levels. Evaluation of an enterprise architecture maturity requires to determine evaluation criteria, to use generally accepted maturity levels of organization's architecture, to conduct survey on interested persons, to assess feedback and ratings. After this maturity evaluation for each criterion and architecture as a whole should be made. If usage of developed perspective architecture of the dairy enterprise is organized properly then completed analysis of its maturity allows to reach fourth managed level of maturity. This maturity level is characterized [19] by ability to collect and analyze the measurements results on which all further activities are based. Measurements are used for efficiency prognosis and they provide better understanding of processes and capabilities. The enterprise architecture documentation is being updated and analyzed systematically. At the same time top management is directly involved in the process of analyzing the organization's architecture.

4. Conclusion

Rapid development and increased digital technologies availability creates favorable conditions for their use in dairy enterprises with purpose of value chain in the production of products tracing, business optimization, productivity and efficiency of production incrementation. This requires business model, strategy, business processes, organizational structure and enterprise information flows to change. In this case architecture of the enterprise becomes basis for determining goals of business, information for its management, IT, business processes and organizational structure.

Developed perspective architecture of dairy enterprise allows to observe the whole enterprise, to demonstrate impact of certain development strategy elements on provided services and business processes as well as to show how they depend on information systems. Presented in this study perspective architecture of dairy enterprise stands as basis for synchronizing all facilities within the organization ensuring the consistency of IT infrastructure and business goals, data collecting, data integrity, data analysis and data consistency which is main support for management decisions.

The IT architecture is based on Internet of Things and Distributed Registry technologies that provide digital monitoring, security, traceability, and analytics services. The architecture model visualizes how objects are perceived and controlled by blockchain and IoT technologies.

With proper organization of its usage developed perspective architecture of dairy farming enterprises allows to reach operated maturity level which corresponds with requirements of all interested parties.

References

[1] Kochetkova O V, Shiryaeva E V, Arkov D P and Matveyev A S 2020 Elaboration of architecture of the enterprise of dairy animal breeding *IOP Conf. Series: Materials Science and Engineering* **873** 012010

[2] Brahim Achour, Malika Belkadi, Idir Filali, Mourad Laghrouche, Mourad Lahdir Image analysis for individual identification and feeding behaviour monitoring of dairy cows based on Convolutional Neural Networks *Biosystems Engineering* **198** 31-49

[3] Berckmans D 2017 General introduction to precision livestock farming *Animal Frontiers* **7(1)** 6-11

[4] Barkema H W, von Keyserlingk M A, Kastelic J P, Lam T J, Luby C, Roy J P, LeBlanc S J, Keefe G P and Kelton D F 2015 Invited review: Changes in the dairy industry affecting dairy cattle health and welfare *J. Dairy Sci.* **98** 7426-7445

[5] Winsten J R, Kerchner, C D, Richardson A, Lichau A and Hyman J M 2010 Trends in the Northeast dairy industry: Large-scale modern confinement feeding and management-intensive grazing *J. Dairy Sci.* **93** 1759-1769

[6] Yu-Chi Tsai, Jih-Tay Hsu, Shih-Torng Ding, Dan Jeric Arcega Rustia and Ta-Te Lin 2020 Assessment of dairy cow heat stress by monitoring drinking behaviour using an embedded imaging system *Biosystems Engineering* **199** 97-108

[7] Hansen M F, Smith M L, Smith L N, Jabbar K A and Forbes D 2018 Automated monitoring of dairy cow body condition, mobility and weight using a single 3D video capture device *Comput. Ind.* **98** 14-22

[8] Neethirajan S 2017 Recent advances in wearable sensors for animal health management *Sensing and Bio-Sensing Research* **12** 15-29

[9] Miguel-Pacheco G G, Kaler J, Remnant J, Cheyne L, Abbott C, French A P, Pridmore T P,

Huxley J N 2014 Behavioural changes in dairy cows with lameness in an automatic milking system *Appl. Animal Behav. Sci.* **150** 1-8

[10] Kruize J W, Wolfert J, Scholten H, Verdouw C N, Kassahun A and Beulens A J M 2016 A reference architecture for Farm Software Ecosystems *Computers and Electronics in Agriculture* **125** 12-28

[11] Yazdanbakhsh O, Zhou Y, Dick S 2017 An intelligent system for livestock disease surveillance *Information Sciences* **378** 26-47

[12] Ferrández-Pastor F J, Garcia-Chamizo J M, Nieto-Hidalgo M, Mora-Pascual J and Mora-Martinez J 2016 Developing ubiquitous sensor network platform using internet of things: application in precision agriculture *Sensors* **16(7)** 1141

[13] Bewley J, Russell R A, Dolecheck K A, Borchers M R, Stone A E, Wadsworth B A and Tsai I-C 2015 Precision dairy monitoring opportunities, limitations, and considerations *Western Dairy Management Conf.* (Reno, USA)

[14] Kochetkova O V and Khoroshavin V S 2019 Research of business processes of dairy animal breeding with purpose of their following automation *Proc. of the Nizhnevolzhsky Agrouniversity Complex: Science and Higher Professional Education* **4(56)** 332-342

[15] Asif Qumer Gill 2015 Agile enterprise architecture modelling: Evaluating the applicabilityand integration of six modelling standards *Information and Software Technologies* **67** 196-206

[16] Breena M, Upton J and Murphy M D 2019 Development of a discrete infrastructure optimization model for economic valuation on dairy farms (DIOMOND) *Computers and Electronics in Agriculture* **156** 508-522

[17] Kethareswaran V 2017 An Indian perspective on the adverse impact of internet of things *Advances in Distributed Computing and Artificial Intelligence J.* **6(4)** 35-40

[18] Kostukova H N 2020 Blockchain and directions of its using in logistics in the conditions of digital transformation *Digital Transformation* **3(12)** 39-44

[19] Gopala Krishna Behara, Prasad Palli 2013 *Maturity Assessment for Enterprise Architecture* https://www.bptrends.com/publicationfiles/01-08-2013-ART-Maturity%20Assessment-Behara.pdf

AUTHOR INDEX

Institute of Physics
Dirac House, Temple Back
Bristol BS1 6BE UK

CURRAN ASSOCIATES INC.
proceedings.com

ISSN: 1755-1307
ISBN 978-1-7138-4947-6

9 781713 849476